Stochastic Mechanics
Random Media
Signal Processing and Image Synthesis
Mathematical Economics and Finance
Stochastic Optimization
Stochastic Control
Stochastic Models in Life Sciences

Applications of Mathematics

Stochastic Modelling
and Applied Probability

55

Edited by   B. Rozovskii
M. Yor

Advisory Board   D. Dawson
D. Geman
G. Grimmett
I. Karatzas
F. Kelly
Y. Le Jan
B. Øksendal
G. Papanicolaou
E. Pardoux

Springer

New York
Berlin
Heidelberg
Hong Kong
London
Milan
Paris
Tokyo

Applications of Mathematics

(continued after index)

G. George Yin   Qing Zhang

Discrete-Time Markov Chains

Two-Time-Scale Methods and Applications

With 31 Figures

 Springer

G. George Yin
Department of Mathematics
Wayne State University
Detroit, MI 48202
USA
gyin@math.wayne.edu

Qing Zhang
Department of Mathematics
University of Georgia
Athens, GA 30602
USA
qingz@math.uga.edu

Managing Editors

B. Rozovskii
University of Southern California
Department of Mathematics
Kaprielian Hall KAP 108
3620 S. Vermont Avenue
Los Angeles, CA 90089
USA
rozovski@math.usc.edu

M. Yor
Laboratoire de Probabilités et Modèles Aléatoires
Université de Paris VI
175, rue du Chevaleret
75013 Paris, France

ISBN 978-1-4419-1955-7   e-ISBN 978-0-387-26871-2

Mathematics Subject Classification (2000): 60J10, 34E05, 93E20

Library of Congress Cataloging-in-Publication Data
Yin, George, 1954–
    Discrete-Time Markov chains: Two-Time-Scale Methods and Applications / George
Yin, Qing Zhang.
        p. cm. — (Applications of mathematics)
    Includes bibliographical references and index.
    (alk. paper)
        1. Markov processes.   I. Zhang, Qing, 1959–   II. Title.   III. Series.
QA274.7.Y57 2005
519.2'33—dc22

Printed on acid-free paper.

9 8 7 6 5 4 3 2 1

springeronline.com

To Our Parents

Contents

Preface

This book focuses on two-time-scale Markov chains in discrete time. Our motivation stems from existing and emerging applications in optimization and control of complex systems in manufacturing, wireless communication, and financial engineering. Much of our effort in this book is devoted to designing system models arising from various applications, analyzing them via analytic and probabilistic techniques, and developing feasible computational schemes. Our main concern is to reduce the inherent system complexity. Although each of the applications has its own distinct characteristics, all of them are closely related through the modeling of uncertainty due to jump or switching random processes.

One of the salient features of this book is the use of multi-time scales in Markov processes and their applications. Intuitively, not all parts or components of a large-scale system evolve at the same rate. Some of them change rapidly and others vary slowly. The different rates of variations allow us to reduce complexity via decomposition and aggregation. It would be ideal if we could divide a large system into its smallest irreducible subsystems completely separable from one another and treat each subsystem independently. However, this is often infeasible in reality due to various physical constraints and other considerations. Thus, we have to deal with situations in which the systems are only nearly decomposable in the sense that there are weak links among the irreducible subsystems, which dictate the occasional regime changes of the system. An effective way to treat such near decomposability is time-scale separation. That is, we set up the systems as if there were two time scales, fast vs. slow.

Following the time-scale separation, we use singular perturbation methodology to treat the underlying systems. Here singular perturbation is interpreted in a broad sense, including both deterministic singular perturbation methods and stochastic averaging. As a consequence, our results may also be divided into analytic and probabilistic. Although the original systems are in discrete time, they are closely related to certain continuous-time systems. To bring them into the framework of continuous-time dynamic systems enables us to use many techniques from the available toolboxes.

This book provides a systematic approach to two-time-scale Markovian systems. We show that the idea of decomposition and aggregation can be made rigorous by deriving asymptotic results of suitably scaled processes. Using the aggregated processes, we can then proceed to study, for example, control and optimization problems such as Markov decision processes, linear quadratic regulator modulated by a Markov chain, and many other hybrid dynamic systems. By deriving a limit problem associated with that of the original system and using the optimal or near-optimal control of the limit system, we then construct controls of the original systems and show such controls are nearly optimal.

Most of the book are an outgrowth of our recent research. Several chapters are concerned with various applications involving two-time scales. The common focus of these chapters is on the reduction of dimensionality of the underlying dynamic systems.

This book is written for applied mathematicians, operations researchers, applied probabilists, control scientists, and financial engineers. It presents results that relate stochastic models, systems theory, and applications in manufacturing, reliability, queueing systems, and stochastic financial markets. Selected materials from this book can also be used in a graduate level course on stochastic processes and applications.

We are very grateful to those people who have helped to make the publication of this book possible. We express our deep gratitude to Wendell Fleming and Harold Kushner, to whom we owe a great intellectual debt. As our mentors, they introduced the stochastic world to us and they have encouraged us throughout each step of our careers. It has been our privilege to work with Rafail Khasminskii on a number of research projects, from whom we have learned much about singular perturbations and Markov processes. We express our special appreciation to Grazyna Badowski, Subhra Dey, Cristina Ion, Vikram Krishnamurthy, Ruihua Liu, Yuanjin Liu, Hongchuan Yang, Jiongmin Yong, Kewen Yin, Hanqin Zhang, and Xunyu Zhou, who have worked with us on various projects related to Markovian systems. The presentation and exposition have much benefited from the comments by Ruihua Liu and Yuanjin Liu, and by three anonymous reviewers, who read early versions of the drafts and offered many insightful comments. Our thanks go to the series editor Boris Rozovsky for his encouragement, time, and consideration. Our thanks also go to Springer senior editor Achi Dosanjh for her assistance and help. We thank the production manager

MaryAnn Brickner and the Springer professionals for their work in finalizing the book. During the years of our study, the research was supported in part by the National Science Foundation, the Office of Naval Research, and the Defense Advanced Research Projects Agency. Their continuing support is greatly appreciated.

Detroit, Michigan   George Yin
Athens, Georgia   Qing Zhang

Conventions

We clarify the numbering system and cross-reference conventions to be used in the book. Equations are numbered consecutively within a chapter. For example, (3.10) indicates the tenth equation in Chapter 3. Corollaries, definitions, examples, lemmas, propositions, remarks, and theorems are numbered sequentially throughout each chapter. For example, Definition 4.1, Theorem 4.2, Corollary 4.3, etc. Assumptions are marked consecutively within a chapter, e.g., (A6.1) is the first listed assumption in Chapter 6. For cross reference either within the chapter or to another chapter, an equation is identified by the chapter number and the equation number; similar conventions are used for theorems, remarks, assumptions, etc.

Throughout the book, we assume that all deterministic processes are Borel measurable and all stochastic processes are measurable with respect to a given filtration. The ith component of a vector $z \in \mathbb{R}^r$ is denoted by z^i. Occasionally, we also use the notion of partitioned vector and write, for example, $v = (v^1, \ldots, v^l)$, where each v^i is a subvector of appropriate dimension. A subscript generally denotes either a finite or an infinite sequence. However, the ε-dependence of a sequence is designated in the superscript. To assist the reader, we provide a glossary of symbols to be used in the subsequent chapters.

Glossary of Symbols

A'	transpose of a matrix (or a vector) A
B^c	complement of a set B
$\mathrm{Cov}(\xi)$	covariance of a random variable ξ
$\mathbb{C}$	space of complex numbers
$C([0, T]; S)$	space of S-valued continuous functions on $[0, T]$
C_L^2	space of functions with bounded derivatives up to the second order and Lipschitz second derivatives
$D([0, T]; S)$	space of S-valued functions being right continuous and having left-hand limits
$E\xi$	expectation of a random variable ξ
$\mathcal{F}$	σ-algebra
$\{\mathcal{F}_t\}$	filtration $\{\mathcal{F}_t, t \geq 0\}$
I	identity matrix of suitable dimension
I_A	indicator function of a set A
K	generic positive constant with convention $K + K = K$ and $KK = K$
$\mathcal{M}$	state space of a Markov chain
$\mathcal{M}_i$	sub-state space of the ith ergodic class
$\mathcal{M}_*$	sub-state space of the transient class
$N(x)$	neighborhood of x
$O(y)$	function of y such that $\sup_y |O(y)|/|y| < \infty$
$O_1(y)$	function of y such that $\sup_y |O_1(y)|/|y| \leq 1$
$P(\xi \in \cdot)$	probability distribution of a random variable ξ
P or P_k	transition matrix with entries p^{ij} or p_k^{ij}
Q or $Q(t)$	generator of a Markov chain

$\overline{Q}$	$= \mathrm{diag}(\nu^1, \ldots, \nu^{l_0})Q\widetilde{\mathbb{1}}$
$\overline{Q}_*$	$= \mathrm{diag}(\nu^1, \ldots, \nu^{l_0})(Q^{11}\widetilde{\mathbb{1}} + Q^{12}A_*)$
$Qf(\cdot)(i)$	$= \sum_{j \neq i} q^{ij}(f(j) - f(i))$ where $Q = (q^{ij})$
$\mathbb{R}^r$	r-dimensional real Euclidean space
$S(r)$ or S_r	ball centered at the origin with radius r
a^+	$= \max\{a, 0\}$ for a real number a
a^-	$= \max\{-a, 0\}$ for a real number a
a.s.	almost surely
$\langle a, b \rangle$	inner product of vectors a and b
$a_1 \wedge \cdots \wedge a_l$	$= \min\{a_1, \ldots, a_l\}$ for $a_i \in \mathbb{R}$, $i = 1, \ldots, l$
$a_1 \vee \cdots \vee a_l$	$= \max\{a_1, \ldots, a_l\}$ for $a_i \in \mathbb{R}$, $i = 1, \ldots, l$
$(a^1, \ldots, a^l) > 0$	$a^1 > 0, \ldots, a^l > 0$
$(a^1, \ldots, a^l) \geq 0$	$a^1 \geq 0, \ldots, a^l \geq 0$
$\mathrm{diag}(A^1, \ldots, A^l)$	diagonal matrix of blocks $A^1, \ldots, A^l$
$\exp(Q)$	e^Q for argument Q
f_x or $\nabla_x f$	gradient of a function f w.r.t. x
i	pure imaginary number with $\mathrm{i}^2 = -1$
i.i.d.	independent and identically distributed
k, k_1, k_2 etc.	discrete time
l_0	total number of recurrent classes in a partitioned Markov chain
m_0	total number of states of a Markov chain
$\log x$	natural logarithm of x
$o(y)$	a function of y such that $\lim_{y \to 0} o(y)/|y| = 0$
p_k^ε	$(P(\alpha_k^\varepsilon = 1), \ldots, P(\alpha_k^\varepsilon = m_0)) \in \mathbb{R}^{1 \times m_0}$
$p^\varepsilon(t)$	$(P(\alpha^\varepsilon(t) = 1), \ldots, P(\alpha^\varepsilon(t) = m_0)) \in \mathbb{R}^{1 \times m_0}$
$\{s_{i1}, \ldots, s_{im_i}\}$	$= \mathcal{M}_i$
$\{s_{*1}, \ldots, s_{*m_*}\}$	$= \mathcal{M}_*$
$\mathrm{tr}(A)$	trace of matrix A
w.p.1	with probability one
$\lfloor x \rfloor$	integer part of x
$\|y\|_T$	$= \max_{i,j} \sup_{0 \leq t \leq T} |y^{ij}(t)|$, where $y = (y^{ij}) \in \mathbb{R}^{r_1 \times r_2}$
$(\Omega, \mathcal{F}, P)$	probability space
α_k or α_k^ε	discrete-time Markov chain
$\alpha(t)$ or $\alpha^\varepsilon(t)$	continuous-time Markov chain or interpolation of α_k or α_k^ε
δ^{ij}	$= 1$ if $i = j$; $= 0$ otherwise
ε	positive small parameter
$\nu(t)$	quasi-stationary distribution
$\varphi_n(\varepsilon k)$ and $\psi_n(k)$	sequences of $\mathbb{R}^{1 \times m_0}$-valued functions
$\varphi_n^{ij}(\varepsilon k)$ and $\psi_n^{ij}(k)$	ijth entries of $\varphi_n(\varepsilon k)$ and $\psi_n(k)$

$\sigma\{\xi(s) : s \le t\}$	σ-algebra generated by the process $\xi(\cdot)$ up to t
$\xi_n \Rightarrow \xi$	ξ_n converges to ξ weakly
$\mathbb{1}_l$	$= (1, \ldots, 1)' \in \mathbb{R}^{l \times 1}$
$\widetilde{\mathbb{1}}$	$= \mathrm{diag}(\mathbb{1}_{m_1}, \ldots, \mathbb{1}_{m_{l_0}})$
$:=$ or $\overset{\mathrm{def}}{=}$	defined to be equal to
$\square$	end of a proof
$|\cdot|$	norm of an Euclidean space or a function space

Part I

Prologue and Preliminaries

1

Introduction, Overview, and Examples

1.1 Introduction

This book is concerned with discrete-time dynamic systems under uncertainty. It studies systems driven by random processes having memoryless properties, namely, Markov processes. The book focuses largely on Markov processes taking values in a finite set, known as finite-state Markov chains.

Markov chains have been used in modeling physical, biological, social, and engineering systems such as population dynamics, queueing networks, and manufacturing systems. The advances in technology have opened up new domains and provided greater opportunities for their further exploration. Applications of Markovian models have emerged from wireless communications, internet traffic modeling, and financial engineering in recent years. One of the main advantages of using Markovian models is that it is general enough to capture the dominant factors of system uncertainty and, in the meantime, it is mathematically trackable.

Most dynamic systems in the real world are inevitably large and complex, mainly due to their interactions with the numerous subsystems. Rapid progress in technology has also made modeling more challenging. An example is the design of a manufacturing system, in which the rate of production depends on current inventory, the future demand, as well as marketing expenditure etc. The new technology enables a flexible machine to produce a wide variety of products with virtually no setup costs. The various scenarios resulting from this renders the system model more complex.

Since exact or closed-form solutions to such large systems are difficult to obtain, one often has to be contented with approximate solutions. Take optimal control of a dynamic system as an example. Because the precise mathematical models are difficult to establish, near-optimal controls often become a viable, and sometimes the only, alternative. Such near optimality requires much less computational effort and often results in more robust policy to attenuate unwanted disturbances.

One of the central themes of the book is to find approximate solutions for Markovian systems. The key stems from modeling using multiple-time scales. Clearly, different elements in a large system evolve at different rates. Some of them vary rapidly and others change slowly. The dynamic system evolves as if different components used different clocks or time scales. To describe it quantitatively, it is crucial to decide the order of magnitude of rates for different elements through comparisons. We should keep in minds that "fast" vs. "slow" and "long time" vs. "short time" are all relative terms.

In fact, time-scale separation is often inherent in the underlying problems. For instance, equity investors in a stock market can be classified into two categories, long-term investors and short-term investors. The former consider a relatively longtime horizon and make decisions based on weekly or monthly performance of the stock, whereas the latter focus on returns in short-term, daily or an even shorter period. Their time scales are in sharp contrast, an example of inherently different time scales.

Another source of two-time-scale formulation comes from large-scale optimization tasks. Consider an optimization problem of a system that involves random regime changes, which are modeled by a Markov chain. To incorporate all the important factors into the models often results in a large state space of the underlying Markov chain. To reduce the complexity, a hierarchical approach is suggested, which leads to a two-time-scale formulation. The hierarchical approach relies on decomposing the states of the Markov chains into several recurrent classes or possibly several recurrent classes plus a group of transient states. The essence is that within each recurrent class the interactions are strong and among different recurrent classes the interactions are weak.

An effective way to describe the distinct rates of changes is to introduce a small parameter $\varepsilon > 0$ into the system. Note that ε is only used to separate different time scales, so that we can provide asymptotic analysis for small ε. In this book, we consider the case when the small parameter ε enters the transition probability matrix, resulting in an additional small perturbation by a generator. (Henceforth, the terms transition probability matrix, transition matrix, and one-step transition probability matrix are used interchangeably throughout the book.)

To see the rationale of introducing ε in the systems, it is important to remember that ε is merely a parameter that separates different scales of the transition probabilities. For a system, we usually have certain knowledge about the behavior of the physical model, e.g., from historical or experimental data or through empirical frequency counts. Such information is useful to manifest the rates of different time scales of the transition probabilities. We can also numerically decompose a transition matrix. Although the original problem may not involve an ε, we may deliberately introduce a small parameter ε to separate the different scales. For example, consider a transition matrix given by

$$\begin{pmatrix} 0.38 & 0.6 & 0.01 & 0.01 \\ 0.21 & 0.78 & 0.01 & 0 \\ 0 & 0.01 & 0.29 & 0.7 \\ 0.01 & 0.02 & 0.5 & 0.47 \end{pmatrix}.$$

Alternatively, we may rewrite it as

$$\begin{pmatrix} 0.4 & 0.6 & 0 & 0 \\ 0.2 & 0.8 & 0 & 0 \\ 0 & 0 & 0.3 & 0.7 \\ 0 & 0 & 0.5 & 0.5 \end{pmatrix} + 0.01 \begin{pmatrix} -2 & 0 & 1 & 1 \\ 1 & -2 & 1 & 0 \\ 0 & 1 & -1 & 0 \\ 1 & 2 & 0 & -3 \end{pmatrix}.$$

Choose $\varepsilon = 0.01$. Then the transition matrix has the form

$$P^\varepsilon = P + \varepsilon Q. \tag{1.1}$$

The one-step transition matrix P^ε given in (1.1) is the basic form to be used in this book. In (1.1), $P = (p^{ij})$ itself is a transition probability matrix and $Q = (q^{ij})$ is a generator of a continuous-time Markov chain (i.e., $q^{ij} \geq 0$, for $j \neq i$, $\sum_j q^{ij} = 0$ for each i). Much of the subsequent study can be extended to certain time-inhomogeneous Markov chains, where both P and Q vary with respect to the time parameter. Nevertheless, to keep the notation simple, we will mainly tackle the time-homogeneous case throughout the book.

In subsequent chapters, we examine asymptotic properties of such systems as $\varepsilon \to 0$. We derive properties associated with the probability distribution vectors and transition probability matrices. We then further our understanding of suitably scaled occupation measures. To integrate analytic and probabilistic methods allows us to have a comprehensive understanding of the structures of the Markovian models, leading to a systematic treatment for systems involving time-scale separation.

To demonstrate, consider the following schematic illustration. Suppose that the state space of the underlying Markov chain has many states clustered into several groups or classes as depicted in Figure 1.1 (a). Naturally, we will try to lump all the states in each cluster into a "super" state. Figure 1.1 (b) shows the aggregated super-states represented by circles.

To solve many problems arising from applications, a clear understanding of the properties of the Markov chains is of crucial importance. The subject matter we study in this book is at the intersection of singular perturbation theory and stochastic processes. Because of their prevalence, this work focuses on stochastic systems in discrete time involving multi-time scales.

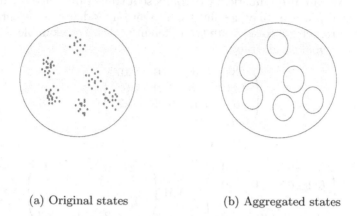

(a) Original states   (b) Aggregated states

FIGURE 1.1. Demonstration of aggregation for Markov chains with many states

Why should we use discrete-time models? In practice, it is convenient to model the underlying system in discrete time mainly because measurements are only available at discrete instants. In addition, simulation and digitization often lead to discrete-time systems. Moreover, owing to the rapid progress in computer control technology, the study of sampled-data systems has become more and more popular. As a result, the planning decisions, strategic policies, and control actions of the underlying systems are often made at discrete times as well. All these make the study of discrete-time systems necessary. Furthermore, many continuous-time models can only be treated via an approximation of discrete-time systems. For example, using a dynamic programming (DP) approach to solve a continuous-time stochastic control problem with Markovian driving noise yields the so-called HJB (Hamilton-Jacobi-Bellman) equation; see Fleming and Rishel [57]. The HJB equations frequently need to be solved numerically via discretization; see Yin and Zhang [158, Chapter 10] and the references therein. In addition, discrete-time systems have distinct features that are very different from their continuous-time counterparts. Because of the aforementioned reasons, this book focuses on discrete-time Markov models.

1.2   Brief Literature Review

For a Markov chain with a finite but large state space, a decomposition approach is often attractive. Ideally, we would like to divide the underlying problem into subproblems that can be solved completely independently; we can then paste together the solutions of the subproblems to obtain the solution to the entire problem. Consider a homogeneous Markov chain. If, for example, the transition matrix is decomposable into several sub-transition matrices (in a diagonal block form), the problem can be solved easily by the aforementioned decomposition methods. Unfortunately, the real world is not ideal. Rather than complete decomposability, one frequently encounters nearly completely decomposable cases. One of the main focuses of this book is to treat such models via the formulation of discrete-time Markovian systems with a two-time-scale approach. Ando and Fisher proposed the so-called nearly completely decomposable matrix models; see Simon and Ando [138]. Such a notion has subsequently been applied in queueing networks for organizing resources in a hierarchical fashion, in computer systems for aggregating memory levels, and in economics for reduction of complexity of large-scale systems; see Courtois [41].

Recent advances in the study of large-scale systems, for example, in production planning, have posed new challenges and provided opportunities for an in-depth understanding of two-time-scale or singularly perturbed Markov chains; see Delebecque and Quadrat [44], Pan and Başar [119], Pervozvanskii and Gaitsgory [123], Phillips and Kokotovic [125], Sethi and Zhang [136], and Yin and Zhang [158], among others. As alluded to previously, for real-world problems, one often faces large-scale systems with uncertainty. Using the idea of hierarchical decomposition and aggregation to deal with a Markovian system enables us to treat a much simpler system with less complexity; see Sethi and Zhang [136] for flexible manufacturing systems, and see also Avramovic, Chow, Kokotovic, Peponides, and Winkelman [8] and Chow, Winkelman, Pai and Sauer [36] for applications to power systems. From a modeling point of view, this amounts to setting up the problems involving different time scales that results in a singularly perturbation formulation. Subsequently, singular perturbation methodology can be used to solve the problem. Here, singular perturbation is interpreted in a broader sense, including both deterministic perturbation methods and stochastic averaging. For general references on singular perturbation methods, we refer the reader to Bogoliubov and Mitropolskii [24], Kevorkian and Cole [80], O'Malley [118], Vasil'eava and Butuzov [145], Wasow [148], and the references therein.

To gain a basic understanding of such systems, it is important to learn the structural properties of the Markov chains. While continuous-time Markovian models were treated in Khasminskii, Yin, and Zhang [85, 86], Pan and Başar [119], Phillips and Kokotovic [125], Sethi and Zhang [136], Yin and Zhang [158], two-time-scale approaches for discrete-time systems were used in various applications in telecommunications, Markov decision processes, control and optimization problems; see Abbad, Filar, and Bielecki [2], Avrachenkov, Filar, and Haviv [7], Blankenship [21], Hoppensteadt and Miranker, [69], Naidu [117], Tse, Gallager and Tsitsiklis [144]. Under a somewhat different setup, averaging of switching and diffusion approximations were analyzed in Anisimov [6].

Two-time-scale stochastic systems have been studied by a host of researchers throughout the years. Khasminskii [81] established a stochastic version of the averaging principle, and brought forward the notion of fast and slow processes in [82]; see also related references in Skorohod [140]. Kushner [96, 97] treated two-time-scale systems in the form of a pair of diffusions and studied control, optimization, and filtering problems, and introduced the notion of near optimality. Recently, Kabanov and Pergamenshchikov [76] considered asymptotic analysis and control of two-time-scale systems. Using analytic methods to tackle probabilistic problems was considered by Papanicolaou [121]. Friedlin and Wentzel [59] examined large deviations of stochastic systems from a random perturbation perspective.

1.3 Motivational Examples

To motivate the subsequent study further, we provide several examples in what follows. They include a manufacturing system, a Markov decision process, a problem arising from discrete optimization, an internet package transmission model, and a parameter identification problem with applications to wireless communication in CDMA/DS (code-division multiple-access/implemented with direct-sequence) signals. We show how a variety of applied problems can be cast into the framework of two-time-scale Markovian systems.

Example 1.1. Consider a manufacturing system in discrete time. Suppose that the system is given by

$$x_{k+1} = x_k + \varepsilon(u_k - z_k), \quad k = 0, 1, 2, \ldots, \quad x_0 \text{ is given}, \tag{1.2}$$

where x_k represents the surplus, u_k is the rate of production, and z_k is the rate of demand at time k. Note that the step size ε in (1.2) is replaceable by a quantity $O(\varepsilon)$ (a quantity that is of the same order as ε). There is a production capacity constraint $0 \leq u_k \leq \alpha_k^\varepsilon$ where α_k^ε, a discrete-time Markov chain (depending on ε) with state space $\mathcal{M} = \{1, \ldots, m_0\}$, represents the capacity of the machine. Our objective is to choose the control subject to both (1.2) and the capacity constraint so that the cost function

$$J^\varepsilon(x, \alpha, u) = E \sum_{k=0}^{T/\varepsilon} \varepsilon g(x_k, u_k, \alpha_k^\varepsilon)$$

is minimized, where $x = x_0$, $\alpha = \alpha_0$, and $u = \{u_k\}$. With the aid of a continuous-time-flow approximation, continuous-time counterparts of such models have been studied extensively in Sethi and Zhang [136] among others. Since in daily operation, the production decision is often made at discrete time, it is also important to examine a model under a discrete-time formulation as given above. What is the rationale for the use of the small parameter $\varepsilon > 0$ in α_k^ε? Suppose that we have a production system consisting of two machines that are subject to breakdown and repair. Assume each of the machines has two states, up (denoted by 1) and down (denoted by 0). Then the system has four states $\{(1,1),(0,1),(1,0),(0,0)\}$. Suppose that the transition probability matrix is

$$\widehat{P} = \begin{pmatrix} 9/20 & 1/2 & 1/20 & 0 \\ 1/2 & 9/20 & 0 & 1/20 \\ 1/20 & 0 & 9/20 & 1/2 \\ 0 & 1/20 & 1/2 & 9/20 \end{pmatrix}.$$

To distinguish the different levels of reliability, we introduce a small parameter ε. Using $\varepsilon = 1/10$ yields

$$\widehat{P} = P^\varepsilon = \begin{pmatrix} 1/2 & 1/2 & 0 & 0 \\ 1/2 & 1/2 & 0 & 0 \\ 0 & 0 & 1/2 & 1/2 \\ 0 & 0 & 1/2 & 1/2 \end{pmatrix} + \varepsilon \begin{pmatrix} -1/2 & 0 & 1/2 & 0 \\ 0 & -1/2 & 0 & 1/2 \\ 1/2 & 0 & -1/2 & 0 \\ 0 & 1/2 & 0 & -1/2 \end{pmatrix}.$$

The transition matrix above is a nearly completely decomposable model. The near-complete decomposability means that the transition matrix above is a sum of decomposable transition matrix with an added perturbation of the order $O(\varepsilon)$.

Example 1.2. Many applications in resource allocation, queueing networks, machine replacement, etc. can be formulated as Markov decision processes (MDPs). Classical treatments of discrete-time-MDP models may be found in Derman [45] and White [149], among others. Consider a discrete-time Markov chain α_k with finite state space $\mathcal{M} = \{1, \ldots, m_0\}$. The transition probability of α_k is action or control dependent, which is denoted by $\alpha_k \sim P(u)$, where u is a control variable. Let Γ denote the control space. We use feedback control $u_k = u(\alpha_k)$, such that $u_k \in \Gamma$, $k \geq 0$. In practical models, the state space $\mathcal{M}$ is large, for example, it is typical for a communication network to have a large number of nodes. A classical approach for solving the MDP problem uses dynamic programming. Nevertheless, such an approach is computationally feasible only if the size of the problem is small or moderate. For large-scale systems, one has to resort to near-optimal schemes.

Consider the Markov chain $\alpha_k = \alpha_k^\varepsilon$ generated by $P^\varepsilon(u)$, where

$$P^\varepsilon(u) = (p^{\varepsilon,ij}(u)) = P(u) + \varepsilon Q(u)$$

is the transition probability matrix, where for each $u \in \Gamma$, $P(u) = (p^{ij}(u))$ is itself an $m_0 \times m_0$-dimensional probability transition matrix, and $Q(u) = (q^{ij}(u))$ is an $m_0 \times m_0$-dimensional generator. Our objective is to find an admissible control u to minimize a cost functional

$$J^\varepsilon(\alpha, u) = E\varepsilon \sum_{k=0}^{\infty} (1 - \beta\varepsilon)^k g(\alpha_k^\varepsilon, u(\alpha_k^\varepsilon)), \tag{1.3}$$

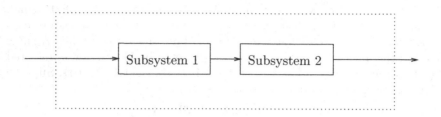

FIGURE 1.2. Two subsystems in tandem

where $\alpha = \alpha_0^\varepsilon$ is the initial state of the chain, $g(\alpha, u)$ is the cost-to-go function, and $\beta > 0$ is a discount factor. In (1.3), the multiplier ε in J^ε acts as a normalizer to ensure the sum to be finite.

To solve the optimal decision making problem requires intensive computation if m_0 is large. To alleviate the computational effort, we search for a near-optimal policy instead. Using asymptotic expansions (to be presented in Chapter 3), we will be able to show that as $\varepsilon \to 0$, the original problem is approximated by a limit problem. The main idea is to lump all the states in each recurrent class into a single state. Thus in the reduced or the limit problem, the underlying process has only l_0 states, where l_0 is the total number of recurrent classes and $m_0 \gg l_0$. Since the computational effort in solving the DP equations depends largely on the total number of the equations to be solved, the required effort in solving the limit problem can be significantly reduced. Then the optimal policy of the limit problem can be used to construct policy for the original problem, which yields a near-optimal policy.

For instance, consider a manufacturing system consisting of two subsystems in tandem; see Figure 1.2. Suppose that each subsystem has ten states denoted by $1, 2, \ldots, 10$. The entire system has 100 states listed as

$$\mathcal{M} = \{s_{1,1}, \ldots, s_{1,10}, \ldots, s_{10,1}, \ldots, s_{10,10}\},$$

where $s_{i,j} = (i, j)$, for $i, j = 1, 2, \ldots, 10$. We can take the control variable u to be a vector representing the rates of preventive maintenance and allocated repair resources. The objective is to choose u to keep the average machine capacity at a reasonable level and to avoid excessive preventive maintenance and repair costs. Let $P^1(u)$ and $P^2(u)$ denote the transition probability matrices of the first and second subsystems, respectively. Consider the situation such that the state of the first system is changing more rapidly than that of the second one. Then the corresponding transition probability matrix can be written as

$$P^\varepsilon(u) = \begin{pmatrix} P^2(u) & & \\ & \ddots & \\ & & P^2(u) \end{pmatrix} + \varepsilon \begin{pmatrix} q^{1,1}(u)I & \cdots & q^{1,10}(u)I \\ & \vdots & \\ q^{10,1}(u)I & \cdots & q^{10,10}(u)I \end{pmatrix},$$

where $Q(u) = (q^{i,j}(u)) = P^1(u) - I$. In this case, the number of the DP equations of the original problem is $m_0 = 100$, while the number for the limit problem (to be specified in Chapter 8) is only $l_0 = 10$.

Example 1.3. Let $\{\eta_m^1\}$ be a Markov chain with state space $\{1, 2\}$ and transition matrix $P^0 = (p^{0,ij})$. Let $\{\eta_k^2\}$ be another Markov chain with state space $\{s_1, s_2\}$ whose transition matrix depends on the value of η_m^1. In fact, for $i = 1, 2$, corresponding to $\eta_m^1 = i$, the transition matrix of η_k^2 is given by P^i. Here η_m^1 is a slow process with $m \in \{n, 2n, 3n, \ldots\}$ for a positive integer n and η_k^2 is a fast process with $k \in \{1, 2, 3, \ldots\}$. Define a joint process $\eta_l = (\eta_m^1, \eta_k^2)$ if $mn \le l < (m+1)n$, for $m = 0, 1, 2, \ldots$ Then the state space of $\{\eta_l\}$ is $\mathcal{M} = \{(1, s_1), (1, s_2), (2, s_1), (2, s_2)\}$. Note that $\{\eta_l\}$ is *not* a Markov chain. However, for a fixed n, the "n-step process" $\{\eta_{mn} : m = 0, 1, 2, \ldots\}$ is a Markov chain with transition matrix

$$\widetilde{P} = \begin{pmatrix} p^{0,11}I & p^{0,12}I \\ p^{0,21}I & p^{0,22}I \end{pmatrix} \begin{pmatrix} (P^1)^n & \\ & (P^2)^n \end{pmatrix},$$

where $P^0 = (p^{0,ij})$ and I is the 2×2 identity matrix. Assume that P^1 and P^2 are irreducible and aperiodic. Then there exist stationary distributions ν^1 and ν^2 corresponding to P^1 and P^2, respectively. Let $\overline{P}^i = (\nu^i, \nu^i)'$, for $i = 1, 2$. For n large enough, we have

$$\widetilde{P} \sim \begin{pmatrix} p^{0,11}I & p^{0,12}I \\ p^{0,21}I & p^{0,22}I \end{pmatrix} \begin{pmatrix} \overline{P}^1 & \\ & \overline{P}^2 \end{pmatrix}.$$

In addition, assume that $P^0 = \exp(\overline{Q})$, where

$$\overline{Q} = \begin{pmatrix} -q^1 & q^1 \\ q^2 & -q^2 \end{pmatrix},$$

for some $q^1 > 0$ and $q^2 > 0$. Let $\alpha_k^{\varepsilon,1}$ and $\alpha_k^{\varepsilon,2}$ be Markov chains such that $\alpha_k^{\varepsilon,1}$ is slowly varying and $\alpha_k^{\varepsilon,2}$ is fast changing. Consider $\alpha_k^\varepsilon = (\alpha_k^{\varepsilon,1}, \alpha_k^{\varepsilon,2})$ with state space $\mathcal{M}$ and transition matrix

$$P^\varepsilon = \begin{pmatrix} P^1 & \\ & P^2 \end{pmatrix} + \varepsilon \begin{pmatrix} -q^1 I & q^1 I \\ q^2 I & -q^2 I \end{pmatrix}.$$

Using Proposition 6.2 and $P^0 = \exp(\overline{Q})$, we can show that

$$(P^\varepsilon)^n \sim \begin{pmatrix} p^{0,11}I & p^{0,12}I \\ p^{0,21}I & p^{0,22}I \end{pmatrix} \begin{pmatrix} \overline{P}^1 & \\ & \overline{P}^2 \end{pmatrix},$$

when $\varepsilon = 1/n$. Thus, with fixed n, α_{mn}^ε has a transition matrix which is close to $\overline{P}$. Consequently, $\alpha_k^{\varepsilon,1}$ can be regarded as an approximation to η_m^1 and $\alpha_k^{\varepsilon,2}$ an approximation to η_k^2. The original two-level structure is recaptured by using the two-time-scale model in a natural way. The advantage of the two-time-scale formulation is that it enables us to simplify the computation via a hierarchical approach. There are many applications involving the two-level formulation. These include, but are not limited to, the multiarmed bandit problem, admission control with buffer management, asset allocation, and employee staffing. We refer the reader to the paper by Chang, Fard, Marcus, and Shayman [33] for further details.

Example 1.4. Consider a discrete optimization problem. Given an objective function $f : \mathcal{M} \to \mathbb{R}$, the problem is to find

$$\iota_* = \arg\min_{\iota \in \mathcal{M}} f(\iota),$$

where $\mathcal{M} = \{1, \ldots, m_0\}$ is the search space and each element of $\mathcal{M}$ is called a design. In the problem of interest, there is no analytic expression of $f(\iota)$ available or the formula is too complicated and/or too difficult to evaluate. Thus we have to solve the problem by using observed values of the objective function. Let $h(\iota)$ be an estimate of $f(\iota)$. We need to use a form of Monte Carlo algorithm to resolve the issue. Let α_k denote the estimate of ι_* at the end of the kth iteration of the algorithm and suppose that $\alpha_k = \iota_k \in \mathcal{M}$. At the $(k+1)$th iteration, another candidate $\widetilde{\alpha}_{k+1}$ is randomly chosen from $\mathcal{M} - \{\iota_k\}$. We assume that there is some $\eta > 0$ such that

$$P(\widetilde{\alpha}_{k+1} = \ell_k | \alpha_k = \iota_k) \geq \eta \tag{1.4}$$

for all k and all $\ell_k \in \mathcal{M} - \{\iota_k\}$. For instance, $\widetilde{\alpha}_{k+1}$ may be randomly chosen as uniformly distributed in $\mathcal{M} - \{\iota_k\}$. Next, we generate M_k independent and identically distributed (i.i.d.) samples for both $h(\iota_k)$ and $h(\ell_k)$. Let $\zeta_k = h(\iota_k) - h(\ell_k)$, and let $h^l(\cdot)$ be the lth sample of $h(\cdot)$. Then we have the i.i.d. samples $\zeta_k^l = h^l(\iota_k) - h^l(\ell_k)$, $l = 1, \ldots, M_k$. To update the estimate of ι_*, use the testing hypotheses

$$H_0 : \ g(\iota_k) - g(\ell_k) < 0,$$
$$H_1 : \ g(\iota_k) - g(\ell_k) > 0$$

based on ζ_k^l, $l = 1, \ldots, M_k$, and set $\alpha_{k+1} = \widetilde{\alpha}_{k+1}$ if H_1 is accepted; $\alpha_{k+1} = \alpha_k$, otherwise. Testing is accomplished by defining a function $t_k : R^{M_k} \to R$ and a test statistic $T_k = t_k(\zeta_k^1, \ldots, \zeta_k^{M_k})$, and by using the decision rule:

$$\alpha_{k+1} = \iota_k \ \text{if} \ T_k < \tau_k;$$
$$= \ell_k \ \text{otherwise.}$$

The threshold τ_k could be different for different k's. The iterations proceed until a stopping criterion is met. The values of the threshold sequence $\{\tau_k\}$ and the sample-size sequence $\{M_k\}$, together with the forms of the test statistics $\{T_k\}$ and the distributions of the samples determine the convergence properties of the algorithm. It is clear that we would like the probability of detection (i.e., the probability that $\alpha_{k+1} = \ell_k$ when H_1 is true) to be large and the probability of false alarm (i.e., the probability that $\alpha_{k+1} = \ell_k$ when H_0 is true) to be small. However, these are conflicting goals and must be balanced in order to have a convergent algorithm.

The algorithm so constructed was shown to be convergent in the sense of w.p.1 (with probability one) in Gong, Kelly, and Zhai [64] by noting that α_k is a Markov chain. Using the two-time-scale approach, the rate of convergence issues can be studied with the help of the asymptotic results presented in this book. It is an interesting application of the singularly perturbed Markov chains under weak irreducibility (to be defined in Chapter 2).

To begin, we partition the state space $\mathcal{M}$ as

$$\mathcal{M} = \{1, 2, \ldots, m_0 - 1\} \cup \{m_0\} = \mathcal{M}_* \cup \mathcal{M}_a,$$

where $\mathcal{M}_*$ contains all of the transient states and $\mathcal{M}_a$ includes the absorbing state $m_0 = \iota_*$. If the Markov chain initially belongs to $\mathcal{M}_*$, it will eventually be absorbed into $\mathcal{M}_a$. Nevertheless, the current situation differs from the classical notion of the absorbing state in Markov chains in that the last row of the transition matrix is not of the form $(0, 0, \ldots, 1)$, but only "close" to it. That is, the state $\{m_0\}$ is not absorbing in the classical sense but is only "approximately absorbing." To accommodate this scenario, we introduce a small perturbation in the transition matrix so that none of the last row of the transition matrix will be 0 and the last element (corresponding to the absorbing state) is close to 1. In addition, it is possible for the transition matrix to be time dependent. Generally, studying the rate of convergence of the underlying algorithm is difficult, which can be resolved by using the techniques of time-scale separation of Markov chains.

Introducing a small parameter $\varepsilon > 0$, we let the transition matrix of the Markov chain $\alpha_k = \alpha_k^\varepsilon$, $k = 0, 1, \ldots, \widetilde{T}/\varepsilon$ for some $\widetilde{T} > 0$, be

$$P^\varepsilon = P + \varepsilon Q, \tag{1.5}$$

where P is a transition matrix of the form

$$P = \begin{pmatrix} p^{11} & \cdots & p^{1, m_0 - 1} & p^{1 m_0} \\ \cdots & \cdots & \cdots & \cdots \\ p^{m_0 - 1, 1} & \cdots & p^{m_0 - 1, m_0 - 1} & p^{m_0 - 1, m_0} \\ 0 & 0 & 0 & 1 \end{pmatrix} \tag{1.6}$$

and Q is a generator of a continuous-time Markov chain. Under such a formulation, the rate of convergence can be obtained. In fact, we have

$$p_k^\varepsilon = \nu + O(\varepsilon + \lambda^k), \quad \text{for some } 0 < \lambda < 1, \tag{1.7}$$

where $\nu = (0, \ldots, 0, 1)$ and p_k^ε denotes the probability distribution vector at time k. The meaning of the small parameter $\varepsilon > 0$ is the ratio of the probability of detection to that of false alarm.

Example 1.5. Wireless technology has provided great opportunities to modern communication applications. In the meantime, owing to their special features of mobility, wireless links usually incur more errors than that do wired channels, which is partially due to such characteristics as path loss, interference, and fading, among others. For mobile radio channels, Rayleigh fading is one of the most commonly used models. A Markov model is used to describe the fading channels; see Wang and Chang [146]. Using the packet as a unit, which consists of a number of bits, signals are transmitted through the channel. Note that a packet sent is correct if all the bits transmitted are correct.

The wireless channel is modeled as a discrete-time Markov chain with two states. That is, the state space of the Markov chain is $\mathcal{M} = \{1, 2\}$, where 1 denotes a "good" state and 2 denotes a "bad" state. No matter which state the channel is in, assume that the sequence of bits transmitted is a sequence of i.i.d. Bernoulli random variables taking values a=correct and b=incorrect. Thus the BER (bit error rate) of the wireless channel can be determined by the channel states. If the channel is in a "good" state, the BER is low, whereas if the channel is in a "bad" state, the BER is high. Assume that the Markov chain is time-inhomogeneous. That is, the transition probability matrix is time varying. Suppose that the transition probability matrix is given by

$$P_k^\varepsilon = I + \varepsilon Q_k, \tag{1.8}$$

where I is a 2×2 identity matrix and Q_k is a sequence of generators. Then we can proceed to analyze the packet transmission error model. The form (1.8) indicates that the transition matrix at any time k is close to an identity. For example, if the channel is currently in a good state, the probability that it will jump to a bad state in the next move is small. However, the term Q_k indicates that such a move is not entirely impossible.

In view of the formulation given in (1.8), the model again involves two time scales. This book presents methods that can be used in analyzing such two-time-scale Markovian models.

Example 1.6. This example is motivated by emerging applications in wireless communications, in particular CDMA signals (see Krishnamurthy and Yin [89] and the references therein). The inherent nature of the underlying system is that it includes a time-varying parameter where the parameter jump changes are not too frequent. Our objective is to track the time-varying signals. Let $\{y_n\}$ be a sequence of real-valued signals representing the observations obtained at time n, and let $\{\theta_n\}$ be the time-varying true parameter, an $\mathbb{R}^r$-valued random process. Suppose that

$$y_n = \varphi_n' \theta_n + e_n, \tag{1.9}$$

where $\varphi_n \in \mathbb{R}^r$ is the regression vector and $\{e_n\}$ (with $e_n \in \mathbb{R}$) is a sequence of zero mean estimation errors.

Suppose that there is a small parameter $\varepsilon > 0$ and that θ_n is a discrete-time homogeneous Markov chain, whose state space is

$$\mathcal{M} = \{\overline{\theta}^1, \ldots, \overline{\theta}^{m_0}\},$$

and whose transition probability matrix is given by

$$P^\varepsilon = I + \varepsilon Q, \tag{1.10}$$

where I is an $\mathbb{R}^{m_0 \times m_0}$ identity matrix and $Q = (q^{ij})$ is an $\mathbb{R}^{m_0 \times m_0}$ dimensional generator of a continuous-time Markov chain. The rationale is that the true parameter θ_n given in (1.9) is time varying, but the variation is small. It hovers about a constant value with occasional jumps to other possible locations.

To track the parameter $\{\theta_n\}$, we use the LMS-type (least mean squares) adaptive filtering procedure and construct a sequence of estimates $\{\widehat{\theta}_n\}$ according to

$$\widehat{\theta}_{n+1} = \widehat{\theta}_n + \mu \varphi_n (y_n - \varphi_n' \widehat{\theta}_n), \tag{1.11}$$

where $\mu > 0$ is a small constant step size.

An important problem is to figure out the bounds on the deviation $\widetilde{\theta}_n = \widehat{\theta}_n - \theta_n$. Under suitable conditions and using asymptotic results of the two-time-scale Markov chains to be presented in this book, we can derive mean squares error bounds for the tracking error sequence $\widetilde{\theta}_n$. The mean squares bounds will assist us further to obtain a weak convergence result of a suitably scaled sequence. In addition, we will be able to find probabilistic bounds on $P(|\widehat{\theta}_n - \theta_n| > \eta)$ for $\eta > 0$.

Example 1.7. This example is concerned with a numerical scheme for the approximation of regime-switching diffusions. Let $\alpha(t)$ be a finite-state Markov chain in continuous time with state space $\mathcal{M} = \{1, \ldots, m_0\}$ and generator Q. Let $f(\cdot, \cdot) : \mathbb{R}^r \times \mathcal{M} \mapsto \mathbb{R}^r$, and $g(\cdot, \cdot) : \mathbb{R}^r \times \mathcal{M} \mapsto \mathbb{R}^{r \times r}$ be appropriate functions. Consider the system of regime-switching diffusions

$$dx(t) = f(x(t), \alpha(t))dt + g(x(t), \alpha(t))dw(t),$$
$$x(0) = x \in \mathbb{R}^r, \ \alpha(0) = \alpha \in \mathcal{M}, \tag{1.12}$$

where $w(\cdot)$ is an r-dimensional standard Brownian motion, and $f(\cdot)$ and $g(\cdot)$ satisfy certain regularity conditions. Assume that the Markov chain $\alpha(\cdot)$ and the Brownian motion $w(\cdot)$ are independent. Assume also that the stochastic differential equation with regime switching (1.12) has a unique solution in distribution for each initial condition.

Frequently, systems of the form (1.12) can only be solved numerically, which makes appropriate discretization and numerical algorithms necessary. One of the discretization techniques used is the Euler-Maruyama approximate solutions; see Kloeden and Platen [88]. Since the Markov chain $\alpha(t)$ has a constant generator Q, it is time homogeneous and its transition probability matrix $P(t) = (p_{ij}(t))$ satisfies the forward equation

$$\dot{P}(t) = P(t)Q, \quad \text{with} \quad P(0) = I. \tag{1.13}$$

Corresponding to the continuous-time Markov chain $\alpha(t)$ generated by Q, choose any positive real number ε. Define $\alpha_n = \alpha(\varepsilon n)$, $n \geq 0$. That is, $\{\alpha_n\}$ is an ε-skeleton of the continuous-time Markov chain $\alpha(t)$; see Chung [38, p.132]. It can be shown that α_n is a discrete-time Markov chain with one-step transition probability matrix

$$P = (p_{ij})_{m_0 \times m_0} = \exp(\varepsilon Q). \tag{1.14}$$

Now, we can proceed to develop a constant-step size Euler-Maruyama approximation for the SDE with regime switching (1.12) by

$$\begin{cases} x_{n+1} = x_n + f(x_n, \alpha_n)\varepsilon + g(x_n, \alpha_n)\Delta w_n, \\ x_0 = y, \ \alpha_0 = \alpha, \end{cases} \tag{1.15}$$

where $\Delta w_n = w(\varepsilon(n+1)) - w(\varepsilon n)$. Note that since $w(\cdot)$ is a standard r-dimensional Brownian motion, it has independent increments. Thus, $\{\Delta w_n\}$ is a sequence of i.i.d. Gaussian random variables such that $E\Delta w_n = 0$ and the covariance is given by εI.

To simplify the calculation further, we take advantage of the appearance of the small parameter ε, and exploit the ε-dependence in the transition matrix. Taking Taylor expansions of the transition matrix in (1.14), we obtain

$$\exp(\varepsilon Q) = \sum_{i=0}^{\infty} \frac{(\varepsilon Q)^i}{i!} = I + \varepsilon Q + O(\varepsilon^2).$$

Thus the first approximation of the transition matrix is given by

$$P^\varepsilon = I + \varepsilon Q, \tag{1.16}$$

which falls in the two-time-scale Markov chain model we are considering in this book. In view of the Taylor expansions above, in lieu of $\exp(\varepsilon Q)$, if we use the transition matrix (1.16), the problem can be much simplified.

The above examples present a multitude of applications involving two-time-scale Markov chains. Another rich source, which is not discussed extensively in this book, comes from queueing systems and networks. Although queues with infinite capacity are frequently treated, there are many important classes of finite-capacity queues; see for example Sharma [137] and the references therein. The two-time-scale formulation can be naturally imbedded in these queueing applications. In addition, for computational purpose, one often has to use finite-capacity queues to approximate queues with infinitely many waiting rooms. Thus, the problem reduces to that of a finite-state Markov chain with a large state space.

1.4 Discrete-Time vs. Continuous-Time Models

There is a close connection between continuous-time, singularly perturbed Markov chains and their discrete-time counterparts. The discussion below, inspired by the ideas exploited in Kumar and Varaiya [91] and motivated by our work in Yin and Zhang [158], reveals the natural connection.

Consider a continuous-time Markov chain $\alpha^\varepsilon(t)$ whose state space and generator are given by $\mathcal{M} = \{1, \ldots, m_0\}$ and $\widetilde{Q}/\varepsilon + \widehat{Q}$, respectively, where $\widetilde{Q}$ and $\widehat{Q}$ are generators of some Markov chains with stationary transition probabilities. Then

$$p^\varepsilon(t) = (P(\alpha^\varepsilon(t) = 1), \ldots, P(\alpha^\varepsilon(t) = m_0)),$$

the probability vector, satisfies the forward equation

$$\dot{p}^\varepsilon(t) = p^\varepsilon(t) \left(\frac{1}{\varepsilon}\widetilde{Q} + \widehat{Q} \right), \ p(0) = p_0.$$

Introduce a new variable $\gamma = t/\varepsilon$. The probability vector $p(\gamma) = p^\varepsilon(t)$ satisfies the rescaled forward equation

$$\dot{p}(\gamma) = \frac{dp(\gamma)}{d\gamma} = p(\gamma)(\widetilde{Q} + \varepsilon\widehat{Q}).$$

Denote $Q^\varepsilon = \widetilde{Q} + \varepsilon\widehat{Q}$, where $\widetilde{Q} = (\tilde{q}^{ij})$ and let

$$q_0 = \max_i(|\tilde{q}^{ii}|) = \max_i \sum_{j \neq i} \tilde{q}^{ij},$$

and fix $q > q_0$. Define

$$\widetilde{P}^\varepsilon = \left(I + \frac{1}{q}\widetilde{Q} \right) + \varepsilon \left(\frac{1}{q}\widehat{Q} \right).$$

Then all entries of $\widetilde{P}^\varepsilon$ are nonnegative and $\widetilde{P}^\varepsilon \mathbb{1}_{m_0} = \mathbb{1}_{m_0}$, for ε small enough. Therefore, $\widetilde{P}^\varepsilon$ is a transition probability matrix, and

$$\dot{p}(\gamma) = p(\gamma)[q(\widetilde{P}^\varepsilon - I)].$$

The formal solution of the above forward equation is

$$p(\gamma) = p_0 \exp(\gamma q(\widetilde{P}^\varepsilon - I))$$
$$= p_0 \exp(-\gamma q) \sum_{j=0}^{\infty} \frac{(\gamma q)^j}{j!} (\widetilde{P}^\varepsilon)^j. \tag{1.17}$$

Consider a discrete-time Markov chain $\widetilde{\alpha}_k^\varepsilon$ having transition matrix $\widetilde{P}^\varepsilon$. Then the corresponding probability vector

$$\widetilde{p}(k) = (P(\widetilde{\alpha}_k^\varepsilon = 1), \dots, P(\widetilde{\alpha}_k^\varepsilon = m_0))$$

with $\widetilde{p}(0) = p_0$ satisfies $\widetilde{p}(k) = p_0(\widetilde{P}^\varepsilon)^k$. This, together with $p(\gamma)$, the solution of the forward equation (1.17), yields

$$p(\gamma) = \sum_{j=0}^{\infty} \exp(-\gamma q) \frac{(q\gamma)^j}{j!} \widetilde{p}(j).$$

Let $\xi(\gamma)$ be a Poisson process with rate q, which is independent of the chain $\widetilde{\alpha}_k^\varepsilon$. Then

$$P(\xi(\gamma) = k) = \exp(-\gamma q) \frac{(q\gamma)^k}{k!}, \ k = 0, 1, 2, \dots$$

Let $\widehat{\alpha}^\varepsilon(\gamma)$ be a continuous-time process obtained from $\xi(\gamma)$ and $\widetilde{\alpha}_k^\varepsilon$ by $\widehat{\alpha}^\varepsilon(\gamma) = \widetilde{\alpha}_{\xi(\gamma)}^\varepsilon$. Then

$$P(\widehat{\alpha}^\varepsilon(\gamma) = i) = \sum_{k=0}^{\infty} P(\widetilde{\alpha}_k^\varepsilon = i) P(\xi(\gamma) = k).$$

This implies

$$P(\widehat{\alpha}^\varepsilon(\gamma) = i) = p(\gamma) = P(\alpha^\varepsilon(\varepsilon\gamma) = i).$$

Concerning the time-scale separation, we note: In a continuous-time setting, we work in a finite time horizon $t \in [0, T]$, whereas in a discrete-time formulation, the time horizon is of the order $O(1/\varepsilon)$, so we work with $0 \leq k \leq \lfloor T/\varepsilon \rfloor$, where $\lfloor T/\varepsilon \rfloor$ denotes the integer part of T/ε. We also note that the time-space separation method considered in the deterministic discrete-time singular perturbation problems by Naidu [117], which mainly dealt with boundary value problems, cannot be carried over to our formulation. Nevertheless, the idea of time-scale separation can still be used and asymptotic expansions can still be constructed for discrete-time Markov chains owing to the work of Hoppensteadt and Miranker [69].

1.5   Organization

This book consists of three parts with a total of fourteen chapters. The first part provides an overview of the book together with mathematical background materials. Part II comprises the development of asymptotic properties of two-time-scale Markov chains, including asymptotic expansions of the probability distribution vectors and probability matrices, structural properties of the Markov chains, and exponential-type large deviation bounds. The development is carried out by examining associated occupation measures. Part III presents several applications in stability, control, optimization, and related fields. In what follows, we give a chapter-by-chapter account of the topics to be covered.

Part I, consisting of Chapters 1 and 2, serves as a prologue. Chapter 1 contains the motivation for the study and a brief review of the literature, together with several illustrative examples. These examples involve multi-time-scale structure and exhibit the scope of the diversity of applications. Some of them will be revisited in the subsequent chapters in more detail.

Mathematical preliminaries are provided in Chapter 2, including the definition of Markov chains, basic notions such as Chapman–Kolmogorov equations, irreducibility, quasi-stationary distributions, Markov chains, martingales, and diffusions. It also collects certain properties of Markov chains, martingales, Gaussian processes, diffusions, and switching diffusions. This chapter can be used as a quick reference. Related readings are suggested at the end of the chapter for further study.

Part II, including Chapters 3–6, is devoted to asymptotic properties of two-time-scale or singularly perturbed Markov chains. It aims to provide an understanding of the intrinsic structural properties of such Markov chains.

In Chapter 3, we study the probability distribution of Markov chains with transition probability $P^\varepsilon = P + \varepsilon Q$, where $\varepsilon > 0$ is a small parameter, P is a transition matrix of a Markov chain, and Q is a generator of a continuous-time Markov chain. To analyze the asymptotics, we develop asymptotic expansions of probability vectors and transition matrices. Including outer expansions and initial layer corrections, the asymptotic expansions are matched through appropriate choices of the initial conditions. After the formal expansions are obtained, we justify the validity of the asymptotic series and obtain the approximation error bounds.

Chapter 4 is concerned with the asymptotic distribution of scaled occupation measures. The results in Chapter 3 are mainly based on analytic techniques, whereas Chapter 4 explores the sample path aspects through the examination of occupation measure. Using the asymptotic expansions, we first obtain mean squares type error bounds on the occupation measures. Assuming the dominating part of the transition matrix can be divided into a number of ergodic classes, we aggregate the states in each recurrent class into a single state. By means of martingale problem formulation, we show that an interpolated sequence of aggregated process converges weakly to a continuous-time Markov chain, whose generator can be explicitly computed. Subsequently, we take a suitable scaling of the occupation measures to obtain limit distribution results. A salient feature of the limit distribution is that it is a mixture of the Gaussian distribution with a Markov chain. That is, the limit turns out to be a switching diffusion process. Different from the usual central limit theorems, the covariance, in fact, involves a function of the limit of the aggregated Markov chain. In addition, the covariance of the limit switching diffusion depends on the asymptotic expansion (more specifically on the initial layer corrections of the transition probability matrices) in an essential way.

The derivations of the results of Chapter 4 such as the switching diffusion limits use essentially a weak convergence analysis. To accommodate the functional invariance theorem and to obtain estimates of rare event probabilities, built on the asymptotic properties, Chapter 5 presents exponential upper bounds for sequences of scaled occupation measures. They are large deviations bounds. These bounds can be conveniently applied to infinite horizon control and optimization problems involving Markov chains, such as controlled dynamic systems whose coefficients are modulated by a Markov chain.

Chapter 6 is an interlude. It summarizes certain results obtained in Chapters 3–5 to be used frequently in the subsequent chapters; a reader may bypass these chapters, if his or her main interests are in control, stability, and other applications covered in the latter parts of the book.

Part III deals with several applications, including stability, Markov decision processes, linear quadratic regulator modulated by a Markov chain, hybrid filtering, the mean variance control problem in financial engineering, production planning, and stochastic approximation. It is divided into seven chapters.

Concerning dynamic systems modulated by Markov chains, many problems in engineering applications require the understanding of the systems' long-term behavior. We present stability analysis of such systems governed by difference equations with regime switching modulated by a two-time-scale Markov chain in Chapter 7. To reduce complexity, one attempts to effectively "replace" the actual systems by a limit system with a simpler structure. To ensure the validity of such a replacement in a longtime horizon, it is crucial that the original system is stable. The fast regime changes and the large state space of the Markov chain make the stability analysis difficult. Under suitable conditions, using the limit dynamic systems and the perturbed Liapunov function methods, we show that if the limit systems are stable, so are the original systems. This justifies the replacement of a complicated original system by its limit from a longtime behavior point of view.

Chapter 8 is concerned with hybrid filtering problems in discrete time. Since numerous problems arising from target tracking, speech recognition, telecommunication, and manufacturing require solutions of filters for a hidden Markov chain, in addition to the random system disturbances and observation noise, we assume that the system under consideration is influenced by a Markov chain with finite state space. We show that a limit filtering problem can be derived in which the underlying Markov chain is replaced by an averaged chain and the system coefficients are averaged out with respect to the stationary measures of each ergodic class. The reduction of complexity is particularly pronounced when the transition matrix of the Markov chain consists of only one ergodic class.

Motivated by applications in resource allocation, queueing networks, machine replacement, and command control, we analyze Markov decision processes whose dynamics are governed by the control-dependent transition matrices $P^{\varepsilon}(u)$ in Chapter 9. A common practice is to use the method of dynamic programming. However, such an approach breaks down when the dimension of the underlying system is too large. Therefore, for large-scale systems, approximate-optimal schemes are necessary. By assuming the transition matrices of the form $P^{\varepsilon}(u) = P(u) + \varepsilon Q(u)$, we use hierarchical decomposition and aggregation to treat large-dimensional systems by examining a limit system that is much simpler with less complexity. One of the interesting aspects is that the limit problem derived is a continuous-time Markov decision problem, although the original one is in discrete time. Using the limit system, we construct controls of the original systems, leading to asymptotic optimality. Both discounted cost and long-run average cost criteria are considered.

Chapter 10 deals with linear quadratic regulator problems that involve regime or configuration switching. We focus on a hybrid linear model consisting of a large number of configurations modulated by a finite-state Markov chain. At any given instance, the system takes one of the possible system configurations, in which the coefficients depend on the state of the underlying Markov chain. Clearly, this model has a greater capability to account for various types of random disturbances in reality. Using hierarchical decomposition and aggregation methodology, we develop asymptotic optimal controls of such systems. Compared with the usual LQ formulation, in lieu of a single Riccati equation, we have to solve a system of Ricatti equations with the total number of equations equal to the number of Markovian states. In this case, the computational effort depends mainly on the number of Riccati equations to be solved. The solution methods presented are a viable alternative for reducing the computational effort. Suppose that corresponding to the original problem we need to solve m_0 Riccati equations (m_0 is equal to the number of states of the Markov chain). In the limit system, we only need to solve l_0 Riccati equations (l_0 is the total number of recurrent classes). If $l_0 \ll m_0$, a substantial reduction of computational effort is achieved.

In Chapter 11, we examine a discrete-time version of Markowitz's mean-variance portfolio selection problem where the market parameters depend on the market mode that jumps among a finite number of states. The

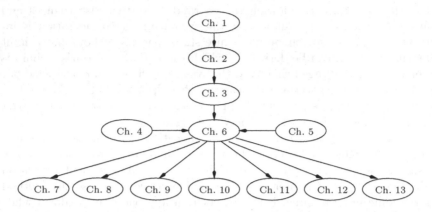

FIGURE 1.3. Relationship and dependence among chapters

random regime switching is delineated by a finite-state Markov chain, based on which a discrete-time Markov modulated portfolio selection model is formulated. Under broad conditions, we derive the limit problem that turns out to be a Markov modulated mean-variance control problem in continuous time, which is a LQG problem with regime switching, and in which the control weight is indefinite. We then proceed to obtain its near-optimal control based on the limit problem.

In Chapter 12, we consider a class of near-optimal production planning problems for discrete-time planning of manufacturing systems. By studying a manufacturing system consisting of a number of machines that produce a number of parts, and assuming that the machines are subject to breakdown and repair, we model the capacity of the machines as a finite-state Markov chain. The goal is to choose the production rates over time so as to minimize an expected cost function. We carry out asymptotic analysis, show that the original problem can be approximated by a limit problem, and construct a near optimal control based on the limit problem.

Chapter 13 is about hybrid stochastic approximation, the study of which is motivated by problems arising from discrete stochastic optimization and tracking time-varying parameters, where the dynamics of the parameters evolve according to a Markov chain. Such problems also arise from emerging applications in wireless communications, for example, adaptive coding in CDMA communication networks. We focus on a class of adaptive algorithms to track the invariant distribution of a conditional Markov chain (conditioned on another Markov chain whose transition probability matrix is "near" identity). That is, the underlying parameter is a Markov chain with infrequent varying dynamics. Our objective is to evaluate the tracking capability of the adaptive stochastic approximation algorithm in terms of mean squares tracking error and asymptotic covariance of the associated limit process.

Finally Chapter 14, providing an appendix, contains a handful of brief discussions, basic notions, and a number of technical results used in the book. The topics include weak convergence, Markov chains, optimal control and HJB equations, and a number of miscellaneous results on convexity, the Arzelá–Ascoli theorem, and the Fredholm alternative etc.

Throughout the rest of the book, each chapter begins with an introduction and an outline, and ends with notes of further remarks, additional literature reviews, and other related matters. To give a "road map" and to show the logical dependence, relationship, and connection among the chapters, we provide a flow chart in Figure 1.3.

2
Mathematical Preliminaries

2.1 Introduction

This chapter provides basic background materials needed in the subsequent chapters of the book. It briefly reviews and summarizes related results of random processes, including Markov chains in both discrete time and continuous time, martingales, Gaussian processes, diffusions, and switching diffusions.

Throughout the book, we work with a probability space $(\Omega, \mathcal{F}, P)$. A collection of σ-algebras $\{\mathcal{F}_t\}$, for $t \geq 0$ or $t = 1, 2, \ldots$, or simply $\mathcal{F}_t$, is called a filtration if $\mathcal{F}_s \subset \mathcal{F}_t$ for $s \leq t$. The $\mathcal{F}_t$ is complete in the sense that it contains all null sets. A probability space $(\Omega, \mathcal{F}, P)$ together with a filtration $\{\mathcal{F}_t\}$ is termed a *filtered probability space*, denoted by $(\Omega, \mathcal{F}, \{\mathcal{F}_t\}, P)$.

2.2 Discrete-Time Markov Chains

Working with discrete time $k \in \{0, 1, \ldots\}$, consider a sequence $\{x_k\}$ of $\mathbb{R}^r$ vectors. If for each k, x_k is a random vector (or an $\mathbb{R}^r$-valued random variable), we call $\{x_k\}$ a stochastic process and write it as $x_k, k = 0, 1, 2, \ldots$, or simply x_k if there is no confusion. A stochastic process is wide-sense (or covariance) stationary, if it has finite second moments, a constant mean, and a covariance that depends only on the time difference. The ergodicity of a stationary sequence $\{x_k\}$ refers to the convergence of the sequence

$$\frac{x_1 + x_2 + \cdots + x_n}{n}$$

to its expectation in the almost sure or some weak sense; see Karlin and Taylor [78, Theorem 5.6, p. 487] for a strong ergodic theorem of a stationary process. A stochastic process x_k is adapted to a filtration $\{\mathcal{F}_k\}$, if for each k, x_k is an $\mathcal{F}_k$-measurable random vector.

Suppose that α_k is a stochastic process taking values in $\mathcal{M}$, which is at most countable (i.e., it is either finite $\mathcal{M} = \{1, 2, \ldots, m_0\}$ or countable $\mathcal{M} = \{1, 2, \ldots\}$). We say that α_k is a Markov chain if

$$
\begin{aligned}
p_{k,k+1}^{ij} &= P(\alpha_{k+1} = j | \alpha_k = i) \\
&= P(\alpha_{k+1} = j | \alpha_0 = i_0, \ldots, \alpha_{k-1} = i_{k-1}, \alpha_k = i),
\end{aligned}
$$

for any $i_0, \ldots, i_{k-1}, i, j \in \mathcal{M}$.

Given i, j, if $p_{k,k+1}^{ij}$ is independent of time k, i.e., $p_{k,k+1}^{ij} = p^{ij}$, we say that α_k has stationary transition probabilities. The corresponding Markov chains are said to be stationary or time-homogeneous or temporally homogeneous or simply homogeneous. In this case, let $P = (p^{ij})$ denote the transition matrix. Denote the n-step transition matrix by $P^{(n)} = (p^{ij,(n)})$, with

$$p^{ij,(n)} = P(x_n = j | x_0 = i).$$

Then $P^{(n)} = (P)^n$. That is, the n-step transition matrix is simply the matrix P to the nth power. Note that

(a) $p^{ij} \geq 0$, $\sum_j p^{ij} = 1$, and

(b) $(P)^{k_1+k_2} = (P)^{k_1}(P)^{k_2}$, for $k_1, k_2 = 1, 2, \ldots$

The last identity is commonly referred to as the Chapman–Kolmogorov equation. In this book, we work with Markov chains with finite state spaces. Thus we confine our discussion to such cases. Certain algebraic properties of Markov chains will be used in the book, some of which are listed next.

Suppose that A is an $r \times r$ square matrix. Denote the collection of eigenvalues of A by Λ. Then the spectral radius of A, denoted by $\rho(A)$, is defined by $\rho(A) = \max_{\lambda \in \Lambda} |\lambda|$. Recall that a matrix with real entries is said to be a positive matrix if it has at least one positive entry and no negative entries. If every entry of A is positive, we call the matrix strictly positive. Similarly, for a vector $x = (x^1, \ldots, x^r)$, by $x \geq 0$, we mean that $x^i \geq 0$ for $i = 1, \ldots, r$; by $x > 0$, we mean that all entries $x^i > 0$.

Let $P = (p^{ij}) \in \mathbb{R}^{m_0 \times m_0}$ be a transition matrix. Clearly, it is a positive matrix. Then $\rho(P) = 1$; see Karlin and Taylor [79, p. 3]. This implies that all eigenvalues of P are on or inside the unit circle.

For a Markov chain α_k, state j is said to be accessible from state i if $p^{ij,(k)} = P(\alpha_k = j | \alpha_0 = i) > 0$ for some $k > 0$. Two states i and j, accessible from each other, are said to communicate. A Markov chain is irreducible if all states communicate with each other. For $i \in \mathcal{M}$, let $d(i)$ denote the period of state i, i.e., the greatest common divisor of all $k \geq 1$ such that $P(\alpha_{k+n} = i | \alpha_n = i) > 0$ (define $d(i) = 0$ if $P(\alpha_{k+n} = i | \alpha_n = i) = 0$ for all k). A Markov chain is called aperiodic if each state has period one. According to Kolmogorov's classification of states, a state i is recurrent if, starting from state i, the probability of returning to state i after some finite time is 1. A state is transient if it is not recurrent. Criteria on recurrence can be found in most standard textbooks of stochastic processes or Markov chains.

Note that (see Karlin and Taylor [79, p. 4]) if P is a transition matrix for a finite-state Markov chain, the multiplicity of the eigenvalue 1 is equal to the number of recurrent classes associated with P. A row vector $\pi = (\pi^1, \ldots, \pi^{m_0})$ with each $\pi^i \geq 0$ is called a stationary distribution of α_k if it is the unique solution to the system of equations

$$\pi P = \pi,$$

$$\sum_i \pi^i = 1.$$

As demonstrated in [79, p. 85], for i in an aperiodic recurrent class, if $\pi^i > 0$, which is the limit of the probability of starting from state i and then entering state i at the nth transition as $n \to \infty$, then for all j in this class of i, $\pi^j > 0$, and the class is termed positive recurrent or strongly ergodic. The following theorem, concerning the spectral gaps, will be used in the asymptotic expansions.

Theorem 2.1. *Let $P = (p^{ij})$ be the transition matrix of an irreducible aperiodic finite-state Markov chain. Then there exist constants $0 < \lambda < 1$ and $c_0 > 0$ such that*

$$|(P)^k - \overline{P}| \leq c_0 \lambda^k \quad \text{for} \quad k = 1, 2, \ldots,$$

where $\overline{P} = \mathbb{1}_{m_0} \pi$, $\mathbb{1}_{m_0} = (1, \ldots, 1)' \in \mathbb{R}^{m_0 \times 1}$, and $\pi = (\pi^1, \ldots, \pi^{m_0})$ is the stationary distribution of α_k. This implies, in particular,

$$\lim_{k \to \infty} P^k = \mathbb{1}_{m_0} \pi.$$

Suppose that α_k is a Markov chain with transition probability matrix P. One of the ergodicity conditions of Markov chains is the Doeblin's condition (see Doob [49, Hypothesis D, p. 192]; see also Meyn and Tweedie [115, p. 391]). Suppose that there is a probability measure μ with the property that for some positive integer n, $0 < \delta < 1$, and $\Delta > 0$, $\mu(A) \leq \delta$ implies that $P^n(x, A) \leq 1 - \Delta$ for all $x \in A$. In the above, $P^n(x, A)$ denotes the transition probability starting from x reaches the set A in n steps. Note that if α_k is a finite-state Markov chain that is irreducible and aperiodic, then the Doeblin's condition is satisfied.

In the subsequent chapters, we often need to treat nonhomogeneous systems of linear equations. Given an $m_0 \times m_0$ irreducible transition matrix P and a vector G, consider

$$F(P - I) = G, \tag{2.1}$$

where F is an unknown vector. Note that zero is an eigenvalue of the matrix $P - I$ and the null space of $P - I$ is spanned by $\mathbb{1}_{m_0}$. Then by the Fredholm alternative (see Lemma 14.36), (2.1) has a solution iff $G\mathbb{1}_{m_0} = 0$, where $\mathbb{1}_{m_0} = (1, \ldots, 1)' \in \mathbb{R}^{m_0 \times 1}$.

Define $Q_c = (P - I \vdots \mathbb{1}_{m_0}) \in \mathbb{R}^{m_0 \times (m_0 + 1)}$. Consider (2.1) together with the condition $F\mathbb{1}_{m_0} = \sum_{i=1}^{m_0} F_i = \widehat{F}$, which may be written as $FQ_c = G_c$ where $G_c = (G \vdots \widehat{F})$. Since for each t, (2.12) has a unique solution, it follows that $Q_c(t)Q_c'(t)$ is a matrix with full rank; therefore, the equation

$$F[Q_cQ_c'] = G_cQ_c' \tag{2.2}$$

has a unique solution, which is given by $G_cQ_c'[Q_cQ_c']^{-1}$. This observation will be used later in this book.

2.3   Discrete-Time Martingales

Many applications involving stochastic processes depend on the concept of martingale. The definition and properties of discrete-time martingales can be found in Breiman [27, Chapter 5], Chung [38, Chapter 9], and Hall and Heyde [67] among others. This section provides a brief review.

Definition 2.2. Suppose that $\{\mathcal{F}_n\}$ is a filtration, and $\{x_n\}$ is a sequence of random variables. The pair $\{x_n, \mathcal{F}_n\}$ is a martingale if for each n,

(a) x_n is $\mathcal{F}_n$-measurable;

(b) $E|x_n| < \infty$;

(c) $E(x_{n+1}|\mathcal{F}_n) = x_n$ w.p.1.

It is a supermartingale (resp. submartingale) if (a) and (b) in the above hold, and

$$E(x_{n+1}|\mathcal{F}_n) \leq x_n \quad (\text{resp. } E(x_{n+1}|\mathcal{F}_n) \geq x_n) \quad \text{w.p.1.}$$

In what follows if the sequence of σ-algebras is clear, we simply say that $\{x_n\}$ is a martingale.

Perhaps the simplest example of a discrete-time martingale is the sum $x_n = \sum_{j=1}^n y_j$ of a sequence of i.i.d. random variables $\{y_n\}$ with zero mean. It is readily seen that

$$E[x_{n+1}|y_1, \ldots, y_n] = E[x_n + y_{n+1}|y_1, \ldots, y_n]$$
$$= x_n + Ey_{n+1} = x_n \quad \text{w.p.1.}$$

The above equation illustrates the defining relation of a martingale.

If $\{x_n\}$ is a martingale, we can define $y_n = x_n - x_{n-1}$, which is known as a martingale difference sequence. Suppose that $\{x_n, \mathcal{F}_n\}$ is a martingale. Then the following properties hold.

(a) Suppose $\varphi(\cdot)$ is an increasing and convex function defined on $\mathbb{R}$, if for each positive integer n, $E|\varphi(x_n)| < \infty$, then $\{\varphi(x_n), \mathcal{F}_n\}$ is a submartingale.

(b) Let τ be a stopping time with respect to $\mathcal{F}_n$ (i.e., an integer-valued random variable such that $\{\tau \leq n\}$ is $\mathcal{F}_n$-measurable for each n). Then $\{x_{\tau \wedge n}, \mathcal{F}_{\tau \wedge n}\}$ is also a martingale.

(c) The martingale inequality (see Kushner [96, p. 3]) states that for each $\lambda > 0$,

$$P\left(\max_{1 \leq j \leq n} |x_j| \geq \lambda\right) \leq \frac{1}{\lambda} E|x_n|,$$
$$E \max_{1 \leq j \leq n} |x_j|^2 \leq 4E|x_n|^2, \quad \text{if } E|x_n|^2 < \infty \text{ for each } n. \tag{2.3}$$

(d) The Doob's inequality (see Hall and Heyde [67, p.15]) states that for each $p > 1$,

$$E^{1/p}|x_n|^p \leq E^{1/p}\left(\max_{1 \leq j \leq n} |x_j|\right)^p \leq qE^{1/p}|x_n|^p,$$

where $p^{-1} + q^{-1} = 1$;

(e) The Burkholder's inequality (see Hall and Heyde [67, p.23]) is: For $1 < p < \infty$, there exist constants K_1 and K_2 such that

$$K_1 E\left|\sum_{j=1}^n y_j^2\right|^{p/2} \leq E|x_n|^p \leq K_2 E\left|\sum_{i=j}^n y_j^2\right|^{p/2},$$

where $y_n = x_n - x_{n-1}$.

Consider a discrete-time Markov chain $\{\alpha_n\}$ with state space $\mathcal{M}$ (either finite or countable) and one-step transition probability matrix $P = (p^{ij})$. Recall that a sequence $\{f(i) : i \in \mathcal{M}\}$ is P-harmonic or right-regular

(Karlin and Taylor [79, p. 48]), if (a) $f(\cdot)$ is a real-valued function such that $f(i) \geq 0$ for each $i \in \mathcal{M}$, and (b)

$$f(i) = \sum_{j \in \mathcal{M}} p^{ij} f(j) \quad \text{for each} \ \ i \in \mathcal{M}. \tag{2.4}$$

If the equality in (2.4) is replaced by $\geq$ (resp. $\leq$), $\{f(i) : i \in \mathcal{M}\}$ is said to be P-superharmonic or right superregular (resp. P-subharmonic or right subregular). Considering $f = (f(i) : i \in \mathcal{M})$ as a column vector, (2.4) can be written as $f = Pf$. Similarly, we can write $f \geq Pf$ for P-superharmonic (resp. $g \leq Pf$ for P-subharmonic). Likewise, $\{f(i) : i \in \mathcal{M}\}$ is said to be P left regular, if (b) above is replaced by

$$f(j) = \sum_{i \in \mathcal{M}} f(i) p^{ij} \quad \text{for each} \ \ j \in \mathcal{M}. \tag{2.5}$$

Similarly, left superregular and subregular functions can be defined.

The following paragraph reveals the natural connection between a martingale and a discrete-time Markov chain. Following the idea presented in Karlin and Taylor [78, p. 241], let $\{f(i) : i \in \mathcal{M}\}$ be a bounded P-harmonic sequence. Define $x_n = f(\alpha_n)$. Then $E|x_n| < \infty$. Moreover, owing to the Markov property,

$$
\begin{aligned}
E(x_{n+1}|\mathcal{F}_n) &= E(f(\alpha_{n+1})|\alpha_n)) \\
&= \sum_{j \in \mathcal{M}} p^{\alpha_n,j} f(j) \\
&= f(\alpha_n) = x_n \ \ \text{w.p.1.}
\end{aligned}
$$

Therefore, $\{x_n, \mathcal{F}_n\}$ is a martingale. Note that if $\mathcal{M}$ is finite, the boundedness of $\{f(i) : i \in \mathcal{M}\}$ is not needed.

As pointed out in Karlin and Taylor [78], one of the widely used ways of constructing martingales is through the utilization of eigenvalues and eigenvectors of a transition matrix. Again, let $\{\alpha_n\}$ be a discrete-time Markov chain with transition matrix P. Recall that a column vector f is a right eigenvector of P associated with an eigenvalue $\lambda \in \mathbb{C}$, if $Pf = \lambda f$. Let f be a right eigenvector of P satisfying $E|f(\alpha_n)| < \infty$ for each n. For $\lambda \neq 0$, define $x_n = \lambda^{-n} f(\alpha_n)$. Then $\{x_n\}$ is a martingale.

2.4 Continuous-Time Martingales and Markov Chains

Denote the space of $\mathbb{R}^r$-valued continuous functions defined on $[0, T]$ by $C([0, T]; \mathbb{R}^r)$, and the space of functions that are right continuous with left-hand limits endowed with the Skorohod topology by $D([0,T];\mathbb{R}^r)$; see Definition 14.2. Consider $x(\cdot) = \{x(t) \in \mathbb{R}^r : t \geq 0\}$. If for each $t \geq 0$, $x(t)$ is an $\mathbb{R}^r$ random vector, we call $x(\cdot)$ a continuous-time stochastic process and write it as $x(t)$, $t \geq 0$, or simply $x(t)$ if there is no confusion.

A process $x(\cdot)$ is *adapted* to a filtration $\{\mathcal{F}_t\}$, if for each $t \geq 0$, $x(t)$ is an $\mathcal{F}_t$-measurable random variable; $x(\cdot)$ is *progressively measurable* if for each $t \geq 0$, the process restricted to $[0,t]$ is measurable with respect to the σ-algebra $\mathcal{B}[0,t] \times \mathcal{F}_t$ in $[0,t] \times \Omega$, where $\mathcal{B}[0,t]$ denotes the Borel sets of $[0,t]$. A progressively measurable process is measurable and adapted, whereas the converse is not generally true. However, any measurable and adapted process with right-continuous sample paths is progressively measurable.

For many applications, we often need to work with a *stopping time*. A stopping time τ on $(\Omega, \mathcal{F}, P)$ with a filtration $\{\mathcal{F}_t\}$ is a nonnegative random variable such that $\{\tau \leq t\} \in \mathcal{F}_t$, for all $t \geq 0$. A stochastic process $\{x(t) : t \geq 0\}$ (real or vector valued) is said to be a martingale on $(\Omega, \mathcal{F}, P)$ with respect to $\{\mathcal{F}_t\}$ if:

(a) For each $t \geq 0$, $x(t)$ is $\mathcal{F}_t$-measurable,

(b) $E|x(t)| < \infty$, and

(c) $E[x(t)|\mathcal{F}_s] = x(s)$ w.p.1 for all $t \geq s$.

If we only say that $x(\cdot)$ is a martingale without specifying the filtration $\mathcal{F}_t$, $\mathcal{F}_t$ is taken to be the natural filtration $\sigma\{x(s) : s \leq t\}$. If there exists a sequence of stopping times $\{\tau_n\}$ such that $0 \leq \tau_1 \leq \tau_2 \leq \cdots \leq \tau_n \leq \tau_{n+1} \leq \cdots$, $\tau_n \to \infty$ w.p.1 as $n \to \infty$, and the process $x^{(n)}(t) := x(t \wedge \tau_n)$ is a martingale, then $x(\cdot)$ is a local martingale.

A jump process is a right-continuous stochastic process with piecewise-constant sample paths. Let $\alpha(\cdot) = \{\alpha(t) : t \geq 0\}$ be a jump process defined on $(\Omega, \mathcal{F}, P)$ taking values in $\mathcal{M}$. Then $\{\alpha(t) : t \geq 0\}$ is a Markov chain with state space $\mathcal{M}$, if

$$P(\alpha(t) = i|\alpha(r) : r \leq s) = P(\alpha(t) = i|\alpha(s)),$$

for all $0 \leq s \leq t$ and $i \in \mathcal{M}$, with $\mathcal{M}$ being either finite or countable.

For any $i, j \in \mathcal{M}$ and $t \geq s \geq 0$, let $p^{ij}(t,s)$ denote the transition probability $P(\alpha(t) = j|\alpha(s) = i)$, and $P(t,s)$ the matrix $(p^{ij}(t,s))$. We name $P(t,s)$ the transition matrix of the Markov chain $\alpha(\cdot)$, and postulate that

$$\lim_{t \to s^+} p^{ij}(t,s) = \delta^{ij},$$

where $\delta^{ij} = 1$ if $i = j$ and 0 otherwise. It follows that for $0 \leq s \leq \varsigma \leq t$,

$$
\begin{cases}
p^{ij}(t,s) \geq 0, \; i,j \in \mathcal{M}, \\
\displaystyle\sum_{j \in \mathcal{M}} p^{ij}(t,s) = 1, \; i \in \mathcal{M}, \\
p^{ij}(t,s) = \displaystyle\sum_{k \in \mathcal{M}} p^{ik}(\varsigma,s)p^{kj}(t,\varsigma), \; i,j \in \mathcal{M}.
\end{cases}
$$

The last identity is usually referred to as the Chapman–Kolmogorov equation. If the transition probability $P(\alpha(t) = j|\alpha(s) = i)$ depends only on $(t - s)$, then $\alpha(\cdot)$ is said to be stationary or it is said to have stationary transition probabilities. In this case, we define $p^{ij}(h) := p^{ij}(s + h, s)$ for any $h \geq 0$. Otherwise, the process is nonstationary. Suppose that $\alpha(t)$ is a continuous-time Markov chain with stationary transition probability $P(t) = (p^{ij}(t))$. It then naturally induces a discrete-time Markov chain. In fact, for each $h > 0$, the transition matrix $(p^{ij}(h))$ is the transition matrix of the discrete-time Markov chain $\alpha_k = \alpha(kh)$, which is called an h-skeleton of the corresponding continuous-time Markov chain in Chung [38, p. 132].

Definition 2.3 (q-Property). A matrix-valued function $Q(t) = (q^{ij}(t))$, for $t \geq 0$, satisfies the q-Property, if

(a) $q^{ij}(t)$ is Borel measurable for all $i, j \in \mathcal{M}$ and $t \geq 0$;

(b) $q^{ij}(t)$ is uniformly bounded. That is, there exists a constant K such that $|q^{ij}(t)| \leq K$, for all $i, j \in \mathcal{M}$ and $t \geq 0$;

(c) $q^{ij}(t) \geq 0$ for $j \neq i$ and $q^{ii}(t) = -\sum_{j \neq i} q^{ij}(t)$, $t \geq 0$.

For any real-valued function f on $\mathcal{M}$ and $i \in \mathcal{M}$, write

$$
Q(t)f(\cdot)(i) = \sum_{j \in \mathcal{M}} q^{ij}(t)f(j) = \sum_{j \neq i} q^{ij}(t)(f(j) - f(i)).
$$

Let us now recall the definition of the generator of a Markov chain.

Definition 2.4 (Generator). A matrix $Q(t)$, $t \geq 0$, is an infinitesimal generator (or in short a generator) of $\alpha(\cdot)$ if it satisfies the q-Property, and for any bounded real-valued function f defined on $\mathcal{M}$

$$
f(\alpha(t)) - \int_0^t Q(\varsigma)f(\cdot)(\alpha(\varsigma))d\varsigma \tag{2.6}
$$

is a martingale.

Remark 2.5. Motivated by the applications we are interested in, a generator is defined for a matrix satisfying the q-Property above, where an additional condition on the boundedness of the entries of the matrix is posed. Different definitions, including other classes of matrices, may be devised as in Chung [38]. To proceed, we give an equivalent condition for a finite-state Markov chain generated by $Q(\cdot)$.

Lemma 2.6. *Let* $\mathcal{M} = \{1, \ldots, m_0\}$. *Then* $\alpha(t) \in \mathcal{M}$, $t \geq 0$, *is a Markov chain generated by* $Q(t)$ *iff*

$$\left(I_{\{\alpha(t)=1\}}, \ldots, I_{\{\alpha(t)=m_0\}}\right) - \int_0^t \left(I_{\{\alpha(\varsigma)=1\}}, \ldots, I_{\{\alpha(\varsigma)=m_0\}}\right) Q(\varsigma)d\varsigma \quad (2.7)$$

is a martingale.

Proof: See Yin and Zhang [158, Lemma 2.4]. □

For any given $Q(t)$ satisfying the q-Property, there exists a Markov chain $\alpha(\cdot)$ generated by $Q(t)$. If $Q(t) = Q$, a constant matrix, the idea of Ethier and Kurtz [55] can be utilized for the construction. For time-varying generator $Q(t)$, we need to use the piecewise-deterministic process approach, described in Davis [42], to define the Markov chain $\alpha(\cdot)$.

Let $0 = \tau_0 < \tau_1 < \cdots < \tau_l < \cdots$ be a sequence of jump times of $\alpha(\cdot)$ such that the random variables τ_1, $\tau_2 - \tau_1$, ..., $\tau_{k+1} - \tau_k$, ... are independent. Let $\alpha(0) = i \in \mathcal{M}$. Then $\alpha(t) = i$ on the interval $[\tau_0, \tau_1)$. The first jump time τ_1 has the probability distribution

$$P(\tau_1 \in B) = \int_B \exp\left\{\int_0^t q^{ii}(s)ds\right\} \left(-q^{ii}(t)\right) dt,$$

where $B \subset [0, \infty)$ is a Borel set. The post-jump location of $\alpha(t) = j$, $j \neq i$, is given by

$$P(\alpha(\tau_1) = j | \tau_1) = \frac{q^{ij}(\tau_1)}{-q^{ii}(\tau_1)}.$$

If $q^{ii}(\tau_1)$ is 0, define $P(\alpha(\tau_1) = j | \tau_1) = 0$, $j \neq i$. Then $P(q^{ii}(\tau_1) = 0) = 0$. In fact, if $B_i = \{t : q^{ii}(t) = 0\}$, then

$$P(q^{ii}(\tau_1) = 0) = P(\tau_1 \in B_i)$$
$$= \int_{B_i} \exp\left\{\int_0^t q^{ii}(s)ds\right\} \left(-q^{ii}(t)\right) dt = 0.$$

In general, $\alpha(t) = \alpha(\tau_l)$ on the interval $[\tau_l, \tau_{l+1})$. The jump time τ_{l+1} has the conditional probability distribution

$$P(\tau_{l+1} - \tau_l \in B_l | \tau_1, \ldots, \tau_l, \alpha(\tau_1), \ldots, \alpha(\tau_l))$$
$$= \int_{B_l} \exp\left\{\int_{\tau_l}^{t+\tau_l} q^{\alpha(\tau_l)\alpha(\tau_l)}(s)ds\right\} \left(-q^{\alpha(\tau_l)\alpha(\tau_l)}(t + \tau_l)\right) dt.$$

The post-jump location of $\alpha(t) = j$, $j \neq \alpha(\tau_l)$ is given by

$$P(\alpha(\tau_{l+1}) = j | \tau_1, \ldots, \tau_l, \tau_{l+1}, \alpha(\tau_1), \ldots, \alpha(\tau_l)) = \frac{q^{\alpha(\tau_l)j}(\tau_{l+1})}{-q^{\alpha(\tau_l)\alpha(\tau_l)}(\tau_{l+1})}.$$

Theorem 2.7. *Suppose that the matrix $Q(t)$ satisfies the q-Property for $t \geq 0$. Then the following statements hold.*

(a) *The process $\alpha(\cdot)$ constructed above is a Markov chain.*

(b) *The process*

$$f(\alpha(t)) - \int_0^t Q(\varsigma)f(\cdot)(\alpha(\varsigma))d\varsigma \tag{2.8}$$

is a martingale for any uniformly bounded function $f(\cdot)$ on $\mathcal{M}$. Thus $Q(t)$ is indeed the generator of $\alpha(\cdot)$.

(c) *The transition matrix $P(t, s)$ satisfies the forward differential equation*

$$\frac{dP(t, s)}{dt} = P(t, s)Q(t), \ t \geq s,$$
$$P(s, s) = I, \tag{2.9}$$

where I is the identity matrix.

(d) *Assume further that $Q(t)$ is continuous in t. Then $P(t, s)$ also satisfies the backward differential equation*

$$\frac{dP(t, s)}{ds} = Q(s)P(t, s), \ t \geq s,$$
$$P(s, s) = I. \tag{2.10}$$

Proof. See Yin and Zhang [158, Theorem 2.5]. □

Suppose that $\alpha(t)$, $t \geq 0$, is a Markov chain generated by an $m_0 \times m_0$ matrix $Q(t)$. The notions of irreducibility and quasi-stationary distribution are given next.

Definition 2.8 (Irreducibility).

(a) A generator $Q(t)$ is said to be weakly irreducible if, for each fixed $t \geq 0$, the system of equations

$$\nu(t)Q(t) = 0,$$
$$\sum_{i=1}^{m_0} \nu^i(t) = 1 \tag{2.11}$$

has a unique solution $\nu(t) = (\nu^1(t), \ldots, \nu^{m_0}(t))$ and $\nu(t) \geq 0$.

(b) A generator $Q(t)$ is said to be irreducible, if for each fixed $t \geq 0$ the systems of equations (2.11) has a unique solution $\nu(t)$ and $\nu(t) > 0$.

By $\nu(t) \geq 0$, we mean that for each $i \in \mathcal{M}$, $\nu^i(t) \geq 0$. Similar interpretation holds for $\nu(t) > 0$. It follows from the definitions above that irreducibility implies weak irreducibility. However, the converse is not true. For example, the generator

$$Q = \begin{pmatrix} -1 & 1 \\ 0 & 0 \end{pmatrix}$$

is weakly irreducible, but it is not irreducible because it contains an absorbing state corresponding to the second row in Q. A moment of reflection reveals that for a two-state Markov chain with generator

$$Q = \begin{pmatrix} -\lambda(t) & \lambda(t) \\ \mu(t) & -\mu(t) \end{pmatrix}$$

the weak irreducibility requires only $\lambda(t) + \mu(t) > 0$, whereas the irreducibility requires that both $\lambda(t)$ and $\mu(t)$ be positive. Such a definition is convenient for many applications (e.g., the manufacturing systems mentioned in Khasminskii, Yin, and Zhang [85, p. 292]).

Definition 2.9 (Quasi-Stationary Distribution). For $t \geq 0$, $\nu(t)$ is termed a quasi-stationary distribution if it is the unique solution of (2.11) satisfying $\nu(t) \geq 0$.

Remark 2.10. While studying homogeneous Markov chains, the stationary distributions play an important role. In the context of nonstationary (non-homogeneous) Markov chains, they are replaced by the quasi-stationary distributions, as defined above.

If $\nu(t) = \nu > 0$, it is termed a stationary distribution. In view of Definitions 2.8 and 2.9, if $Q(t)$ is weakly irreducible, then there is a quasi-stationary distribution. Note that the rank of a weakly irreducible $m_0 \times m_0$ matrix $Q(t)$ is $m_0 - 1$, for each $t \geq 0$. The definition above emphasizes the probabilistic interpretation. An equivalent definition pinpointing the algebraic properties of $Q(t)$ is provided next. One can verify their equivalence using the Fredholm alternative; see Lemma 14.36.

Definition 2.11. A generator $Q(t)$ is said to be weakly irreducible if, for each fixed $t \geq 0$, the system of equations

$$f(t)Q(t) = 0,$$
$$\sum_{i=1}^{m_0} f^i(t) = 0 \tag{2.12}$$

has only the trivial (zero) solution.

2.5   Gaussian, Diffusion, and Switching Diffusion Processes

A Gaussian random vector $x = (x^1, x^2, \ldots, x^r)$ is one whose characteristic function has the form

$$\phi(y) = \exp\left(i\langle y, \mu\rangle - \frac{1}{2}\langle \Sigma y, y\rangle\right),$$

where $\mu \in \mathbb{R}^r$ is a constant vector, $\langle y, \mu\rangle$ is the usual inner product, i denotes the pure imaginary number satisfying $i^2 = -1$, and Σ is a symmetric nonnegative definite $r \times r$ matrix. In the above, μ and Σ are the mean vector and covariance matrix of x, respectively.

Let $x(t)$, $t \geq 0$, be a stochastic process. It is a Gaussian process if for any $0 \leq t_1 < t_2 < \cdots < t_k$ and $k = 1, 2, \ldots$, $(x(t_1), x(t_2), \ldots, x(t_k))$ is a Gaussian vector. A random process $x(\cdot)$ has *independent increments* if for any $0 \leq t_1 < t_2 < \cdots < t_k$ and $k = 1, 2, \ldots$,

$$(x(t_1) - x(0)), \ (x(t_2) - x(t_1)), \ \ldots, \ (x(t_k) - x(t_{k-1}))$$

are independent. A sufficient condition for a process to be Gaussian is given next, whose proof can be found in Skorohod [139, p. 7].

Lemma 2.12. *Suppose that the process $x(\cdot)$ has independent increments and continuous sample paths with probability one. Then $x(\cdot)$ is a Gaussian process.*

An $\mathbb{R}^r$-valued random process for $t \geq 0$ is a Brownian motion, if

(a) $B(0) = 0$ w.p.1;

(b) $B(\cdot)$ is a process with independent increments;

(c) $B(\cdot)$ has continuous sample paths with probability one;

(d) the increments $B(t) - B(s)$ have Gaussian distribution with $E(B(t) - B(s)) = 0$ and $\mathrm{Cov}(B(t), B(s)) = \Sigma|t - s|$ for some nonnegative definite $r \times r$ matrix Σ, where $\mathrm{Cov}(B(t), B(s))$ denotes the covariance.

A process $B(\cdot)$ is said to be a standard Brownian motion if $\Sigma = I$. By virtue of Lemma 2.12, a Brownian motion is necessarily a Gaussian process. For an $\mathbb{R}^r$-valued Brownian motion $B(t)$, let $\mathcal{F}_t = \sigma\{B(s) : s \leq t\}$. Let $h(\cdot)$ be an $\mathcal{F}_t$-measurable process taking values in $\mathbb{R}^{r \times r}$ such that $\int_0^t E|h(s)|^2 ds < \infty$ for all $t \geq 0$. Using $B(\cdot)$ and $h(\cdot)$, one can define a stochastic integral $\int_0^t h(s)dB(s)$ such that it is a martingale with mean 0 and

$$E\left|\int_0^t h(s)dB(s)\right|^2 = \int_0^t E\left[\mathrm{tr}(h(s)h'(s))ds\right].$$

Suppose that $b(\cdot)$ and $\sigma(\cdot)$ are non-random Borel measurable functions. A process $x(\cdot)$ defined as

$$x(t) = x(0) + \int_0^t b(s, x(s))ds + \int_0^t \sigma(s, x(s))dB(s) \qquad (2.13)$$

is called a diffusion. Then $x(\cdot)$ defined in (2.13) is a Markov process in the sense that the Markov property

$$P(x(t) \in A|\mathcal{F}_s) = P(x(t) \in A|x(s))$$

holds for all $0 \le s \le t$ and for any Borel set A. A slightly more general definition allows $b(\cdot)$ and $\sigma(\cdot)$ to be $\mathcal{F}_t$-measurable processes. However, the current definition is sufficient for our purpose.

Associated with the diffusion process, there is an operator $\mathcal{L}$, known as the generator of the diffusion $x(\cdot)$, defined as follows. Let $C^{1,2}$ be the class of real-valued functions on (a subset of) $\mathbb{R}^r \times [0, \infty)$ whose first-order partial derivative with respect to t and the second-order mixed partial derivatives with respect to x are continuous. Define an operator $\mathcal{L}$ on $C^{1,2}$ by

$$\mathcal{L}f(t,x) = \frac{\partial f(t,x)}{\partial t} + \sum_{i=1}^r b^i(t,x)\frac{\partial f(t,x)}{\partial x^i} + \frac{1}{2}\sum_{i,j=1}^r a^{ij}(t,x)\frac{\partial^2 f(t,x)}{\partial x^i \partial x^j}, \quad (2.14)$$

where $A(t,x) = (a^{ij}(t,x)) = \sigma(t,x)\sigma'(t,x)$. The well-known Ito's lemma (see Gihman and Skorohod [62], Kunita and Watanabe [92], and Liptser and Shiryayev [105]) states that

$$df(t, x(t)) = \mathcal{L}f(t, x(t)) + f_x'(t, x(t))\sigma(t, x(t))dB(t),$$

or in its integral form

$$\begin{aligned} f(t, x(t)) - f(0, x(0)) &= \int_0^t \mathcal{L}f(s, x(s))ds \\ &+ \int_0^t f_x'(s, x(s))\sigma(s, x(s))dB(s). \end{aligned}$$

One of the consequences of the Ito's lemma is that

$$M_f(t) = f(t, x(t)) - f(0, x(0)) - \int_0^t \mathcal{L}f(s, x(s))ds$$

is a square integrable $\mathcal{F}_t$-martingale. Conversely, let $x(\cdot)$ be right continuous. Using the notation of martingale problems given by Stroock and Varadhan [143], $x(\cdot)$ is said to be a solution of the martingale problem with operator $\mathcal{L}$ if $M_f(\cdot)$ is a martingale for each $f(\cdot, \cdot) \in C_0^{1,2}$ (the class of $C^{1,2}$ functions with compact support).

Suppose that $\alpha(\cdot)$ is a continuous-time Markov chain with finite-state space $\mathcal{M} = \{1, \ldots, m_0\}$ and generator $Q(t)$ and that $\alpha(\cdot)$ is independent of the standard r-dimensional Brownian motion $B(\cdot)$. Then the process $x(\cdot)$

$$x(t) = x(0) + \int_0^t b(s, x(s), \alpha(s))ds + \int_0^t \sigma(s, x(s), \alpha(s))dB(s)$$

is called a switching diffusion or system of diffusions with regime switching. The corresponding operator is defined as follows. For each $\iota \in \mathcal{M}$ and each $f(\cdot, \cdot, \iota) \in C^{1,2}$,

$$
\begin{aligned}
\mathcal{L}f(t, x, \iota) &= \frac{\partial f(t, x, \iota)}{\partial t} + \sum_{i=1}^r b^i(t, x, \iota)\frac{\partial f(t, x, \iota)}{\partial x^i} \\
&\quad + \frac{1}{2}\sum_{i,j=1}^r a^{ij}(t, x, \iota)\frac{\partial^2 f(t, x, \iota)}{\partial x^i \partial x^j} + Q(t)f(t, x, \cdot)(\iota),
\end{aligned}
\tag{2.15}
$$

where $A(t, x, \iota) = (a^{ij}(t, x, \iota)) = \sigma(t, x, \iota)\sigma'(t, x, \iota)$. Similar to the case of diffusions, with the $\mathcal{L}$ defined in (2.15), for each $i \in \mathcal{M}$ and $f(\cdot, \cdot, i) \in C^{1,2}$, a result known as generalized Ito's lemma (see [19]) reads

$$
\begin{aligned}
df(t, x(t), \alpha(t)) &= \mathcal{L}f(t, x(t), \alpha(t)) \\
&\quad + f'_x(t, x(t), \alpha(t))\sigma(t, x(t), \alpha(t))dB(t),
\end{aligned}
$$

or in its integral form

$$
\begin{aligned}
&f(t, x(t), \alpha(t)) - f(0, x(0), \alpha(0)) \\
&= \int_0^t \mathcal{L}f(s, x(s), \alpha(s))ds + \int_0^t f'_x(s, x(s), \alpha(s))\sigma(s, x(s), \alpha(s))dB(s).
\end{aligned}
$$

In addition,

$$M_f(t) = f(t, x(t), \alpha(t)) - f(0, x(0), \alpha(0)) - \int_0^t \mathcal{L}f(s, x(s), \alpha(s))ds$$

is a martingale. Similar to the case of diffusion processes, we can define the corresponding notion of solution of martingale problem accordingly.

2.6   Notes

A nonmeasure theoretic introduction to stochastic processes can be found in Ross [130]. The two volumes by Karlin and Taylor [78, 79] provide an introduction to discrete-time and continuous-time Markov chains. More advanced treatments can be found in Chung [38] and Revuz [127]. A book that deals exclusively with finite-state Markov chain is Iosifescu [73]. The book of Meyn and Tweedie [115] examines Markov chains and their stability. The connection between generators of Markov processes and martingales is explained in Ethier and Kurtz [55]. An account of piecewise-deterministic processes is in Davis [42]. Results on basic probability theory may be found in Chow and Teicher [37]; theory of stochastic processes can be found in Gihman and Skorohod [62]. More detailed discussions regarding martingales and diffusions can be found in Elliott [54]; in-depth study of stochastic differential equations and diffusion processes can be found in Kunita and Watanabe [92].

Part II

Asymptotic Properties

3
Asymptotic Expansions

3.1   Introduction

This chapter constructs asymptotic expansions of probability distribution vectors and transition matrices for discrete-time Markov chains having two time scales. As alluded to in Chapter 1, many applications require finding optimal or near-optimal controls, carrying out optimization tasks, and/or analyzing stability for dynamic systems modulated by discrete-time Markov chains. In these problems, Markov chains play a crucial role. To be able to find solutions to such problems, the foremost requirement is to have a thorough understanding of the probability structure of the Markov chains. As a first step, it is imperative to learn the properties of the probability distributions of the Markov chains. By scrutinizing the difference equations representing the associated probabilities, we focus on obtaining approximate solutions of the underlying equations.

Using k to denote the discrete time, we work with the time horizon $k = 0, 1, \ldots, \lfloor T/\varepsilon \rfloor$ for some $T > 0$, where $\lfloor z \rfloor$ denotes the integer part of a real number z. For notational simplicity, we will often suppress the floor function symbol $\lfloor \cdot \rfloor$ and write T/ε in lieu of $\lfloor T/\varepsilon \rfloor$. In view of the discussion of Chapter 1, we consider the following model. For a small parameter $\varepsilon > 0$, let α_k^ε be a discrete-time Markov chain depending on ε and having finite state space $\mathcal{M} = \{1, \ldots, m_0\}$ and transition matrix

$$P_k^\varepsilon = P_k + \varepsilon Q_k, \tag{3.1}$$

where for each k, P_k is a transition probability matrix and $Q_k = (q_k^{ij})$ is a generator of a continuous-time Markov chain (i.e., $q_k^{ij} \geq 0$, for $j \neq i$, $\sum_j q_k^{ij} = 0$ for each k, i). We are interested in the behavior of the Markov chain α_k^ε for $k \leq \lfloor T/\varepsilon \rfloor = O(1/\varepsilon)$. Let p_k^ε denote the probability vector

$$p_k^\varepsilon = (P(\alpha_k^\varepsilon = 1), \ldots, P(\alpha_k^\varepsilon = m_0)) \in \mathbb{R}^{1 \times m_0}.$$

Assuming that the initial probability p_0^ε is independent of ε, i.e., $p_0^\varepsilon = p_0 = (p_0^1, \ldots, p_0^{m_0})$ such that $p_0^i \geq 0$ for $i = 1, \ldots, m_0$ and $p_0 \mathbb{1}_{m_0} = \sum_{i=1}^{m_0} p_0^i = 1$. It is well known that p_k^ε is a solution of the vector-valued difference equation

$$p_{k+1}^\varepsilon = p_k^\varepsilon P_k^\varepsilon, \ k = 0, 1, \ldots, \lfloor T/\varepsilon \rfloor,$$
$$p_0^\varepsilon = p_0. \tag{3.2}$$

With such a structure, we wish to answer the question: What is the limit probability distribution as $\varepsilon \to 0$ for p_k^ε given by (3.2) with $0 \leq k \leq \lfloor T/\varepsilon \rfloor$? We are equally interested in finding limit properties of the associated equations for k-step transition probability matrices

$$\widetilde{P}_{k+1}^\varepsilon = \widetilde{P}_k^\varepsilon P_k^\varepsilon, \ k = 0, 1, \ldots, \lfloor T/\varepsilon \rfloor,$$
$$\widetilde{P}_0^\varepsilon = I \in \mathbb{R}^{m_0 \times m_0}. \tag{3.3}$$

Note that when P_k^ε is independent of k, $\widetilde{P}_k^\varepsilon = (P^\varepsilon)^k$, the kth power of matrix P^ε.

Equations (3.2) and (3.3) are the discrete-time analog of the forward equations for continuous-time Markov chains. In studying continuous-time systems, our focus was on the generators of the corresponding Markov chains in Yin and Zhang [158]. For discrete-time Markov chains, the basic element taking a similar role to a generator in a continuous-time Markov chain is the transition matrix, also known as one-step transition matrix, which is what our study will be focused on. Concentrating on the transition probabilities, this chapter is devoted to obtaining matched asymptotic expansions; see, for instance, Hoppensteadt and Miranker [69], Kevorkian and Cole [80], Pan and Başar [119], Pervozvanskii and Gaitsgory [123], Yin and Zhang [158] among others.

3.1.1   Why Do We Need Asymptotic Expansions?

Before proceeding further, we take a pause and answer the question, why are asymptotic expansions needed? The examples given in Chapter 1 all involve discrete-time Markov chains with two time scales. The solutions of these problems depend on the asymptotic properties of the probability distributions. Iterating on (3.2) yields the desired probability vectors,

$$p_{k+1}^\varepsilon = p_0 P_0^\varepsilon P_1^\varepsilon \cdots P_k^\varepsilon. \tag{3.4}$$

However, the matrix product involved in (3.4) is merely a representation; its evaluation is by no means simple. Even if the matrices P_k and Q_k are independent of k, evaluating the power $(P+\varepsilon Q)^k$ could still be a nontrivial task when the dimensions of the transition matrix and the generator are large. On the other hand, for many problems arising from a wide variety of applications, the exact solutions may not be easily computable due to their complexity, whereas an approximate solution is often as valuable as the exact solution from a practical point of view. Based on such observations, our study begins with the construction of the asymptotic expansions of the probability vectors and transition matrices.

Consider (3.2). A naive thought might be that since $\varepsilon \to 0$, we can simply drop the εQ_k term. This, however, is incorrect. The reader may wish to convince himself or herself by considering $P_k^\varepsilon = P^\varepsilon = P + \varepsilon Q$ with

$$P = \begin{pmatrix} P^1 & 0 \\ 0 & P^1 \end{pmatrix}, \quad P^1 = \begin{pmatrix} 0.7 & 0.3 \\ 0.2 & 0.8 \end{pmatrix},$$

and Q being a 4×4 generator. In this case, we cannot drop εQ when $\varepsilon \to 0$, $k \to \infty$, but εk remains bounded away from zero.

One of our primary motivations of studying two-time-scale Markov chains is the reduction of complexity for control and optimization of large-scale systems. The main ideas are based on the replacement of the actual system by a reduced-order or limit system in which the coefficients of the original system are averaged out with respect to the invariant measures. To obtain optimal or nearly optimal controls, one often needs to use asymptotic properties of a sequence of occupation measures to figure out the limit systems, which depend mainly on the asymptotic expansions.

In Chapter 4, we shall study asymptotic distributions of scaled occupation measures. To obtain such results, the asymptotic expansions play a vital role. To illustrate, consider a singularly perturbed Markov chain with P given by (3.1). Suppose that for simplicity, $P_k = P$ is irreducible and aperiodic and $Q_k = Q$. Denote the stationary distribution corresponding to P by $\nu = (\nu^1, \ldots, \nu^{m_0}) \in \mathbb{R}^{1 \times m_0}$. It will be shown that

$$\sqrt{\varepsilon} \sum_{i=0}^{k} [I_{\{\alpha_k^\varepsilon = i\}} - \nu^i], \quad \text{for } k = O(1/\varepsilon),$$

has a limit normal distribution. Moreover, the limit variance depends on the initial layer expansion terms. Thus without the asymptotic expansions, such a central limit result will be virtually impossible to obtain.

Asymptotic expansions are also needed for state aggregations. Consider a Markov chain α_k^ε with a finite state space $\mathcal{M}$ consisting of a number of recurrent classes. Suppose that there are l_0 recurrent classes $\mathcal{M}_i$ for $i \leq l_0$. We can aggregate all of the states in each class $\mathcal{M}_i$ as one state to get an aggregated process. That is, define $\overline{\alpha}_k^\varepsilon$ by $\overline{\alpha}_k^\varepsilon = i$ if $\alpha_k^\varepsilon \in \mathcal{M}_i =$

$\{s_{i1}, \ldots, s_{im_i}\}$. Define a sequence of scaled occupation measures by $n_k^\varepsilon = (n_k^{\varepsilon, ij})$ with

$$n_k^{\varepsilon, ij} = \sqrt{\varepsilon} \sum_{k_1=0}^{k} [I_{\{\alpha_{k_1}^\varepsilon = s_{ij}\}} - \nu^{ij} I_{\{\overline{\alpha}_{k_1}^\varepsilon = i\}}], \quad \text{for } k \le \lfloor T/\varepsilon \rfloor,$$

and the continuous-time interpolation $n^\varepsilon(t)$ by

$$n^\varepsilon(t) = n_k^\varepsilon, \quad \text{for } t \in [\varepsilon k, \varepsilon(k+1)),$$

where ν^{ij} denotes the jth component of the stationary measure ν^i. Under suitable conditions, we will derive limit process $n^\varepsilon(\cdot)$, show it converges weakly to a switching diffusion process, and derive exponential-type upper bounds of the form

$$E \exp\left(\frac{K_T}{(T+1)^3} |n^\varepsilon(t)|\right) \le K,$$

for some $K, K_T > 0$. In carrying out these tasks, the asymptotic expansions are indispensable. More of these will be said in Chapters 4 and 5.

3.1.2   Outline of the Chapter

For ease of presentation, in the technical development throughout the book, we will mainly concentrate on time-homogeneous Markov chains, in which $P_k = P$ and $Q_k = Q$ given in (3.1) are constant matrices not depending on k. A large portion of the results can be extended to certain time-inhomogeneous Markov chains.

The rest of the chapter is arranged as follows. Section 3.2 presents the formulation and the conditions needed. Section 3.3 proceeds with the study of constructing asymptotic expansions when the transition matrix is irreducible for motivational purposes and for introducing the basic techniques to be used. For this simple and easy to understand case, the matching of initial layer corrections with the outer expansions is relatively straightforward. It allows us to present the main steps involved in obtaining the formal asymptotic expansions. Section 3.4 takes up the issue of constructing asymptotic expansions when multiple irreducible classes are involved. The problem becomes more involved; care must be taken for the matching. The formal expansions are obtained in Section 3.4, whereas error bounds are derived in Section 3.5. In Section 3.6, we present several examples to demonstrate how the asymptotic expansions can be constructed. Finally, we place the proofs of several technical results in Section 3.7, and close the chapter with some further remarks in Section 3.8.

3.2 Formulation and Conditions

Being a dominating force in P^ε, the structure of P is important. In a finite-state Markov chain, there is at least one recurrent state (i.e., not all states are transient). In fact, (see, for example, Iosifescu [73, p. 94]) any transition probability matrix of a finite-state Markov chain with stationary transition probabilities can be put into the form of either

$$
P = \text{diag}(P^1, \ldots, P^{l_0}) = \begin{pmatrix} P^1 & & & \\ & P^2 & & \\ & & \ddots & \\ & & & P^{l_0} \end{pmatrix}, \tag{3.5}
$$

or

$$
P = \begin{pmatrix} P^1 & & & & \\ & P^2 & & & \\ & & \ddots & & \\ & & & P^{l_0} & \\ P^{*,1} & P^{*,2} & \cdots & P^{*,l_0} & P^* \end{pmatrix}, \tag{3.6}
$$

where each P^i is a transition matrix within the ith recurrent class $\mathcal{M}_i$ for $i \le l_0$, and the last row $(P^{*,1}, \cdots P^{*,l_0}, P^*)$ in (3.6) is a result from the transient states. Here and henceforth, $\text{diag}(A^1, \ldots, A^j)$ denotes a block-diagonal matrix with matrix entries $A^1, \ldots, A^j$ having appropriate dimensions.

If the Markov chain has no transient states, by appropriately rearranging the states, the transition matrix can always be written as (3.5), and the chain is referred to as recurrent. The matrix P alone does not allow any transitions from recurrent class i to recurrent class j for $i \ne j$. The term εQ, however, facilitates the transitions among different recurrent classes. Nevertheless, compared with the transitions dictated by P, the transitions attributed to εQ represent "weak" interactions. If there are transient states, by rearrangement, P will be of the form (3.6). Note that $P^{*,i}$, $i = 1, \ldots, l_0$, are the transition probabilities from the transient states to the recurrent states, and P^* is the transition probabilities within the transient states. To proceed, we need the following condition.

(A3.1) Consider $P^\varepsilon = P + \varepsilon Q$, where $\varepsilon > 0$ is a small parameter, P is a transition matrix (i.e., $P = (p^{ij})$ with $p^{ij} \ge 0$ and $\sum_j p^{ij} = 1$ for each i) given by (3.5) or (3.6), and Q is a generator (i.e., $Q = (q^{ij})$ with $q^{ij} \ge 0$ if $i \ne j$ and $\sum_j q^{ij} = 0$ for each i) of a continuous-time Markov chain. For P given by (3.5) or (3.6), and for each $i \le l_0$, P^i is transition probability matrix that is irreducible and aperiodic. In addition, for P given by (3.6), P^* is a matrix having all of its eigenvalues inside the unit circle.

For the case of inclusion of transient states, the assumption above implies that P^*, corresponding to transition probabilities within the class of transient states, after an extended time will be negligible.

3.2.1   Decomposition and Subspaces

For P given in (3.5), the state space of the Markov chain $\mathcal{M}$ admits a decomposition of the form

$$\mathcal{M} = \mathcal{M}_1 \cup \cdots \cup \mathcal{M}_{l_0}$$
$$= \{s_{11}, \ldots, s_{1m_1}\} \cup \cdots \cup \{s_{l_0 1}, \ldots, s_{l_0 m_{l_0}}\}, \tag{3.7}$$

whereas for P given in (3.6), the state space $\mathcal{M}$ is decomposable into the following form

$$\mathcal{M} = \mathcal{M}_1 \cup \cdots \cup \mathcal{M}_{l_0} \cup \mathcal{M}_*$$
$$= \{s_{11}, \ldots, s_{1m_1}\} \cup \cdots \cup \{s_{l1}, \ldots, s_{lm_{l_0}}\} \cup \{s_{*1}, \ldots, s_{*m_*}\}, \tag{3.8}$$

with $m_0 = m_1 + m_2 + \cdots + m_{l_0} + m_*$. The subspaces $\mathcal{M}_i$ for $i = 1, \ldots, l_0$ consist of recurrent states belonging to l_0 different ergodic classes, and the subspace $\mathcal{M}_*$ consists of transient states.

3.2.2   Asymptotic Expansions

For future use, for an appropriate function $f : \mathbb{R}^{1 \times m_0} \mapsto \mathbb{R}^{1 \times m_0}$, define an operator $\mathcal{L}^\varepsilon$ as

$$\mathcal{L}^\varepsilon f(k) = f(k+1) - f(k) P_k^\varepsilon. \tag{3.9}$$

We construct matched asymptotic expansions of the form

$$p_k^\varepsilon = U_{n_0}^\varepsilon(\varepsilon k) + V_{n_0}^\varepsilon(k) + e_{n_0}^\varepsilon(k), \tag{3.10}$$

where $e_{n_0}^\varepsilon(k)$ represents the approximation error, and

$$U_{n_0}^\varepsilon(\varepsilon k) = \sum_{j=0}^{n_0} \varepsilon^j \varphi_j(\varepsilon k), \quad V_{n_0}^\varepsilon(k) = \sum_{j=0}^{n_0} \varepsilon^j \psi_j(k). \tag{3.11}$$

The term $U^\varepsilon(\varepsilon k)$ is the outer expansion and $V^\varepsilon(k)$ the initial-layer correction. After these terms are found, we then prove the formal asymptotic series has the desired approximation error bounds.

3.3   Asymptotic Expansions: Irreducible Case

In this section, we consider a simple case with an irreducible Markov chain. We present asymptotic expansions of the probability vectors and transition matrices under the condition of irreducibility. This section is mainly for motivational purpose; it also outlines the main steps in constructing the asymptotic series without much of the undue technical complication. It is informative and instructive to work out this less complicated situation first. We begin with the difference equation in (3.2), assume that the transition matrix P^ε is given by (3.1) and that P is irreducible and aperiodic. We will show in what follows that p_k^ε given in (3.2) converges to $\nu = (\nu^1, \ldots, \nu^{m_0})$, the stationary distribution. Note that the convergence takes place as $\varepsilon \to 0$, $k \to \infty$, but εk remains to be bounded. For future reference, we state the desired asymptotic expansions in the following remark.

Remark 3.1. Suppose that $P^\varepsilon = P + \varepsilon Q$ and that P is irreducible. For an integer $n_0 > 0$, for some $T > 0$, and for any $0 \le k \le \lfloor T/\varepsilon \rfloor$, as $\varepsilon \to 0$ and $k \to \infty$, $p_k^\varepsilon \to \nu$, where ν is the stationary distribution corresponding to P. Moreover, there exist two sequences $\{\varphi_i(t)\}_{i=0}^{n_0}$, $0 \le t \le T$, and $\{\psi_i(k)\}_{i=0}^{n_0}$ such that $|\psi_i(k)| \le K\lambda_0^k$ for some $0 < \lambda_0 < 1$, that $\varphi_i(\cdot)$ for $i = 0, \ldots, n_0$ are sufficiently smooth, and that

$$\sup_{0 \le k \le \lfloor T/\varepsilon \rfloor} \left| p_k^\varepsilon - \sum_{i=0}^{n_0} \varepsilon^i \varphi_i(\varepsilon k) - \sum_{i=0}^{n_0} \varepsilon^i \psi_i(k) \right| = O(\varepsilon^{n_0+1}).$$

Since P and Q are constant matrices, the k-step transition matrix associated with P^ε depends only on the time difference k and can be denoted by $(P^\varepsilon)^k$. Then

$$(P^\varepsilon)^k = \overline{P} + \Psi_0(k) + \varepsilon \Phi_1(\varepsilon k) + \varepsilon \Psi_1(k) + O(\varepsilon^2), \qquad (3.12)$$

such that $\overline{P} = \mathbb{1}\nu$, a matrix having identical rows, with each row being the stationary distribution associated with the transition matrix P. Moreover, $|\Psi_i(k)| \le K\lambda_0^k$ for some $K > 0$ and $i = 0, 1$. In the above and hereafter, $K > 0$ represents a generic positive constant; its values may change for different uses. It is understood that the convention $K + K = K$ and $KK = K$ is used.

The λ_0 given above is related to the absolute value of the "largest" non-unity eigenvalues of P. Note that any transition matrix always has an eigenvalue 1. The aperiodicity implies that no eigenvalues other than 1 are on the unit circle. It can be shown that both $\varphi_1(t)$ and $\Phi_1(t)$ are independent of the initial conditions and $\Phi_1(t)$ has identical rows.

To obtain the asymptotic series, the first step is to construct the formal expansions, the second step derives the decay properties of the initial layer corrections, and the third step validates the asymptotic expansions by providing error estimates. In this section, we only concern ourselves with the formal asymptotic expansions. Validations of the asymptotic expansions for general cases are given in Section 3.7.

Note that in (3.10), there are two time scales, fast time scale $k \leq \lfloor T/\varepsilon \rfloor$ and slow time scale εk. Define a new variable $t = \varepsilon k$. Our task is to construct an approximation to the solution of (3.2). Using a Taylor expansion of $\varphi_i(t + \varepsilon)$ about t, we obtain

$$\varphi_i(t + \varepsilon) - \varphi_i(t) = \sum_{j=1}^{n_0+1-i} \frac{\varepsilon^j}{j!} \frac{d^j \varphi_i(t)}{dt^j} + O(\varepsilon^{n_0+2-i}), \quad 0 \leq i \leq n_0,$$

$$\varphi_{n_0+1}(t + \varepsilon) - \varphi_{n_0+1}(t) = \varepsilon \frac{d\varphi_{n_0+1}(t)}{dt} + O(\varepsilon^2).$$

(3.13)

In view of (3.10) and (3.11), substituting $U_{n_0+1}^\varepsilon(t) + V_{n_0+1}^\varepsilon(k)$ into

$$\mathcal{L}^\varepsilon [U_{n_0+1}^\varepsilon(t) + V_{n_0+1}^\varepsilon(k)] = 0,$$

for the outer expansion terms, we obtain

$$\varphi_0(t)(P - I) = 0,$$
$$\varphi_i(t)(P - I) = \widetilde{\xi}_{i-1}(t), \quad \text{for } 1 \leq i \leq n_0 + 1,$$

(3.14)

where

$$\widetilde{\xi}_{i-1} \stackrel{\text{def}}{=} \sum_{j=1}^{i} \frac{1}{j!} \frac{d^j \varphi_{i-j}(t)}{dt^j} - \varphi_{i-1}(t)Q, \quad 1 \leq i \leq n_0 + 1.$$

Define an augmented matrix $Q_c = ((P - I) \vdots \mathbb{1}_{m_0})$. The system of homogeneous equations

$$\begin{cases} \varphi_0(t)(P - I) = 0, \\ \varphi_0(t)\mathbb{1}_{m_0} = 1 \end{cases}$$

can be rewritten as

$$\varphi_0(t)Q_c = (0'_{m_0} \vdots 1),$$

(3.15)

where $0_{m_0} \in \mathbb{R}^{m_0 \times 1}$ with all entries being 0. Note that $Q_c Q_c'$ is an $m_0 \times m_0$ matrix. The irreducibility implies that $Q_c Q_c'$ has full rank, i.e., $\text{rank}[Q_c Q_c'] = m_0$. The unique solution of (3.15) is given by

$$\varphi_0(t) = (0'_{m_0} \vdots 1)Q_c'[Q_c Q_c']^{-1}.$$

(3.16)

Using exactly the same technique, we obtain the solutions of the rest of systems of nonhomogeneous equations as

$$\varphi_i(t) = \widetilde{\xi}_{i-1}(t)Q_c'[Q_c Q_c']^{-1}, \quad \text{for } 1 \leq i \leq n_0 + 1.$$

(3.17)

Note that these $\varphi_i(t)$'s are obtained in a consecutive manner. Thus, when we solve the equation for $\varphi_i(t)$, $\varphi_j(t)$ for $j \leq i - 1$ have been found and $\widetilde{\xi}_{i-1}(t)$ is a known function.

The outer expansions provide a good approximation for t to be away from 0. However, it does not satisfy the initial condition. To compensate, we need to use the initial layer corrections. To do so, in view of (3.10) and (3.11), considering the expansion terms $\psi_i(k)$ in $\mathcal{L}^\varepsilon[U^\varepsilon_{n_0+1}(t) + V^\varepsilon_{n_0+1}(k)] = 0$ leads to

$$\psi_0(k+1) = \psi_0(k)P,$$
$$\psi_i(k+1) = \psi_i(k)P + \psi_{i-1}(k)Q, \quad 0 < i \le n_0 + 1. \tag{3.18}$$

To ensure the match of the outer expansions and the initial layer corrections, we choose the initial conditions to be

$$\varphi_0(0) + \psi_0(0) = p_0,$$
$$\varphi_i(0) + \psi_i(0) = 0, \ 1 \le i \le n_0 + 1.$$

Then it is easily seen that

$$\psi_0(k+1) = (p_0 - \varphi_0(0))(P)^{k+1},$$
$$\psi_i(k+1) = -\varphi_i(0)(P)^{k+1} + \sum_{j=0}^{k} \psi_{i-1}(j)Q(P)^{k-j}, \ 0 < i \le n_0 + 1. \tag{3.19}$$

Since P is irreducible and aperiodic, all of its non-unity eigenvalues are inside the unit circle. Noting the orthogonality $[p_0 - \varphi_0(0)]\overline{P} = 0$, it can be seen that

$$|\psi_0(k)| = |(p_0 - \varphi_0(0))((P)^k - \overline{P})| \le K\lambda_0^k,$$

for some $K > 0$ and some $0 < \lambda_0 < 1$. Similarly, noting $\varphi_i(0)\mathbb{1}_{m_0} = 0$, $\psi_i(k)\mathbb{1}_{m_0} = 0$, and (3.19), we have

$$|\psi_i(k+1)| = \left| -\varphi_i(0)(P)^{k+1} + \sum_{j=0}^{k} \psi_{i-1}(j)Q(P)^{k-j} \right|$$
$$\le |\varphi_i(0)[(P)^{k+1} - \mathbb{1}_{m_0}\nu]| + \left| \sum_{j=0}^{k} \psi_{i-1}(j)Q[(P)^{k-j} - \mathbb{1}_{m_0}\nu] \right|$$
$$\le K\lambda_0^{k+1} + K\sum_{j=0}^{k} \lambda_0^j \lambda_0^{k-j}$$
$$\le K\lambda_0^{k+1} + K\sqrt{k}(\sqrt{\lambda_0})^k \le K\lambda_1^k,$$

where $0 < \lambda_0 \le \lambda_1 = \sqrt{\lambda_0} < 1$, in which, we have used $k\sqrt{\lambda_0^k} \le K$, it then follows that $|\psi_1(k)| \le K\lambda_0^k$. Thus all the outer expansions and initial layer correction terms have been found and the initial layer terms decay geometrically (or exponentially) fast.

As for the derivation of the corresponding result for transition matrices, we seek expansions of the form

$$\Phi_0(t) + \varepsilon\Phi_1(t) + \Psi_0(k) + \varepsilon\Psi_1(k) + \widetilde{e}_1^\varepsilon(k),$$

with $t = \varepsilon k$, and $\widetilde{e}_1^\varepsilon(k)$ represents the error. To illustrate, consider $\Phi_0(t)$ and $\Psi_0(k)$. Then it is easily seen that $\Phi_0(t)(P-I) = 0$. Use $\Phi_0^i(t)$ to denote the ith row in $\Phi_0(t)$. Then

$$\Phi_0^i(t)(P-I) = 0 \text{ and } \Phi_0^i(t)\mathbb{1}_{m_0} = 1.$$

The system above has a unique solution $\Phi_0^i(t) = \nu$, since P is irreducible and aperiodic. Thus $\Phi_0^i(t) = \nu$ is independent of t for each $i \in \mathcal{M}$. Similarly, we can obtain $\Phi_1(t)$. Thus

$$\overline{P} = \mathbb{1}_{m_0}\nu = \begin{pmatrix} \nu^1 & \cdots & \nu^{m_0} \\ \cdots & \cdots & \cdots \\ \nu^1 & \cdots & \nu^{m_0} \end{pmatrix},$$

$$\Phi_1(t) = \mathbb{1}_{m_0}\varphi_1(t) = \begin{pmatrix} \varphi_1^1(t) & \cdots & \varphi_1^{m_0}(t) \\ \cdots & \cdots & \cdots \\ \varphi_1^1(t) & \cdots & \varphi_1^{m_0}(t) \end{pmatrix}.$$

$$(3.20)$$

As for the initial layer corrections, we have

$$\Psi_0(k) = \Psi_0(k-1)P,$$
$$\Psi_0(0) = I - \overline{P}. \tag{3.21}$$

$$\Psi_1(k) = \Psi_1(k-1)P + \Psi_0(k-1)Q,$$
$$\Psi_0(0) = -\Phi_0(0). \tag{3.22}$$

We proceed to verify the geometric decay property. Similar to the asymptotic expansions of p_k^ε, in view of (3.21), since $(I - \overline{P})\overline{P} = 0$,

$$|\Psi_0(k)| = |(I - \overline{P})(P)^k| = |(I - \overline{P})((P)^k - \overline{P})|$$
$$\leq |I - \overline{P}||(P)^k - \overline{P}| \leq K\lambda^k,$$

for some $K > 0$ and some $0 < \lambda < 1$. Similar calculations and estimates can be used for higher-order expansions to obtain full asymptotic expansions of the k-step transition matrix.

3.4 Asymptotic Expansions: General Case

This section is devoted to the asymptotic expansion of the solution of the difference equation (3.2) with the matrix P given by (3.5) or (3.6). In what follows, the detailed development and the verbatim proofs are given for the more complex case–inclusion of transient states. When P is given by (3.5), the argument is even simpler. Our effort to follow lies in the construction of $U_{n_0}^\varepsilon(\varepsilon k)$ and $V_{n_0}^\varepsilon(k)$ and the proof of $e_{n_0}^\varepsilon(k) = O(\varepsilon^{n_0+1})$ uniformly in $0 \le k \le \lfloor T/\varepsilon \rfloor$. The quantity n_0 is the order of the asymptotic expansion. For time-varying transition probability matrices, the n_0 is related to the smoothness of the transition functions for time-varying Markov chains. For time-homogeneous transition matrices, (3.10) holds for any positive integer n_0. That is, effectively, we can have an asymptotic series in lieu of a polynomial in terms of ε. Our study is inspired and motivated by the work of Hoppensteadt and Miranker [69]. Non-separable two-time methods are used in that paper, whereas our asymptotic expansion uses a separable form. As demonstrated for the continuous-time models, the separable form of the asymptotic expansion is advantageous in many applications and is useful for the study of occupation measures and related limit problems. Note that in Hoppensteadt and Miranker [69], linear systems are treated; it is required the matrices P and Q commute, and P be invertible. These assumptions appear to be restrictive and are not suitable for transition probability matrices; they are not needed in our approach. Furthermore, our results can also be extended to time-inhomogeneous (time-dependent) matrices P_k and Q_k. In what follows, the formal expansion is obtained in this section, and the justification of the expansion is provided in the next section.

Again, we emphasize that in (3.10), there are two time scales, the fast scale $k \le \lfloor T/\varepsilon \rfloor$ for some $T > 0$ and the slow time scale εk. For convenience, define a new time variable $t = \varepsilon k$. In view of the slow-time variable t, we may write the Taylor expansion of $\varphi_j(t + \varepsilon)$ as in (3.13). Substituting $U_{n_0+1}^\varepsilon(t) + V_{n_0+1}^\varepsilon(k)$ defined in (3.11) into (3.2), using (3.13), and comparing the coefficients of like powers of ε^i result in

$$\varphi_0(t)(P - I) = 0,$$

$$\varphi_1(t)(P - I) = \frac{d\varphi_0(t)}{dt} - \varphi_0(t)Q,$$

$$\dots \tag{3.23}$$

$$\varphi_{n_0+1}(t)(P - I) = \sum_{j=1}^{n_0+1} \frac{1}{j!} \frac{d^j \varphi_{n_0+1-j}(t)}{dt^j} - \varphi_{n_0}(t)Q,$$

and

$$\psi_0(k + 1) = \psi_0(k)P,$$

$$\psi_1(k + 1) = \psi_1(k)P + \psi_0(k)Q,$$

$$\dots \tag{3.24}$$

$$\psi_{n_0+1}(k + 1) = \psi_{n_0+1}(k)P + \psi_{n_0}(k)Q.$$

We proceed to find $\varphi_i(t)$ and $\psi_i(k)$. As is usual with singular perturbation method, for $i \leq n_0$, $\varphi_i(t)$ does not satisfy the initial conditions in general. To obtain the matched asymptotic expansions, care must be taken to match the outer expansions with the initial layer corrections. Compared with the case of irreducible matrix P, the problem is more delicate to handle now. We illustrate the features in what follows.

3.4.1 Constructing $\varphi_0(\varepsilon k)$ and $\psi_0(k)$

Henceforth, for a vector v, we use a partitioned form

$$v = (v^{11}, \ldots, v^{1m_1}, \ldots, v^{l_0 1}, \ldots, v^{l_0 m_{l_0}}, v^{*1}, \ldots, v^{*m_*})$$
$$= (v^1, \ldots, v^{l_0}, v^*),$$

where $v^i \in \mathbb{R}^{1 \times m_i}$ for each $i = 1, \ldots, l_0, *$. Using this notation and the partitioned form of the matrix P, for the leading term in the outer expansion, we obtain

$$\varphi_0^i(t)(P^i - I) + \varphi_0^*(t)P^{*,i} = 0, \quad 1 \leq i \leq l_0,$$
$$\varphi_0^*(t)(P^* - I) = 0. \tag{3.25}$$

Here and hereafter, I always denotes the identity matrix of appropriate dimensions. For instance, in $P^i - I$, I is an $m_i \times m_i$ identity matrix, and in $P^* - I$, I is an $m_* \times m_*$ identity matrix. We suppress the m_i dependence when there is no confusion from the context.

Since P^* is a nonnegative matrix having all of its eigenvalues inside the unit circle, $P^* - I$ is invertible. As a result, the last equation in (3.25) has only the trivial solution $\varphi_0^*(t) = 0$. Since $\varphi_0(t)$ represents the limit probability vector, the limit probability of the transient states disappears, which is expected. Substituting this into the rest of the l_0 equations in (3.25) yields that

$$\varphi_0^i(t)(P^i - I) = 0, \quad 1 \leq i \leq l_0.$$

Since for each $i = 1, \ldots, l_0$, P^i is irreducible,

$$\varphi_0^i(t)(P^i - I) = 0, \quad \varphi_0^i(t)\mathbb{1}_{m_i} = \theta_0^i(t), \tag{3.26}$$

is uniquely solvable, where $\theta_0^i(t)$ satisfying $\theta_0^i(t) \geq 0$ and $\sum_{i=1}^{l_0} \theta_0^i(t) = 1$ is to be determined. Write the leading term as

$$\varphi_0(t) = (\theta_0^1(t), \ldots, \theta_0^{l_0}(t), 0'_{m_*}) \mathrm{diag}(\nu^1, \ldots, \nu^{l_0}, 0_{m_* \times m_*}),$$

where $0_{m_*} \in \mathbb{R}^{m_* \times 1}$ is a zero vector, $0_{m_* \times m_*}$ is an $m_* \times m_*$ zero matrix, and for $i = 1, \ldots, l_0$, ν^i is the stationary distribution corresponding to the transition matrix P^i. Our task now is to find $\theta_0(t)$.

To proceed, define $\widetilde{\mathbb{1}}$ and $\widetilde{\mathbb{1}}_*$ as

$$\widetilde{\mathbb{1}} = \mathrm{diag}(\mathbb{1}_{m_1}, \ldots, \mathbb{1}_{m_{l_0}}) = \begin{pmatrix} \mathbb{1}_{m_1} & & \\ & \ddots & \\ & & \mathbb{1}_{m_{l_0}} \end{pmatrix} \quad \text{and}$$

$$\widetilde{\mathbb{1}}_* = \begin{pmatrix} \widetilde{\mathbb{1}} \\ A_* & 0_{m_* \times m_*} \end{pmatrix} = \begin{pmatrix} \mathbb{1}_{m_1} & & & \\ & \ddots & & \\ & & \mathbb{1}_{m_{l_0}} & \\ a^1 & \cdots & a^{l_0} & 0_{m_* \times m_*} \end{pmatrix}, \tag{3.27}$$

where $A_* = (a^1, \ldots, a^{l_0}) \in \mathbb{R}^{m_* \times (m_1 + \cdots + m_{l_0})}$ with

$$a^i = -(P^* - I)^{-1} P^{*,i} \mathbb{1}_{m_i}, \quad i = 1, \ldots, l_0. \tag{3.28}$$

Note that $a^i \geq 0$ for $i = 1, \ldots, l_0$ (i.e., all of their components are nonnegative; see Chapter 2 for the notation). Moreover, using

$$P^* \mathbb{1}_{m_*} + \sum_{i=1}^{l_0} P^{*,i} \mathbb{1}_{m_i} = \mathbb{1}_{m_*},$$

it is easily verified that

$$\sum_{i=1}^{l_0} a^i = \mathbb{1}_{m_*}. \tag{3.29}$$

Therefore, a^i represents the probability of transition from the transient states to the ith recurrent class $\mathcal{M}_i$.

It is clear that $(P - I)\widetilde{\mathbb{1}}_* = 0$. This orthogonality condition will be used in the subsequent development. Multiplying the second equation in (3.23) by $\widetilde{\mathbb{1}}_*$ results in

$$\varphi_0(t) Q \widetilde{\mathbb{1}}_* - \frac{d\varphi_0(t)}{dt} \widetilde{\mathbb{1}}_* = 0. \tag{3.30}$$

Partition Q as $Q = \begin{pmatrix} Q^{11} & Q^{12} \\ Q^{21} & Q^{22} \end{pmatrix}$, where $Q^{11} \in \mathbb{R}^{(m_0 - m_*) \times (m_0 - m_*)}$, $Q^{12} \in \mathbb{R}^{(m_0 - m_*) \times m_*}$, $Q^{21} \in \mathbb{R}^{m_* \times (m_0 - m_*)}$, $Q^{22} \in \mathbb{R}^{m_* \times m_*}$, and

$$\overline{Q}_* = \mathrm{diag}(\nu^1, \ldots, \nu^{l_0})(Q^{11}\widetilde{\mathbb{1}} + Q^{12}(a^1, \ldots, a^{l_0})). \tag{3.31}$$

Using $\theta_0(t) = (\theta_0^1(t), \ldots, \theta_0^{l_0}(t)) \in \mathbb{R}^{1 \times l_0}$ and (3.26), (3.30) becomes

$$\frac{d}{dt}\theta_0(t) = \theta_0(t)\overline{Q}_*. \tag{3.32}$$

Noting that p_0 is independent of ε, $p_0 = (p_0^1, \ldots, p_0^{l_0}, p_0^*)$ (where $p_0^i \in \mathbb{R}^{1 \times m_i}$) and examining $p_0 \widetilde{\mathbb{1}}_*$, we choose the initial condition $\theta_0^i(0)$ for each $i = 1, \ldots, l_0$ as

$$\theta_0^i(0) = p_0^i \mathbb{1}_{m_i} + p_0^* a^i, \quad \text{or equivalently} \quad \theta_0(0) = p_0 \widetilde{\mathbb{1}}_*.$$

It follows from (3.29) that

$$\theta_0^1(0) + \cdots + \theta_0^{l_0}(0) = 1.$$

Thus (3.32) is uniquely solvable, so $\varphi_0(t)$ is completely determined.

Next, examining (3.24), the solution of $\psi_0(k)$ is given by

$$\psi_0(k) = \psi_0(0)(P)^k.$$

To ensure the match of the outer expansion and the initial layer correction, we choose the initial conditions so that

$$\varphi_0(0) + \psi_0(0) = p_0. \tag{3.33}$$

Remark 3.2. To proceed, we deduce an ergodicity result. The essence is an estimate derivable from Perron and Frobenius's work on nonnegative matrices (see Iosifescu [73, p. 51]). For each $i \le l_0$, the stationary distribution ν^i is a left eigenvector of P^i and $\mathbb{1}_{m_i}$ is a right eigenvector of P^i. By virtue of Billingsley [18, p. 167 and p. 168] (or Iosifescu [73, p. 123]), as $k \to \infty$, $(P^i)^k \to \mathbb{1}_{m_i}\nu^i$ and $|(P^i)^k - \mathbb{1}_{m_i}\nu^i| \le K\lambda^k$ for some $0 < \lambda < 1$. Note that similar to (3.20), $\mathbb{1}_{m_i}\nu^i$ has identical rows containing the stationary distribution ν^i. Define

$$\widetilde{\nu}_* = \widetilde{\mathbb{1}}_*\mathrm{diag}(\nu^1, \ldots, \nu^{l_0}, 0_{m_* \times m_*}) \in \mathbb{R}^{m_0 \times m_0}. \tag{3.34}$$

The following lemma demonstrates that $(P)^k$ converges to $\widetilde{\nu}_*$ geometrically fast, proof of which is deferred until Section 3.6 in order not to disrupt the flow of presentation.

Lemma 3.3. *For some $0 < \lambda < 1$ and for $k \ge 0$, $|P^k - \widetilde{\nu}_*| \le K\lambda^k$.*

By virtue of Lemma 3.3, we proceed to establish the geometric decay property of $\psi_0(k)$. Note that (3.33) implies $\psi_0(0)\widetilde{\mathbb{1}}_* = 0$. It follows that $(p_0 - \varphi_0(0))\widetilde{\nu}_* = 0$. Thus,

$$\psi_0(k) = (p_0 - \varphi_0(0))P^k = (p_0 - \varphi_0(0))(P^k - \widetilde{\nu}_*).$$

Lemma 3.3 then implies that for some $0 < \lambda < 1$ and for $k \ge 0$,

$$|\psi_0(k)| \le K\lambda^k.$$

The leading terms $\varphi_0(\varepsilon k)$ and $\psi_0(k)$ in the asymptotic expansion have been constructed together with the verification of the decay property of $\psi_0(k)$.

Remark 3.4 Note that $\varphi_0(t)$ is the limit of the probability vector p_k^ε. It has the form $(\theta_0^1(t)\nu^1, \ldots, \theta_0^{l_0}(t)\nu^{l_0}, 0'_{m_*})$, where $0'_{m_*}$ is an m_*-dimensional zero row vector. In view of (3.8), for the transient part, $P(\alpha_k^\varepsilon = s_{*j}) \to 0$. As for the recurrent part, for each $i = 1, \ldots, l_0$, the limit distribution is of the form $\theta_0^i(t)\nu^i$, which has the meaning of total probability and is the limit of

$$P(\alpha_k^\varepsilon \in \mathcal{M}_i)P(\alpha_k^\varepsilon = s_{ij}|\alpha_k^\varepsilon \in \mathcal{M}_i).$$

3.4.2   Constructing $\varphi_n(t)$ and $\psi_n(k)$ for $n \geq 1$

The construction relies on the solutions of (3.23) and (3.24). It is done inductively. Suppose that $\varphi_j(t)$ and $\psi_j(k)$, for $j = 0, 1, \ldots, n - 1$, have been constructed such that $\varphi_j(t)$ are sufficiently smooth and $\psi_j(k)$ decay exponentially fast. We proceed to obtain the nth terms. For the outer expansion, consider

$$\varphi_n(t)(P - I) = \sum_{j=1}^{n} \frac{1}{j!} \frac{d^j \varphi_{n-j}(t)}{dt^j} - \varphi_{n-1}(t)Q \overset{\text{def}}{=} \xi_{n-1}(t). \qquad (3.35)$$

Again, we use the partitioned vector

$$\varphi_n(t) = (\varphi_n^1(t), \varphi_n^2(t), \ldots, \varphi_n^{l_0}(t), \varphi_n^*(t)).$$

Then equation (3.35) can be rewritten as

$$\begin{aligned} \varphi_n^i(t)(P^i - I) + \varphi_n^*(t)P^{*,i} &= \xi_{n-1}^i(t), \quad i = 1, \ldots, l_0, \\ \varphi_n^*(t)(P^* - I) &= \xi_{n-1}^*(t). \end{aligned} \qquad (3.36)$$

The solution of the last equation above is

$$\varphi_n^*(t) = \xi_{n-1}^*(t)(P^* - I)^{-1}. \qquad (3.37)$$

This leads to

$$\varphi_n^i(t)(P^i - I) = \xi_{n-1}^i(t) - \xi_{n-1}^*(t)(P^* - I)^{-1}P^{*,i}. \qquad (3.38)$$

Owing to the Fredholm alternative (see Lemma 14.36), this equation has a solution if and only if the right-hand side is orthogonal to $\mathbb{1}_{m_i}$, which is easily verified by a direct computation. Consequently, the solution can be written as

$$\varphi_n^i(t) = \theta_n^i(t)\nu^i + \widetilde{\varphi}_n^i(t), \qquad (3.39)$$

where $\widetilde{\varphi}_n^i(t)$ denotes a particular solution of (3.38). Substituting $\varphi_n^i(t)$ given in (3.39) into (3.35) yields the differential equation

$$\frac{d}{dt}\theta_n(t) = \theta_n(t)\overline{Q}_* + \widehat{\xi}_n(t), \qquad (3.40)$$

where

$$\theta_n(t) = (\theta_n^1(t), \ldots, \theta_n^{l_0}(t)),$$

and $\widehat{\xi}_n(t)$ is the vector consists of the first l_0 components of

$$-\frac{d}{dt}\widetilde{\varphi}_n(t)\widetilde{\mathbb{1}}_* + \widetilde{\varphi}_n(t)Q\widetilde{\mathbb{1}}_* - \sum_{j=2}^{n+1} \frac{1}{j!} \frac{d^j}{dt^j}\varphi_{n+1-j}(t)\widetilde{\mathbb{1}}_*.$$

Using the defining equation of $\psi_n(k)$, we obtain

$$\psi_n(k+1) = \psi_n(0)P^{k+1} + \sum_{j=0}^{k} \psi_{n-1}(j)QP^{k-j}. \qquad (3.41)$$

Next we determine the initial conditions $\theta_n^i(0)$ through the matching of $\varphi_n(\varepsilon k)$ with $\psi_n(k)$. Choose the initial values such that

$$\psi_n(0) = -\varphi_n(0). \qquad (3.42)$$

Examining (3.41), we demand that $\psi_n(k) \to 0$ as $k \to \infty$. This implies that

$$\psi_n(0)\tilde{\nu}_* = -\sum_{j=0}^{\infty} \psi_{n-1}(j)Q\tilde{\nu}_* \overset{\text{def}}{=} -\bar{s}_{n-1}\tilde{\nu}_*. \qquad (3.43)$$

Remark 3.5. Since $\varphi_j(t)$ and $\psi_j(k)$ for $j \le n-1$ have been found, $\bar{s}_{n-1}$ is a known vector. By virtue of the definition of $\tilde{\nu}_*$ in (3.34), (3.43) leads to

$$(\psi_n^1(0)\mathbb{1}_{m_1}\nu^1 + \psi_n^*(0)a^1\nu^1, \dots, \psi_n^{l_0}\mathbb{1}_{m_{l_0}}\nu^{l_0} + \psi_n^*(0)a^{l_0}\nu^{l_0}, 0_{m_*})$$
$$= (\bar{s}_{n-1}^1 \mathbb{1}_{m_1}\nu^1 + \bar{s}_{n-1}^* a^1\nu^1, \dots, \bar{s}_{n-1}^{l_0}\mathbb{1}_{m_{l_0}}\nu^{l_0} + \bar{s}_{n-1}^* a^{l_0}\nu^{l_0}, 0_{m_*}).$$

Since $\varphi_n^*(t)$ can be obtained from (3.36), $\psi_n^*(0)$ can be obtained from (3.42). Note that $\mathbb{1}_{m_i}\nu^i$ has identical rows $(\nu^{i1}, \dots, \nu^{im_i})$. Although (3.43) contains m_0 equations, there are only l_0 unknowns, namely, $\psi_n^i(0)\mathbb{1}_{m_i}$. To determine the initial conditions, we first solve for these l_0 unknowns. Then we obtain $\theta_n^i(0) = -\psi_n^i(0)\mathbb{1}_{m_i}$. Once we obtain $\theta_n^i(0)$, $\varphi_n(t)$ is completely specified, so $\varphi_n(0)$ is found. Finally, we choose $\psi_n(0)$ by using (3.42), and complete the construction of the formal expansions.

We detail the steps outlined in the above remark as follows. First,

$$\psi_n^i(0)\mathbb{1}_{m_i} + \psi_n^*(0)a^i = -[\bar{s}_{n-1}^i \mathbb{1}_{m_i} + \bar{s}_{n-1}^* a^i].$$

Owing to (3.37), and $\varphi_n^*(0) = -\psi_n^*(0)$,

$$\psi_n^i(0)\mathbb{1}_{m_i} = -[\bar{s}_{n-1}^i \mathbb{1}_{m_i} + \bar{s}_n^* a^i] + \xi_{n-1}^*(0)(P^* - I)^{-1}a^i.$$

We choose $\theta_n^i(0) = -\psi_n^i(0)\mathbb{1}_{m_i}$ for $i = 1, \dots, l_0$. Using $\theta_n^i(0)$ to determine $\varphi_n(0)$ as in (3.39) and then determine $\psi_n(0)$ by means of (3.42). Thus the initial conditions are uniquely determined and the construction of the asymptotic expansion is completed. We summarize the development so far into the following theorem.

Theorem 3.6. *Suppose that Condition (A3.1) holds for P given by (3.6). Then the asymptotic expansions can be constructed via (3.23) and (3.24) together with the specification of the initial data (3.42) and (3.43) such that for $n = 0, \dots, n_0$, $\varphi_n(t)$ are sufficiently smooth and $|\psi_n(k)| \le K\lambda_0^k$ for some $0 < \lambda_0 < 1$.*

Remark 3.7. In what follows, for notational simplicity, we will not distinguish different λ's. We will simply write λ as a generic constant satisfying $0 < \lambda < 1$.

3.5 Error Bounds

Up to now, we have constructed the asymptotic expansions formally. To validate the expansions, we must derive the desired error bounds. Recall the operator $\mathcal{L}^\varepsilon$ defined in (3.9) and the nth-order approximation error $e_n^\varepsilon(k)$ given by

$$e_n^\varepsilon(k) = \sum_{j=0}^{n} \varepsilon^j \varphi_j(\varepsilon k) + \sum_{j=0}^{n} \varepsilon^j \psi_j(k) - p_k^\varepsilon, \quad \text{for } n = 0, \dots, n_0 + 1,$$

where p_k^ε is the solution of (3.2). Our plan is as follows. We first obtain a lemma that establishes a bound on $\mathcal{L}^\varepsilon e_n^\varepsilon(k)$. The second step presents an auxiliary lemma based on a non-homogeneous equation. Combining the two steps yields the desired error estimates.

3.5.1 Estimates on $\mathcal{L}^\varepsilon e_n^\varepsilon(k)$

Note that in view of the definition of $\mathcal{L}^\varepsilon$ and by virtue of Theorem 3.6, for each $n = 0, \dots, n_0 + 1$,

$$\mathcal{L}^\varepsilon e_n^\varepsilon(k) = \mathcal{L}^\varepsilon \left(\sum_{j=0}^{n} \varepsilon^j \varphi_j(\varepsilon k) + \sum_{j=0}^{n} \varepsilon^j \psi_j(k) \right).$$

Owing to (3.23) and (3.24), we have

$$\mathcal{L}^\varepsilon \left(\sum_{j=0}^{n} \varepsilon^j \varphi_j(\varepsilon k) + \sum_{j=0}^{n} \varepsilon^j \psi_j(k) \right)$$

$$= \sum_{j=0}^{n} \varepsilon^j [\varphi_j(\varepsilon(k+1)) - \varphi_j(\varepsilon k)(P + \varepsilon Q)] \qquad (3.44)$$

$$+ \sum_{j=0}^{n} \varepsilon^j [\psi_j(k+1) - \psi_j(k)(P + \varepsilon Q)].$$

Using (3.24), it is easily checked that

$$\sum_{j=0}^{n} \varepsilon^j [\psi_j(k+1) - \psi_j(k)(P + \varepsilon Q)] = -\varepsilon^{n+1} \psi_n(k) Q.$$

By virtue of the defining relation (3.23), detailed estimates yield

$$\sum_{j=0}^{n} \varepsilon^j [\varphi_j(\varepsilon(k+1)) - \varphi_j(\varepsilon k)(P + \varepsilon Q)]$$

$$= \sum_{j=0}^{n} \varepsilon^j [\varphi_j(\varepsilon(k+1)) - \varphi_j(\varepsilon k)] - \sum_{j=0}^{n} \varepsilon^j \varphi_j(\varepsilon k)((P - I) + \varepsilon Q)$$

$$= \sum_{j=0}^{n} \varepsilon^j \sum_{i=1}^{n+1-j} \left[\frac{\varepsilon^i}{i!} \frac{d^i \varphi_j(\varepsilon k)}{dt^i} + O(\varepsilon^{n+2-i}) \right]$$

$$- \sum_{j=0}^{n} \varepsilon^j \varphi_j(\varepsilon k)((P - I) + \varepsilon Q)$$

$$= O(\varepsilon^{n+1}).$$

(3.45)

Thus the following lemma is obtained.

Lemma 3.8. *Under the conditions of Theorem 3.6, we have* $\mathcal{L}^\varepsilon e_n^\varepsilon(k) = O(\varepsilon^{n+1})$, *for each* $n = 0, \ldots, n_0 + 1$.

3.5.2   An Estimate of $e_n^\varepsilon(k)$

The main result here is recorded in the following lemma.

Lemma 3.9. *Suppose that, for some* $n > 0$,

$$\mathcal{L}^\varepsilon e(k) = O(\varepsilon^{n+1}), \ e(0) = 0.$$

Then the solution of the above initial value problem satisfies

$$\sup_{0 \le k \le \lfloor T/\varepsilon \rfloor} |e(k)| = O(\varepsilon^n).$$

Proof. Since $e(k)$ is the solution of the difference equation

$$e(k) = e(k-1)P^\varepsilon + O(\varepsilon^{n+1}),$$

with initial data $e(0) = 0$, the solution can be written as

$$e(k) = \sum_{j=0}^{k-1} O(\varepsilon^{n+1})(P^\varepsilon)^j, \ \text{for } k = 0, \ldots, \lfloor T/\varepsilon \rfloor, \qquad (3.46)$$

and as a result

$$\sup_{0 \le k \le \lfloor T/\varepsilon \rfloor} |e(k)| \le O(\varepsilon^{n+1})\lfloor T/\varepsilon \rfloor = O(\varepsilon^n)$$

as desired.   □

With the preparation of Lemmas 3.8 and 3.9, we are ready to derive the desired error estimates. The result is stated in the next theorem, and the proof is postponed until the next section.

Theorem 3.10. *Under the conditions of Theorem* 3.6,

$$\sup_{0 \le k \le \lfloor T/\varepsilon \rfloor} |e_{n_0}^{\varepsilon}(k)| = O(\varepsilon^{n_0+1}).$$

Note that Theorems 3.6 and 3.10 also hold for P given by (3.5), and the analysis is simpler.

3.5.3 Expansions of k-step Transition Matrices

Parallel to the asymptotic development of the probability vector p_k^{ε}, we may also obtain asymptotic expansions of k-step transition matrices. Recall the defining equation (3.3). As stated earlier that when $P_k = P$ and $Q_k = Q$, the k-step transition matrix $\widetilde{P}_k^{\varepsilon} = (P^{\varepsilon})^k$, owing to the homogeneity of the transition matrices. Note that (3.3) can be considered as a special case of (3.2) since the ith row of the matrix is a solution of (3.2) with a specific initial condition (the ith unit vector).

We seek asymptotic expansions of the form

$$\widetilde{U}_j^{\varepsilon}(\varepsilon k) = \sum_{i=0}^{j} \varepsilon^i \Phi_i(\varepsilon k), \quad \widetilde{V}_j^{\varepsilon}(k) = \sum_{i=0}^{j} \varepsilon^i \Psi_i(k), \quad j = 1, \dots, n_0. \tag{3.47}$$

Proceeding exactly as we did in the previous cases, substituting the above expansion into the equation $(P^{\varepsilon})^{k+1} = (P^{\varepsilon})^k P^{\varepsilon}$ and comparing coefficients of like powers of ε, we obtain

$$\Phi_0(t)(P - I) = 0,$$
$$\Phi_{i+1}(t)(P - I) = \sum_{j=1}^{i+1} \frac{1}{j!} \frac{d^j \Phi_{i+1-j}(t)}{dt^j} - \Phi_i(t)Q, \quad i = 0, \dots, n_0, \tag{3.48}$$

and

$$\Psi_0(k+1) = \Psi_0(k)P,$$
$$\Psi_{i+1}(k+1) = \Psi_{i+1}(k)P + \Psi_i(k)Q, \quad i = 0, \dots, n_0, \tag{3.49}$$

with the initial data

$$\Phi_0(0) + \Psi_0(0) = I,$$
$$\Phi_i(0) = -\Psi_i(0), \quad i = 1, \dots, n_0 + 1. \tag{3.50}$$

To find these $\Phi_n(t)$'s, we consider, for example,

$$\Phi_0(t) = (\Phi_0^1(t), \dots, \Phi_0^{l_0}(t), \Phi_0^*(t))$$

where $\Phi_0^i(t) \in \mathbb{R}^{m_0 \times m_i}$ for $i = 1, \ldots, l_0, *$. Then we obtain

$$\Phi_0^i(t)(P^i - I) + \Phi_0^* P^{*,i} = 0, \quad i = 1, \ldots, l_0,$$
$$\Phi_0^*(t)(P^* - I) = 0.$$

Similar to the asymptotic expansions for p_k^ε, $\Phi_0^*(t) = 0$, so we can proceed to solve $\Phi_0^i(t)(P^i - I) = 0$ for $i = 1, \ldots, l_0$. Denote

$$\Theta_*(t) = \mathrm{diag}(\Theta(t), I_{m_* \times m_*}) \in \mathbb{R}^{(l_0 + m_*) \times (l_0 + m_*)},$$

where $\Theta(t) = (\theta^{ij}(t)) \in \mathbb{R}^{l_0 \times l_0}$ satisfies

$$\frac{d\Theta(t)}{dt} = \Theta(t)\overline{Q}, \quad \Theta(0) = I. \tag{3.51}$$

Or equivalently, $\Theta_*(t)$ satisfies

$$\frac{d\Theta_*(t)}{dt} = \Theta_*(t)\overline{Q}_*, \quad \Theta(0) = I.$$

Detailed argument leads to

$$\Phi_0(t) = \widetilde{\mathbb{1}}_* \Theta_*(t) \widetilde{\nu}_* = \begin{pmatrix} \widetilde{\mathbb{1}}\Theta(t)\nu & 0_{(m_0 - m_*) \times (m_0 - m_*)} \\ A_* \Theta(t)\nu & 0_{m_* \times m_*} \end{pmatrix}, \tag{3.52}$$

where A_* is defined in (3.27). For the initial layer correction term, using the initial condition $\Psi_0(0) = I - \Phi_0(0)$, similar to Theorem 3.6, we can show that there is a $0 < \lambda < 1$ such that $|P^k - \widetilde{\nu}_*| \leq K\lambda^k$ and, consequently,

$$|\Psi_0(k)| = |(I - \Phi_0(0))(P^k - \widetilde{\nu}_*)| \leq K\lambda^k.$$

Working our way to higher-order terms, the same methods of proof as in Theorems 3.6 and 3.10 yield the following theorem.

Theorem 3.11. *Under* (A3.1), *asymptotic expansions* $\widetilde{U}_j^\varepsilon(\varepsilon k)$ *and* $\widetilde{V}_j^\varepsilon(k)$ *for* $j = 1, \ldots, n_0$ *can be constructed such that* $\Phi_i(t)$ *are sufficiently smooth and* $\Psi(k)$ *decay exponentially fast. Moreover,*

$$\sup_{0 \leq k \leq \lfloor T/\varepsilon \rfloor} |(P^\varepsilon)^k - \widetilde{U}_{n_0}^\varepsilon(\varepsilon k) - \widetilde{V}_{n_0}^\varepsilon(k)| = O(\varepsilon^{n_0 + 1}). \tag{3.53}$$

Remark 3.12. The theorem above is concerned with time-homogeneous discrete-time Markov chains exclusively. When the transition matrices are time dependent (i.e., in (3.1), P_k or Q_k or both are k-dependent), the situation is more complex and care needs to be exercised in obtaining the approximation of the k-step transition matrices; see Yin and Zhang [158, pp. 82–84] for a continuous-time analog.

3.6   Examples

This section presents several examples. In these examples, the discrete-time Markov chains all involve two time scales. We demonstrate how to construct the asymptotic expansions of their probability vectors and transition matrices.

Example 3.13. In Chapter 1, Examples 1.5 and 1.6, we have encountered such Markov chains α_k^ε whose transition matrix is given by $P^\varepsilon = I + \varepsilon Q$. Note that the identity matrix may be written as $I = \text{diag}(1, \ldots, 1)$, so $P = I$ can be regarded as a transition matrix consisting of m_0 ergodic classes, each of which is an absorbing state. Using the results in Theorems 3.6 and 3.10, we obtain that

$$p_k^\varepsilon = \varphi_0(\varepsilon k) + O(\varepsilon), \ 0 \le k \le T/\varepsilon,$$

where $\varphi_0(t)$ satisfies the ordinary differential equation

$$\frac{d\varphi_0(t)}{dt} = \varphi_0(t)Q, \ \varphi_0(0) = p_0.$$

Thus the 0th-order initial layer term $\psi_0(k) = 0$ for $0 \le k \le T/\varepsilon$ due to the fact $\psi_0(0) = 0$. It is interesting to note that a naive thought may suggest dropping the εQ in the limit, but such a conclusion in fact is not correct. The absence of the 0th-order initial layer term is a distinct characteristic tied up with the absorbing states. For a continuous-time analog, we refer the reader to Yin and Zhang [158, Section 6.3].

Example 3.14. Suppose that α_k^ε is a discrete-time Markov chain with state space $\mathcal{M}$ given in (3.7) and transition matrix $P^\varepsilon = P + \varepsilon Q$. Define N_k to be the counting process that counts the number of transitions of the Markov chain α_k^ε up to k. Denote

$$f(k+1) = E(N_{k+1} - N_k) \ \text{ with } \ f(0) = 0,$$

which is referred to as the frequency of the transition and is the mean number of transitions in the $(k+1)$ step. Then it is easily seen that

$$f(k+1) = \sum_{i=i}^{l_0} \sum_{j=1}^{m_i} P(\alpha_k^\varepsilon = s_{ij})(1 - p^{\varepsilon, ij, ij}),$$

where

$$p^{\varepsilon, ij, ij} = P(\alpha_{k+1}^\varepsilon = s_{ij} | \alpha_k^\varepsilon = s_{ij})$$

denotes the transition probability that the Markov chain will remain in state s_{ij} during the $(k+1)$st transition. Now consider the case that the transition probability matrix P^ε of the Markov chain α_k^ε has the form $P^\varepsilon =$

$P+\varepsilon Q$ with P given by (3.5). Now all the functions of interests are indexed by ε. Consequently, using the asymptotic expansions in Theorem 3.10, we obtain

$$f^\varepsilon(k+1) = \sum_{i=1}^{l_0} \sum_{j=1}^{m_i} \theta_0^i(\varepsilon k)\nu^{ij}(1 - p^{ij,ij}) + O(\varepsilon), \quad k = O(1/\varepsilon),$$

where $p^{ij,ij}$ is defined as that of $p^{\varepsilon,ij,ij}$ but corresponding to the transition matrix P not P^ε, and ν^{ij} denotes the jth component of ν^i the stationary distribution of P^i. Furthermore, we can approximate the mean number of transitions from $\mathcal{M}_{i_1} = \{s_{i_11}, \ldots, s_{i_1 m_{i_1}}\}$ to $\mathcal{M}_{i_2} = \{s_{i_21}, \ldots, s_{i_2 m_{i_2}}\}$ for $i_1 \neq i_2$ as

$$f_{\mathcal{M}_{i_1}\mathcal{M}_{i_2}}(k+1) = \sum_{j_1=1}^{m_{i_1}} \sum_{j_2=1}^{m_{i_2}} \theta_0^{i_1}(\varepsilon k)\nu^{i_1 j_1} p^{i_1 j_1, i_2 j_2} + O(\varepsilon), \quad k = O(1/\varepsilon).$$

Example 3.15. Suppose that α_k^ε is a discrete-time Markov chain whose transition probability matrix is given by (3.1) with P given by (3.5). Suppose that (A3.1) holds. Then by virtue of Theorems 3.6 and 3.10, we can construct asymptotic expansions of p_k^ε, the solution of (3.2), and

$$p_k^\varepsilon = \text{diag}(\nu^1, \ldots, \nu^{l_0})\theta(t) + O(\varepsilon + \lambda^k),$$

for some $0 < \lambda < 1$, where $\theta(t) \in \mathbb{R}^{1 \times l_0}$ satisfies the differential equation

$$\frac{d\theta(t)}{dt} = \theta(t)\overline{Q}, \quad \theta(0) = p_0\widetilde{\mathbb{1}}.$$

Suppose that in addition that P_k^ε is irreducible. Consider

$$J^\varepsilon = \limsup_{N \to \infty} \frac{1}{N}E \sum_{k=0}^{N-1} g(\alpha_k^\varepsilon),$$

for a suitable Borel measurable function $g(\cdot)$. Then it is well known that a Borel function $h^\varepsilon(\cdot)$ satisfies the so-called Poisson equation

$$h^\varepsilon(i) + J^\varepsilon = g(i) + \sum_{j=1}^{m_0} p^{\varepsilon,ij} h^\varepsilon(j),$$

or equivalently,

$$J^\varepsilon = g(i) + \sum_{j=1}^{m_0} (p^{\varepsilon,ij} - \delta^{ij})h^\varepsilon(j), \quad i = 1, \ldots, m_0, \qquad (3.54)$$

where $\delta^{ij} = 1$ if $i = j$ and is 0 otherwise. It will be shown in Chapter 8 that associated with (3.54), there is a limit problem

$$J^0 = \limsup_{T \to \infty} E \int_0^T g(\overline{\alpha}(t))dt,$$

where $\overline{\alpha}(t)$ is a continuous-time Markov chain with state space $\overline{\mathcal{M}} = \{1, \ldots, l_0\}$ and generator $\overline{Q}$. In fact, much more complicated Markov decision processes involving control variables will be considered there. Note that Poisson equations have been used extensively in the context of stochastic dynamic programming (see Ross [131]) and stochastic approximation (see Metivier and Priouret [114]), among others.

Example 3.16. Let α_k^ε be a discrete-time Markov chain whose transition probability matrix is given by (3.1) with P given by (3.6) and Q being a generator of a continuous-time Markov chain. Assume that (A3.1) holds. Then using Theorems 3.6 and 3.10, we can approximate the expectation $Ef(\alpha_k^\varepsilon)$ for an arbitrary measurable function $f(\cdot)$. Note that

$$Ef(\alpha_k^\varepsilon) = \sum_{i=1}^{l_0} \sum_{j=1}^{m_i} f(s_{ij}) P(\alpha_k^\varepsilon = s_{ij}).$$

The first approximation is to replace the probability $P(\alpha_k^\varepsilon = s_{ij})$ by the leading term in the asymptotic expansion. A moment of reflection reveals that we can, in fact, find the asymptotic expansions of the expectation with desired order of accuracy. In fact, we have

$$Ef(\alpha_k^\varepsilon) = \sum_{i=1}^{l_0} \sum_{j=1}^{m_i} \sum_{\ell=0}^{n} \varepsilon^\ell f(s_{ij})[\varphi_\ell^{ij}(t) + \psi_\ell^{ij}(k)] + O(\varepsilon^{n+1}),$$

and the error bounds hold uniformly for $0 \leq k \leq T/\varepsilon$, where $t = \varepsilon k$, $\varphi_\ell^{ij}(t)$ and $\psi_\ell^{ij}(k)$ denote the jth components in the ith blocks $\varphi_\ell^i(t)$ and $\psi_\ell^i(k)$, respectively.

3.7 Proofs of Results

This section contains the proofs of a number of technical results in the construction of the asymptotic expansions and the asymptotic validation.

Proof of Lemma 3.3. Consider the following difference equation

$$w(k+1) = w(k)P, \quad w(0) = w_0.$$

Rewriting $w(k)$ in terms of the partitioned vectors, we arrive at

$$w^i(k+1) = w^i(k)P^i + w^*(k)P^{*,i}, \quad i = 1, \ldots, l_0,$$
$$w^*(k+1) = w^*(k)P^*. \tag{3.55}$$

The solution of the last equation in (3.55) is $w^*(k+1) = w_0^*[(P^*)^{k+1}]$. Substituting this into the rest of the equations in (3.55) leads to

$$w^i(k+1) = w_0^i(P^i)^{k+1} + w_0^* \sum_{j=0}^{k} (P^*)^j P^{*,i} (P^i)^{k-j}.$$

A simple calculation shows that

$$\sum_{j=0}^{\infty} (P^*)^j = (I - P^*)^{-1} = -(P^* - I)^{-1}$$

since P^* has all of its eigenvalues inside the unit circle. By (A3.1) and Remark 3.2, for each $i = 1, \ldots, l_0$, $(P^i)^k \to \mathbb{1}_{m_i} \nu^i$ as $k \to \infty$,

$$\left| \sum_{j=0}^{k} (P^*)^j P^{*,i} (P^i)^{k-j} \right| \leq \sum_{j=0}^{\infty} |(P^*)^j P^{*,i} \mathbb{1}_{m_i} \nu^i|$$

$$+ \sum_{j=0}^{k} |(P^*)^j P^{*,i} [(P^i)^{k-j} - \mathbb{1}_{m_i} \nu^i]|$$

$$\leq K \sum_{j=0}^{\infty} |(P^*)^j P^{*,i} \mathbb{1}_{m_i} \nu^i| + K < \infty.$$

Note that

$$w_0^i \mathbb{1}_{m_i} \nu^i + w_0^* \sum_{j=0}^{\infty} (P^*)^j P^{*,i} \mathbb{1}_{m_i} \nu^i = (w_0^1, \ldots, w_0^{l_0}, w_0^*) \tilde{\nu}_*,$$

where $\tilde{\nu}_*$ is defined in (3.34). To obtain the desired estimate, we work with each partitioned vector. For each $i = 1, \ldots, l_0$,

$$w^i(k+1) - \left(w_0^i \mathbb{1}_{m_i} \nu^i + w_0^* \sum_{j=0}^{\infty} (P^*)^j P^{*,i} \mathbb{1}_{m_i} \nu^i \right)$$

$$= w_0^i[(P^i)^k - \mathbb{1}_{m_i} \nu^i] + w_0^* \sum_{j=0}^{k} (P^*)^j P^{*,i}[(P^i)^{k-j} - \mathbb{1}_{m_i} \nu^i]$$

$$- w_0^* \sum_{j=k+1}^{\infty} (P^*)^j P^{*,i} \mathbb{1}_{m_i} \nu^i.$$

The condition on P^* implies

$$\left| w_0^* \sum_{j=k+1}^{\infty} (P^*)^j P^{*,i} \mathbb{1}_{m_i} \nu^i \right| \leq K |w_0^*| \lambda_*^{k+1}.$$

Since for each $i = 1, \ldots, l_0$, P^i is irreducible and aperiodic, by virtue of Remark 3.2, $|(P^i)^k - \mathbb{1}_{m_i} \nu^i| \leq K \lambda_i^k$, for some $\lambda_i > 0$. Choose $\lambda = \max(\lambda_1, \ldots, \lambda_{l_0}, \lambda_*)$. The desired estimate follows. $\square$

Proof of Theorem 3.6. The smoothness of the outer-expansion terms $\varphi_n(\cdot)$ can be verified by examining the solutions directly. As for the geometric decay property of $\psi_n(k)$, in view of (3.41), we have

$$\psi_n(k+1) = \psi_n(0)(P^{k+1} - \widetilde{\nu}_*) + \sum_{j=0}^{k} \psi_{i-1}(j)Q(P^{k-j} - \widetilde{\nu}_*)$$

$$- \sum_{j=k+1}^{\infty} \psi_{n-1}(j)Q\widetilde{\nu}_*.$$

(3.56)

Upon using $k(\sqrt{\lambda})^k \leq K$, for each $k = 0, \ldots, \lfloor T/\varepsilon \rfloor$, and by virtue of Lemma 3.3,

$$|\psi_n(k+1)| \leq K\lambda^{k+1} + K\sum_{j=0}^{k} \lambda^j \lambda^{k-j} + K\lambda^{k+1}$$

$$\leq (K+k)\lambda^k \leq K(\sqrt{\lambda})^k \leq K\lambda_0^k,$$

where $0 < \lambda < \sqrt{\lambda} = \lambda_0 < 1$. □

Proof of Theorem 3.10. By virtue of Lemma 3.8,

$$\mathcal{L}^\varepsilon e_{n_0+1}^\varepsilon(k) = O(\varepsilon^{n_0+2}).$$

Applying Lemma 3.9 to $e_{n_0+1}^\varepsilon(k)$ leads to

$$\sup_{0 \leq k \leq \lfloor T/\varepsilon \rfloor} |e_{n_0+1}^\varepsilon(k)| = O(\varepsilon^{n_0+1}).$$

Since

$$e_{n_0+1}^\varepsilon(k) = e_{n_0}^\varepsilon(k) + \varepsilon^{n_0+1}\varphi_{n_0+1}(\varepsilon k) + \varepsilon^{n_0+1}\psi_{n_0+1}(k),$$

the boundedness of $\varphi_{n_0+1}(\varepsilon k)$ and $\psi_{n_0+1}(k)$ then implies

$$e_{n_0+1}^\varepsilon(k) = e_{n_0}^\varepsilon(k) + O(\varepsilon^{n_0+1}),$$

and hence

$$e_{n_0}^\varepsilon(k) = e_{n_0+1}^\varepsilon(k) + O(\varepsilon^{n_0+1})$$

$$= O(\varepsilon^{n_0+1}) + O(\varepsilon^{n_0+1})$$

$$= O(\varepsilon^{n_0+1}).$$

The proof of the theorem is thus concluded. □

3.8 Notes

The subject matter discussed in this chapter is at the intersection of singular perturbation and Markov processes, both of which are covered widely in the literature. Singular perturbation theory (see, for example, Bender and Orszag [12] or O'Malley [118] for a general introduction) may be traced back to the study of Prandtl at the beginning of the 20th century, whose work established the foundation of boundary layer theory. The methods have been extensively used in, for example, various branches of physics, including statistical mechanics, solid state physics, chemical physics, and molecular biophysics. For related applications in control theory and optimization, we refer the reader to Phillips and Kokotovic [125], Pan and Başar [119], Pervozvanskii and Gaitsgori [123], and Yin and Zhang [158]. Discrete-time singularly perturbed Markov chains have been considered by Abbad, Filar, and Bielecki [2], Blankenship [21], and Delebecque and Quadrat [44], among others. The idea of two-time-scale expansions has also been found in emerging applications in communication theory (see Tse, Gallager, and Tsitsiklis [144], among others), which opens up new avenues for diverse applications.

Parallel to the advances in the deterministic theory, there is a stochastic version of the averaging methods. It began with a seminal paper by Khasminskii [81], continued with the large deviations approach of Friedlin and Wentzell [59], and extended to the martingale averaging methods of Kushner [97].

Much of our initial study was geared toward applications of manufacturing systems, in which the systems under consideration are of a hybrid type and jump processes are used in the models. The recent book of Sethi and Zhang contains a wide range of applications of production planning and optimal control problems. Our systematic study began with the work of Khasminskii, Yin, and Zhang [85, 86]; our subsequent effort has been devoted to singularly perturbed systems involving Markov chains and related control problems; see Zhang, Liu, and Yin [177], Yin and Zhang [158, 159, 160], Yin, Zhang, and Badowski [162, 163, 164], Yin, Zhang, and Liu [166], Yin, Zhang, Yang, and Yin [168], Zhang [172, 173], Zhang and Yin [179, 180, 181], and the references therein. Related work in singularly perturbed diffusions and singularly perturbed switching diffusions can be found in Il'in, Khasminskii, and Yin [71, 72], Khasminskii and Yin [84], and Yin [154].

4
Occupation Measures

4.1 Introduction

The asymptotic expansions obtained in Chapter 3 rely purely on analytic techniques and the procedure is deterministic in nature. This demonstrates how the probability distributions can be approximated by the corresponding equilibrium distributions. To further understand asymptotic properties of the two-time-scale Markov chains, in this chapter, we examine sequences of occupation measures of the model considered in Chapter 3. Suppose that $T > 0$ and $\varepsilon > 0$ is a small parameter, and that for $0 \leq k \leq \lfloor T/\varepsilon \rfloor$, α_k^ε is a discrete-time Markov chain with state space $\mathcal{M} = \{1, \ldots, m_0\}$ and transition matrix

$$P^\varepsilon = P + \varepsilon Q, \tag{4.1}$$

where P is a transition matrix of a discrete-time Markov chain and Q is a generator of a continuous-time Markov chain.

To motivate our study, let us begin with a relatively simple case, in which P given in (4.1) is irreducible and aperiodic. Using the indicator function I_A of a set A, the expression

$$\varepsilon \sum_{l=0}^{k-1} I_{\{\alpha_l^\varepsilon = i\}} \quad \text{for each } i \in \mathcal{M} \tag{4.2}$$

represents the duration that the Markov chain spends in state i before time $k - 1$ and hence is termed an occupation measure. Clearly, it is a random process. Note that the above summation is scaled by ε, which is needed to make the sum well defined (finite) in the limit as $\varepsilon \to 0$. By virtue of Billingsley [18, Example 2, p. 167], α_k^ε is ϕ-mixing and hence is strongly ergodic. Consequently, as $\varepsilon \to 0$, we have that

$$\varepsilon \sum_{l=0}^{T/\varepsilon-1} I_{\{\alpha_l^\varepsilon = i\}} = T\left[\frac{1}{(T/\varepsilon)} \sum_{l=0}^{T/\varepsilon-1} I_{\{\alpha_l^\varepsilon = i\}} \right] \tag{4.3}$$
$$\to T\nu^i \text{ w.p.1,}$$

where ν^i is the ith component of the stationary distribution ν corresponding to P. The above assertion is nothing but a strong law of large numbers. In the above and henceforth, we use the convention that T/ε represents its integer part. Define sequences of centered occupation measures by

$$\pi_k^{\varepsilon,i} = \varepsilon \sum_{l=0}^{k-1} [I_{\{\alpha_l^\varepsilon = i\}} - \nu^i] \text{ for } i \in \mathcal{M}, \ 0 \le k \le T/\varepsilon. \tag{4.4}$$

Both the asymptotic expansions developed in Chapter 3 for the probability distribution of α_k^ε and the strong law of large numbers (4.3) for the occupation measures indicate that somehow α_k^ε can be approximated by its stationary distribution ν in a suitable sense and $\pi_{T/\varepsilon}^{\varepsilon,i} \to 0$ w.p.1 as $\varepsilon \to 0$. How close is such an approximation? How fast does $\pi_{T/\varepsilon}^{\varepsilon,i} \to 0$? Is there any central-limit-theorem type result? Answers to these questions along with mean squares estimates and structural properties of the underlying Markov chains are the central theme of this chapter. The question of assessing the quality of approximation of α_k^ε by ν from a stochastic point of view can be addressed by considering suitably scaled sequences. Note that if k in (4.4) is independent of ε, the result is trivial. Thus we are mainly interested in the case that $k = O(1/\varepsilon)$. For this reason, k may be written as k_ε since it is ε dependent. Roughly, the discussion above tells us that the occupation measures (4.2) can be approximated by a deterministic quantity. Such an approximation crucially depends on the fact that α_k^ε is ϕ-mixing.

If the Markov chain has transition matrix of the form (4.1) with P given by

$$P = \text{diag}(P^1, \ldots, P^{l_0}) = \begin{pmatrix} P^1 & & \\ & \ddots & \\ & & P^{l_0} \end{pmatrix}, \tag{4.5}$$

where $P^i \in \mathbb{R}^{m_i \times m_i}$ are transition matrices and $\sum_{i=1}^{l_0} m_i = m_0$, or

$$P = \begin{pmatrix} P^1 & & & \\ & P^2 & & \\ & & \ddots & \\ & & & P^{l_0} \\ P^{*,1} & P^{*,2} & \cdots & P^{*,l_0} & P^* \end{pmatrix}, \tag{4.6}$$

where P^i are transition matrices for each $i \leq l_0$ and $(P^{*,1}, \ldots, P^{*,l_0}, P^*)$ in (4.6) corresponds to the transient states. Note that the mixing condition will no longer hold if P is given by (4.5) or (4.6). This can be seen by examining the asymptotic expansions for the probability vectors and transition matrices. One of the consequences is that (4.2) cannot be approximated by a deterministic function. That is, in treating a centered occupation measure similar to that of (4.4), the centering term must be a random process. This makes matters substantially more difficult and leads to a more involved limit process.

We arrange the rest of the chapter as follows. Section 4.2 gives a motivational example and illustrates how to obtain the desired limit results via an examination of a simpler case under irreducible P. Section 4.3 addresses the issues when the transition matrix P is given by either (4.5) or (4.6). We will derive functional limit theorems which yield a limit switching diffusion process. Section 4.4 includes the full proof of the result for the case that P is given in (4.5). With the detailed proofs for P given by (4.5) and definitions of related quantities for P given by (4.6), the reader can carry out the extensions to include the transient state cases. Finally, Section 4.5 provides notes and further references.

4.2 Irreducible Chains

This section provides motivation for the subsequent technical development. To make the discussion simple and easily accessible, we use a relatively simpler model, namely, P being irreducible and aperiodic.

4.2.1 A Motivational Example

Let α_k^ε be a Markov chain with state space $\mathcal{M} = \{1, 2\}$ and transition probability matrix (4.1) such that P and Q are given by

$$P = \begin{pmatrix} 0.55 & 0.45 \\ 0.4 & 0.6 \end{pmatrix} \quad \text{and} \quad Q = \begin{pmatrix} -0.6 & 0.6 \\ 0.5 & -0.5 \end{pmatrix}.$$

It is easily seen that P is irreducible and $\nu = (0.4706, 0.5294)$ is the corresponding stationary distribution. For demonstration purposes, let us plot the sample path of α_k^ε with $\varepsilon = 0.01$ and $\alpha_0^\varepsilon = 1$ in Figure 4.1. Strictly speaking, it should have been a scatter plot on lattice points only, however, for a better visualization, we delineated the paths as if they were continuous in time.

Examining (4.4), we would expect the scaled sequence $n_k^{\varepsilon,i} = \pi_k^{\varepsilon,i}/\sqrt{\varepsilon}$ for $i = 1, 2$ to behave like a normal distribution. To study the distribution of $n_k^{\varepsilon,i}$, we first fix the time k at $k = 1,000$ and calculate $\pi_k^{\varepsilon,i}/\sqrt{\varepsilon}$ using 1000 replicates. The different random seeds yield a frequency distribution–a histogram that is a function of the underlying sample point ω.

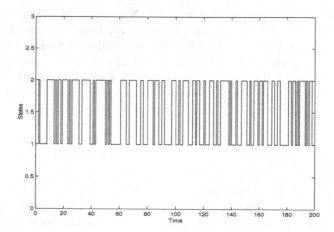

FIGURE 4.1. Sample path of Markov chain with $\varepsilon = 0.01$

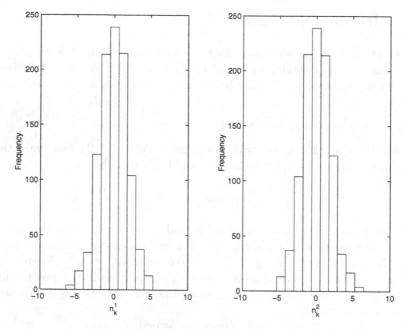

FIGURE 4.2. Distribution

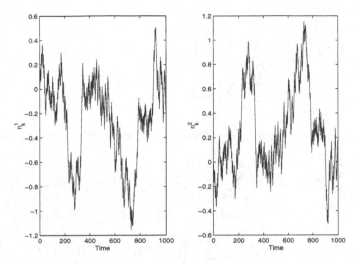

FIGURE 4.3. Sample paths of scaled occupation measures

Figure 4.2 presents the frequency distribution of the Markov chain given in Figure 4.1. The histogram on the left-hand side is for the first component, whereas that of the second component is given on the right. Next, we focus on one realization (keeping ω fixed) and trace out the sample paths of $n_k^{\varepsilon,i}$ for a fixed sample point ω. The sample paths are shown in Figure 4.3, where the left-hand side graph displays the sample path of $n_k^{\varepsilon,1}$ and the right-hand side graph shows that of $n_k^{\varepsilon,2}$, respectively. Numerical experiments using various P^ε and with different initial conditions have shown similar results.

4.2.2   Discussion

The above numerical experiments confirm our intuition that the scaled occupation measures are approximately normally distributed. To proceed, we demonstrate how such a result can be obtained for the simpler case when P is irreducible and aperiodic. The discussion is by no means detailed since we will provide a verbatim proof for the more general cases later. This part only serves the purpose of illustrating the main ideas.

We first obtain a mean-squares estimate. Note that using the asymptotic expansions, we have

$$P(\alpha_k^\varepsilon = i) = \nu^i + O(\varepsilon + \lambda^k), \ k \geq 0,$$
$$P(\alpha_{k_2}^\varepsilon = i | \alpha_{k_1}^\varepsilon) = \nu^i + O(\varepsilon + \lambda^{k_2-k_1}) \ \text{ for } \ k_2 \geq k_1 \geq 0.$$

Let $\xi_{k_1 k_2} = P(\alpha_{k_1}^\varepsilon = i, \alpha_{k_2}^\varepsilon = i) - \nu^i P(\alpha_{k_1}^\varepsilon = i) - \nu^i P(\alpha_{k_2}^\varepsilon = i) + (\nu^i)^2$.

Then we have

$$
\begin{aligned}
E(\pi_k^{\varepsilon,i})^2 &= \varepsilon^2 \sum_{k_1=0}^{k-1} \sum_{k_2=0}^{k-1} E\left(I_{\{\alpha_{k_1}^\varepsilon=i\}} - \nu^i\right)\left(I_{\{\alpha_{k_2}^\varepsilon=i\}} - \nu^i\right) \\
&= \varepsilon^2 \sum_{k_1=0}^{k-1} \sum_{k_2=0}^{k-1} \xi_{k_1 k_2}.
\end{aligned}
$$

Using $\xi_{k_1 k_2}$, we need only consider the partial sum with $k_2 \geq k_1$. Therefore,

$$
\begin{aligned}
\varepsilon^2 \sum_{k_1=0}^{k-1} \sum_{k_2=0}^{k-1} \xi_{k_1 k_2} &\leq K\varepsilon^2 \sum_{k_1=0}^{k-1} \sum_{k_2 \geq k_1}^{k-1} \Big[P(\alpha_{k_1}^\varepsilon = i) P(\alpha_{k_2}^\varepsilon = i | \alpha_{k_1}^\varepsilon = i) \\
&\qquad -\nu^i P(\alpha_{k_1}^\varepsilon = i) - \nu^i P(\alpha_{k_2}^\varepsilon = i) + (\nu^i)^2 \Big] \\
&\leq K\varepsilon^2 \sum_{k_1=0}^{k-1} \sum_{k_2 \geq k_1}^{k-1} \Big[(\nu^i + O(\varepsilon + \lambda^{k_1}))(\nu^i + O(\varepsilon + \lambda^{k_2-k_1})) \\
&\qquad -(\nu^i)^2 + O(\varepsilon + \lambda^{k_1} + \lambda^{k_2}) \Big].
\end{aligned}
$$

(4.7)

Noting that

$$
\sum_{l=0}^{k-1} \lambda^l = \frac{1-\lambda^k}{1-\lambda} \leq \frac{1}{1-\lambda} = O(1)
$$

and $k \leq T/\varepsilon$, we obtain

$$
\sup_{0 \leq k \leq T/\varepsilon} E(\pi_k^{\varepsilon,i})^2 = O(\varepsilon).
$$

To proceed, define a sequence of the scaled occupation measures as

$$
\begin{aligned}
n_k^{\varepsilon,i} &= \sqrt{\varepsilon} \sum_{l=0}^{k-1} [I_{\{\alpha_l^\varepsilon=i\}} - \nu^i], \ i \in \mathcal{M}, \\
n_k^\varepsilon &= (n_k^{\varepsilon,1}, \ldots, n_k^{\varepsilon,m_0}) \in \mathbb{R}^{1 \times m_0}.
\end{aligned}
$$

(4.8)

Define a sequence of piecewise constant functions by

$$
n^\varepsilon(t) = n_k^\varepsilon \ \text{for} \ t \in [\varepsilon k, \varepsilon k + \varepsilon).
$$

Note that $n^\varepsilon(t) = (\pi^{\varepsilon,1}(t)/\sqrt{\varepsilon}, \ldots, \pi^{\varepsilon,m_0}(t)/\sqrt{\varepsilon})$, where

$$
\pi^{\varepsilon,i}(t) = \pi_k^{\varepsilon,i} \ \text{for} \ t \in [\varepsilon k, \varepsilon k + \varepsilon).
$$

We next characterize the limit of $n^\varepsilon(\cdot)$. To do so, we divide the work into several steps.

Step 1: Calculate the limit mean of $n_\varepsilon(\cdot)$. In view of the asymptotic expansions, we have

$$En^{\varepsilon,i}(t) = \sqrt{\varepsilon} \sum_{l=0}^{t/\varepsilon-1} O(\lambda^l) + \sum_{l=0}^{t/\varepsilon-1} O(\varepsilon^{3/2})$$

$$= O(\sqrt{\varepsilon}), \quad \text{for each } i \in \mathcal{M}.$$

Thus, $En^\varepsilon(t) \to 0$ as $\varepsilon \to 0$.

Step 2: Compute the covariance of the limit process $n(\cdot)$. We begin with the examination of $E\left(n^{\varepsilon,i}(t)n^{\varepsilon,j}(t)\right)$ for $i,j \in \mathcal{M}$. Using the asymptotic expansions, detailed calculation reveals that

$$
\begin{aligned}
&E\left(n^{\varepsilon,i}(t)n^{\varepsilon,j}(t)\right) \\
&= \varepsilon \sum_{k_2=0}^{t/\varepsilon-1} \sum_{k_1=0}^{k_2} P(\alpha_{k_1}^\varepsilon = i)[P(\alpha_{k_2}^\varepsilon = j|\alpha_{k_1}^\varepsilon = i) - \nu^j] \\
&\quad + \varepsilon \sum_{k_1=0}^{t/\varepsilon-1} \sum_{k_2=0}^{k_1-1} P(\alpha_{k_2}^\varepsilon = j)[P(\alpha_{k_1}^\varepsilon = i|\alpha_{k_2}^\varepsilon = j) - \nu^i] + o(1),
\end{aligned}
\tag{4.9}
$$

where $o(1) \to 0$ as $\varepsilon \to 0$ uniformly in $t \in [0,T]$. Again, by virtue of the asymptotic expansions with substitution $l = k_2 - k_1$, interchanging the orders of summations, and using the Dirichlet summation formula,

$$
\begin{aligned}
&\varepsilon \sum_{k_2=0}^{t/\varepsilon-1} \sum_{k_1=0}^{k_2} P(\alpha_{k_1}^\varepsilon = i)[P(\alpha_{k_2}^\varepsilon = j|\alpha_{k_1}^\varepsilon = i) - \nu^j] \\
&\quad = \varepsilon\nu^i \sum_{l=0}^{t/\varepsilon-1} (t/\varepsilon - l)\psi_0^{ij}(l) + o(1),
\end{aligned}
\tag{4.10}
$$

and

$$
\begin{aligned}
&\varepsilon \sum_{k_1=0}^{t/\varepsilon-1} \sum_{k_2=0}^{k_1-1} P(\alpha_{k_2}^\varepsilon = j)[P(\alpha_{k_1}^\varepsilon = i|\alpha_{k_2}^\varepsilon = j) - \nu^i] \\
&\quad = \varepsilon\nu^j \sum_{l=1}^{t/\varepsilon-1} (t/\varepsilon - l)\psi_0^{ji}(l) + o(1),
\end{aligned}
\tag{4.11}
$$

where $\Psi_0(l) = (\psi_0^{ij}(l))$ is the zeroth-order initial layer correction term in the asymptotic expansions (3.47), and $o(1) \to 0$ as $\varepsilon \to 0$ uniformly in $t \in [0,T]$. Thus we need only find the limits on the last lines of (4.10) and (4.11).

By means of the asymptotic expansions, we have

$$\varepsilon \sum_{k_1=0}^{t/\varepsilon-1} \sum_{k_2 \geq k_1} \psi_0^{ij}(k_2 - k_1) = O(1),$$

$$\sum_{k_1=0}^{t/\varepsilon-1} \sum_{k_2 \geq k_1} \psi_0^i(k_1)\psi_1^{ij}(k_2 - k_1) = O(1),$$

$$\varepsilon \sum_{k_1=0}^{t/\varepsilon-1} \sum_{k_2 \geq k_1} \psi_0^i(k_1) = O(1),$$

and the bounds hold uniform in $t \in [0, T]$. Noting

$$\sum_{l=0}^{1/\varepsilon-1} \psi_0^{ij}(l) = \varepsilon \sum_{l=0}^{t/\varepsilon-1} \frac{t}{\varepsilon}\psi_0^{ij}(l) \to t \sum_{l=0}^{\infty} \psi_0^{ij}(l) \quad \text{as} \quad \varepsilon \to 0,$$

and using the geometric decay property of $\Psi_0(l)$ and the error estimates

$$\varepsilon \left| \sum_{l=0}^{t/\varepsilon-1} l\psi_0^{ij}(l) \right| \leq \varepsilon \sum_{l=0}^{t/\varepsilon-1} l\lambda^l = \lambda\varepsilon \sum_{l=1}^{t/\varepsilon-1} l\lambda^{l-1}$$

$$\leq \lambda\varepsilon \frac{d}{d\lambda} \sum_{l=0}^{\infty} \lambda^l \to 0 \quad \text{as} \quad \varepsilon \to 0,$$

we can show that

$$\varepsilon \sum_{l=0}^{t/\varepsilon-1} (t/\varepsilon - l)\psi_0^{ij}(l) \to t \sum_{l=0}^{\infty} \psi_0^{ij}(l),$$

$$\varepsilon \sum_{l=0}^{t/\varepsilon-1} (t/\varepsilon - l)\psi_0^{ji}(l) \to t \sum_{l=0}^{\infty} \psi_0^{ji}(l). \tag{4.12}$$

Therefore, the limit covariance is given by At, with $A = (a^{ij})$ satisfying

$$A = \sum_{l=0}^{\infty} [\text{diag}(\nu^1, \ldots, \nu^{m_0})\Psi(l) + \Psi'(l)\text{diag}(\nu^1, \ldots, \nu^{m_0})].$$

Step 3: Verify the tightness of $\{n^\varepsilon(\cdot)\}$ in $D([0, T]; \mathbb{R}^{m_0})$, the space of $\mathbb{R}^{m_0}$-valued functions that are right continuous and have left-hand limits, endowed with the Skorohod topology (see Definition 14.2). It can be shown that $n^\varepsilon(\cdot)$ is ϕ-mixing with an exponential mixing rate, and

$$E|n^\varepsilon(t + s) - n^\varepsilon(t)|^{2k} = O(s^k), \quad \text{for} \quad k = 1, 2.$$

Step 4: As a consequence of Step 3, the limit $n(\cdot)$ has continuous paths w.p.1. Using characteristic functions, we can show $n(\cdot)$ has independent increments. Then Theorem 2.12 implies that $n(\cdot)$ is necessarily a Gaussian process with desired mean and covariance as calculated in Steps 1 and 2.

4.3   Recurrent States

4.3.1   Formulation and Preliminary

For definiteness, in what follows, we assume that P in (4.1) takes the form (4.5). Then the state space $\mathcal{M}$ can be written as

$$\mathcal{M} = \{s_{11}, \ldots, s_{1m_1}\} \cup \{s_{21}, \ldots, s_{2m_2}\} \cup \ldots \cup \{s_{l_0 1}, \ldots, s_{l_0 m_{l_0}}\}$$
$$= \mathcal{M}_1 \cup \mathcal{M}_2 \cup \cdots \cup \mathcal{M}_{l_0}, \tag{4.13}$$

with $m_0 = m_1 + m_2 + \cdots + m_{l_0}$. The subspace $\mathcal{M}_i$, for each $i = 1, \ldots, l_0$, consists of recurrent states belonging to the ith ergodic class. To proceed, we state the conditions needed first.

(A4.1) P^ε, P, and P^i for $i \leq l_0$ are one-step transition probability matrices such that for each $i \leq l_0$, P^i is irreducible and aperiodic.

In Chapter 3, we obtained asymptotic expansions of the probability vector and the associated transition probabilities. We recapture these expansions up to the second order in the next theorem.

Theorem 4.1. *Under* (A4.1), *the following assertions hold:*

(a) *For the probability distribution vector* $p_k^\varepsilon = (P(\alpha_k^\varepsilon = s_{ij})) \in \mathbb{R}^{1 \times m_0}$, *we have*

$$p_k^\varepsilon = \theta(\varepsilon k) \mathrm{diag}(\nu^1, \ldots, \nu^{l_0}) + O(\varepsilon + \lambda^k) \tag{4.14}$$

for some λ *with* $0 < \lambda < 1$, *where* ν^i *is the stationary distribution corresponding to the transition matrix* P_i, *and* $\theta(t) = (\theta^1(t), \ldots, \theta^{l_0}(t)) \in \mathbb{R}^{1 \times l_0}$ *satisfies*

$$\frac{d\theta(t)}{dt} = \theta(t)\overline{Q}, \quad \theta(0) = p_0 \widetilde{\mathbb{1}},$$

where

$$\overline{Q} = \mathrm{diag}(\nu^1, \ldots, \nu^{l_0}) Q \widetilde{\mathbb{1}},$$
$$\widetilde{\mathbb{1}} = \mathrm{diag}(\mathbb{1}_{m_1}, \ldots, \mathbb{1}_{m_{l_0}}). \tag{4.15}$$

(b) *For* $k \leq T/\varepsilon$, *the k-step transition probability matrix* $(P^\varepsilon)^k$ *satisfies*

$$(P^\varepsilon)^k = \Phi(t) + \varepsilon\widehat{\Phi}(t) + \Psi(k) + \varepsilon\widehat{\Psi}(k) + O\left(\varepsilon^2\right), \tag{4.16}$$

where

$$\Phi(t) = \widetilde{\mathbb{1}}\Theta(t)\mathrm{diag}(\nu^1, \ldots, \nu^{l_0}),$$
$$\frac{d\Theta(t)}{dt} = \Theta(t)\overline{Q}, \quad \Theta(0) = I. \tag{4.17}$$

Moreover, $\Phi(t)$ *and* $\widehat{\Phi}(t)$ *are uniformly bounded in* $[0, T]$ *and* $\Psi(k)$ *and* $\widehat{\Psi}(k)$ *decay exponentially, i.e.,* $|\Psi(k)| + |\widehat{\Psi}(k)| \leq \lambda^k$ *for some* $0 < \lambda < 1$.

Remark 4.2. We reiterate the following items, which are to be used in subsequent discussions.

(a) In view of the asymptotic expansion, we have

$$(P^\varepsilon)^k = \Phi(\varepsilon k) + O(\varepsilon + \lambda^k).$$

(b) Recall that $\Phi(t) + \varepsilon\widehat{\Phi}(t)$ need not satisfy the initial condition. To compensate, the initial layer correction terms $\Psi(k)$ and $\widehat{\Psi}(k)$ are so selected that $\Phi(0) + \Psi(0) = I$ and $\widehat{\Phi}(0) + \widehat{\Psi}(0) = 0$.

(c) For the initial layer term,

$$\Psi(k) = \Psi(0)P^k.$$

In view of (4.16) and (4.17), since $\Theta(0) = I$,

$$\Psi(k) = (I - \widetilde{1}\text{diag}(\nu^1, \ldots, \nu^{l_0}))P^k$$
$$= \text{diag}((I_{m_1} - 1_{m_1}\nu^1)(P^1)^k, \ldots, (I_{m_{l_0}} - 1_{m_{l_0}}\nu^{l_0})(P^{l_0})^k),$$
$$(4.18)$$

where I_{m_i} is an $m_i \times m_i$ identity matrix. Thus $\Psi(k)$ is again of block-diagonal form.

4.3.2 Aggregation

In view of the decomposability of $\mathcal{M}$, we define an aggregated process $\overline{\alpha}_k^\varepsilon$ of α_k^ε by

$$\overline{\alpha}_k^\varepsilon = i \text{ if } \alpha_k^\varepsilon \in \mathcal{M}_i \text{ for } i = 1, \ldots, l_0. \qquad (4.19)$$

In order to achieve the complexity reduction, by means of aggregation in (4.19), we reduced the total number of states of the random process from m_0 to l_0. If $l_0 \ll m_0$, there will be significant savings in computational effort for applications arising in control and optimization. Note that the aggregated process is generally non-Markov.

To proceed, define continuous-time interpolations by

$$\alpha^\varepsilon(t) = \alpha_k^\varepsilon, \text{ and } \overline{\alpha}^\varepsilon(t) = \overline{\alpha}_k^\varepsilon \text{ for } t \in [\varepsilon k, \varepsilon(k+1)), \qquad (4.20)$$

and denote by $D([0,T]; \mathcal{M})$ the space of functions that are defined on $[0,T]$ taking values in $\mathcal{M}$ and that are right continuous and have left limits endowed with the Skorohod topology. We shall show that $\overline{\alpha}^\varepsilon(\cdot)$, the interpolation of $\overline{\alpha}_k^\varepsilon$, has a weak limit that is a continuous-time Markov chain. The result is stated below. Its proof is postponed until Section 4.5.

Theorem 4.3. *Assume (A4.1). Then as $\varepsilon \to 0$, $\overline{\alpha}^\varepsilon(\cdot)$ converges weakly to $\overline{\alpha}(\cdot)$, a Markov chain generated by $\overline{Q}$ given by (4.15).*

As was illustrated in the introduction, if P in (4.1) is reducible, one cannot approximate the occupation measures by a deterministic function. What can one do regarding the approximation now? An appropriate choice comes from suitable centering by a random process and an aggregated process similar to (4.4). For $k = 0, \ldots, T/\varepsilon$, $i = 1, \ldots, l_0$, and $j = 1, \ldots, m_i$, define sequences of occupation measures by

$$\pi_k^{\varepsilon,ij} = \varepsilon \sum_{l=0}^{k-1} \left(I_{\{\alpha_l^\varepsilon = s_{ij}\}} - \nu^{ij} I_{\{\overline{\alpha}_l^\varepsilon = i\}} \right),$$

$$\pi_k^\varepsilon = (\pi_k^{\varepsilon,ij}) \in \mathbb{R}^{1 \times m_0}. \tag{4.21}$$

We can write π_k^ε recursively as a difference equation

$$\pi_{k+1}^\varepsilon = \pi_k^\varepsilon + \varepsilon W(\alpha_k^\varepsilon), \tag{4.22}$$

where

$$W(\alpha) = (w^{ij}(\alpha)) \in \mathbb{R}^{1 \times m_0}, \quad \text{with}$$

$$w^{ij}(\alpha) = I_{\{\alpha = s_{ij}\}} - \nu^{ij} I_{\{\alpha \in \mathcal{M}_i\}}. \tag{4.23}$$

Define continuous-time interpolations

$$\pi^{\varepsilon,ij}(t) = \pi_k^{\varepsilon,ij} \text{ for } t \in [k\varepsilon, (k+1)\varepsilon),$$

$$\pi^\varepsilon(t) = (\pi^{ij}(t)) \in \mathbb{R}^{1 \times m_0}. \tag{4.24}$$

Remark 4.4. For the case when P is irreducible, we have demonstrated that the sequence of scaled occupation measures verify a mixing condition. This mixing property no longer holds for the multi-ergodic class cases. One of the reasons is that unlike the case of irreducible P given by (4.5), neither $\Phi(t)$ nor $\widehat{\Phi}(t)$ has identical rows. Nevertheless, the following mean squares estimates still hold.

Theorem 4.5. *Assume* (A4.1). *Then for* $i = 1, \ldots, l_0$, $j = 1, \ldots, m_i$,

$$\sup_{0 \leq k \leq T/\varepsilon} E|\pi_k^{\varepsilon,ij}|^2 = O(\varepsilon) \quad \text{and} \quad \sup_{t \in [0,T]} E|\pi^{\varepsilon,ij}(t)|^2 = O(\varepsilon).$$

4.3.3 Asymptotic Distribution

This section obtains the switching diffusion limit for a sequence of suitably scaled occupation measures. For each $i = 1, \ldots, l_0$, $j = 1, \ldots, m_i$, and each $0 < k \leq T/\varepsilon$, define sequences of normalized occupation measures

$$n_k^{\varepsilon,ij} = \sqrt{\varepsilon} \sum_{l=0}^{k-1} w^{ij}(\alpha_l^\varepsilon) = \frac{1}{\sqrt{\varepsilon}} \pi_k^{\varepsilon,ij},$$

$$n_k^\varepsilon = (n_k^{\varepsilon,ij}) \in \mathbb{R}^{1 \times m_0}. \tag{4.25}$$

Then, we have the following difference equation

$$n_{k+1}^\varepsilon = n_k^\varepsilon + \sqrt{\varepsilon} W(\alpha_k^\varepsilon). \qquad (4.26)$$

Define a continuous-time interpolation

$$n^\varepsilon(t) = n_k^\varepsilon \text{ for } t \in [\varepsilon k, \varepsilon k + \varepsilon). \qquad (4.27)$$

Our objective is to show that $n^\varepsilon(\cdot)$ so defined converges weakly to a switching diffusion process. In fact, it is necessary to treat the pair of processes $(n^\varepsilon(\cdot), \overline{\alpha}^\varepsilon(\cdot))$ together. Following the weak convergence analysis approach, we first show that this sequence is tight in $D([0, T]; \mathbb{R}^{m_0} \times \mathcal{M})$, the space of functions that are defined on $[0, T]$ taking values in $\mathbb{R}^{m_0} \times \mathcal{M}$ and that are right continuous have left limits endowed with the Skorohod topology. Then we characterize its limit by identifying it as a solution of a martingale problem with appropriate operator $\mathcal{L}$. The procedure is carried out by proving a series of lemmas. First, we present Lemma 4.6 as a preparation, whose proof is contained in Section 4.5. Then Lemma 4.7 establishes the tightness of the sequence. The proofs are moved to Section 4.5 for ease of presentation.

Lemma 4.6. *Assume* (A4.1). *Let* $\mathcal{F}_t^\varepsilon = \sigma\{\alpha^\varepsilon(s) : s \le t\}$. *Then*

(a) $\displaystyle\sup_{0 \le t \le t+s \le T} E[(n^\varepsilon(t+s) - n^\varepsilon(t)) | \mathcal{F}_t^\varepsilon] = O(\sqrt{\varepsilon})$;

(b) *For each* $i = 1, \ldots, l_0$ *and* $j = 1, \ldots, m_i$, *and for any* $0 < s \le \delta$,

$$E[(n^{\varepsilon, ij}(t+s) - n^{\varepsilon, ij}(t))^2 | \mathcal{F}_t^\varepsilon] = O(\delta).$$

With the help of Lemma 4.6, we have the following result.

Lemma 4.7. *Under the conditions of Lemma* 4.6, $\{n^\varepsilon(\cdot), \overline{\alpha}^\varepsilon(\cdot)\}$ *is tight in* $D([0, T]; \mathbb{R}^{m_0} \times \mathcal{M})$.

Since $\{n^\varepsilon(\cdot), \overline{\alpha}^\varepsilon(\cdot)\}$ is tight, by Prohorov's theorem, we can extract a weakly convergent subsequence. Choose such a subsequence with limit $(n(\cdot), \overline{\alpha}(\cdot))$, and still denote the subsequence by $(n^\varepsilon(\cdot), \overline{\alpha}^\varepsilon(\cdot))$ for notational simplicity. To obtain the desired result, we need only show that the limit $(n(\cdot), \overline{\alpha}(\cdot))$ solves a martingale problem with operator $\mathcal{L}$ given by

$$\mathcal{L}f(x, i) = \frac{1}{2} \sum_{j_1=1}^{m_i} \sum_{j_2=1}^{m_i} a^{j_1 j_2}(i) \frac{\partial^2 f(x, i)}{\partial x^{ij_1} \partial x^{ij_2}} + \overline{Q} f(x, \cdot)(i), \qquad (4.28)$$

for $i = 1, \ldots, l_0$, where $A(i) = (a^{j_1 j_2}(i))$ is symmetric and nonnegative definite to be specified later. The symmetry and the nonnegative definiteness of $(a^{j_1 j_2}(i))$ can be verified. Moreover, the values of these $a^{j_1 j_2}(i)$'s are determined as follows. (In the following, Lemma 4.8 identifies the limit covariance, and Lemma 4.9 presents the uniqueness of the associated martingale problem.)

Lemma 4.8. *For each $i = 1, \ldots, l_0$, the covariance of the limit process is $S(i) = \Sigma(i)\Sigma'(i)$, where*

$$\Sigma(i) = \mathrm{diag}(0_{m_1 \times m_1}, \ldots, 0_{m_{i-1} \times m_{i-1}}, \sigma(i), 0_{m_{i+1} \times m_{i+1}}, \ldots, 0_{m_{l_0} \times m_{l_0}}), \tag{4.29}$$

$\sigma(i) \in \mathbb{R}^{m_i \times m_i}$ *satisfying* $\sigma(i)\sigma'(i) = A(i) = (a^{j_1 j_2}(i))$, *and*

$$A(i) = \nu^{i,\mathrm{diag}} \sum_{k=0}^{\infty} \Psi(k, i) + \sum_{k=0}^{\infty} \Psi'(k, i)\nu^{i,\mathrm{diag}}, \tag{4.30}$$

with $\nu^{i,\mathrm{diag}} = \mathrm{diag}(\nu^{i1}, \ldots, \nu^{im_i}) \in \mathbb{R}^{m_i \times m_i}$,

$$\Psi(k+1, i) = \Psi(k, i)P^i,$$

$$\Psi(0, i) = I - \mathbb{1}_{m_i}\nu^i.$$

Lemma 4.9. *A martingale problem associated with operator $\mathcal{L}$ defined in (4.28) has a unique solution.*

We next state the main result. Its proof will be obtained by proving a series of lemmas and is relegated to Section 4.5 as well.

Theorem 4.10. *Assume (A4.1). Then $(n^\varepsilon(\cdot), \overline{\alpha}^\varepsilon(\cdot))$ converges weakly to $(n(\cdot), \overline{\alpha}(\cdot))$ such that the limit is the solution of the martingale problem with operator $\mathcal{L}$ given by (4.28).*

Remark 4.11. Note that $\overline{\alpha}(\cdot)$ is a Markov chain generated by $\overline{Q}$. Define a stochastic process

$$n^0(t) = \int_0^t (\Sigma(\overline{\alpha}(s))dw(s))', \tag{4.31}$$

where $w(\cdot)$ is a standard Brownian motion and $\Sigma(\alpha)$ is given in Lemma 4.8. Then for each $i = 1, \ldots, l_0$, and for any $f(\cdot, i)$ that is a real-valued function with bounded derivatives up to the second order and with Lipschitz continuous second derivative,

$$f(n^0(t), \overline{\alpha}(t)) - \int_0^t \mathcal{L}f(n^0(s), \overline{\alpha}(s))ds \text{ is a martingale.}$$

Therefore, $(n^0(\cdot), \overline{\alpha}(\cdot))$ is a solution of the martingale problem with operator $\mathcal{L}$. By virtue of Lemma 4.9, the uniqueness of the martingale problem with operator $\mathcal{L}$ then implies that $(n^0(\cdot), \overline{\alpha}(\cdot))$ has the same distribution as that of $(n(\cdot), \overline{\alpha}(\cdot))$. The process given in (4.31) has both diffusive behavior as well as a switching property, hence it is a switching diffusion.

As far as the proof is concerned, for the recurrent case, we will use a perturbed test function approach. In fact, it is possible to prove the desired result via direct averaging. Such an approach will be used in Chapter 10. The reason that we use the current setup is to give the reader a broad range of methods for related problems.

4.4 Inclusion of Transient States

In this section, we consider the occupation measure when the underlying Markov chain has transient states. We only state the corresponding results. Their proofs can be obtained in a manner similar to that used for recurrent case.

Let α_k^ε be a Markov chain whose transition probability matrix is given by (4.1) with P having the form (4.6). Replace (A4.1) by (A4.2).

(A4.2) P^ε, P, and P^i for $i \leq l_0$ are one-step transition probability matrices such that for each $i \leq l_0$, P^i is irreducible and aperiodic. All the eigenvalues of P^* are inside the unit circle.

Partition the matrix Q as

$$Q = \begin{pmatrix} Q^{11} & Q^{12} \\ Q^{21} & Q^{22} \end{pmatrix}, \tag{4.32}$$

where

$$Q^{11} \in \mathbb{R}^{(m_0-m_*) \times (m_0-m_*)}, \; Q^{12} \in \mathbb{R}^{(m_0-m_*) \times m_*},$$

$$Q^{21} \in \mathbb{R}^{m_* \times (m_0-m_*)}, \;\; \text{and} \; Q^{22} \in \mathbb{R}^{m_* \times m_*}.$$

Write

$$\overline{Q}_* = \operatorname{diag}(\nu^1, \ldots, \nu^{l_0})(Q^{11}\widetilde{\mathbb{1}} + Q^{12}A_*), \tag{4.33}$$

where

$$A_* = (a^1, \ldots, a^{l_0}) \in \mathbb{R}^{m_* \times l_0}, \;\; \text{with}$$

$$a^i = -(P^* - I)^{-1}P^{*,i}\mathbb{1}_{m_i}, \;\text{for } i = 1, \ldots, l_0. \tag{4.34}$$

Remark 4.12. Note that $a^i = (a^{i,1}, \ldots, a^{i,m_*})' \in \mathbb{R}^{m_* \times 1}$ and that $a^{i,j}$ is the probability of entering the ith recurrent class starting from transient state s_{*j}. Thus, $a^{i,j} \geq 0$ and $\sum_{i=1}^{l_0} a^{i,j} = 1$ (i.e., starting from transient state s_{*j}, the probabilities of transitions from transient states to recurrent classes add up to 1).

Define

$$\widetilde{\mathbb{1}}_* = \begin{pmatrix} \widetilde{\mathbb{1}} & 0_{(m_0-m_*) \times m_*} \\ A_* & 0_{m_* \times m_*} \end{pmatrix},$$

$$\nu_* = \operatorname{diag}(\nu^1, \ldots, \nu^{l_0}, 0_{m_* \times m_*});$$

define sequences of centered occupation measures by

$$\pi_k^{\varepsilon,ij} = \begin{cases} \varepsilon \sum_{l=0}^{k-1} (I_{\{\alpha_l^\varepsilon = s_{ij}\}} - \nu^{ij} I_{\{\overline{\alpha}_l^\varepsilon = i\}}), & \text{for } i = 1, \ldots, l_0, \\ \varepsilon \sum_{l=0}^{k-1} I_{\{\alpha_l^\varepsilon = s_{*j}\}}, & \text{for } i = *; \end{cases} \tag{4.35}$$

define the vector π_k^ε, and its continuous-time interpolation $\pi^\varepsilon(t)$. Consider an aggregated process

$$\overline{\alpha}_k^\varepsilon = \begin{cases} i, & \text{if } \alpha_k^\varepsilon \in \mathcal{M}_i, \\ U_j, & \text{if } \alpha_k^\varepsilon = s_{*j}, \end{cases} \tag{4.36}$$

where U_j is given by

$$U_j = I_{\{0 \le U \le a^{1,j}\}} + 2I_{\{a^{1,j} < U \le a^{1,j}+a^{2,j}\}} + \cdots + l_0 I_{\{a^{1,j}+\cdots+a^{l_0-1,j} < U \le 1\}},$$

and U is a random variable uniformly distributed on $[0, 1]$, independent of the Markov chain α_k^ε.

Remark 4.13. We have aggregated the states in each recurrent class into one state. This is similar to the treatment of the recurrent case. If the chain is in a transient state, then it will eventually enter one of the recurrent classes. Starting from each of the transient states, the process may jump to one of the l_0 possible locations. This is similar to the generation of discrete random variables with l_0 possible values, which leads to definition (4.36).

Theorem 4.14. *Under (A4.2), the following assertions hold:*

(a) *The probability vector* $p_k^\varepsilon = \theta(\varepsilon k)\text{diag}(\nu^1, \ldots, \nu^{l_0}, 0_{m_*}) + O(\varepsilon + \lambda^k)$, *where* $0_{m_*} \in \mathbb{R}^{1 \times m_*}$ *and* $\theta(t) = (\theta^1(t), \ldots, \theta^{l_0}(t)) \in \mathbb{R}^{1 \times l_0}$ *satisfies*

$$\frac{d\theta(t)}{dt} = \theta(t)\overline{Q}_*, \quad \theta^i(0) = p^i(0)\mathbb{1}_{m_i} - p^*(0)a^i.$$

The transition matrix satisfies

$$P^\varepsilon(\varepsilon k_0, \varepsilon k) = \Phi(t_0, t) + \Psi(k_0, k) + \varepsilon\widehat{\Phi}(t_0, t) + \varepsilon\widehat{\Psi}(k_0, k) + O(\varepsilon^2),$$

for some λ *with* $0 < \lambda < 1$, *where*

$$\Phi(t_0, t) = \widetilde{\mathbb{1}}_*\Theta_*(t_0, t)\nu_* \tag{4.37}$$

with

$$\Theta_*(t_0, t) = \text{diag}(\Theta(t_0, t), I_{m_* \times m_*})$$

where $\Theta(t_0, t) = (\theta^{ij}(t_0, t))$ *satisfies the differential equation*

$$\frac{\partial \Theta(t_0, t)}{\partial t} = \Theta(t_0, t)\overline{Q}_*(t), \quad \Theta(t_0, t_0) = I.$$

(b) *For each* $j = 1, \ldots, m_i$,

$$\sup_{t \in [0,T]} E|\pi^{\varepsilon, ij}(t)|^2 = \begin{cases} O(\varepsilon), & \text{for } i = 1, \ldots, l_0, \\ O(\varepsilon^2), & \text{for } i = *; \end{cases}$$

(c) $\overline{\alpha}^\varepsilon(\cdot)$ *converges weakly to* $\overline{\alpha}(\cdot)$, *a Markov chain generated by* $\overline{Q}_*$;

(d) *Lemma 4.6 continues to hold;*

(e) *Theorem 4.10 continues to hold and for each $i \in \overline{\mathcal{M}}$, the limit operator is given by*

$$\mathcal{L}f(x,i) = \frac{1}{2} \sum_{j_1=1}^{m_i} \sum_{j_2=1}^{m_i} a^{j_1 j_2}(i) \frac{\partial^2 f(x,i)}{\partial x^{i j_1} \partial x^{i j_2}} + \overline{Q}_* f(x,\cdot)(i).$$

4.5   Proofs of Results

In this section, we provide proofs of Lemmas 4.3, 4.6, 4.7, 4.8, and 4.9, and Theorems 4.5, 4.10, and 4.14.

Proof of Lemma 4.3. First we prove that $\overline{\alpha}^\varepsilon(\cdot)$ is tight. Recall that $\mathcal{F}_t^\varepsilon$ and $\mathcal{F}_k$ are the σ-algebras generated by $\{\alpha^\varepsilon(u), u \leq t\}$ and $\{\alpha_l^\varepsilon, l \leq k\}$, respectively, and use ν^{ij} to denote the jth component of the stationary distribution ν^i. In view of the interpolation (4.20) and the Markov property of the discrete-time Markov chain,

$$
\begin{aligned}
E[(\overline{\alpha}^\varepsilon(t+s) - \overline{\alpha}^\varepsilon(s))^2 | \mathcal{F}_s^\varepsilon] &= E[(\overline{\alpha}^\varepsilon(t+s) - \overline{\alpha}^\varepsilon(s))^2 | \alpha^\varepsilon(r), r \leq s] \\
&= E[(\overline{\alpha}_{(t+s)/\varepsilon}^\varepsilon - \overline{\alpha}_{s/\varepsilon}^\varepsilon)^2 | \alpha_1^\varepsilon, \dots, \alpha_{s/\varepsilon}^\varepsilon] \\
&= E[(\overline{\alpha}_{(t+s)/\varepsilon}^\varepsilon - \overline{\alpha}_{s/\varepsilon}^\varepsilon)^2 | \alpha_{s/\varepsilon}^\varepsilon].
\end{aligned}
$$

Note that

$$\{\overline{\alpha}_k^\varepsilon = i\} = \bigcup_{j=1}^{m_i} \{\alpha_k^\varepsilon = s_{ij}\}.$$

It then follows that

$$
\begin{aligned}
E[(\overline{\alpha}^\varepsilon(t+s) &- \overline{\alpha}^\varepsilon(s))^2 | \alpha_{s/\varepsilon}^\varepsilon = s_{ij}] \\
&= \sum_{p=1}^{l_0} (p-i)^2 P(\overline{\alpha}_{(t+s)/\varepsilon}^\varepsilon = p | \alpha_{s/\varepsilon}^\varepsilon = s_{ij}) \\
&\leq l_0^2 \sum_{p \neq i} P(\overline{\alpha}_{(t+s)/\varepsilon}^\varepsilon = p | \alpha_{s/\varepsilon}^\varepsilon = s_{ij}) \\
&= l_0^2 \sum_{p \neq i} \sum_{p_1=1}^{m_p} P(\alpha_{(t+s)/\varepsilon}^\varepsilon = s_{pp_1} | \alpha_{s/\varepsilon}^\varepsilon = s_{ij}) \\
&= l_0^2 \sum_{p \neq i} \sum_{p_1=1}^{m_p} \nu^{pp_1} \theta^{ip}(t) + O(\varepsilon + \lambda^{t/\varepsilon}) \\
&= l_0^2 \sum_{p \neq i} \theta^{ip}(t) + O(\varepsilon + \lambda^{t/\varepsilon}).
\end{aligned}
$$

Note that the next to the last line above follows from Theorem 4.1. Since $\lim_{t\to 0}\theta^{ip}(t)=0$ for $i\neq p$, there is a $\gamma^\varepsilon(t)\geq 0$ that is $\mathcal{F}_s^\varepsilon$-measurable such that

$$|E[(\overline{\alpha}^\varepsilon(t+s)-\overline{\alpha}^\varepsilon(s))^2|\mathcal{F}_s^\varepsilon]|\leq \gamma^\varepsilon(t), \quad \text{and} \quad \lim_{t\to 0}\limsup_{\varepsilon\to 0} E\gamma^\varepsilon(t)=0.$$

Thus $\overline{\alpha}^\varepsilon(\cdot)$ is tight by the tightness criterion in Theorem 14.12.

To proceed, it suffices to prove the convergence of the finite dimensional distributions of $\overline{\alpha}^\varepsilon(\cdot)$ to $\overline{\alpha}(\cdot)$. For any $0\leq t_1 < t_2 < \ldots < t_\iota \leq T$, and $i_1,\ldots,i_\iota \in \overline{\mathcal{M}}=\{1,\ldots,l_0\}$,

$$
\begin{aligned}
&P\left(\overline{\alpha}^\varepsilon(t_\iota)=i_\iota,\ldots,\overline{\alpha}^\varepsilon(t_1)=i_1\right)\\
&= P\left(\overline{\alpha}_{t_\iota/\varepsilon}^\varepsilon=i_\iota,\ldots,\overline{\alpha}_{t_1/\varepsilon}^\varepsilon=i_1\right)\\
&= P\left(\alpha_{t_\iota/\varepsilon}^\varepsilon\in\mathcal{M}_{i_\iota},\ldots,\alpha_{t_1/\varepsilon}^\varepsilon\in\mathcal{M}_{i_1}\right)\\
&= \sum_{j_1=1}^{m_{i_1}}\cdots\sum_{j_\iota=1}^{m_{i_\iota}} P\left(\alpha_{t_\iota/\varepsilon}^\varepsilon=s_{i_\iota j_\iota},\ldots,\alpha_{t_1/\varepsilon}^\varepsilon=s_{i_1 j_1}\right)\\
&= \sum_{j_1=1}^{m_{i_1}}\cdots\sum_{j_\iota=1}^{m_{i_\iota}} P(\alpha_{t_1/\varepsilon}^\varepsilon=s_{i_1 j_1})\prod_{\kappa=2}^{\iota}P(\alpha_{t_\kappa/\varepsilon}^\varepsilon=s_{i_\kappa j_\kappa}|\alpha_{t_{\kappa-1}/\varepsilon}^\varepsilon=s_{i_{\kappa-1}j_{\kappa-1}})\\
&\longrightarrow \sum_{j_1=1}^{m_{i_1}}\cdots\sum_{j_\iota=1}^{m_{i_\iota}}\nu^{i_1 j_1}\theta^{i_1}(t_1)\prod_{\kappa=2}^{\iota}\nu^{i_\kappa j_\kappa}\theta^{i_{\kappa-1}i_\kappa}(t_\kappa-t_{\kappa-1}) \quad \text{as } \varepsilon\to 0\\
&= \theta^{i_1}(t_1)\prod_{\kappa=2}^{\iota}\theta^{i_{\kappa-1}i_\kappa}(t_\kappa-t_{\kappa-1})\\
&= P\left(\overline{\alpha}(t_\iota)=i_\iota,\ldots,\overline{\alpha}(t_1)=i_1\right).
\end{aligned}
$$

In the above, going from the fourth line to the fifth line, we used the Markov property; the sixth line is a consequence of the asymptotic expansions; the seventh line follows from $\sum_{j=1}^{m_i}\nu^{ij}=1$. Thus, $\overline{\alpha}^\varepsilon(\cdot)\to\overline{\alpha}(\cdot)$ in distribution, and the proof is completed. $\square$

Proof of Theorem 4.5. We verify the second inequality; the proof of the first inequality can be obtained similarly. For any $t\in[0,T]$, in view of the interpolation, $\pi^{\varepsilon,ij}(t)=\pi_{t/\varepsilon}^{\varepsilon,ij}$. Thus

$$
\begin{aligned}
E|\pi^{\varepsilon,ij}(t)|^2 &= E\left(\varepsilon\sum_{l=0}^{t/\varepsilon-1}[I_{\{\alpha_l^\varepsilon=s_{ij}\}}-\nu^{ij}I_{\{\overline{\alpha}_l^\varepsilon=i\}}]\right)^2\\
&= \varepsilon^2 E\sum_{l=0}^{t/\varepsilon-1}\sum_{p=0}^{t/\varepsilon-1}\zeta^{lp},
\end{aligned}
\tag{4.38}
$$

where

$$\zeta^{lp} = I_{\{\alpha_l^\varepsilon = s_{ij}, \alpha_p^\varepsilon = s_{ij}\}} - \nu^{ij} I_{\{\overline{\alpha}_l^\varepsilon = i, \alpha_p^\varepsilon = s_{ij}\}}$$
$$- \nu^{ij} I_{\{\alpha_l^\varepsilon = s_{ij}, \overline{\alpha}_p^\varepsilon = i\}} + \nu^{ij} \nu^{ij} I_{\{\overline{\alpha}_l^\varepsilon = i, \overline{\alpha}_p^\varepsilon = i\}}.$$

(4.39)

Note that ζ^{lp} depends on ε and s_{ij}, but for simplicity, we suppress the dependence of ε and s_{ij}. It follows that

$$E|\pi_{t/\varepsilon}^{\varepsilon, ij}|^2 = \varepsilon^2 \left(\sum_{p<l} E\zeta^{lp} + \sum_{l<p} E\zeta^{lp} + \sum_{l=p} E\zeta^{lp} \right).$$

There are three cases to consider.
   Case (i) $p < l$:

$$P(\alpha_l^\varepsilon = s_{ij}, \alpha_p^\varepsilon = s_{ij}) = P(\alpha_l^\varepsilon = s_{ij}|\alpha_p^\varepsilon = s_{ij})P(\alpha_p^\varepsilon = s_{ij}),$$
$$P(\overline{\alpha}_l^\varepsilon = i, \alpha_p^\varepsilon = s_{ij}) = P(\overline{\alpha}_l^\varepsilon = i|\alpha_p^\varepsilon = s_{ij})P(\alpha_p^\varepsilon = s_{ij}),$$
$$P(\alpha_l^\varepsilon = s_{ij}, \overline{\alpha}_p^\varepsilon = i) = \sum_{j_2=1}^{m_i} P(\alpha_l^\varepsilon = s_{ij}|\alpha_p^\varepsilon = s_{ij_2})P(\alpha_p^\varepsilon = s_{ij_2}),$$
$$P(\overline{\alpha}_l^\varepsilon = i, \overline{\alpha}_p^\varepsilon = i) = \sum_{j_2=1}^{m_i} P(\overline{\alpha}_l^\varepsilon = i|\alpha_p^\varepsilon = s_{ij_2})P(\alpha_p^\varepsilon = s_{ij_2}).$$

By Theorem 4.1, we have

$$P(\alpha_l^\varepsilon = s_{ij}|\alpha_p^\varepsilon = s_{ij}) = \nu^{ij}\theta^{ii}(\varepsilon(l-p)) + O(\varepsilon + \lambda^{l-p}),$$
$$P(\overline{\alpha}_l^\varepsilon = i|\alpha_p^\varepsilon = s_{ij}) = \sum_{j_1=1}^{m_i} P(\alpha_l^\varepsilon = s_{ij_1}|\alpha_p^\varepsilon = s_{ij})$$
$$= \sum_{j_1=1}^{m_i} \nu^{ij_1}\theta^{ii}(\varepsilon(l-p)) + O(\varepsilon + \lambda^{l-p})$$
$$= \theta^{ii}(\varepsilon(l-p)) + O(\varepsilon + \lambda^{l-p}).$$

Note that

$$P(\alpha_l^\varepsilon = s_{ij}|\alpha_p^\varepsilon = s_{ik_2}) = \nu^{ij}\theta^{ii}(\varepsilon(l-p)) + O(\varepsilon + \lambda^{l-p}).$$

It follows that

$$P(\alpha_l^\varepsilon = s_{ij}, \overline{\alpha}_p^\varepsilon = i) = \nu^{ij}\theta^{ii}(\varepsilon(l-p)) \sum_{j_2=1}^{m_i} P(\alpha_p^\varepsilon = s_{ij_2}) + O(\varepsilon + \lambda^{l-p}),$$

$$P(\overline{\alpha}_l^\varepsilon = i, \overline{\alpha}_p^\varepsilon = i) = \sum_{j_1=1}^{m_i} \sum_{j_2=1}^{m_i} P(\alpha_l^\varepsilon = s_{ij_1} | \alpha_p^\varepsilon = s_{ij_2}) P(\alpha_p^\varepsilon = s_{ij_2})$$

$$= \sum_{j_1=1}^{m_i} \sum_{j_2=1}^{m_i} \nu^{ij_1}\theta^{ii}(\varepsilon(l-p)) P(\alpha_p^\varepsilon = s_{ij_2}) + O(\varepsilon + \lambda^{l-p})$$

$$= \theta^{ii}(\varepsilon(l-p)) \sum_{j_2=1}^{m_i} P(\alpha_p^\varepsilon = s_{ij_2}) + O(\varepsilon + \lambda^{l-p}).$$

Collecting the above estimates, we arrive at

$$E\zeta^{lp} = O(\varepsilon + \lambda^{l-p}) \text{ for } p < l.$$

Case (ii) $p < l$: By symmetry and the calculation in Case (i),

$$E\zeta^{lp} = O(\varepsilon + \lambda^{p-l}) \text{ for } l < p.$$

Case (iii) $p = l$: In this case, since $E\zeta^{ll}$ is bounded,

$$\varepsilon^2 \sum_{l=0}^{t/\varepsilon-1} E\zeta^{ll} = O(\varepsilon).$$

Moreover, note that

$$\sum_{p=0}^{t/\varepsilon-1} \sum_{l=0}^{p-1} \lambda^{p-l} = \frac{\lambda}{1-\lambda} \left[\frac{t}{\varepsilon} - \frac{1-\lambda^{t/\varepsilon-1}}{1-\lambda} \right] = O(1/\varepsilon),$$

and similarly

$$\sum_{l=0}^{t/\varepsilon-1} \sum_{p=0}^{l-1} \lambda^{l-p} = O(1/\varepsilon).$$

Therefore, we have

$$E|\pi^{\varepsilon,ij}(t)|^2 = \varepsilon^2 \left[\sum_{p=0}^{t/\varepsilon-1} \sum_{l=0}^{p-1} O(\varepsilon + \lambda^{p-l}) + \sum_{l=0}^{t/\varepsilon-1} \sum_{p=0}^{l-1} O(\varepsilon + \lambda^{l-p}) \right] + O(\varepsilon)$$

$$= \varepsilon^2 (t/\varepsilon)^2 O(\varepsilon) + \varepsilon^2 O(1/\varepsilon) + O(\varepsilon) = O(\varepsilon).$$

Furthermore, the bound holds uniformly in $t \in [0, T]$. $\square$

Proof of Lemma 4.6. Note that for any fixed i, j, $s \geq 0$, and $t \geq 0$,

$$
E[(n^{\varepsilon,ij}(t+s) - n^{\varepsilon,ij}(t))|\mathcal{F}_t^\varepsilon]
$$
$$
= E[(n^{\varepsilon,ij}(t+s) - n^{\varepsilon,ij}(t))|\alpha^\varepsilon(u), u \leq t]
$$
$$
= E[(n^{\varepsilon,ij}_{(t+s)/\varepsilon} - n^{\varepsilon,ij}_{t/\varepsilon})|\alpha_l^\varepsilon, \, l \leq t/\varepsilon]
$$
$$
= E[(n^{\varepsilon,ij}_{(t+s)/\varepsilon} - n^{\varepsilon,ij}_{t/\varepsilon})|\alpha_{t/\varepsilon}^\varepsilon]
$$
$$
= \sqrt{\varepsilon} \sum_{l=t/\varepsilon}^{(t+s)/\varepsilon-1} E[I_{\{\alpha_l^\varepsilon = s_{ij}\}} - \nu^{ij} I_{\{\overline{\alpha}_l = i\}}|\alpha_{t/\varepsilon}^\varepsilon]
$$
$$
= \sqrt{\varepsilon} \sum_{l=t/\varepsilon}^{(t+s)/\varepsilon-1} [P(\alpha_l^\varepsilon = s_{ij}|\alpha_{t/\varepsilon}^\varepsilon) - \nu^{ij} P(\overline{\alpha}_l = i|\alpha_{t/\varepsilon}^\varepsilon)].
$$

Owing to the asymptotic expansions given in Theorem 4.1,

$$
P(\alpha_l^\varepsilon = s_{ij}|\alpha_{t/\varepsilon}^\varepsilon) - \nu^{ij} P(\overline{\alpha}_l = i|\alpha_{t/\varepsilon}^\varepsilon)
$$
$$
= \sum_{i_0=1}^{l_0} \sum_{j_0=1}^{m_{i_0}} \left(P(\alpha_l^\varepsilon = s_{ij}|\alpha_{t/\varepsilon}^\varepsilon = s_{i_0 j_0}) - \nu^{ij} P(\overline{\alpha}_l = i|\alpha_{t/\varepsilon}^\varepsilon = s_{i_0 j_0}) \right)
$$
$$
\times I_{\{\alpha_{t/\varepsilon}^\varepsilon = s_{i_0 j_0}\}}
$$
$$
= O(\varepsilon + \lambda^{l-(t/\varepsilon)}).
$$

Thus, we have

$$
E[(n^{\varepsilon,ij}(t+s) - n^{\varepsilon,ij}(t))|\mathcal{F}_t^\varepsilon]
$$
$$
= \sqrt{\varepsilon} \sum_{l=t/\varepsilon}^{(t+s)/\varepsilon-1} O(\varepsilon + \lambda^{l-(t/\varepsilon)}) = O(\sqrt{\varepsilon}).
$$

Moreover, the above estimates hold uniformly in $0 \leq t \leq t + s \leq T$. This implies (a).

To prove (b), note that by the Markov property,

$$
E[(n^{\varepsilon,ij}(t+s) - n^{\varepsilon,ij}(t))^2|\mathcal{F}_t^\varepsilon]
$$
$$
= E\left[\left(n^{\varepsilon,ij}_{(t+s)/\varepsilon} - n^{\varepsilon,ij}_{t/\varepsilon} \right)^2 \Big| \alpha_{t/\varepsilon}^\varepsilon \right]
$$
$$
= \varepsilon E\left[\left(\sum_{l=t/\varepsilon}^{(t+s)/\varepsilon-1} \left(I_{\{\alpha_l^\varepsilon = s_{ij}\}} - \nu^{ij} I_{\{\overline{\alpha}_l = i\}} \right) \right)^2 \Big| \alpha_{t/\varepsilon}^\varepsilon \right]
$$
$$
= \varepsilon \sum_{l=t/\varepsilon}^{(t+s)/\varepsilon-1} \sum_{p=t/\varepsilon}^{(t+s)/\varepsilon-1} E\Big[I_{\{\alpha_l^\varepsilon = s_{ij}, \alpha_p^\varepsilon = s_{ij}\}} - \nu^{ij} I_{\{\overline{\alpha}_l = i, \alpha_p^\varepsilon = s_{ij}\}}
$$
$$
- \nu^{ij} I_{\{\alpha_l^\varepsilon = s_{ij}, \overline{\alpha}_p = i\}} + \nu^{ij}\nu^{ij} I_{\{\overline{\alpha}_l = i, \overline{\alpha}_p = i\}})|\alpha_{t/\varepsilon}^\varepsilon \Big].
$$

Using ζ^{lp} defined in (4.39),

$$E\left[\left(n_{(t+s)/\varepsilon}^{\varepsilon,ij} - n_{t/\varepsilon}^{\varepsilon,ij}\right)^2 \Big| \alpha_{t/\varepsilon}^{\varepsilon}\right] = \varepsilon \sum_{l=t/\varepsilon}^{(t+s)/\varepsilon-1} \sum_{p=t/\varepsilon}^{(t+s)/\varepsilon-1} E\left[\zeta^{lp} \Big| \alpha_{t/\varepsilon}^{\varepsilon}\right],$$

and

$$E\left[\zeta^{lp} \Big| \alpha_{t/\varepsilon}^{\varepsilon}\right] = \sum_{i_0=1}^{l_0} \sum_{j_0=1}^{m_{i_0}} E\left[\zeta^{lp} \Big| \alpha_{s/\varepsilon}^{\varepsilon} = s_{i_0 j_0}\right] I_{\{\alpha_{t/\varepsilon}^{\varepsilon}=s_{i_0 j_0}\}}.$$

As in the proof of Theorem 4.1, we can show that

$$E\left[\zeta^{lp} \Big| \alpha_{t/\varepsilon}^{\varepsilon} = s_{i_0 j_0}\right] = \begin{cases} O(\varepsilon + \lambda^{l-p}) & \text{for } t/\varepsilon \le p < l \le (t+s)/\varepsilon - 1, \\ O(\varepsilon + \lambda^{p-l}) & \text{for } t/\varepsilon \le l < p \le (t+s)/\varepsilon - 1, \\ O(1), & \text{for } t/\varepsilon \le l = p \le (t+s)/\varepsilon - 1. \end{cases}$$

Therefore,

$$E\left[\left(n_{(t+s)/\varepsilon}^{\varepsilon,ij} - n_{t/\varepsilon}^{\varepsilon,ij}\right)^2 \Big| \alpha_{t/\varepsilon}^{\varepsilon}\right]$$

$$= \varepsilon \sum_{l=t/\varepsilon}^{(t+s)/\varepsilon-1} \sum_{p=t/\varepsilon}^{l-1} O(\varepsilon + \lambda^{l-p}) + \varepsilon \sum_{p=t/\varepsilon}^{(t+s)/\varepsilon-1} \sum_{l=t/\varepsilon}^{p-1} O(\varepsilon + \lambda^{p-l})$$

$$+ \varepsilon \sum_{l=t/\varepsilon}^{(t+s)/\varepsilon-1} O(1)$$

$$= \varepsilon O(s/\varepsilon) = O(s).$$

This completes the proof. □

Proof of Lemma 4.7. In view of the weak convergence of the process $\overline{\alpha}^{\varepsilon}(\cdot)$ (Theorem 4.3), it suffices to prove the tightness of $\{n^{\varepsilon}(\cdot)\}$. By Lemma 4.6 (b),

$$\lim_{\delta \to 0} \limsup_{\varepsilon \to 0} E\left(E|n^{\varepsilon}(t+s) - n^{\varepsilon}(t)|^2 \Big| \mathcal{F}_t^{\varepsilon}\right) = \lim_{\delta \to 0} O(\delta) = 0.$$

The assertion then follows from the tightness criterion Lemma 14.12. □

Sketch of Proof of Lemma 4.8. The proof is similar to that of Yin and Zhang [158, pp. 200–203]. A short outline is provided below. Let us begin with the calculation in a single block i ($i = 1, \ldots, l_0$). Corresponding to this block, $W(\alpha)$ in the definition of occupation measures becomes $I_{\{\alpha=s_{ij}\}} - \nu^{ij}$ (since $\alpha \in \mathcal{M}_i$, $I_{\{\alpha \in \mathcal{M}_i\}} = 1$). Consider a singularly perturbed discrete-time Markov chain with transition matrix $P^{\varepsilon} = P^i + \varepsilon \check{Q}^i$, where P^i is irreducible and aperiodic and $\check{Q}^i$ is a generator of a continuous-time

Markov chain. Note that similar to the continuous-time counter part, to calculate the covariance, the actual form of $\varepsilon \check{Q}^i$ is asymptotically unimportant (see Yin and Zhang [158, p. 203]). Let $n_k^{\varepsilon,i}$ and $n^{\varepsilon,i}(t)$ be the scaled occupation measure and its continuous-time interpolation, respectively. Since the underlying Markov chain has a finite state space, it is φ-mixing. Using Theorem 4.1 and Theorem 4.5, detailed estimates yield (4.30).

Now consider P having the form (4.5). It can be rewritten as

$$P = \text{diag}(P^1, 0_{m_2 \times m_2}, \ldots, 0_{m_{l_0} \times m_{l_0}})$$
$$+ \cdots + \text{diag}(0_{m_1 \times m_1}, \ldots, 0_{m_{l_0-1} \times m_{l_0-1}}, P^{l_0}).$$

The argument in the preceding paragraph indicates that if $\alpha \in \mathcal{M}_i$, $\Sigma(i)$ is given by (4.29) as desired.   □

Proof of Lemma 4.9. The argument is similar to that of Yin and Zhang [158, p. 199 and p. 200] (see also related problems in Lemma 14.20). By virtue of Lemma 14.8, it suffices to verify the uniqueness in distribution of $(n(t), \bar{\alpha}(t))$ for each $t \in [0, T]$. Consider the function

$$\widetilde{\phi}(x, l) = \exp(\mathbf{i}(x\theta + \theta_0 l)),$$

for each positive integer l, $x \in \mathbb{R}^{1 \times m_0}$, $\theta \in \mathbb{R}^{m_0 \times 1}$, $\theta_0 \in \mathbb{R}$, and $\mathbf{i}^2 = -1$. Note that $x\theta$ above is just the usual inner product. Define

$$\phi^{ij}(t) = E[I_{\{\bar{\alpha}(t)=i\}} \widetilde{\phi}(n(t), j)] \quad \text{for} \quad i, j = 1, \ldots, l_0.$$

Taking partials with respect to the variables x^{j_1} and x^{j_2} and using the coupling due to the presence of generator $\overline{Q}$, we obtain

$$\phi^{ij}(t) - \phi^{ij}(0) - \int_0^t \left\{ \sum_{j_1, j_2=1}^{m_i} a^{j_1 j_2}(i)(-\theta^{ij_1} \theta^{ij_2}) \phi^{ij}(s) \right. \tag{4.40}$$
$$\left. + \sum_{j_0=1}^{l_0} \bar{q}^{j_0 i}(s) \phi^{j_0 j}(s) \right\} ds = 0,$$

where $\phi^{ij}(0) = EI_{\{\bar{\alpha}(0)=i\}} \widetilde{\phi}(0, j)$. Let

$$\phi(t) = (\phi^{ij}(t), \ i, j = 1, \ldots, l_0).$$

Then (4.40) becomes

$$\phi(t) = \phi(0) + \int_0^t \phi(s) G(s) ds,$$

where $\phi(0) = (\phi^{ij}(0))$ and $G(t)$ is a matrix-valued function defined by the integrand of (4.40). The equation for $\phi(t)$ is a linear ordinary differential equation, thus it has a unique solution. Hence, $\phi(t)$ is uniquely determined. As a result,

$$E \exp\left(\mathrm{i}\{n(t)\theta + \theta_0\overline{\alpha}(t)\}\right)$$
$$= \sum_{j=1}^{l_0} E\left(I_{\{\overline{\alpha}(t)=j\}} \exp\left(\mathrm{i}\{n(t)\theta + j\theta_0\}\right)\right)$$

is uniquely determined for all $\theta \in \mathbb{R}^{m_0 \times 1}$ and $\theta_0 \in \mathbb{R}$. Therefore, the distribution of $(n(t), \overline{\alpha}(t))$ is uniquely determined by the well-known uniqueness and inversion formula for characteristic functions (see Theorem 14.40). $\square$

Proof of Theorem 4.10. To proceed, we prove the weak convergence of $(n^\varepsilon(\cdot), \overline{\alpha}^\varepsilon(\cdot))$ by using Theorem 14.19. The essence is to apply the perturbed test function methods. For an appropriate function $g(\cdot)$, define the operator $\mathcal{L}^\varepsilon$ by

$$\mathcal{L}^\varepsilon g(n_k^\varepsilon, \overline{\alpha}_k^\varepsilon) = \frac{1}{\varepsilon} E_k[g(n_{k+1}^\varepsilon, \overline{\alpha}_{k+1}^\varepsilon) - g(n_k^\varepsilon, \overline{\alpha}_k^\varepsilon)], \qquad (4.41)$$

where E_k denotes the conditional expectation with respect to $\mathcal{F}_k$, the σ-algebra generated by $\{\alpha_l^\varepsilon, l \leq k\}$. We will construct a perturbed test function f^ε and show that all conditions in Theorem 14.19 are satisfied. We also obtain the representation of the limit operator and the limit covariance matrix. Hence the weak convergence result follows.

For each $i = 1, \ldots, l_0$, let $f(\cdot, i)$ be any real-valued function with bounded derivatives up to the second order such that the second derivatives are Lipschitz continuous. Define

$$\overline{f}(x, \alpha) = \sum_{i=1}^{l_0} f(x, i) I_{\{\alpha \in \mathcal{M}_i\}} = \begin{cases} f(x, 1), & \text{if } \alpha \in \mathcal{M}_1, \\ \vdots \\ f(x, l_0), & \text{if } \alpha \in \mathcal{M}_{l_0}. \end{cases} \qquad (4.42)$$

Definition (4.42) allows us to replace $f(n_k^\varepsilon, \overline{\alpha}_k^\varepsilon)$ by $\overline{f}(n_k^\varepsilon, \alpha_k^\varepsilon)$. Denote

$$\chi_k^\varepsilon = (I_{\{\alpha_k^\varepsilon = s_{ij}\}}) \in \mathbb{R}^{1 \times m_0},$$
$$\overline{\chi}_k^\varepsilon = (I_{\{\overline{\alpha}_k^\varepsilon = 1\}}, \ldots, I_{\{\overline{\alpha}_k^\varepsilon = l_0\}}) \in \mathbb{R}^{1 \times l_0},$$
$$\widehat{\nu} = \mathrm{diag}(\nu^1, \ldots, \nu^{l_0}) \in \mathbb{R}^{l_0 \times m_0},$$
$$\overline{F}(x) = \begin{pmatrix} f(x, 1)\mathbb{1}_{m_1} \\ f(x, 2)\mathbb{1}_{m_2} \\ \vdots \\ f(x, l_0)\mathbb{1}_{m_{l_0}} \end{pmatrix} \in \mathbb{R}^{m_0 \times 1}, \text{ and } F(x) = \begin{pmatrix} f(x, 1) \\ \vdots \\ f(x, l_0) \end{pmatrix} \in \mathbb{R}^{l_0 \times 1}.$$

$$(4.43)$$

In view of the block-diagonal structure of the transition matrix P, it is easy to see that $(P - I)\overline{f}(x, \cdot)(\alpha) = 0$. That is, $\overline{f}(x, \alpha)$ is orthogonal to

$P - I$, so $(P - I)\overline{F}(x) = 0$ by virtue of (4.43). Moreover, we obtain

$$
\begin{aligned}
\varepsilon \mathcal{L}^\varepsilon \overline{f}(n_k^\varepsilon, \alpha_k^\varepsilon) &= E_k[\overline{f}(n_{k+1}^\varepsilon, \alpha_{k+1}^\varepsilon) - \overline{f}(n_k^\varepsilon, \alpha_k^\varepsilon)] \\
&= E_k[\overline{f}(n_{k+1}^\varepsilon, \alpha_{k+1}^\varepsilon) - \overline{f}(n_{k+1}^\varepsilon, \alpha_k^\varepsilon)] \\
&\quad + E_k[\overline{f}(n_{k+1}^\varepsilon, \alpha_k^\varepsilon) - \overline{f}(n_k^\varepsilon, \alpha_k^\varepsilon)].
\end{aligned}
$$

Using Taylor expansions, the last term in the above equation can be written as

$$
\begin{aligned}
&E_k[\overline{f}(n_{k+1}^\varepsilon, \alpha_k^\varepsilon) - \overline{f}(n_k^\varepsilon, \alpha_k^\varepsilon)] \\
&= \sqrt{\varepsilon} W(\alpha_k^\varepsilon) \overline{f}_x(n_k^\varepsilon, \alpha_k^\varepsilon) + \frac{\varepsilon}{2} W(\alpha_k^\varepsilon) \overline{f}_{xx}(n_k^\varepsilon, \alpha_k^\varepsilon) W'(\alpha_k^\varepsilon) + e_k^{\varepsilon,1},
\end{aligned}
$$

$$(4.44)$$

where $E|e_k^{\varepsilon,1}| = o(\varepsilon)$ uniformly in $k \leq T/\varepsilon$, and $\overline{f}_x(x, \alpha) \in \mathbb{R}^{m_0 \times 1}$ denotes the gradient w.r.t. the variable x. Note that $W(\alpha)\overline{f}_x(x, \alpha)$ in (4.44) is the usual inner product of two vectors. To estimate the first term in (4.44), we have

$$
\begin{aligned}
&E_k[\overline{f}(n_{k+1}^\varepsilon, \alpha_{k+1}^\varepsilon) - \overline{f}(n_{k+1}^\varepsilon, \alpha_k^\varepsilon)] \\
&= \sum_{i_1=1}^{l_0} \sum_{j_1=1}^{m_{i_1}} E_k \left[\sum_{i=1}^{l_0} \sum_{j=1}^{m_i} \overline{f}(n_{k+1}^\varepsilon, s_{ij}) P(\alpha_{k+1}^\varepsilon = s_{ij} | \alpha_k^\varepsilon = s_{i_1 j_1}) \right. \\
&\qquad\qquad\qquad\qquad \left. - \overline{f}(n_{k+1}^\varepsilon, s_{i_1 j_1}) \right] I_{\{\alpha_k^\varepsilon = s_{i_1 j_1}\}} \\
&= \chi_k^\varepsilon (P^\varepsilon - I) E_k \overline{F}(n_{k+1}^\varepsilon) \\
&= \chi_k^\varepsilon (P - I + \varepsilon Q) E_k \overline{F}(n_{k+1}^\varepsilon) \\
&= \varepsilon \chi_k^\varepsilon Q E_k \overline{F}(n_{k+1}^\varepsilon) \\
&= \varepsilon \chi_k^\varepsilon Q \overline{F}(n_k^\varepsilon) + e_k^{\varepsilon,2} \\
&= \varepsilon Q \overline{f}(n_k^\varepsilon, \cdot)(\alpha_k^\varepsilon) + e_k^{\varepsilon,2},
\end{aligned}
$$

$$(4.45)$$

where $\sup_{0 < k \leq T/\varepsilon} E|e_k^{\varepsilon,2}| = o(\varepsilon)$. Thus

$$
\begin{aligned}
\varepsilon \mathcal{L}^\varepsilon \overline{f}(n_k^\varepsilon, \alpha_k^\varepsilon) &= \varepsilon Q \overline{f}(n_k^\varepsilon, \cdot)(\alpha_k^\varepsilon) + \frac{\varepsilon}{2} W(\alpha_k^\varepsilon) \overline{f}_{xx}(n_k^\varepsilon, \alpha_k^\varepsilon) W'(\alpha_k^\varepsilon) \\
&\quad + \sqrt{\varepsilon} W(\alpha_k^\varepsilon) \overline{f}_x(n_k^\varepsilon, \alpha_k^\varepsilon) + e_k^{\varepsilon,1} + e_k^{\varepsilon,2}.
\end{aligned}
$$

$$(4.46)$$

For the purpose of averaging, we introduce several perturbations. The rationale is that these perturbations should be small in magnitude and should result in the desired cancellation. On the interval $t \in [k\varepsilon, k\varepsilon + \varepsilon)$, define the first perturbed test function by

$$
f_1^\varepsilon(x, t) = \sqrt{\varepsilon} \left(\sum_{l=k}^{T/\varepsilon} E_k W(\alpha_l^\varepsilon) \right) \overline{f}_x(x, \alpha_k^\varepsilon).
$$

$$(4.47)$$

We claim that $E|f_1^\varepsilon(n_k^\varepsilon, \varepsilon k)| = o(1)$, as $\varepsilon \to 0$. Noting that $\overline{f}_x(n_k^\varepsilon, \alpha_k^\varepsilon)$ is bounded and $\mathcal{F}_k$-measurable, definition (4.47), the Markov property, and the Cauchy–Schwarz inequality yield

$$
\begin{aligned}
E|f_1^\varepsilon(n_k^\varepsilon, \varepsilon k)| &= \sqrt{\varepsilon} E \left| \left(\sum_{l=k}^{T/\varepsilon} E_k W(\alpha_l^\varepsilon) \right) \overline{f}_x(n_k^\varepsilon, \alpha_k^\varepsilon) \right| \\
&\leq \sqrt{\varepsilon} E^{1/2} \left| \sum_{l=k}^{T/\varepsilon} E\left(\chi_l^\varepsilon - \overline{\chi}_l^\varepsilon \widehat{\nu} \Big| \alpha_k^\varepsilon \right) \right|^2 E^{1/2} |\overline{f}_x(n_k^\varepsilon, \alpha_k^\varepsilon)|^2 \\
&\leq K \sqrt{\varepsilon} E^{1/2} \left| \sum_{l=k}^{T/\varepsilon} E\left(\chi_l^\varepsilon - \overline{\chi}_l^\varepsilon \widehat{\nu} \Big| \alpha_k^\varepsilon \right) \right|^2.
\end{aligned}
$$

A closer look at the summand in the above leads to

$$
\begin{aligned}
&\sum_{i_1=1}^{l_0} \sum_{j_1=1}^{m_{i_1}} \left(P(\alpha_l^\varepsilon = s_{ij} | \alpha_k^\varepsilon = s_{i_1 j_1}) - \nu^{ij} \sum_{j_2=1}^{m_i} P(\alpha_l^\varepsilon = s_{ij_2} | \alpha_k^\varepsilon = s_{i_1 j_1}) \right) \\
&= \sum_{i_1=1}^{l_0} \sum_{j_1=1}^{m_{i_1}} \left(\nu^{ij} \theta^{i_1 i} - \nu^{ij} \left(\sum_{j_2=1}^{m_i} \nu^{ij_2} \right) \theta^{i_1 i} \right) + O(\varepsilon + \lambda^{l-k}) \\
&= O(\varepsilon + \lambda^{l-k}).
\end{aligned}
$$

Note that

$$
\sqrt{\varepsilon} \sum_{l=k}^{T/\varepsilon} O\left(\varepsilon + \lambda^{l-k} \right) = O(\sqrt{\varepsilon}) \to 0 \quad \text{as } \varepsilon \to 0.
$$

It follows that

$$
E|f_1^\varepsilon(n_k^\varepsilon, \varepsilon k)| = o(1) \quad \text{as } \varepsilon \to 0.
$$

Write

$$
f_1^\varepsilon(n_k^\varepsilon, \varepsilon k) = \sqrt{\varepsilon} W(\alpha_k^\varepsilon) \overline{f}_x(n_k^\varepsilon, \alpha_k^\varepsilon) + \sqrt{\varepsilon} \left(\sum_{l=k+1}^{T/\varepsilon} E_k W(\alpha_l^\varepsilon) \right) \overline{f}_x(n_k^\varepsilon, \alpha_k^\varepsilon).
$$

Then we have

$$
\begin{aligned}
E_k f_1^\varepsilon(n_{k+1}^\varepsilon, &\varepsilon(k+1)) - f_1^\varepsilon(n_k^\varepsilon, \varepsilon k) \\
&= -\sqrt{\varepsilon} W(\alpha_k^\varepsilon) \overline{f}_x(n_k^\varepsilon, \alpha_k^\varepsilon) \\
&+ \sqrt{\varepsilon} \sum_{l=k+1}^{T/\varepsilon} E_k W(\alpha_l^\varepsilon) [\overline{f}_x(n_{k+1}^\varepsilon, \alpha_{k+1}^\varepsilon) - \overline{f}_x(n_{k+1}^\varepsilon, \alpha_k^\varepsilon)] \\
&+ \sqrt{\varepsilon} \sum_{l=k+1}^{T/\varepsilon} E_k W(\alpha_l^\varepsilon) \overline{f}_x(n_{k+1}^\varepsilon, \alpha_k^\varepsilon) - \sqrt{\varepsilon} \sum_{l=k+1}^{T/\varepsilon} E_k W(\alpha_l^\varepsilon) \overline{f}_x(n_k^\varepsilon, \alpha_k^\varepsilon).
\end{aligned}
$$

$$
(4.48)
$$

By virtue of (4.45), the third line of (4.48) contributes a negligible term. It follows that

$$
\begin{aligned}
\varepsilon \mathcal{L}^\varepsilon f_1^\varepsilon(n_k^\varepsilon, \varepsilon k) \\
= E_k f_1^\varepsilon(n_{k+1}^\varepsilon, \varepsilon(k+1)) - f_1^\varepsilon(n_k^\varepsilon, \varepsilon k) \\
= -\sqrt{\varepsilon} W(\alpha_k^\varepsilon) \overline{f}_x(n_k^\varepsilon, \alpha_k^\varepsilon) + e_k^{\varepsilon,3} \\
+\varepsilon \left(\sum_{l=k+1}^{T/\varepsilon} E_k W(\alpha_l^\varepsilon) \overline{f}_{xx}(n_k^\varepsilon, \alpha_k^\varepsilon) \right) W'(\alpha_k^\varepsilon),
\end{aligned}
\tag{4.49}
$$

where $\sup_{0 \le k \le T/\varepsilon} E|e_k^{\varepsilon,3}| = o(\varepsilon)$.

To proceed, define (for $\varepsilon k = t$),

$$
\begin{aligned}
f_2^\varepsilon(x,t) \\
= \frac{\varepsilon}{2} \sum_{l=k}^{T/\varepsilon} [E_k W(\alpha_l^\varepsilon) \overline{f}_{xx}(x, \alpha_k^\varepsilon) W'(\alpha_k^\varepsilon) - EW(\alpha_l^\varepsilon) \overline{f}_{xx}(x, \alpha_k^\varepsilon) W'(\alpha_k^\varepsilon)],
\end{aligned}
$$

$$
\begin{aligned}
f_3^\varepsilon(x,t) \\
= \varepsilon \sum_{p=k}^{T/\varepsilon} \sum_{l=p+1}^{T/\varepsilon} [E_p W(\alpha_l^\varepsilon) \overline{f}_{xx}(x, \alpha_p^\varepsilon) W'(\alpha_p^\varepsilon) - EW(\alpha_l^\varepsilon) \overline{f}_{xx}(x, \alpha_p^\varepsilon) W'(\alpha_p^\varepsilon)],
\end{aligned}
$$

$$
f_4^\varepsilon(x,t) = \varepsilon \sum_{l=k}^{T/\varepsilon} E_k \left(\chi_l^\varepsilon - \overline{\chi}_l^\varepsilon \widehat{\nu} \right) Q \widetilde{\mathbb{1}} F(x),
$$

(4.50)

where χ_l^ε, $\overline{\chi}_l^\varepsilon$, and $\widehat{\nu}$ are defined in (4.43). Note that $Q\widetilde{\mathbb{1}}F(x) = Q\overline{F}(x)$. Similar to the previous estimates for $f_1^\varepsilon(n_k, \varepsilon k)$, by virtue of Theorem 4.5, it can be verified that

$$
\sup_{0 \le k \le T/\varepsilon} E|f_{i_1}^\varepsilon(n_k^\varepsilon, \varepsilon k)| \to 0, \quad \text{as } \varepsilon \to 0 \text{ for } i_1 = 2, 3, 4.
\tag{4.51}
$$

Moreover, detailed computation reveals that

$$
\begin{aligned}
\varepsilon \mathcal{L} f_2^\varepsilon(n_k^\varepsilon, \varepsilon k) = -\frac{\varepsilon}{2} \Big[& W(\alpha_k^\varepsilon) \overline{f}_{xx}(n_k^\varepsilon, \alpha_k^\varepsilon) W'(\alpha_k^\varepsilon) \\
& -EW(\alpha_k^\varepsilon) \overline{f}_{xx}(n_k^\varepsilon, \alpha_k^\varepsilon) W'(\alpha_k^\varepsilon) \Big] + e_k^{\varepsilon,4},
\end{aligned}
$$

$$
\begin{aligned}
\varepsilon \mathcal{L} f_3^\varepsilon(n_k^\varepsilon, \varepsilon k) = -\varepsilon \sum_{l=k+1}^{T/\varepsilon} \Big[& E_k W(\alpha_l^\varepsilon) \overline{f}_{xx}(n_k^\varepsilon, \alpha_k^\varepsilon) W'(\alpha_k^\varepsilon) \\
& -EW(\alpha_l^\varepsilon) \overline{f}_{xx}(n_k^\varepsilon, \alpha_k^\varepsilon) W'(\alpha_k^\varepsilon) \Big] + e_k^{\varepsilon,5},
\end{aligned}
\tag{4.52}
$$

$$
\varepsilon \mathcal{L}^\varepsilon f_4^\varepsilon(n_k^\varepsilon, \varepsilon k) = -\varepsilon(\chi_k^\varepsilon - \overline{\chi}_k^\varepsilon \widehat{\nu}) Q\overline{F}(n_k^\varepsilon) + e_k^{\varepsilon,6},
$$

and that

$$\sup_{0 \le k \le T/\varepsilon} E|e_k^{\varepsilon,\iota}| = o(\varepsilon) \quad \text{as} \quad \varepsilon \to 0, \quad \text{for} \quad \iota = 4,5,6.$$

Define

$$f^\varepsilon(x, \varepsilon k) = \overline{f}(x, \alpha_k^\varepsilon) + \sum_{\iota=1}^{4} f_\iota^\varepsilon(x, \varepsilon k), \tag{4.53}$$

where $f_\iota^\varepsilon(\cdot)$, $\iota = 1,2,3,4$ are as specified in (4.47) and (4.50). The preceding estimates of $E|f_\iota^\varepsilon(x, \varepsilon k)|$ lead to

$$E|f^\varepsilon(n_k^\varepsilon, \varepsilon k) - f(n_k^\varepsilon, \overline{\alpha}_k^\varepsilon)| = E|f^\varepsilon(n_k^\varepsilon, \varepsilon k) - \overline{f}(n_k^\varepsilon, \alpha_k^\varepsilon)| \to 0 \text{ as } \varepsilon \to 0. \tag{4.54}$$

In addition, we obtain

$$\begin{aligned}
\mathcal{L}^\varepsilon f^\varepsilon(n_k^\varepsilon, \varepsilon k) &= \overline{Q}f(n_k^\varepsilon, \cdot)(\overline{\alpha}_k^\varepsilon) + \frac{1}{2} EW(\alpha_k^\varepsilon)\overline{f}_{xx}(n_k^\varepsilon, \alpha_k^\varepsilon)W'(\alpha_k^\varepsilon) \\
&\quad + \frac{1}{2} \sum_{l=k+1}^{T/\varepsilon} EW(\alpha_l^\varepsilon)\overline{f}_{xx}(n_k^\varepsilon, \alpha_k^\varepsilon)W'(\alpha_k^\varepsilon) \\
&\quad + \frac{1}{2} \sum_{l=k+1}^{T/\varepsilon} EW(\alpha_k^\varepsilon)\overline{f}_{xx}(n_k^\varepsilon, \alpha_k^\varepsilon)W'(\alpha_l^\varepsilon) + e_k^\varepsilon,
\end{aligned} \tag{4.55}$$

where

$$e_k^\varepsilon = \sum_{\iota=1}^{6} e_k^{\varepsilon,\iota} \quad \text{and} \quad \sup_{0 \le k \le T/\varepsilon} E|e_k^\varepsilon| = o(\varepsilon) \quad \text{as} \quad \varepsilon \to 0.$$

To proceed, we need to evaluate the limit of the next to the last term in (4.55) (the term of the second line of (4.55) can be treated similarly). Note that

$$\begin{aligned}
&\sum_{l=k+1}^{T/\varepsilon} EW(\alpha_k^\varepsilon)\overline{f}_{xx}(n_k^\varepsilon, \alpha_k^\varepsilon)W'(\alpha_l^\varepsilon) \\
&= \sum_{l=k+1}^{T/\varepsilon} E\text{tr}\left(W'(\alpha_l^\varepsilon)W(\alpha_k^\varepsilon)\overline{f}_{xx}(n_k^\varepsilon, \alpha_k^\varepsilon)\right),
\end{aligned} \tag{4.56}$$

where $\text{tr}(A) = \sum_i a_{ii}$ denotes the trace of a square matrix A as usual. To get the desired limit, using $x = n^\varepsilon(t) = n_k^\varepsilon$, it suffices to examine the preceding expression with $\overline{f}_{xx}(n_k^\varepsilon, \alpha_k^\varepsilon)$ replaced by $\overline{f}_{xx}(x, \alpha_k^\varepsilon)$ for each fixed x. To this end, for $l > k$,

$$\begin{aligned}
&\text{tr}\left(W'(\alpha_l^\varepsilon)W(\alpha_k^\varepsilon)\overline{f}_{xx}(x, \alpha_k^\varepsilon)\right) \\
&= \text{tr}\left(W'(\alpha_l^\varepsilon)W(\alpha_k^\varepsilon)\sum_{\iota=1}^{l_0} f_{xx}(x, \iota)I_{\{\alpha_k^\varepsilon \in \mathcal{M}_\iota\}}\right) \\
&= \sum_{i_1=1}^{l_0}\sum_{j_1=1}^{m_{i_1}}\sum_{i_2=1}^{l_0}\sum_{j_2=1}^{m_{i_2}} w^{i_1 j_1}(\alpha_l^\varepsilon)w^{i_2 j_2}(\alpha_k^\varepsilon)\frac{\partial^2 f(x, i_2)}{\partial x^{i_2 j_2}\partial x^{i_1 j_1}}.
\end{aligned} \tag{4.57}$$

Similar to (4.38) and (4.39), we obtain four terms as

$$Ew^{i_1j_1}(\alpha_l^\varepsilon)w^{i_2j_2}(\alpha_k^\varepsilon)\frac{\partial^2 f(x,i_2)}{\partial x^{i_2j_2}\partial x^{i_1j_1}} = \rho_1 + \rho_2 + \rho_3 + \rho_4, \qquad (4.58)$$

where

$$\rho_1 = P(\alpha_l^\varepsilon = s_{i_1j_1}, \alpha_k^\varepsilon = s_{i_2j_2})\frac{\partial^2 f(x,i_2)}{\partial x^{i_2j_2}\partial x^{i_1j_1}},$$

$$\rho_2 = -\nu^{i_1j_1}P(\alpha_l^\varepsilon \in \mathcal{M}_{i_1}, \alpha_k^\varepsilon = s_{i_2j_2})\frac{\partial^2 f(x,i_2)}{\partial x^{i_2j_2}\partial x^{i_1j_1}},$$

$$\rho_3 = -\nu^{i_2j_2}P(\alpha_l^\varepsilon = s_{i_1j_1}, \alpha_k^\varepsilon \in \mathcal{M}_{i_2})\frac{\partial^2 f(x,i_2)}{\partial x^{i_2j_2}\partial x^{i_1j_1}},$$

$$\rho_4 = \nu^{i_1j_1}\nu^{i_2j_2}P(\alpha_l^\varepsilon \in \mathcal{M}_{i_1}, \alpha_k^\varepsilon \in \mathcal{M}_{i_2})\frac{\partial^2 f(x,i_2)}{\partial x^{i_2j_2}\partial x^{i_1j_1}}.$$

We claim that under (A4.1),

$$\sup_{\varepsilon,k}\sum_{l=k+1}^{T/\varepsilon}\left|E\mathrm{tr}\left(W'(\alpha_l^\varepsilon)W(\alpha_k^\varepsilon)\overline{f}_{xx}(n_k^\varepsilon,\alpha_k^\varepsilon)\right)\right| < \infty. \qquad (4.59)$$

To prove (4.59), working with (4.58) and using Theorem 4.1, straightforward although detailed estimates reveal that

$$\rho_1 = \left[P(\alpha_k^\varepsilon = s_{i_2j_2})[\nu^{i_1j_1}\theta^{i_2i_1}(\varepsilon(l-k)) + \psi^{i_2j_2,i_1j_1}(l-k)\right.$$
$$\left. +\varepsilon\widehat{\varphi}^{i_2j_2,i_1j_1}(\varepsilon(l-k)) + \varepsilon\widehat{\psi}^{i_2j_2,i_1j_1}(l-k)] + O(\varepsilon^2)\right]\frac{\partial^2 f(x,i_2)}{\partial x^{i_2j_2}\partial x^{i_1j_1}},$$

$$\rho_2 = -\left[\nu^{i_1j_1}P(\alpha_k^\varepsilon = s_{i_2j_2})\sum_{j_3=1}^{m_{i_1}}[\nu^{i_1j_3}\theta^{i_2i_1}(\varepsilon(l-k)) + \psi^{i_2j_2,i_1j_3}(l-k)\right.$$
$$\left. +\varepsilon\widehat{\varphi}^{i_2j_2,i_1j_3}(\varepsilon(l-k)) + \varepsilon\widehat{\psi}^{i_2j_2,i_1j_3}(l-k)] + O(\varepsilon^2)\right]\frac{\partial^2 f(x,i_2)}{\partial x^{i_2j_2}\partial x^{i_1j_1}},$$

$$\rho_3 = -\left[\nu^{i_2j_2}\sum_{j_4=1}^{m_{i_2}}P(\alpha_k^\varepsilon = s_{i_2j_4})[\nu^{i_1j_1}\theta^{i_2i_1}(\varepsilon(l-k)) + \psi^{i_2j_4,i_1j_1}(l-k)\right.$$
$$\left. +\varepsilon\widehat{\varphi}^{i_2j_4,i_1j_1}(\varepsilon(l-k)) + \varepsilon\widehat{\psi}^{i_2j_4,i_1j_1}(l-k)] + O(\varepsilon^2)\right]\frac{\partial^2 f(x,i_2)}{\partial x^{i_2j_2}\partial x^{i_1j_1}},$$

$$\rho_4 = \left[\nu^{i_1 j_1} \nu^{i_2 j_2} \sum_{j_3=1}^{m_{i_1}} \sum_{j_4=1}^{m_{i_2}} P(\alpha_k^\varepsilon = s_{i_2 j_4})[\nu^{i_1 j_3} \theta^{i_2 i_1}(\varepsilon(l-k)) \right.$$
$$\left. + \psi^{i_2 j_4, i_1 j_3}(l-k) + \varepsilon \widehat{\varphi}^{i_2 j_4, i_1 j_3}(\varepsilon(l-k)) + \varepsilon \widehat{\psi}^{i_2 j_4, i_1 j_3}(l-k)] \right.$$
$$\left. + O(\varepsilon^2) \right] \frac{\partial^2 f(x, i_2)}{\partial x^{i_2 j_2} \partial x^{i_1 j_1}}.$$

Using the asymptotic expansions, similar to the proof of Theorem 4.5, it is easily seen that the sum of the leading terms in $\sum_\iota \rho_\iota$ (the first terms on the right-hand side of ρ_ι, $\iota = 1, \ldots, 4$) is 0. Since $\Psi(\cdot)$ and $\widehat{\Psi}(\cdot)$ decay geometrically, for each i, j, p, q,

$$\sum_{l=k+1}^{T/\varepsilon} \sum_{i,j,p,q} \psi^{ij,pq}(l-k) < \infty,$$
$$\sum_{l=k+1}^{T/\varepsilon} \sum_{i,j,p,q} \widehat{\psi}^{ij,pq}(l-k) < \infty.$$

In addition, by using the asymptotic expansions, it follows that

$$\varepsilon \sum_{l=k+1}^{T/\varepsilon} |\widehat{\varphi}(\varepsilon(l-k))| \le K\varepsilon(T/\varepsilon) = KT < \infty.$$

As a result, we have

$$\sum_{l=k+1}^{T/\varepsilon} \sum_{i_1, j_2, i_2, j_2} \sum_{\iota=1}^{4} |\rho_\iota| < \infty.$$

Thus (4.59) is proved.

Equation (4.59) together with (4.48) and (4.52) yields that

$$\sup_{\varepsilon, k} |\mathcal{L}^\varepsilon f^\varepsilon(n_k^\varepsilon, \varepsilon k)| < \infty.$$

Moreover,

$$\lim_{\varepsilon \to 0} \sum_{l=k+1}^{T/\varepsilon} E\left[\mathrm{tr}\left(W'(\alpha_l^\varepsilon) W(\alpha_k^\varepsilon) \overline{f}_{xx}(n_k^\varepsilon, \alpha_k^\varepsilon) \right) \right] \quad \text{exists.} \tag{4.60}$$

Using (4.59) and the estimates obtained thus far,

$$\lim_{\varepsilon \to 0} E\left| \frac{E_k f^\varepsilon(k\varepsilon + \varepsilon) - f^\varepsilon(k\varepsilon)}{\varepsilon} - \mathcal{L}f(n^\varepsilon(k\varepsilon), \overline{\alpha}^\varepsilon(k\varepsilon)) \right| = 0, \tag{4.61}$$

where the limit operator is given by (4.28). Up to now, all the conditions in Theorem 14.19 have been verified. Thus, by using that theorem, the proof of Theorem 4.10 is concluded.

Remark 4.15. If the transition matrix P given in (4.5) is irreducible (i.e., the associated Markov chain has only one single ergodic class), (4.60) follows from the mixing properties, because the correlations between $W(\alpha_l^\varepsilon)$ and $W(\alpha_k^\varepsilon)$ decay exponentially fast. This can also be seen from the discussion in the previous section concerning the mixing condition. If P consists of multi-ergodic blocks, (4.60) indicates that for fixed k, although the correlations between $W(\alpha_l^\varepsilon)$ and $W(\alpha_k^\varepsilon)$ may not decay as fast as the single ergodic class case, they are summable.

Proof of Theorem 4.14. The proof of Part (a) is similar to what was before, with the modification of including the transient states. The proof of Part (e) uses Parts (a)–(d) and detailed estimates similar to the previous section. Thus we will only prove Parts (b)–(d).

To prove Part (b), we note that for $i = 1, \ldots, l_0$, the proof is the same as before. As for $i = *$ (for the transient states),

$$E|\pi^{\varepsilon,*j}(t)|^2 = E|\pi_{t/\varepsilon}^{\varepsilon,*j}|^2 = E\left(\varepsilon \sum_{l=0}^{t/\varepsilon-1} I_{\{\alpha_l^\varepsilon = s_{*j}\}}\right)^2$$

$$= \varepsilon^2 \sum_{l=0}^{t/\varepsilon-1} \sum_{p=0}^{t/\varepsilon-1} P(\alpha_l^\varepsilon = s_{*j}, \alpha_p^\varepsilon = s_{*j}).$$

When $p < l$, we have

$$\sum_{l=0}^{t/\varepsilon-1} \sum_{p=0}^{l-1} P(\alpha_l^\varepsilon = s_{*j}, \alpha_p^\varepsilon = s_{*j})$$

$$= \sum_{l=0}^{t/\varepsilon-1} \sum_{p=0}^{l-1} P(\alpha_l^\varepsilon = s_{*j} | \alpha_p^\varepsilon = s_{*j}) P(\alpha_p^\varepsilon = s_{*j})$$

$$= \sum_{l=0}^{t/\varepsilon-1} \sum_{p=0}^{l-1} O(\lambda^{l-p}) O(\lambda^p)$$

$$\leq K \sum_{l=0}^{t/\varepsilon-1} (l\lambda^{l/2})(\lambda^{1/2})^l = O(1).$$

By symmetry, for $l < p$, we have

$$\sum_{p=0}^{t/\varepsilon-1} \sum_{l=0}^{p-1} P(\alpha_l^\varepsilon = s_{*j}, \alpha_p^\varepsilon = s_{*j}) = O(1).$$

As for $p = l$, it follows from the asymptotic expansion

$$\sum_{l=0}^{t/\varepsilon-1} P(\alpha_l^\varepsilon = s_{*j}) = O(1).$$

Thus the desired result follows.

To prove Part (c), define

$$\overline{\chi}^\varepsilon(\overline{\alpha}^\varepsilon(t)) = (I_{\{\overline{\alpha}^\varepsilon(t)=1\}}, \ldots, I_{\{\overline{\alpha}^\varepsilon(t)=l_0\}}).$$

We show that $\{\overline{\chi}^\varepsilon(\cdot)\}$ is tight. To do so, we first prove that

$$\limsup_{\varepsilon \to 0} E[\overline{\chi}^\varepsilon(\overline{\alpha}^\varepsilon(t+s)) - \overline{\chi}^\varepsilon(\overline{\alpha}^\varepsilon(t))|\mathcal{F}_t^\varepsilon] \le \gamma(s), \qquad (4.62)$$

for some function $\gamma(\cdot)$ such that $E\gamma(s) \to 0$ as $s \to 0$. By the Markov property, and the definition of $\alpha^\varepsilon(\cdot)$,

$$E[\overline{\chi}^\varepsilon(\overline{\alpha}^\varepsilon(t+s)) - \overline{\chi}^\varepsilon(\overline{\alpha}^\varepsilon(t))|\mathcal{F}_t^\varepsilon] = E[\overline{\chi}^\varepsilon(\overline{\alpha}_{(t+s)/\varepsilon}^\varepsilon) - \overline{\chi}^\varepsilon(\overline{\alpha}_{t/\varepsilon}^\varepsilon)|\alpha_{t/\varepsilon}^\varepsilon]$$

Moreover, when $\alpha_{t/\varepsilon}^\varepsilon = s_{ij}$ and $i \ne *$,

$$E[I_{\{\overline{\alpha}_{t/\varepsilon}^\varepsilon=p\}}|\alpha_{t/\varepsilon}^\varepsilon = s_{ij}] = \delta_{ip} = \begin{cases} 1 & \text{if } i = p, \\ 0 & \text{if } i \ne p. \end{cases}$$

Therefore,

$$\begin{aligned}
&E[I_{\{\overline{\alpha}_{(t+s)/\varepsilon}^\varepsilon=p\}} - I_{\{\overline{\alpha}_{t/\varepsilon}^\varepsilon=p\}}|\alpha_{t/\varepsilon}^\varepsilon = s_{ij}] \\
&= P(\overline{\alpha}_{(t+s)/\varepsilon}^\varepsilon = p|\alpha_{t/\varepsilon}^\varepsilon = s_{ij}) - \delta_{ip} \\
&= P(\alpha_{(t+s)/\varepsilon}^\varepsilon \in \mathcal{M}_p|\alpha_{t/\varepsilon}^\varepsilon = s_{ij}) \\
&\quad + \sum_{j_1=1}^{m_*} P(\alpha_{(t+s)/\varepsilon}^\varepsilon = s_{*j_1}|\alpha_{t/\varepsilon}^\varepsilon = s_{ij})P(U_{j_1} = p) - \delta_{ip} \\
&= \theta^{ip}(s) + O(\varepsilon + \lambda^{s/\varepsilon}) - \delta_{ip}.
\end{aligned}$$

Since $\Theta(t) = (\theta^{ij}(t)) \to I$ as $t \to 0$,

$$\limsup_{\varepsilon \to 0} E[I_{\{\overline{\alpha}_{(t+s)/\varepsilon}^\varepsilon=p\}} - I_{\{\overline{\alpha}_{t/\varepsilon}^\varepsilon=p\}}|\alpha_{t/\varepsilon}^\varepsilon = s_{ij}] = \theta^{ip}(s) - \delta_{ip} \le \widetilde{\gamma}(s), \quad (4.63)$$

and $\widetilde{\gamma}(s) \to 0$ as $s \to 0$. Thus in this case, (4.62) is verified.

When $\alpha_{t/\varepsilon}^\varepsilon = s_{*j}$, since U_j is independent of $\alpha^\varepsilon(\cdot)$,

$$\begin{aligned}
&E[I_{\{\overline{\alpha}_{(t+s)/\varepsilon}^\varepsilon=p\}} - I_{\{\overline{\alpha}_{t/\varepsilon}^\varepsilon=p\}}|\alpha_{t/\varepsilon}^\varepsilon = s_{*j}] \\
&= P(\overline{\alpha}_{(t+s)/\varepsilon}^\varepsilon = p|\alpha_{t/\varepsilon}^\varepsilon = s_{*j}) - P(\overline{\alpha}_{t/\varepsilon}^\varepsilon = p|\alpha_{t/\varepsilon}^\varepsilon = s_{*j}) \\
&= P(\alpha_{(t+s)/\varepsilon}^\varepsilon \in \mathcal{M}_p|\alpha_{t/\varepsilon}^\varepsilon = s_{*j}) \\
&\quad + \sum_{j_1=1}^{m_*} P(\alpha_{(t+s)/\varepsilon}^\varepsilon = s_{*j_1}|\alpha_{t/\varepsilon}^\varepsilon = s_{*j})P(U_{j_1} = p) - P(U_j = p) \\
&= O(s) + O(\varepsilon + \lambda^{s/\varepsilon}) = O(s + \varepsilon + \lambda^{s/\varepsilon}).
\end{aligned}$$

We obtain

$$\limsup_{\varepsilon \to 0} E[\overline{\chi}^{\varepsilon}(\overline{\alpha}^{\varepsilon}(t+s)) - \overline{\chi}^{\varepsilon}(\overline{\alpha}^{\varepsilon}(t))|\mathcal{F}_s^{\varepsilon}] = O(s).$$

Thus (4.62) is again verified.

To proceed, we use (4.62) to prove the desired tightness. Since $I_A^2 = I_A$ for any indicator function of the set A, we have

$$E[|\overline{\chi}^{\varepsilon}(\overline{\alpha}^{\varepsilon}(t+s)) - \overline{\chi}^{\varepsilon}(\overline{\alpha}^{\varepsilon}(t))|^2|\mathcal{F}_t^{\varepsilon}]$$

$$= E[\overline{\chi}^{\varepsilon}(\overline{\alpha}^{\varepsilon}(t+s))\overline{\chi}^{\varepsilon,\prime}(\overline{\alpha}^{\varepsilon}(t+s)) - \overline{\chi}^{\varepsilon}(\overline{\alpha}^{\varepsilon}(t+s))\overline{\chi}^{\varepsilon,\prime}(\overline{\alpha}^{\varepsilon}(t))$$

$$\quad - \overline{\chi}^{\varepsilon}(\overline{\alpha}^{\varepsilon}(t))\overline{\chi}^{\varepsilon,\prime}(\overline{\alpha}^{\varepsilon}(t+s)) + \overline{\chi}^{\varepsilon}(\overline{\alpha}^{\varepsilon}(t))\overline{\chi}^{\varepsilon,\prime}(\overline{\alpha}^{\varepsilon}(t))|\mathcal{F}_t^{\varepsilon}]$$

$$= \sum_{p=1}^{l_0} E[I_{\{\overline{\alpha}^{\varepsilon}(t+s)=p\}} - 2I_{\{\overline{\alpha}^{\varepsilon}(t+s)=p\}}I_{\{\overline{\alpha}^{\varepsilon}(t)=p\}} + I_{\{\overline{\alpha}^{\varepsilon}(t)=p\}}|\mathcal{F}_t^{\varepsilon}].$$

(4.64)

Since $\sum_{p=1}^{l_0} I_{\{\overline{\alpha}^{\varepsilon}(t+s)=p\}} = 1$ and $\sum_{p=1}^{l_0} I_{\{\overline{\alpha}^{\varepsilon}(t)=p\}} = 1$,

$$\sum_{p=1}^{l_0} E[I_{\{\overline{\alpha}^{\varepsilon}(t+s)=p\}} - 2I_{\{\overline{\alpha}^{\varepsilon}(t+s)=p\}}I_{\{\overline{\alpha}^{\varepsilon}(t)=p\}} + I_{\{\overline{\alpha}^{\varepsilon}(t)=p\}}|\mathcal{F}_t^{\varepsilon}]$$

$$= 2\left[1 - \sum_{p=1}^{l_0} E[I_{\{\overline{\alpha}^{\varepsilon}(t+s)=p\}}I_{\{\overline{\alpha}^{\varepsilon}(t)=p\}}|\mathcal{F}_t^{\varepsilon}]\right]$$

$$= 2\left[1 - \sum_{p=1}^{l_0} E[I_{\{\overline{\alpha}^{\varepsilon}(t+s)=p\}} - I_{\{\overline{\alpha}^{\varepsilon}(t)=p\}}|\mathcal{F}_t^{\varepsilon}]I_{\{\overline{\alpha}^{\varepsilon}(t)=p\}} - \sum_{p=1}^{l_0} I_{\{\overline{\alpha}^{\varepsilon}(t)=p\}}\right]$$

$$= 2\sum_{p=1}^{l_0} E[I_{\{\overline{\alpha}^{\varepsilon}(t+s)=p\}} - I_{\{\overline{\alpha}^{\varepsilon}(t)=p\}}|\mathcal{F}_t^{\varepsilon}]I_{\{\overline{\alpha}^{\varepsilon}(t)=p\}}.$$

By using (4.62), the above estimates, and (4.64),

$$\limsup_{\varepsilon \to 0} E[|\overline{\chi}^{\varepsilon}(\overline{\alpha}^{\varepsilon}(t+s)) - \overline{\chi}^{\varepsilon}(\overline{\alpha}^{\varepsilon}(t))|^2|\mathcal{F}_t^{\varepsilon}] \le \gamma(s)$$

such that $\gamma(s) \to 0$ as $s \to 0$. Thus

$$\lim_{s \to 0}\limsup_{\varepsilon \to 0} E\left(E[|\overline{\chi}^{\varepsilon}(\overline{\alpha}^{\varepsilon}(t+s)) - \overline{\chi}^{\varepsilon}(\overline{\alpha}^{\varepsilon}(t))|^2|\mathcal{F}_t^{\varepsilon}]\right) = 0,$$

and $\{\overline{\chi}^{\varepsilon}(\cdot)\}$ is tight. Noting that $\overline{\alpha}^{\varepsilon}(t) = \sum_{i=1}^{l_0} iI_{\{\overline{\alpha}^{\varepsilon}(t)=i\}}$, the tightness of $\{\overline{\alpha}^{\varepsilon}(\cdot)\}$ thus follows. The convergence of the finite-dimensional distributions is proved in a similar way as was done in Theorem 4.10.

Let

$$w^{ij}(\alpha) = \begin{cases} I_{\{\alpha=s_{ij}\}} - \nu^{ij}I_{\{\alpha \in M_i\}}, & \text{for } i = 1, \ldots, l_0, \\ I_{\{\alpha=s_{*j}\}}, & \text{for } i = *, \end{cases}$$

(4.65)

and consider the normalized occupation measure

$$n_k^{\varepsilon,ij} = \sqrt{\varepsilon} \sum_{l=0}^{k-1} w^{ij}(\alpha_l^\varepsilon).$$

Define n_k^ε and the continuous-time interpolations as in the previous case.

To prove Part (d), for fixed i, j, and $i = 1, \ldots, l_0$ by using the martingale property,

$$E[(n^{\varepsilon,ij}(t+s) - n^{\varepsilon,ij}(t)|\mathcal{F}_t^\varepsilon] = E[n_{(t+s)/\varepsilon}^{\varepsilon,ij} - n_{t/\varepsilon}^{\varepsilon,ij}|\alpha_{t/\varepsilon}^\varepsilon]$$
$$= \sqrt{\varepsilon} \sum_{l=t/\varepsilon}^{(t+s)/\varepsilon-1} E[w^{ij}(\alpha_l^\varepsilon)|\alpha_{t/\varepsilon}^\varepsilon].$$

If $\alpha \notin \mathcal{M}_*$, a similar argument as in the previous case leads to

$$E[w^{ij}(\alpha_l^\varepsilon)|\alpha_{t/\varepsilon}^\varepsilon] = O(\varepsilon + \lambda^{l-(t/\varepsilon)}).$$

If $\alpha = s_{*j}$, then we have for $i \neq *$,

$$E[w^{ij}(\alpha_l^\varepsilon)|\alpha_{t/\varepsilon}^\varepsilon = s_{*j}]$$
$$= P(\alpha_l^\varepsilon = s_{ij}|\alpha_{t/\varepsilon}^\varepsilon = s_{*j}) - \nu^{ij}(\alpha_l^\varepsilon \in \mathcal{M}_i|\alpha_{t/\varepsilon}^\varepsilon = s_{*j})$$
$$= O(\varepsilon + \lambda^{l-(t/\varepsilon)}).$$

and for $i = *$,

$$E[w^{ij}(\alpha_l^\varepsilon)|\alpha_{t/\varepsilon}^\varepsilon] = P(\alpha_l^\varepsilon = s_{*j}|\alpha_{t/\varepsilon}^\varepsilon) = O(\varepsilon + \lambda^{l-(t/\varepsilon)}).$$

Thus $E[w^{ij}(\alpha_l^\varepsilon)|\alpha_{t/\varepsilon}^\varepsilon] = O(\varepsilon + \lambda^{l-(t/\varepsilon)})$ and Part (a) is proved.

To prove Part (b), using ζ^{lp} defined in (4.39), consider

$$\zeta^{lp,ij} = \begin{cases} \zeta^{lp}, & \text{for } i = 1, \ldots, l_0, \\ I_{\{\alpha_l^\varepsilon = s_{*j}\}} I_{\{\alpha_p^\varepsilon = s_{*j}\}}, & \text{for } i = *. \end{cases}$$

Again by the Markov property

$$E[(n^{\varepsilon,ij}(t+s) - n^{\varepsilon,ij}(t))^2|\mathcal{F}_t^\varepsilon] = E[(n_{(t+s)/\varepsilon}^{\varepsilon,ij} - n_{t/\varepsilon}^{\varepsilon,ij})^2|\alpha_{t/\varepsilon}^\varepsilon]$$
$$= \varepsilon \sum_{l=t/\varepsilon}^{(t+s)/\varepsilon-1} \sum_{p=t/\varepsilon}^{(t+s)/\varepsilon-1} E[\zeta^{lp,ij}|\alpha_{t/\varepsilon}^\varepsilon].$$

$$(4.66)$$

As before, we obtain

$$E[\zeta^{lp,ij}|\alpha_{t/\varepsilon}^\varepsilon] = \begin{cases} O(\varepsilon + \lambda^{l-p}), & \text{for } t/\varepsilon \leq p < l \leq (t+s)/\varepsilon - 1, \\ O(\varepsilon + \lambda^{p-l}), & \text{for } t/\varepsilon \leq l < p \leq (t+s)/\varepsilon - 1, \\ O(1), & \text{for } t/\varepsilon \leq l = p \leq (t+s)/\varepsilon - 1. \end{cases}$$

By virtue of (4.66),

$$\varepsilon \sum_{l=t/\varepsilon}^{(t+s)/\varepsilon-1} \sum_{p=t/\varepsilon}^{(t+s)/\varepsilon-1} E[\zeta^{rp,ij}|\alpha_{t/\varepsilon}^{\varepsilon}] = O(s).$$

Thus Part (d) is proved.   □

4.6   Notes

Classical central limit theorems concerning Markov chains may be found in Dobrushin [48], for instance. Diffusion approximation arising from the context of differential equations first appeared in Khasminskii [81]. Diffusion approximations for wideband noise were treated in Kushner [96, 97]. Functional central limit theorems and structural properties of continuous-time Markov chains with two-time scales were considered in Yin and Zhang [158], Yin, Zhang, and Badowski [162], and Zhang and Yin [179, 180]. Under discrete-time setup, the case with irreducible matrices was dealt with in Yin, Zhang, Yang, and Yin [168]. One of the main difficulties we run into is that for the more general setup with P given in (4.5) or (4.6), the transition matrix is not irreducible, and as a result, the usual mixing conditions cannot be verified. The main part of the results of this chapter is based on Yin, Zhang, and Badowski [165]. That paper also contains more general cases for non-homogeneous Markov chains.

5

Exponential Bounds

5.1 Introduction

This chapter is concerned with exponential upper bounds for scaled sequences of occupation measures. In Chapters 3 and 4, we examined asymptotic properties of singularly perturbed Markov chains in discrete time, including asymptotic expansions of the probability vectors and transition matrices, aggregations of the underlying processes, and the switching diffusion limit of a sequence of scaled occupation measures. In applications, it is also useful to provide estimates of rare event probabilities. In this chapter, we focus on the exponential type bounds for singularly perturbed systems. Such a study is important to further understand system stability and to facilitate the calculation of probabilities beyond the normal deviation range.

The rest of the chapter is arranged as follows. Section 5.2 begins with the formulation of the problem. Section 5.3 presents the main results. First we deal with the case in which the fast changing part of the transition probability matrix is irreducible, a necessary step for investigating properties of Markov chains with more complex structures. Then we consider the case in which the fast-changing transition matrix is decomposable into l_0 sub-transition matrices corresponding to decomposing the state space into l_0 irreducible classes. Finally, we treat the case when transient states are included. Section 5.4 gives examples of applications of these error bounds. Extensions that incorporate time dependence and that include transient states are given in Section 5.5. Detailed proofs are contained in Section 5.6. The chapter closes with some further remarks.

5.2   Formulation

Suppose T is a positive real number, $\varepsilon > 0$ is a small parameter, and α_k^ε, for $0 \le k \le \lfloor T/\varepsilon \rfloor$, is a discrete-time Markov chain with finite state space $\mathcal{M} = \{1, \ldots, m_0\}$, where $\lfloor z \rfloor$ denotes the integer part of a real number z. For notational simplicity, as was done in the previous chapters, we often simply write T/ε in lieu of $\lfloor T/\varepsilon \rfloor$ in what follows. Consider a discrete-time Markov chain α_k^ε with stationary transition probability matrix P^ε given by

$$P^\varepsilon = P + \varepsilon Q, \tag{5.1}$$

where $P = (p^{ij})$ is a transition probability matrix, and $Q = (q^{ij})$ is a generator.

We are interested in cases when a transition probability matrix of a finite-state Markov chain can be put into the form

$$P = \mathrm{diag}(P^1, \ldots, P^{l_0}), \tag{5.2}$$

where for each $i \le l_0$, P^i is a transition matrix within the ith recurrent class. A Markov chain with transition matrix given by (5.2) consists of l_0 recurrent classes. Let $\mathcal{M}_i = \{s_{i1}, \ldots, s_{im_i}\}$ denote the states corresponding to the ith block P^i. Then the state space of α_k^ε can be put into the form $\mathcal{M} = \mathcal{M}_1 \cup \cdots \cup \mathcal{M}_{l_0}$. In addition to the recurrent chains given above, we will also examine the case when transient states are included.

5.3   Main Results

We present the results in several steps. In what follows, we first derive a result when P is an irreducible matrix. Using the bounds for such irreducible chains, we then work on Markov chain α_k^ε with transition matrix P consisting of l_0 recurrent classes given by (5.2). Finally, we extend the results to include transient states.

5.3.1   Irreducible Chains

Here, we suppose that the Markov chain has a transition matrix given by (5.1) with P being irreducible and aperiodic. We aim to obtain exponential error bounds, which are needed for studying problems with a more general transition matrix P. It is also interesting in its own right.

Let $\mathcal{M} = \{1, \ldots, m_0\}$ be the state space, and $\beta_k = \mathrm{diag}(\beta_k^1, \ldots, \beta_k^{m_0})$, where $\{\beta_k^i\}$ are bounded sequences of real numbers. For each $i = 1, \ldots, m_0$, define

$$n_k^{\varepsilon,i} = \sqrt{\varepsilon} \sum_{l=0}^{k-1} (I_{\{\alpha_l^\varepsilon = i\}} - \nu^i)\beta_l^i, \tag{5.3}$$

and
$$n_k^\varepsilon = (n_k^{\varepsilon,1}, \ldots, n_k^{\varepsilon,m_0})$$
$$= \sqrt{\varepsilon} \sum_{l=0}^{k-1} (I_{\{\alpha_l^\varepsilon=1\}} - \nu^1, \ldots, I_{\{\alpha_l^\varepsilon=m_0\}} - \nu^{m_0})\beta_l.$$

In view of Theorem 4.1, $(P^\varepsilon)^k \to \overline{P} = \mathbb{1}_{m_0}\nu$, as $\varepsilon \to 0$ and $k = O(1/\varepsilon)$, where $\mathbb{1}_{m_0} = (1, \ldots, 1)' \in \mathbb{R}^{m_0 \times 1}$ and $\nu = (\nu^1, \ldots, \nu^{m_0})$ is the stationary distribution of P. Moreover, for $0 \le k \le T/\varepsilon$ for some finite $T > 0$, the k-step transition matrix $(P^\varepsilon)^k$ satisfies $(P^\varepsilon)^k - \overline{P} = O(\varepsilon + \lambda^k)$ for some $0 < \lambda < 1$. Denote the least upper bound of

$$\frac{(P^\varepsilon)^k - \overline{P}}{\varepsilon + \lambda^k}$$

by K_T for $0 \le k \le T/\varepsilon$. For convenience, introduce the notation $O_1(\cdot)$ as a "normalized" order symbol in that $O_1(y)$ is a function of y such that $|O_1(y)|/|y| \le 1$. We then have

$$(P^\varepsilon)^k - \overline{P} = (P + \varepsilon Q)^k - \overline{P}$$
$$= K_T O_1(\varepsilon + \lambda^k), \quad \text{for } 0 \le k \le T/\varepsilon. \tag{5.4}$$

For a vector $v = (v^i)$, we use $|\cdot|$ to denote the max norm

$$|v| = \max_i |v^i|,$$

and given a matrix $A = (a^{ij})$, we use the norm

$$|A| = \max_{i,j} |a^{ij}|.$$

Similarly, for a vector-valued sequence $z_k \in \mathbb{R}^{1 \times m_0}$ and a matrix-valued sequence $A_k \in \mathbb{R}^{m \times m}$, we use

$$|z|_T = \sup_{0 \le k \le T/\varepsilon} |z_k|, \qquad |A|_T = \sup_{0 \le k \le T/\varepsilon} |A_k|, \tag{5.5}$$

respectively.

Theorem 5.1. *Assume that P is irreducible and aperiodic. Let c_T be a constant such that*

$$0 \le c_T \le \frac{1-\lambda}{K_T(|\beta|_T + 1)}. \tag{5.6}$$

Then there exist an $\varepsilon_0 > 0$ and a constant K such that for all $0 \le \varepsilon \le \varepsilon_0$, the following error bound holds: for $1 \le k \le T/\varepsilon$,

$$E \exp\left(\frac{c_T}{(T+1)^{\frac{3}{2}}} |n_k^\varepsilon|\right) \le K, \tag{5.7}$$

where K is a constant independent of ε and T and λ is as given in Theorem 4.1. Moreover, for any $0 < \theta < 1$, we have

$$E \sup_{1 \le k \le T/\varepsilon} \exp\left(\frac{\theta c_T}{(T+1)^{\frac{3}{2}}} |n_k^\varepsilon|\right) \le K. \tag{5.8}$$

5.3.2   Recurrent Chains

This section is devoted to the exponential bounds for singularly perturbed Markov chains, in which the fast-changing part of the transition probability matrix includes l_0 recurrent classes. That is, all states are recurrent. In this case, the state space is decomposed to

$$\begin{aligned}
\mathcal{M} &= \mathcal{M}_1 \cup \mathcal{M}_2 \cup \cdots \cup \mathcal{M}_{l_0} \\
&= \{s_{11}, \ldots, s_{1m_1}\} \cup \{s_{21}, \ldots, s_{2m_2}\} \cup \cdots \cup \{s_{l_0 1}, \ldots, s_{l_0 m_{l_0}}\}.
\end{aligned} \tag{5.9}$$

Let $\{\beta_k\}$ be a sequence of diagonal matrices with real entries such that

$$\beta_k = \mathrm{diag}\left(\beta_k^{11}, \ldots, \beta_k^{1m_1}, \ldots, \beta_k^{l_0 1}, \ldots, \beta_k^{l_0 m_{l_0}}\right).$$

Define

$$\begin{aligned}
n_k^\varepsilon &= \sqrt{\varepsilon} \sum_{l=0}^{k-1} \Big(I_{\{\alpha_l^\varepsilon = s_{11}\}} - \nu^{11} I_{\{\overline{\alpha}_l^\varepsilon = 1\}}, \ldots, I_{\{\alpha_l^\varepsilon = s_{1m_1}\}} - \nu^{1m_1} I_{\{\overline{\alpha}_l^\varepsilon = 1\}}, \\
&\quad \ldots, I_{\{\alpha_l^\varepsilon = s_{l1}\}} - \nu^{l_0 1} I_{\{\overline{\alpha}_l^\varepsilon = l_0\}}, \ldots, I_{\{\alpha_l^\varepsilon = s_{l_0 m_{l_0}}\}} - \nu^{l_0 m_{l_0}} I_{\{\overline{\alpha}_l^\varepsilon = l_0\}} \Big) \beta_l.
\end{aligned}$$

Let c_T be a constant such that

$$0 \le c_T \le \frac{1 - \lambda}{K_T(|\beta|_T + 1)},$$

where

$$K_T = \sup_{\varepsilon, k} \frac{(P^\varepsilon)^k - \Phi(\varepsilon k)}{\varepsilon + \lambda^k}.$$

The exponential bounds for the recurrent chains are given as follows.

Theorem 5.2. *Assume that P^ε, P, and P^i for $i \le l_0$ are transition probability matrices such that for each $i \le l$, P^i is irreducible and aperiodic. Then there exist an $\varepsilon_0 > 0$ and a constant K such that for all $0 \le \varepsilon \le \varepsilon_0$, the following error bound holds: For $1 \le k \le T/\varepsilon$,*

$$E \exp\left(\frac{c_T}{(T+1)^{\frac{3}{2}}} |n_k^\varepsilon|\right) \le K. \tag{5.10}$$

Moreover, given $0 < \theta < 1$, there exist an $\varepsilon_0 > 0$ and a constant K such that, for all $0 \le \varepsilon \le \varepsilon_0$, we have

$$E \sup_{1 \le k \le T/\varepsilon} \exp \left(\frac{\theta c_T}{(T+1)^{\frac{3}{2}}} |n_k^\varepsilon| \right) \le K. \tag{5.11}$$

In obtaining a limit theorem for asymptotic distribution of n_k^ε, an important step involves proving the scaled sequence of being bounded in probability. With the exponential bounds obtained, such a probability bound can be readily obtained. In addition, moment bounds can also be obtained. They are stated in the next two corollaries with proofs in Section 5.5.

Corollary 5.3. *Suppose that the conditions of Theorem 5.2 are fulfilled. Then for any $\delta > 0$, there exists a $K_\delta > 0$ such that for all $0 \le k \le T/\varepsilon$,*

$$P \left(\sup_{k \le T/\varepsilon} |n_k^\varepsilon| > K_\delta \right) < \delta. \tag{5.12}$$

Corollary 5.4. *Under the conditions of Theorem 5.2, for any $0 < l < \infty$,*

$$E \sup_{k \le T/\varepsilon} |n_k^\varepsilon|^l \le \frac{K(T+1)^{\frac{3l}{2}} l!}{(c_T)^l}, \tag{5.13}$$

where K is given in Theorem 5.2.

5.3.3 Markov Chains with Transient States

Let $P^\varepsilon = P + \varepsilon Q$ and P given by (4.6). Define a diagonal matrix

$$\beta_k = \text{diag} \left(\beta_k^{11}, \dots, \beta_k^{1m_1}, \dots, \beta_k^{l_0 1}, \dots, \beta_k^{l_0 m_{l_0}}, \beta_k^{*1}, \dots, \beta_k^{*m_*} \right).$$

Also define

$$n_k^\varepsilon = \sqrt{\varepsilon} \sum_{l=0}^{k-1} \left(I_{\{\alpha_l^\varepsilon = s_{11}\}} - \nu^{11} I_{\{\overline{\alpha}_l^\varepsilon = 1\}}, \dots, I_{\{\alpha_l^\varepsilon = s_{1m_1}\}} - \nu^{1m_1} I_{\{\overline{\alpha}_l^\varepsilon = 1\}}, \right.$$
$$\dots, I_{\{\alpha_l^\varepsilon = s_{l1}\}} - \nu^{l_0 1} I_{\{\overline{\alpha}_l^\varepsilon = l_0\}}, \dots, I_{\{\alpha_l^\varepsilon = s_{l_0 m_{l_0}}\}} - \nu^{l_0 m_{l_0}} I_{\{\overline{\alpha}_l^\varepsilon = l_0\}},$$
$$\left. I_{\{\alpha_k^\varepsilon = s_{*1}\}}, \dots, I_{\{\alpha_k^\varepsilon = s_{*m_*}\}} \right) \beta_l.$$

We present the exponential bounds of the scaled occupation measures when the transient states are included.

Theorem 5.5 *Assume P^ε, P, and P^i for $i = 1, \dots, l_0$ are transition probability matrices such that each P^i is irreducible and aperiodic. Moreover, suppose that all the eigenvalues of P^* are inside the unit circle. Then the conclusion of Theorem 5.2 continues to hold.*

5.4   Applications

This section presents two applications of the exponential bound results. It includes tightness, moment bounds, and asymptotic normality.

Example 5.6.   Assume that the conditions of Theorem 5.1 hold. As an application of the exponential bounds, we can re-derive a central limit result. For simplicity, consider a fixed $i \in \mathcal{M}$, and set

$$N^\varepsilon = \frac{\sqrt{\varepsilon}}{\sqrt{T}} \sum_{k=0}^{T/\varepsilon - 1} (I_{\{\alpha_k^\varepsilon = i\}} - \nu^i),$$

where we have suppressed the i-dependence in N^ε.

Under the conditions of Theorem 5.1, N^ε converges in distribution to a normal random vector with mean 0 and covariance matrix $\sigma^2 I$, where

$$\sigma^2 = \nu^i \psi^{ii}(0) + 2\nu^i \sum_{k=1}^{\infty} \psi^{ii}(k),$$

with the $\psi^{ij}(k)$ being the initial layer correction term as given in Theorem 4.1. In view of Theorem 5.1, the characteristic function

$$G^\varepsilon(z) = E \exp(\mathrm{i} N^\varepsilon z)$$

exists for all $z \in \mathbb{R}$, where i is the imaginary number satisfying $\mathrm{i}^2 = -1$. Since P is irreducible, α_k^ε is a ϕ-mixing process with an exponential mixing rate. Using the mixing inequality, we can then verify $G^\varepsilon(z) \to \exp(-\sigma^2 z^2/2)$ as $\varepsilon \to 0$, and hence conclude that N^ε has a normal limit distribution. Note that the above example is a simple illustration only. In Chapter 4, by defining $n^\varepsilon(t) = n_k^\varepsilon$ for $t \in [\varepsilon k, \varepsilon k + \varepsilon)$, we showed that $n^\varepsilon(\cdot)$ converges weakly to a diffusion process or a switching diffusion process depending on whether P given in (5.1) is irreducible or has the form (5.2).

Example 5.7.   In production planning with long-run average cost problems, a finite-state Markov chain is often used to characterize the underlying machine capacity and demand rate processes. Suppose that $\{\alpha_k^\varepsilon\}$ is a function of the machine state process representing the production capacity, and let τ be the first time when the sum $\sum_{k=0}^{n} (\alpha_k^\varepsilon - z) = \mu_0$ for given z and μ_0, where z is a constant demand rate and μ_0 is a measure of the accumulative difference of the demand and the capacity. It is useful to provide estimates on the first and second moments of τ in terms of μ_0. The exponential bound obtained in Theorem 5.2 is crucial in deriving such finite moments; see Sethi et al. [134] for details.

5.5   Proofs of Results

Proof of Theorem 5.1. For notational convenience, let

$$\widehat{n}_k^\varepsilon = n_k^\varepsilon - \sqrt{\varepsilon}(\chi_0^\varepsilon - \nu) + \sqrt{\varepsilon}(\chi_k^\varepsilon - \nu). \tag{5.14}$$

Note that the χ_i is bounded, so are

$$\sqrt{\varepsilon}(\chi_0^\varepsilon - \nu) = O(\sqrt{\varepsilon}) \ \text{ and } \ \sqrt{\varepsilon}(\chi_k^\varepsilon - \nu) = O(\sqrt{\varepsilon}).$$

It suffices to show Theorem 5.1 with n_k^ε replaced by $\widehat{n}_k^\varepsilon$. The rest of the proof is divided into seven steps.

Step 1: (Verifying martingale property of a suitable sequence). Define

$$\begin{aligned}
\chi_k^\varepsilon &= (I_{\{\alpha_k^\varepsilon=1\}}, \ldots, I_{\{\alpha_k^\varepsilon=m_0\}}) \in \mathbb{R}^{1 \times m_0} \\
w_k^\varepsilon &= \chi_k^\varepsilon - \chi_0^\varepsilon - \sum_{l=0}^{k-1} \chi_l^\varepsilon(P + \varepsilon Q - I).
\end{aligned} \tag{5.15}$$

Denote $\Delta w_k^\varepsilon = w_{k+1}^\varepsilon - w_k^\varepsilon$. It is easily shown that

$$|\Delta w_k^\varepsilon| \le 1.$$

A simple calculation reveals that

$$\chi_{k+1}^\varepsilon = \chi_k^\varepsilon + \chi_k^\varepsilon(P + \varepsilon Q - I) + \Delta w_k^\varepsilon. \tag{5.16}$$

It can be verified that

$$w_k \overline{P} = \chi_k^\varepsilon \overline{P} - \chi_0^\varepsilon \overline{P} - \sum_{l=0}^{k-1} \chi_l^\varepsilon (P + \varepsilon Q - I)\overline{P} = 0. \tag{5.17}$$

Thus w_k is orthogonal to $\overline{P}$. Denote by $\mathcal{F}_k$ the σ-algebra generated by $\{\alpha_l^\varepsilon : l \le k\}$. We next show that Δw_k^ε is a martingale difference sequence with respect to $\mathcal{F}_k$. In fact, for each fixed $j \in \mathcal{M}$, by the Markov property,

$$\begin{aligned}
E(I_{\{\alpha_{k+1}^\varepsilon=j\}}|\mathcal{F}_k) &= E(I_{\{\alpha_{k+1}^\varepsilon=j\}}|\alpha_k^\varepsilon) \\
&= \sum_{i=1}^{m_0} I_{\{\alpha_k^\varepsilon=i\}} E(I_{\{\alpha_{k+1}^\varepsilon=j\}}|\alpha_k^\varepsilon = i) \\
&= \sum_{i=1}^{m_0} I_{\{\alpha_k^\varepsilon=i\}} p^{\varepsilon,ij},
\end{aligned}$$

so

$$\begin{aligned}
E(\chi_{k+1}^\varepsilon|\mathcal{F}_k) &= \left(\sum_{i=1}^{m_0} I_{\{\alpha_k^\varepsilon=i\}} p^{\varepsilon,i1}, \ldots, \sum_{i=1}^{m_0} I_{\{\alpha_k^\varepsilon=i\}} p^{\varepsilon,im_0} \right) \\
&= \chi_k^\varepsilon(P + \varepsilon Q).
\end{aligned}$$

As a result, it follows that

$$E(\Delta w_k | \mathcal{F}_k) = E(\chi^\varepsilon_{k+1} | \mathcal{F}_k) - \chi^\varepsilon_k(P + \varepsilon Q) = 0. \tag{5.18}$$

That is, the martingale difference property is verified.
Step 2: (Preliminary estimates). By virtue of (5.16),

$$\chi^\varepsilon_{k+1} = \chi^\varepsilon_0(P + \varepsilon Q)^{k+1} + \sum_{l=0}^{k} \Delta w^\varepsilon_l (P + \varepsilon Q)^{k-l}. \tag{5.19}$$

Noting that $\chi^\varepsilon_0 \overline{P} = \nu$ and the orthogonality of w_k and $\overline{P}$ given by (5.17), we arrive at

$$\chi^\varepsilon_{k+1} - \nu = \chi^\varepsilon_0[(P + \varepsilon Q)^{k+1} - \overline{P}] + \sum_{l=0}^{k} \Delta w^\varepsilon_l \{[(P + \varepsilon Q)^{k-l} - \overline{P}] + \overline{P}]\}$$

$$= \chi^\varepsilon_0 \rho_{k+1} + \sum_{l=0}^{k} \Delta w^\varepsilon_l \rho_{k-l},$$

$$\tag{5.20}$$

where

$$\rho_i = [(P + \varepsilon Q)^i - \overline{P}]. \tag{5.21}$$

Let $1 \le \kappa_\varepsilon \le T/\varepsilon$. It follows from (5.20),

$$\sum_{k=1}^{\kappa_\varepsilon} [\chi^\varepsilon_k - \nu]\beta_k = \sum_{k=0}^{\kappa_\varepsilon-1} [\chi^\varepsilon_{k+1} - \nu]\beta_{k+1}$$

$$= \sum_{k=0}^{\kappa_\varepsilon-1} \chi^\varepsilon_0 \rho_{k+1}\beta_{k+1} + \sum_{k=0}^{\kappa_\varepsilon-1} \left(\sum_{l=0}^{k} \Delta w^\varepsilon_l \rho_{k-l}\right)\beta_{k+1}. \tag{5.22}$$

Considering the last term in (5.22) and interchanging the order of summations, we obtain

$$\sum_{k=0}^{\kappa_\varepsilon-1} \left(\sum_{l=0}^{k} \Delta w^\varepsilon_l \rho_{k-l}\right)\beta_{k+1} = \sum_{l=0}^{\kappa_\varepsilon-1} \Delta w^\varepsilon_l \widetilde{\rho}_l,$$

where

$$\widetilde{\rho}_l = \sum_{l_1=0}^{\kappa_\varepsilon-1-l} \rho_{l_1}\beta_{l_1+l+1}.$$

Equation (5.4) implies that

$$|\widetilde{\rho}_l| \le K_T \left(T + \frac{1}{1-\lambda}\right)|\beta|_T, \quad \text{for all } l \le T/\varepsilon.$$

It follows from (5.22) that

$$|\widehat{n}^\varepsilon_{\kappa_\varepsilon}| = \sqrt{\varepsilon} \left| \sum_{k=0}^{\kappa_\varepsilon-1} (\chi^\varepsilon_{k+1} - \nu)\beta_{k+1} \right|$$

$$\leq \sqrt{\varepsilon} \sum_{k=0}^{\kappa_\varepsilon-1} |\chi^\varepsilon_0 \rho_{k+1}\beta_{k+1}| + \sqrt{\varepsilon} \left| \sum_{l=0}^{\kappa_\varepsilon-1} \Delta w^\varepsilon_l \tilde{\rho}_l \right|.$$

In view of (5.4), we have

$$\sum_{k=0}^{\kappa_\varepsilon-1} |\chi^\varepsilon_0 \rho_{k+1}\beta_{k+1}| \leq |\beta|_T K_T \sum_{k=0}^{\kappa_\varepsilon-1} (\varepsilon + \lambda^{k+1}) \tag{5.23}$$

$$\leq |\beta|_T K_T \left(T + \frac{\lambda}{1-\lambda} \right).$$

Thus,

$$\exp\left(\frac{c_T \sqrt{\varepsilon}}{(T+1)^{\frac{3}{2}}} \sum_{k=0}^{\kappa_\varepsilon-1} |\chi^\varepsilon_0 \rho_{k+1}\beta_{k+1}| \right)$$

$$\leq \exp\left(\frac{c_T \sqrt{\varepsilon}}{(T+1)^{\frac{3}{2}}} |\beta|_T K_T \left(T + \frac{\lambda}{1-\lambda} \right) \right) \leq e,$$

for ε small enough.

Step 3: (Exponential estimates). Let $d_k = (d^1_k, \ldots, d^{m_0}_k)$ with

$$d^i_k = \sum_{l=0}^{k-1} \left(\frac{c_T}{T+1} \right) \Delta w^{\varepsilon,i}_l \tilde{\rho}_l, \ i \in M,$$

where $\Delta w^{\varepsilon,i}_l$ denotes the ith component of Δw^ε_l. To complete the proof of inequality (5.4), it suffices to show

$$E \exp\left(\frac{\sqrt{\varepsilon}}{\sqrt{T+1}} |d_{\kappa_\varepsilon}| \right) \leq K. \tag{5.24}$$

Note that

$$E e^\xi \leq e + (e-1) \sum_{l=1}^{\infty} e^l P(\xi \geq l), \tag{5.25}$$

for any nonnegative random variable ξ. Moreover, for each $a > 0$, using the maximum norm, we have

$$P(|d_{\kappa_\varepsilon}| \geq a) = P\left(\bigcup_{i=1}^{m_0} |d^i_{\kappa_\varepsilon}| \geq a \right) \leq \sum_{i=1}^{m_0} P(|d^i_{\kappa_\varepsilon}| \geq a). \tag{5.26}$$

In view of these inequalities, (taking $\xi = (\sqrt{\varepsilon}/\sqrt{T+1})|d_{\kappa_\varepsilon}|$ in (5.25) and $a = l\sqrt{T+1}/\sqrt{\varepsilon}$ in (5.26)), it suffices to show

$$\sum_{l=1}^{\infty} e^l P\left(|d^i_{\kappa_\varepsilon}| \geq \frac{l\sqrt{T+1}}{\sqrt{\varepsilon}} \right) < \infty, \ \text{for each } i \in \{1, \ldots, m_0\}. \tag{5.27}$$

In fact, for $0 \leq k \leq T/\varepsilon$,

$$|\Delta d_k^i| = |d_{k+1}^i - d_k^i| = \left| \left(\frac{c_T}{T+1} \right) \Delta w_k^{\varepsilon,i} \tilde{\rho}_k \right| \leq 1.$$

Given $i \in \{1, \ldots, m_0\}$ and $\zeta > 0$, define

$$q_k = 1 + \zeta \sum_{l=0}^{k-1} q_l \Delta d_l^i, \quad q_0 = 1.$$

Note that ζ is a parameter to be chosen later to fit our needs. Then, $\{q_k\}$ is a martingale and $E q_k = 1$. Moreover,

$$q_{k+1} - q_k = \zeta q_k \Delta d_k^i.$$

It follows that

$$q_k = \prod_{l=0}^{k-1} (1 + \zeta \Delta d_l^i).$$

Using the condition $|\Delta d_l^i| \leq 1$, it is easy to show that

$$1 + (\Delta d_l^i)\zeta \geq \exp\left((\Delta d_l^i)\zeta - \overline{\kappa}\zeta^2 \right),$$

for some $\overline{\kappa} > 0$. This implies that

$$q_k \geq \prod_{l=0}^{k-1} \exp\left((\Delta d_l^i)\zeta - \overline{\kappa}\zeta^2 \right) = \exp\left(d_k^i \zeta - \overline{\kappa} k \zeta^2 \right), \quad \text{for } k \leq T/\varepsilon.$$

Note that

$$P\left(|d_{\kappa_\varepsilon}^i| \geq \frac{l\sqrt{T+1}}{\sqrt{\varepsilon}} \right)$$
$$= P\left(d_{\kappa_\varepsilon}^i \geq \frac{l\sqrt{T+1}}{\sqrt{\varepsilon}} \right) + P\left(-d_{\kappa_\varepsilon}^i \geq \frac{l\sqrt{T+1}}{\sqrt{\varepsilon}} \right).$$

We need only consider the first term because the second one can be treated similarly. We have

$$P\left(d_{\kappa_\varepsilon}^i \geq \frac{l\sqrt{T+1}}{\sqrt{\varepsilon}} \right) \leq P\left(q_{\kappa_\varepsilon} \geq \exp\left(\frac{l\sqrt{T+1}}{\sqrt{\varepsilon}}\zeta - \overline{\kappa}\kappa_\varepsilon\zeta^2 \right) \right)$$
$$\leq \exp\left(-\frac{l\sqrt{T+1}}{\sqrt{\varepsilon}}\zeta + \overline{\kappa}\kappa_\varepsilon\zeta^2 \right).$$

Now choose $\zeta = 2\sqrt{\varepsilon}/\sqrt{T+1}$. It follows that

$$-\left(\frac{l\sqrt{T+1}}{\sqrt{\varepsilon}} \right)\zeta + \overline{\kappa}\kappa_\varepsilon\zeta^2 \leq 4\overline{\kappa} - 2l.$$

Therefore,

$$\sum_{l=1}^{\infty} e^l e^{4\bar{\kappa}-2l} = e^{4\bar{\kappa}} \sum_{l=1}^{\infty} e^{-l} = \frac{e^{4\bar{\kappa}}}{e-1} < \infty,$$

which implies (5.27). Hence, (5.24) follows, so does (5.7). We next prove (5.8).

Step 4: (Near martingale property). We claim that for ε small enough,

$$|E[\widehat{n}_{t/\varepsilon}^{\varepsilon,i}|\mathcal{F}_{s/\varepsilon}] - \widehat{n}_{s/\varepsilon}^{\varepsilon,i}| \leq O(\sqrt{\varepsilon}), \text{ for all } \omega \in \Omega \text{ and } 0 \leq s \leq t \leq T. \quad (5.28)$$

Here $O(\sqrt{\varepsilon})$ is deterministic. To see this, note that for all $i_0 \in \mathcal{M}$,

$$E\left[\sum_{l=s/\varepsilon+1}^{t/\varepsilon} (I_{\{\alpha_l^\varepsilon=i\}} - \nu^i)\beta_l \middle| \alpha_{s/\varepsilon}^\varepsilon = i_0\right]$$

$$= \sum_{l=s/\varepsilon+1}^{t/\varepsilon} (E[I_{\{\alpha_l^\varepsilon=i\}}|\alpha_{s/\varepsilon}^\varepsilon = i_0] - \nu^i)\beta_l$$

$$= \sum_{l=s/\varepsilon+1}^{t/\varepsilon} (P(\alpha_l^\varepsilon = i|\alpha_{s/\varepsilon}^\varepsilon = i_0) - \nu^i)\beta_l$$

$$= \sum_{l=s/\varepsilon+1}^{t/\varepsilon} O(\varepsilon + \lambda^{l-s/\varepsilon}) = O(1).$$

So, (5.28) follows.

Step 5: (Near submartingale property). We show that, for each $a > 0$,

$$E[\exp(a|\widehat{n}_{t/\varepsilon}^{\varepsilon,i}|)|\mathcal{F}_{s/\varepsilon}] \geq \exp(a|\widehat{n}_{s/\varepsilon}^{\varepsilon,i}|)(1 + O(\sqrt{\varepsilon})), \quad (5.29)$$

for $0 \leq s \leq t \leq T$.

First of all, note that $\phi(x) = |x|$ is a convex function. Noting that $O(\sqrt{\varepsilon}) = -O(\sqrt{\varepsilon})$, we have

$$E[|\widehat{n}_{t/\varepsilon}^{\varepsilon,i}| \,|\mathcal{F}_{s/\varepsilon}] \geq |\widehat{n}_{s/\varepsilon}^{\varepsilon,i}| + \phi_+(\widehat{n}_{s/\varepsilon}^{\varepsilon,i})E[\widehat{n}_{t/\varepsilon}^{\varepsilon,i} - \widehat{n}_{s/\varepsilon}^{\varepsilon,i}|\mathcal{F}_{s/\varepsilon}]$$

$$\geq |\widehat{n}_{s/\varepsilon}^{\varepsilon,i}| - |\phi_+(\widehat{n}_{s/\varepsilon}^{\varepsilon,i})|E[|\widehat{n}_{t/\varepsilon}^{\varepsilon,i} - \widehat{n}_{s/\varepsilon}^{\varepsilon,i}||\mathcal{F}_{s/\varepsilon}]$$

$$= |\widehat{n}_{s/\varepsilon}^{\varepsilon,i}| - O(\sqrt{\varepsilon})$$

$$= |\widehat{n}_{s/\varepsilon}^{\varepsilon,i}| + O(\sqrt{\varepsilon}),$$

where $\phi_+(x)$ is the right-hand derivative of $\phi(x)$ and is bounded by 1.

Moreover, note that e^{ax} is also convex. It follows that

$$E[\exp(a|\widehat{n}^{\varepsilon,i}_{t/\varepsilon}|)|\mathcal{F}_{s/\varepsilon}]$$
$$\geq \exp(a|\widehat{n}^{\varepsilon,i}_{s/\varepsilon}|) + a\exp(a|\widehat{n}^{\varepsilon,i}_{s/\varepsilon}|)E[|\widehat{n}^{\varepsilon,i}_{t/\varepsilon}| - |\widehat{n}^{\varepsilon,i}_{s/\varepsilon}|\,|\mathcal{F}_{s/\varepsilon}]$$
$$\geq \exp(a|\widehat{n}^{\varepsilon,i}_{s/\varepsilon}|) - a\exp(a|\widehat{n}^{\varepsilon,i}_{s/\varepsilon}|)E[|\widehat{n}^{\varepsilon,i}_{t/\varepsilon} - \widehat{n}^{\varepsilon,i}_{s/\varepsilon}|\,|\mathcal{F}_{s/\varepsilon}]$$
$$= \exp(a|\widehat{n}^{\varepsilon,i}_{s/\varepsilon}|)(1 - O(\sqrt{\varepsilon}))$$
$$= \exp(a|\widehat{n}^{\varepsilon,i}_{s/\varepsilon}|)(1 + O(\sqrt{\varepsilon})).$$

Step 6: (Optional sampling). Let

$$x^{\varepsilon}_k = \exp(a|\widehat{n}^{\varepsilon,i}_k|) \quad \text{for} \quad a > 0. \tag{5.30}$$

Then, for any $\mathcal{F}_k$ stopping time $\tau \leq T/\varepsilon$,

$$E[x^{\varepsilon}_{T/\varepsilon}|\mathcal{F}_\tau] \geq x^{\varepsilon}_\tau(1 + O(\sqrt{\varepsilon})). \tag{5.31}$$

Note that $\tau \in \{0, 1, 2, \ldots, T/\varepsilon\}$. Taking $t = T$ and $s = \varepsilon k$ in (5.29), we have

$$E[x^{\varepsilon}_{T/\varepsilon}|\mathcal{F}_k] \geq x^{\varepsilon}_k(1 + O(\sqrt{\varepsilon})).$$

For all $A \in \mathcal{F}_\tau$, we have $A \cap \{\tau = k\} \in \mathcal{F}_k$. Therefore,

$$\int_{A \cap \{\tau=k\}} x^{\varepsilon}_{T/\varepsilon}dP \geq \left(\int_{A \cap \{\tau=k\}} x^{\varepsilon}_\tau dP\right)(1 + O(\sqrt{\varepsilon})).$$

Thus

$$\int_A x^{\varepsilon}_{T/\varepsilon}dP \geq \left(\int_A x^{\varepsilon}_\tau dP\right)(1 + O(\sqrt{\varepsilon})),$$

and (5.31) follows.

Step 7: (Submartingale inequality). Take $a = c_T/(T+1)^{\frac{3}{2}}$ in Step 6. Then for ε small enough, there exists K such that

$$P\left(\sup_{k \leq T/\varepsilon} x^{\varepsilon}_k \geq x\right) \leq \frac{K}{x}, \tag{5.32}$$

for all $x > 0$.

In fact, let $\tau = \min\{k : x^{\varepsilon}_k \geq x, \ k \leq T/\varepsilon\}$. We adopt the convention and take $\tau = \infty$ if $\{k : x^{\varepsilon}_k \geq x\} = \emptyset$. Then we have

$$Ex^{\varepsilon}_{T/\varepsilon} \geq (Ex^{\varepsilon}_{(T/\varepsilon)\wedge\tau})(1 + O(\sqrt{\varepsilon})),$$

and we can write

$$Ex^{\varepsilon}_{(T/\varepsilon)\wedge\tau} = Ex^{\varepsilon}_\tau I_{\{\tau \leq T/\varepsilon\}} + Ex^{\varepsilon}_{T/\varepsilon}I_{\{\tau > T/\varepsilon\}} \geq Ex^{\varepsilon}_\tau I_{\{\tau \leq T/\varepsilon\}}.$$

Moreover, in view of the definition of τ, we have

$$Ex_\tau^\varepsilon I_{\{\tau \le T/\varepsilon\}} \ge xP(\tau \le T/\varepsilon) = xP\left(\sup_{k \le T/\varepsilon} x_k^\varepsilon \ge x\right).$$

It follows that

$$P\left(\sup_{k \le T/\varepsilon} x_k^\varepsilon \ge x\right) \le \frac{Ex_{T/\varepsilon}^\varepsilon}{(1 + O(\sqrt{\varepsilon}))x} \le \frac{K}{x}.$$

Thus, (5.32) follows.

To complete the proof of (5.8), note that for $0 < \theta < 1$,

$$E\exp\left(\frac{\theta c_T}{(T+1)^{\frac{3}{2}}} \sup_{k \le T/\varepsilon} |\widehat{n}_{t/\varepsilon}^{\varepsilon,i}|\right) = E \sup_{k \le T/\varepsilon} (x_k^\varepsilon)^\theta.$$

It follows that

$$\begin{aligned}
E \sup_{k \le T/\varepsilon} (x_k^\varepsilon)^\theta &= \int_0^\infty P\left(\sup_{k \le T/\varepsilon} (x_k^\varepsilon)^\theta \ge x\right) dx \\
&\le 1 + \int_1^\infty P\left(\sup_{k \le T/\varepsilon} (x_k^\varepsilon)^\theta \ge x\right) dx \\
&\le 1 + \int_1^\infty P\left(\sup_{k \le T/\varepsilon} x_k^\varepsilon \ge x^{1/\theta}\right) dx \\
&\le 1 + \int_1^\infty Kx^{-1/\theta} dx < \infty.
\end{aligned}$$

The desired exponential bound is obtained. □

The following result is a direct consequence of Steps 1–3. It will be used in the proof of Theorem 5.2.

Lemma 5.8. *If, for any sequence $\{\rho_k\}$ satisfying $|\rho_k| \le K_T O_1(\varepsilon + \lambda^k)$, then for*

$$0 \le c_T \le \frac{1 - \lambda}{K_T(|\beta|_T + 1)},$$

and $1 \le \kappa_\varepsilon \le T/\varepsilon$, we have

$$E\exp\left(\frac{c_T\sqrt{\varepsilon}}{(T+1)^{\frac{3}{2}}}\left|\sum_{k=0}^{\kappa_\varepsilon - 1}\left(\sum_{l=0}^{k}\Delta w_l^\varepsilon \rho_{k-l}\right)\beta_{k+1}\right|\right) \le K.$$

Proof of Theorem 5.2. We only verify (5.10). The proof of (5.11) is similar to that of (5.8).

Define

$$\begin{aligned}
\chi_k^\varepsilon &= (I_{\{\alpha_k^\varepsilon = s_{11}\}}, \ldots, I_{\{\alpha_k^\varepsilon = s_{1m_1}\}}, \ldots, I_{\{\alpha_k^\varepsilon = s_{l_0 1}\}}, \ldots, I_{\{\alpha_k^\varepsilon = s_{l_0 m_{l_0}}\}}), \\
\widetilde{\chi}_k^\varepsilon &= \chi_k^\varepsilon \widetilde{\mathbb{1}}, \quad \widetilde{\mathbb{1}} = \text{diag}(\mathbb{1}_{m_1}, \ldots, \mathbb{1}_{m_{l_0}}), \\
\overline{\chi}_k^\varepsilon &= \widetilde{\chi}_k^\varepsilon \widehat{\nu}, \quad \widehat{\nu} = \text{diag}(\nu^1, \ldots, \nu^{l_0}),
\end{aligned}$$

$$(5.33)$$

and define Δw_k^ε and w_k^ε as in (5.15) and (5.16).

Using the definitions given in (5.33), we have

$$\chi_{k+1}^\varepsilon = \chi_0^\varepsilon (P + \varepsilon Q)^{k+1} + \sum_{l=0}^{k} \Delta w_l^\varepsilon (P + \varepsilon Q)^{k-l}. \tag{5.34}$$

Let

$$\eta_{k+1}^\varepsilon = \chi_0^\varepsilon [(P + \varepsilon Q)^{k+1} - \Phi(\varepsilon(k+1))]$$
$$+ \sum_{l=0}^{k} \Delta w_l^\varepsilon [(P + \varepsilon Q)^{k-l} - \Phi(\varepsilon(k-l))].$$

Then,

$$\chi_{k+1}^\varepsilon = \chi_0^\varepsilon \Phi(\varepsilon(k+1)) + \sum_{l=0}^{k} \Delta w_l^\varepsilon \Phi(\varepsilon(k-l)) + \eta_{k+1}^\varepsilon. \tag{5.35}$$

On the other hand, multiplying from the right on both sides of (5.35) by $\tilde{\mathbb{1}}$, using $\tilde{\chi}_k^\varepsilon = \chi_k^\varepsilon \tilde{\mathbb{1}}$, $\Phi(\varepsilon k)\tilde{\mathbb{1}} = \tilde{\mathbb{1}}\Theta(\varepsilon k)$ (see Theorem 4.1), and noting that $\tilde{\nu}\tilde{\mathbb{1}}$ equals to the $l_0 \times l_0$ identity matrix, we have

$$\tilde{\chi}_{k+1}^\varepsilon = \chi_0^\varepsilon \Phi(\varepsilon(k+1))\tilde{\mathbb{1}} + \sum_{l=0}^{k} \Delta w_l^\varepsilon \Phi(\varepsilon(k-l))\tilde{\mathbb{1}} + \eta_{k+1}^\varepsilon \tilde{\mathbb{1}}$$
$$= \chi_0^\varepsilon \tilde{\mathbb{1}}\Theta(\varepsilon(k+1)) + \sum_{l=0}^{k} \Delta w_l^\varepsilon \tilde{\mathbb{1}}\Theta(\varepsilon(k-l)) + \eta_{k+1}^\varepsilon \tilde{\mathbb{1}}.$$

Thus, we arrive at

$$\overline{\chi}_{k+1}^\varepsilon = \chi_0^\varepsilon \Phi(\varepsilon(k+1)) + \sum_{l=0}^{k} \Delta w_l^\varepsilon \Phi(\varepsilon(k-l)) + \eta_{k+1}^\varepsilon \tilde{\mathbb{1}}\tilde{\nu}. \tag{5.36}$$

Combining (5.35) and (5.36), we obtain

$$\chi_{k+1}^\varepsilon - \overline{\chi}_{k+1}^\varepsilon = \eta_{k+1}^\varepsilon (I - \tilde{\mathbb{1}}\tilde{\nu}). \tag{5.37}$$

In view of Theorem 4.1, we have

$$|(P + \varepsilon Q)^k - \Phi(\varepsilon k)| \leq K_T O_1(\varepsilon + \lambda^k), \text{ for } 0 \leq k \leq T/\varepsilon.$$

Let

$$\rho_k = ((P + \varepsilon Q)^k - \Phi(\varepsilon k))(I - \tilde{\mathbb{1}}\tilde{\nu}).$$

Then

$$|\rho_k| \leq K_T O_1(\varepsilon + \lambda^k).$$

We have

$$\left| \sum_{k=1}^{\kappa_\varepsilon} [\chi_k^\varepsilon - \overline{\chi}_k^\varepsilon] \beta_k \right| \leq \left| \sum_{k=0}^{\kappa_\varepsilon-1} \chi_0^\varepsilon \rho_{k+1} \beta_{k+1} \right| + \left| \sum_{k=0}^{\kappa_\varepsilon-1} \left(\sum_{l=0}^{k} \Delta w_l^\varepsilon \rho_{k-l} \right) \beta_{k+1} \right|.$$

Following Lemma 5.8 and (5.23), we obtain

$$\exp\left(\frac{\sqrt{\varepsilon} c_T}{(T+1)^{\frac{3}{2}}} \left| \sum_{k=1}^{\kappa_\varepsilon} [\chi_k^\varepsilon - \overline{\chi}_k^\varepsilon] \beta_k \right| \right) \leq K.$$

The desired result then follows. □

Proof of Corollary 5.3. By virtue of the Markov inequality and Theorem 5.2,

$$P\left(\sup_{k \leq T/\varepsilon} |n_k^\varepsilon| > K_\delta \right) \leq P\left(\sup_{k \leq T/\varepsilon} \exp\left(\frac{c_T}{(T+1)^{\frac{3}{2}}} |n_k^\varepsilon| \right) > \exp\left(\frac{c_T K_\delta}{(T+1)^{\frac{3}{2}}} \right) \right)$$

$$\leq \exp\left(-\frac{c_T K_\delta}{(T+1)^{\frac{3}{2}}} \right) E \sup_{k \leq T/\varepsilon} \exp\left(\frac{c_T}{(T+1)^{\frac{3}{2}}} |n_k^\varepsilon| \right).$$

Since

$$E \sup_{k \leq T/\varepsilon} \exp\left(\frac{c_T}{(T+1)^{\frac{3}{2}}} |n_k^\varepsilon| \right) \leq K,$$

we obtain that

$$P\left(\sup_{k \leq T/\varepsilon} |n_k^\varepsilon| > K_\delta \right) \leq K \exp\left(-\frac{c_T K_\delta}{(T+1)^{\frac{3}{2}}} \right). \tag{5.38}$$

To make the right-hand side of (5.38) be less than δ, it suffices to have

$$K_\delta > (T+1)^{\frac{3}{2}} \log(K/\delta)/c_T,$$

which yields (5.12). □

Proof of Corollary 5.4. To verify this assertion, note that for any $z \in \mathbb{R}$, we have $\exp(|z|) \geq |z|^l/l!$. Thus

$$\sup_{k \leq T/\varepsilon} |n_k^\varepsilon|^l \leq l! \sup_{k \leq T/\varepsilon} \exp\left(\frac{c_T}{(T+1)^{\frac{3}{2}}} |n_k^\varepsilon| \right) \frac{(T+1)^{\frac{3l}{2}}}{c_T^l}.$$

Taking expectation and using Theorem 5.2, we obtain (5.13). □

Proof of Theorem 5.5. In view of the developments in Chapter 3 and Chapter 4 for the case of inclusion of transient states, Theorem 4.14 is in force. That is, we also obtain the desired asymptotic expansions, the mean squares estimates of the occupation measures, and the weak convergence of the aggregated process. Define

$$\widetilde{\mathbb{1}}_* = \begin{pmatrix} \mathbb{1} & 0_{(m-m_*)\times m_*} \\ A_* & 0_{m_*\times m_*} \end{pmatrix},$$
$$\nu_* = \operatorname{diag}(\nu^1,\ldots,\nu^{l_0},0_{m_*\times m_*}), \tag{5.39}$$

$$\chi_k^\varepsilon = (I_{\{\alpha_k^\varepsilon=s_{11}\}},\ldots,I_{\{\alpha_k^\varepsilon=s_{1m_1}\}},\ldots,I_{\{\alpha_k^\varepsilon=s_{l_01}\}},\ldots,I_{\{\alpha_k^\varepsilon=s_{l_0m_{l_0}}\}},$$
$$I_{\{\alpha_k^\varepsilon=s_{*1}\}},\ldots,I_{\{\alpha_k^\varepsilon=s_{*m_*}\}}),$$
$$\widetilde{\chi}_k^\varepsilon = \chi_k^\varepsilon\widetilde{\mathbb{1}}_*, \qquad \overline{\chi}_k^\varepsilon = \widetilde{\chi}_k^\varepsilon\nu_*, \tag{5.40}$$

and define Δw_k^ε and w_k^ε as in the recurrent case.

To obtain the desired exponential upper bounds, recall the definitions (5.39) and (5.40). Using $v = (v^1,v^2,\ldots,v^{l_0},v^*)$, the notation of a partitioned vector, with $v^i \in \mathbb{R}^{1\times m_i}$, we can readily see that

$$\chi_k^\varepsilon = (I_k^\varepsilon\mathbb{1}_{m_1} + I_k^{\varepsilon,*}a^1,\ldots,I_k^{\varepsilon,l_0}\mathbb{1}_{m_{l_0}} + I_k^{\varepsilon,*}a^{l_0},0_{m_*}),$$
$$\overline{\chi}_k^\varepsilon = ((I_k^\varepsilon\mathbb{1}_{m_1} + I_k^{\varepsilon,*}a^1)\nu^1,\ldots,(I_k^{\varepsilon,l_0}\mathbb{1}_{m_{l_0}} + I_k^{\varepsilon,*}a^{l_0})\nu^{l_0},0_{m_*}),$$

where 0_{m_*} is an $\mathbb{R}^{1\times m_*}$ zero vector. As in (5.37), we can derive

$$\chi_{k+1}^\varepsilon - \overline{\chi}_{k+1}^\varepsilon = \eta_{k+1}^\varepsilon(I - \widetilde{\mathbb{1}}_*\nu_*), \tag{5.41}$$

where

$$\eta_{k+1}^\varepsilon = \chi_0^\varepsilon[(P^\varepsilon)^{k+1} - \Phi(\varepsilon(k+1))] + \sum_{l=0}^{k}\Delta w_l^\varepsilon[(P^\varepsilon)^{k-l} - \Phi(\varepsilon(k-l))].$$

The rest of the development follows from the same line of argument as that of Theorem 5.2. $\square$

5.6   Notes

This chapter is concerned with exponential-type bounds for sequences of scaled and centered occupation measures. The results are useful for the development of infinite horizon stochastic control problems and for the further study of hybrid systems involving singularly perturbed Markov chains. This chapter is based on our recent work in Zhang and Yin [183]. In dealing with continuous-time, two-time-scale Markov chains, exponential error bounds were obtained in Zhang and Yin [179]. Such bounds were used in developing asymptotic normality for the fast-varying irreducible generators. Related results on continuous-time problems can be found in Yin and Zhang [158, Chapter 7] and Yin, Zhang, and Liu [166].

6
Interim Summary and Extensions

6.1 Introduction

In the previous chapters, we have developed asymptotic expansions of probability vectors and transition matrices, obtained asymptotic distributions of scaled sequences of occupation measures, and their exponential error bounds. The remaining chapters will cover control, stability, filtering, and stochastic approximation problems arising from numerous applications. To be more specific, we will study stability of Markov-modulated dynamic systems, Markov decision processes, linear quadratic regulators, hybrid filtering, mean variance control, production planning, and stochastic approximations. We will make use of the results obtained thus far. This chapter serves as a link between the theory covered in the previous chapters and applications to be presented in the following chapters. It is an interlude and provides a self-contained summary and extensions to models with nonstationary Markov chains. A reader who is mainly interested in applications may skip Chapters 3–5 and utilize this chapter as a user's guide.

The rest of the chapter is arranged as follows. In the next section, we recapture the basic models and conditions. Section 6.3 deals with chains with recurrent states. Section 6.4 takes care of the case when transient states are also included in addition to the recurrent classes of states. Section 6.5 discusses the case where the transition matrices are time dependent. Throughout the chapter, the conditions needed along with the results (in the format of propositions) will be presented; the proofs, having been presented essentially in Chapters 3–5, will not be repeated, however.

6.2   The Model

Suppose T is a positive number, $\varepsilon > 0$ is small parameter, and α_k^ε, for $k = 0, 1, \ldots, \lfloor T/\varepsilon \rfloor$, is a discrete-time Markov chain with finite state space $\mathcal{M} = \{1, \ldots, m_0\}$, where $\lfloor z \rfloor$ denotes the integer part of a real number z. For notational simplicity, we often simply write T/ε in lieu of $\lfloor T/\varepsilon \rfloor$.

Suppose the transition probability matrix P^ε of α_k^ε is given by

$$P^\varepsilon = P + \varepsilon Q, \tag{6.1}$$

where P itself is a transition matrix and $Q = (q^{ij})$ is a generator of a continuous-time Markov chain, i.e., for each $i \neq j$, $q^{ij} \geq 0$ and for each i, $\sum_j q^{ij} = 0$ or $Q\mathbb{1}_{m_0} = 0$. Recall that $\mathbb{1}_l \in \mathbb{R}^{l \times 1}$ denotes a column vector with all components being 1.

In (6.1), P is the dominating force for ε sufficiently small. In accordance with the well-known Kolmogorov classification of states, in a finite-state Markov chain, there is at least one recurrent state, and either all states are recurrent or there is also a collection of transient states in addition to the recurrent states. Furthermore, any transition probability matrix of a finite-state Markov chain can be put into the form of (see, for example, Iosifescu [73, p. 94]) either

$$P = \operatorname{diag}(P^1, \ldots, P^{l_0}) = \begin{pmatrix} P^1 & & & \\ & P^2 & & \\ & & \ddots & \\ & & & P^{l_0} \end{pmatrix}, \tag{6.2}$$

or

$$P = \begin{pmatrix} P^1 & & & & \\ & P^2 & & & \\ & & \ddots & & \\ & & & P^{l_0} & \\ P^{*,1} & P^{*,2} & \cdots & P^{*,l_0} & P^* \end{pmatrix}, \tag{6.3}$$

where for each $i = 1, \ldots, l_0$, P^i is a transition matrix within the ith recurrent class $\mathcal{M}_i = \{s_{i1}, \ldots, s_{im_i}\}$; and the last row $(P^{*,1}, \ldots, P^{*,l_0}, P^*)$ in (6.3) corresponds to the transient states $\mathcal{M}_* = \{s_{*1}, \ldots, s_{*m_*}\}$. A Markov chain with transition matrix given by (6.2) consists of l_0 recurrent classes, whereas a Markov chain with transition matrix (6.3) has l_0 recurrent classes plus m_* transient states. In the next section, we concentrate on the transition matrix P^ε with P specified by (6.2). Then in Section 6.4, we treat certain models with P given by (6.3); in Section 6.5, we consider time-inhomogeneous chains having transition matrices $P^\varepsilon(\varepsilon k) = P(\varepsilon k) + \varepsilon Q(\varepsilon k)$ with $P(\varepsilon k)$ having the partitioned form (6.3).

Remark 6.1. To use the two-time-scale approach in practice, one often needs to convert a given transition probability matrix P into the form of

(6.1). In this connection, an example was given in Section 1.1; see also the algorithm of conversion to canonical form for continuous-time generators in Yin and Zhang [158, Section 3.6]. The reduction to canonical form may also be derived from the decomposition outlined by Avramovic; see Phillips and Kokotovic [125] for more details. Such a decomposition of P may be non-unique. Nevertheless, this will not affect the applicability of the approach because the non-uniqueness only leads to more than one near-optimal solution in the context of control and optimization and any near-optimal control will be good enough for practical purposes.

6.3 Recurrent Chains

Suppose that α_k^ε has transition probabilities (6.1) with P having a block diagonal form (6.2). For each $k = 0, 1, \ldots, T/\varepsilon$, the probability vector

$$p_k^\varepsilon = (P(\alpha_k^\varepsilon = 1), \ldots, P(\alpha_k^\varepsilon = m_0)) \in \mathbb{R}^{1 \times m_0}$$

satisfies the vector-valued difference equation

$$p_{k+1}^\varepsilon = p_k^\varepsilon P^\varepsilon, \quad p_0^\varepsilon = p_0, \tag{6.4}$$

where p_0 is the initial probability distribution, i.e., $p_0^i \geq 0$ and $p_0 \mathbb{1}_{m_0} = \sum_{i=1}^{m_0} p_0^i = 1$, which is assumed to be independent of ε. To proceed, we make the following assumption.

(HR) P^ε and P are transition probability matrices. Moreover, for $i = 1, \ldots, l_0$, P^i are irreducible and aperiodic transition probability matrices.

Here, HR stands for "homogeneous and recurrent." Let ν^i be the stationary distribution associated with P^i. That is, $\nu^i = (\nu^{i1}, \ldots, \nu^{im_i})$ is the only positive solution to

$$\nu^i = \nu^i P^i \quad \text{and} \quad \nu^i \mathbb{1}_{m_i} = 1, \quad \text{for} \ i = 1, \ldots, l_0.$$

We next define an aggregated process. Given $k \leq T/\varepsilon$ and $i = 1, \ldots, l_0$, define $\overline{\alpha}_k^\varepsilon = i$ if $\alpha_k^\varepsilon \in \mathcal{M}_i$. Moreover, for $k \leq T/\varepsilon$, $i = 1, \ldots, l_0$, and $j = 1, \ldots, m_i$, define a sequence of occupation measures

$$\pi_k^{\varepsilon, ij} = \varepsilon \sum_{l=0}^{k-1} \left(I_{\{\alpha_l^\varepsilon = s_{ij}\}} - \nu^{ij} I_{\{\overline{\alpha}_l^\varepsilon = i\}} \right),$$
$$\pi_k^\varepsilon = \left(\pi_k^{\varepsilon, 11}, \ldots, \pi_k^{\varepsilon, 1m_1}, \ldots, \pi_k^{\varepsilon, l_0 1}, \ldots, \pi_k^{\varepsilon, l_0 m_{l_0}} \right). \tag{6.5}$$

Furthermore, define $\overline{\alpha}^\varepsilon(t)$ to be the continuous-time interpolation of α_k^ε. That is,

$$\overline{\alpha}^\varepsilon(t) = \overline{\alpha}_k^\varepsilon \quad \text{for} \ t \in [\varepsilon k, \varepsilon k + \varepsilon).$$

Let

$$\overline{Q} = \text{diag}(\nu^1, \ldots, \nu^{l_0})Q\widetilde{\mathbb{1}}, \tag{6.6}$$

where $\widetilde{\mathbb{1}} = \text{diag}(\mathbb{1}_{m_1}, \ldots, \mathbb{1}_{m_{l_0}})$.

Proposition 6.2. *Assume condition* (HR). *Then the following assertions hold:*

(a) *For the probability distribution vector* p_k^ε, *we have*

$$p_k^\varepsilon = \theta(\varepsilon k)\text{diag}(\nu^1, \ldots, \nu^{l_0}) + O(\varepsilon + \lambda^k) \tag{6.7}$$

for some λ *with* $0 < \lambda < 1$, *where* $\theta(t) = (\theta^1(t), \ldots, \theta^{l_0}(t)) \in \mathbb{R}^{1 \times l_0}$ *satisfies*

$$\frac{d\theta(t)}{dt} = \theta(t)\overline{Q}, \ \theta^i(0) = p_0^i \mathbb{1}_{m_i}.$$

(b) *For* $k \leq T/\varepsilon$, *the* k-*step transition probability matrix* $(P^\varepsilon)^k$ *satisfies*

$$(P^\varepsilon)^k = \Phi(\varepsilon k) + \varepsilon\widehat{\Phi}(\varepsilon k) + \Psi(k) + \varepsilon\widehat{\Psi}(k) + O\left(\varepsilon^2\right), \tag{6.8}$$

where

$$\Phi(t) = \widetilde{\mathbb{1}}\Theta(t)\text{diag}(\nu^1, \ldots, \nu^{l_0}),$$
$$\frac{d\Theta(t)}{dt} = \Theta(t)\overline{Q}, \ \Theta(0) = I. \tag{6.9}$$

Moreover, $\Phi(t)$ *and* $\widehat{\Phi}(t)$ *are uniformly bounded in* $[0, T]$ *and* $\Psi(k)$ *and* $\widehat{\Psi}(k)$ *decay exponentially fast, i.e.,* $|\Psi(k)| + |\widehat{\Psi}(k)| = O(\lambda^k)$ *for some* $0 < \lambda < 1$.

(c) *For* $i = 1, \ldots, l_0, \ j = 1, \ldots, m_i$,

$$\sup_{0 \leq k \leq T/\varepsilon} E|\pi_k^{\varepsilon, ij}|^2 = O(\varepsilon).$$

(d) $\overline{\alpha}^\varepsilon(\cdot)$ *converges weakly to* $\overline{\alpha}(\cdot)$, *which is a continuous-time Markov chain with generator* $\overline{Q}$.

Parts (a) and (b) were proved in Theorems 3.10 and 3.11 and Parts (c) and (d) were obtained in Theorems 4.5 and 4.3, respectively.

Remark 6.3. Using the exponential error bound estimate in (6.12), one can obtain a stronger version of Part (c) as follows:

$$E \sup_{0 \leq k \leq T/\varepsilon} |\pi_k^{\varepsilon, ij}|^2 = O(\varepsilon).$$

We next consider the scaled occupation measures and their asymptotic distribution and exponential bounds. Let

$$n_k^\varepsilon = (n_k^{\varepsilon,11}, \ldots, n_k^{\varepsilon,1m_1}, \ldots, n_k^{\varepsilon,l_0 1}, \ldots, n_k^{\varepsilon,l_0 m_{l_0}})$$

denote a sequence of the scaled occupation measures with

$$n_k^{\varepsilon,ij} = \frac{1}{\sqrt{\varepsilon}} \pi_k^{\varepsilon,ij}.$$

Define a sequence of piecewise constant functions by

$$n^\varepsilon(t) = n_k^\varepsilon \text{ for } t \in [\varepsilon k, \varepsilon k + \varepsilon) \text{ and } \pi^\varepsilon(t) = \pi_k^\varepsilon \text{ for } t \in [\varepsilon k, \varepsilon k + \varepsilon).$$

Note that $n^\varepsilon(t) = \pi^\varepsilon(t)/\sqrt{\varepsilon}$.

For each $i = 1, \ldots, l_0$, let

$$A(i) = \nu^{i,\mathrm{diag}} \sum_{k=0}^{\infty} \Psi(k,i) + \sum_{k=0}^{\infty} \Psi'(k,i)\nu^{i,\mathrm{diag}}, \qquad (6.10)$$

where $\nu^{i,\mathrm{diag}} = \mathrm{diag}(\nu^{11}, \ldots, \nu^{im_i}) \in \mathbb{R}^{m_i \times m_i}$ and

$$\Psi(k+1,i) = \Psi(k,i)P^i,$$

$$\Psi(0,i) = I - \mathbb{1}_{m_i}\nu^i.$$

Then, $A(i) = (a^{j_1 j_2}(i))$ is symmetric and nonnegative definite.

Proposition 6.4. *Assume condition* (HR). *Then the following assertions hold.*

(a) $(n^\varepsilon(\cdot), \overline{\alpha}^\varepsilon(\cdot))$ *converges weakly to* $(n(\cdot), \overline{\alpha}(\cdot))$ *such that the limit is the unique solution of the martingale problem with operator*

$$\mathcal{L}f(x,i) = \frac{1}{2} \sum_{j_1=1}^{m_i} \sum_{j_2=1}^{m_i} a^{j_1 j_2}(i) \frac{\partial^2 f(x,i)}{\partial x^{ij_1} \partial x^{ij_2}} + \overline{Q}f(x,\cdot)(i),$$

for $i = 1, \ldots, l_0$.

(b) *Let* c_T *be a constant such that*

$$0 \le c_T \le \frac{1-\lambda}{2K_T},$$

where

$$K_T = \sup_{\varepsilon,k} \frac{(P^\varepsilon)^k - \Phi(\varepsilon k)}{\varepsilon + \lambda^k}.$$

Then there exist an $\varepsilon_0 > 0$ and a constant K such that for all $0 \leq \varepsilon \leq \varepsilon_0$ and for $1 \leq k \leq T/\varepsilon$,

$$E \exp\left(\frac{c_T}{(T+1)^{\frac{3}{2}}} |n_k^\varepsilon|\right) \leq K. \tag{6.11}$$

Moreover, for any $0 < \theta < 1$, there exist an $\varepsilon_0 > 0$ and a constant K such that for all $0 \leq \varepsilon \leq \varepsilon_0$, we have

$$E \sup_{1 \leq k \leq T/\varepsilon} \exp\left(\frac{\theta c_T}{(T+1)^{\frac{3}{2}}} |n_k^\varepsilon|\right) \leq K. \tag{6.12}$$

Furthermore, let $\beta_k = (\beta_k^{11}, \ldots, \beta_k^{1m_1}, \ldots, \beta_k^{l_0 1}, \ldots, \beta_k^{l_0 m_{l_0}})$. Then both (6.11) and (6.12) hold with $n_k^{\varepsilon, ij}$ replaced by $n_k^{\varepsilon, ij} \beta_k^{ij}$ for c_T satisfying

$$0 \leq c_T \leq \frac{1 - \lambda}{K_T(|\beta|_T + 1)}.$$

Part (a) was derived in Chapter 4 and Part (b) was proved in Chapter 5.

6.4   Inclusion of Transient States

In this section, we treat the model with transient states in addition to the l_0 ergodic classes. We consider the Markov chain with transition matrix P^ε with P given by (6.3). This together with the result of the last section takes care of most of the practical concerns of finite-state Markov chains. To proceed, we modify condition (HR) slightly.

(HT) P^ε, P, and P^i for $i = 1, \ldots, l_0$ are transition probability matrices such that each P^i is irreducible and aperiodic. Moreover, all the eigenvalues of P^* are inside the unit circle.

Under condition (HT), the matrix $P^* - I$ is invertible. Let

$$a^i = -(P^* - I)^{-1} P^{*,i} \mathbb{1}_{m_i}, \text{ for } i = 1, \ldots, l_0, \tag{6.13}$$

and $A_* = (a^1, \ldots, a^{l_0})$. Define

$$\widetilde{\mathbb{1}}_* = \begin{pmatrix} \widetilde{\mathbb{1}} & 0_{(m-m_*) \times m_*} \\ A_* & 0_{m_* \times m_*} \end{pmatrix},$$

$$\nu_* = \text{diag}(\nu^1, \ldots, \nu^{l_0}, 0_{m_* \times m_*}). \tag{6.14}$$

Partition the matrix Q as

$$Q = \begin{pmatrix} Q^{11} & Q^{12} \\ Q^{21} & Q^{22} \end{pmatrix}, \tag{6.15}$$

where

$$Q^{11} \in \mathbb{R}^{(m-m_*)\times(m-m_*)}, \ Q^{12} \in \mathbb{R}^{(m-m_*)\times m_*},$$
$$Q^{21} \in \mathbb{R}^{m_*\times(m-m_*)}, \ \text{and} \ Q^{22} \in \mathbb{R}^{m_*\times m_*}.$$

Write

$$\overline{Q}_* = \text{diag}(\nu^1, \ldots, \nu^{l_0})(Q^{11}\widetilde{\mathbb{1}} + Q^{12}A_*),$$
$$\overline{G} = \nu_* Q\widetilde{\mathbb{1}}_* = \text{diag}(\overline{Q}_*, 0_{m_*\times m_*}). \tag{6.16}$$

Define sequences of scaled occupation measures by

$$\pi_k^{\varepsilon,ij} = \begin{cases} \varepsilon \sum_{l=0}^{k-1} (I_{\{\alpha_l^\varepsilon=s_{ij}\}} - \nu^{ij}I_{\{\overline{\alpha}_l^\varepsilon=i\}}), & \text{for } i = 1, \ldots, l_0, \\ & \text{and } j = 1, \ldots, m_i, \\ \varepsilon \sum_{l=0}^{k-1} I_{\{\alpha_l^\varepsilon=s_{*j}\}}, & \text{for } i = *, \\ & \text{and } j = 1, \ldots, m_*. \end{cases} \tag{6.17}$$

Consider an aggregated process given by

$$\overline{\alpha}_k^\varepsilon = \begin{cases} i, & \text{if } \alpha_k^\varepsilon \in \mathcal{M}_i, \\ U_j, & \text{if } \alpha_k^\varepsilon = s_{*j}, \end{cases} \tag{6.18}$$

where U_j is given by

$$U_j = I_{\{0 \le U \le a^{1,j}\}} + 2I_{\{a^{1,j} < U \le a^{1,j}+a^{2,j}\}} + \cdots + l_0 I_{\{a^{1,j}+\cdots+a^{l_0-1,j} < U \le 1\}},$$

and U is a random variable uniformly distributed on $[0,1]$, independent of α_k^ε. Define $\overline{\alpha}^\varepsilon(t) = \overline{\alpha}_k^\varepsilon$ for $t \in [\varepsilon k, \varepsilon k + \varepsilon)$.

Proposition 6.5. *Under* (HT), *the following assertions hold:*

(a) $p_k^\varepsilon = (\theta(\varepsilon k)\text{diag}(\nu^1, \ldots, \nu^{l_0}), 0'_{m_*}) + O(\varepsilon + \lambda^k)$, *where* $0'_{m_*} \in \mathbb{R}^{1\times m_*}$ *and* $\theta(t) = (\theta^1(t), \ldots, \theta^{l_0}(t)) \in \mathbb{R}^{1\times l_0}$ *satisfies*

$$\frac{d\theta(t)}{dt} = \theta(t)\overline{Q}_*, \quad \theta^i(0) = p^i(0)\mathbb{1}_{m_i} + p^*(0)a^i.$$

(b) *The k-step transition matrix satisfies*

$$(P^\varepsilon)^k = \Phi(\varepsilon k) + \Psi(k) + \varepsilon\widehat{\Phi}(\varepsilon k) + \varepsilon\widehat{\Psi}(k) + O(\varepsilon^2), \ \text{for } k \le T/\varepsilon,$$

for some λ with $0 < \lambda < 1$, where

$$\Phi(t) = \widetilde{\mathbb{1}}_*\Theta_*(t)\nu_* \ \text{with} \ \Theta_*(t) = \text{diag}(\Theta(t), I_{m_*\times m_*}), \tag{6.19}$$

and $\Theta(t) = (\theta^{ij}(t))$ *satisfies the differential equation*

$$\frac{d\Theta(t)}{dt} = \Theta(t)\overline{Q}_*, \quad \Theta(0) = I.$$

(c) *For each* $j = 1, \ldots, m_i$,

$$\sup_{0 \le k \in T/\varepsilon} E|\pi_k^{\varepsilon,ij}|^2 = \begin{cases} O(\varepsilon), & \text{for } i = 1, \ldots, l_0, \\ O(\varepsilon^2), & \text{for } i = *. \end{cases}$$

(d) $\overline{\alpha}^\varepsilon(\cdot)$ *converges weakly to* $\overline{\alpha}(\cdot)$, *which is a continuous-time Markov chain with state space* $\{1, \ldots, l_0\}$ *and generator* $\overline{Q}_*$.

Note that the error bound $O(\varepsilon + \lambda^k)$ depends on T. Such dependence can be relaxed by allowing k to be unbounded. The following lemma is an extension of this kind, which will be used in Chapter 9.

Proposition 6.6. *Assume condition* (HT). *Then there exist positive constants* K *and* $0 < \lambda < 1$ *(both independent of* ε *and* k*) such that for* $j = 1, \ldots, m_i$, $i = 1, \ldots, l_0$,

$$\left| P(\alpha_k^\varepsilon = s_{ij}) - \nu^{ij} \theta^i(\varepsilon k) \right| \le K(k\varepsilon^2 + \varepsilon + \lambda^k), \tag{6.20}$$

and for $j = 1, \ldots, m_*$,

$$P(\alpha_k^\varepsilon = s_{*j}) \le K(k\varepsilon^2 + \varepsilon + \lambda^k), \tag{6.21}$$

where $\theta^i(\varepsilon k)$, $i = 1, \ldots, l_0$, *can be obtained from the solution* $\theta^i(t)$ *of the following system of differential equations*

$$\begin{cases} \dfrac{d}{dt} \left(\theta^1(t), \ldots, \theta^{l_0}(t) \right) = \left(\theta^1(t), \ldots, \theta^{l_0}(t) \right) \overline{Q}_*, \\ \theta^i(0) = p_0^i \mathbb{1}_{m_i} + p_0^* a^i. \end{cases} \tag{6.22}$$

Let $\beta_k = \left(\beta_k^{11}, \ldots, \beta_k^{1m_1}, \ldots, \beta_k^{l_0 1}, \ldots, \beta_k^{l_0 m_{l_0}}, \beta_k^{*1}, \ldots, \beta_k^{*m_*} \right)$. Define

$$n_k^{\varepsilon,ij} = \frac{1}{\sqrt{\varepsilon}} \pi_k^{\varepsilon,ij} \beta_k^{ij}.$$

We obtain the exponential bounds of the scaled occupation measures when transient states are included.

Proposition 6.7. *Assume condition* (HT). *Let* c_T *be a constant such that*

$$0 \le c_T \le \frac{1 - \lambda}{K_T(|\beta|_T + 1)},$$

where

$$K_T = \sup_{\varepsilon, k} \frac{(P^\varepsilon)^k - \Phi(\varepsilon k)}{\varepsilon + \lambda^k}.$$

Then there exists an $\varepsilon_0 > 0$ *and a constant* K *such that for all* $0 \le \varepsilon \le \varepsilon_0$ *and for* $1 \le k \le T/\varepsilon$,

$$E \exp\left(\frac{c_T}{(T+1)^{\frac{3}{2}}} |n_k^\varepsilon| \right) \le K. \tag{6.23}$$

Moreover, for any $0 < \theta < 1$, there exists an $\varepsilon_0 > 0$ and a constant K such that for all $0 \leq \varepsilon \leq \varepsilon_0$, we have

$$E \sup_{1 \leq k \leq T/\varepsilon} \exp\left(\frac{\theta c_T}{(T+1)^{\frac{3}{2}}}|n_k^\varepsilon|\right) \leq K. \qquad (6.24)$$

6.5   Nonstationary Markov Chains

Let α_k^ε be a nonstationary Markov chain with a time-dependent transition matrix P^ε, i.e.,

$$p_{k+1}^\varepsilon = p_k^\varepsilon P^\varepsilon(k).$$

Given $P(t)$ and $Q(t)$ for $0 \leq t \leq T$, assume $P^\varepsilon(k)$ is of the form

$$P^\varepsilon(k) = P(\varepsilon k) + \varepsilon Q(\varepsilon k), \qquad (6.25)$$

with

$$P(\varepsilon k) = \begin{pmatrix} P^1(\varepsilon k) & & & & \\ & P^2(\varepsilon k) & & & \\ & & \ddots & & \\ & & & P^{l_0}(\varepsilon k) & \\ P^{*,1}(\varepsilon k) & P^{*,2}(\varepsilon k) & \cdots & P^{*,l_0}(\varepsilon k) & P^*(\varepsilon k) \end{pmatrix}.$$

We assume the following conditions.

(NH1) For each $t \in [0, T]$, $P^\varepsilon(t)$, $P(t)$, and $P^i(t)$ for $i = 1, \ldots, l_0$ are transition probability matrices such that each $P^i(t)$ is irreducible and aperiodic. There exists an $m_* \times m_*$ nonsingular matrix $B(t)$ and constant matrices P^* and $P^{*,i}$ satisfying

$$P^*(t) - I = B(t)(P^* - I) \text{ and } P^{*,i}(t) = B(t)P^{*,i},$$

for $i = 1, \ldots, l_0$. All the eigenvalues of $P^*(t)$ are inside the unit circle.

(NH2) On $[0, T]$, the matrix-valued function $P(t)$ is twice continuously differentiable and $Q(t)$ is Lipschitz continuous.

It follows from condition (NH1) and (NH2), that the following quantities

$$a^i(t) = -(P^*(t) - I)^{-1} P^{*,i}(t) \mathbb{1}_{m_i}$$
$$= -(P^* - I) P^{*,i} \mathbb{1}_{m_i} = a^i, \quad \text{for } i = 1, \ldots, l_0,$$

are independent of time with $a^i(t) = a^i$. Define $A_* = (a^1, \ldots, a^{l_0}) \in \mathbb{R}^{m_* \times l_0}$. As in (6.15), partition the matrix $Q(t)$ so that

$$Q(t) = \begin{pmatrix} Q^{11}(t) & Q^{12}(t) \\ Q^{21}(t) & Q^{22}(t) \end{pmatrix}. \tag{6.26}$$

Write

$$\begin{aligned}
\overline{Q}_*(t) &= \mathrm{diag}(\nu^1(t), \ldots, \nu^{l_0}(t))(Q^{11}(t)\tilde{\mathbb{1}} + Q^{12}(t)A_*), \\
\overline{G}(t) &= \nu_*(t)Q(t)\tilde{\mathbb{1}}_* = \mathrm{diag}(\overline{Q}_*(t), 0_{m_* \times m_*}),
\end{aligned} \tag{6.27}$$

where $\nu_*(t) = \mathrm{diag}(\nu^1(t), \ldots, \nu^{l_0}(t), 0_{m_* \times m_*})$. Define sequences of centered occupation measures by

$$\pi_k^{\varepsilon, ij} = \begin{cases} \varepsilon \sum_{l=0}^{k-1} (I_{\{\alpha_l^\varepsilon = s_{ij}\}} - \nu^{ij}(\varepsilon l)I_{\{\overline{\alpha}_l^\varepsilon = i\}}), & \text{for } i = 1, \ldots, l_0, \\ & \text{and } j = 1, \ldots, m_i, \\ \varepsilon \sum_{l=0}^{k-1} I_{\{\alpha_l^\varepsilon = s_{*j}\}}, & \text{for } i = *. \\ & \text{and } j = 1, \ldots, m_*. \end{cases} \tag{6.28}$$

Define the corresponding aggregated process $\overline{\alpha}_k^\varepsilon$ as in (6.18). Note that the k-step transition probability matrix $P^\varepsilon(k, j)$ for $0 \leq j \leq k$ takes the form

$$P^\varepsilon(k, j) = P(\varepsilon j)P(\varepsilon(j+1)) \cdots P(\varepsilon k),$$
$$P^\varepsilon(k+1, k) = I.$$

The following results can be derived as in Chapters 3 and 4; see also Yin, Zhang, and Badowski [165].

Proposition 6.8. *Under* (NH1) *and* (NH2), *the following assertions hold:*

(a) $p_k^\varepsilon = (\theta(\varepsilon k)\mathrm{diag}(\nu^1(\varepsilon k), \ldots, \nu^{l_0}(\varepsilon k)), 0'_{m_*}) + O(\varepsilon + \lambda^k)$, *where* $0'_{m_*} \in \mathbb{R}^{1 \times m_*}$ *and* $\theta(t) = (\theta^1(t), \ldots, \theta^{l_0}(t)) \in \mathbb{R}^{1 \times l_0}$ *satisfies*

$$\frac{d\theta(t)}{dt} = \theta(t)\overline{Q}_*(t), \quad \theta^i(0) = p^i(0)\mathbb{1}_{m_i} - p^*(0)a^i.$$

(b) *For* $k \leq T/\varepsilon$, *the transition matrix satisfies*

$$P^\varepsilon(k, k_0) = \Phi(\varepsilon k, \varepsilon k_0) + \Psi(k, k_0) + \varepsilon\widehat{\Phi}(\varepsilon k, \varepsilon k_0) + \varepsilon\widehat{\Psi}(k, k_0) + O(\varepsilon^2),$$

for some λ *with* $0 < \lambda < 1$, *where*

$$\Phi(t, t_0) = \tilde{\mathbb{1}}_* \Theta_*(t, t_0)\nu_*(t) \quad \text{with} \quad \Theta_*(t, t_0) = \mathrm{diag}(\Theta(t, t_0), I_{m_* \times m_*}), \tag{6.29}$$

where $\Theta(t, t_0) = (\theta^{ij}(t, t_0))$ *satisfies the differential equation*

$$\frac{\partial\Theta(t, t_0)}{\partial t} = \Theta(t, t_0)\overline{Q}_*(t), \quad \Theta(t_0, t_0) = I.$$

(c) *For each* $j = 1, \ldots, m_i$,

$$\sup_{0 \le k \in T/\varepsilon} E|\pi_k^{\varepsilon,ij}|^2 = \begin{cases} O(\varepsilon), & for\ i = 1, \ldots, l_0, \\ O(\varepsilon^2), & for\ i = *. \end{cases}$$

(d) $\overline{\alpha}^\varepsilon(\cdot)$ *converges weakly to* $\overline{\alpha}(\cdot)$, *which is a continuous-time Markov chain having generator* $\overline{Q}_*(t)$.

(e) *Proposition 6.7 holds with*

$$K_T = \sup_{\varepsilon,k,k_0} \frac{P^\varepsilon(k,k_0) - \Phi(\varepsilon k, \varepsilon k_0)}{\varepsilon + \lambda^{k-k_0}}.$$

6.6 Notes

Structures of transition probability matrices of the forms (6.2) and (6.3) for finite-state Markov chains are in Iosifescu [73], among others. For results concerning two-time-scale Markov chains with nonstationary transition probabilities, see Yin, Zhang, and Badowski [165]. The continuous-time counterparts can be found in Yin and Zhang [158].

Part III

Applications

7

Stability of Dynamic Systems

7.1 Introduction

This chapter focuses on stability of dynamic systems with regime switching in discrete time. Suppose that the underlying systems are modeled by difference equations and/or difference equations subject to an additional exogenous random noise input source. Different from the traditional setup, the dynamics of the systems are subject to regime changes that are modulated by discrete-time Markov chains. We aim to investigate the long-term behavior of the system characterized by Liapunov functions.

Due to modeling requirements and/or the complex structure of the underlying systems, frequently, the Markov chain is either inherent of two time scales or has a large state space naturally divisible into subspaces leading to a time-scale separation. To tackle such systems (e.g., to find their optimal controls), computation of solutions is often deemed infeasible. To overcome such difficulties, by noting the high contrast of rates of transitions and introducing a small parameter $\varepsilon > 0$, we can formulate the problems as systems subject to singularly perturbed Markov chains.

In the previous chapters, we have focused on the structural properties of the underlying Markov chains. Using those results, we can reduce system complexity through decomposition and aggregation, which is accomplished by showing that the underlying system yields a limit system whose coefficients are averaged out with respect to the invariant measure of the Markov chain. Such an approach is also useful for finding optimal controls of dynamic systems, which will be dealt with in later chapters of the book. The essence of these approaches is to note the connection of the original system with a limit system. The technique of comparing original and limit systems will also be adopted in this chapter.

Up to now, we have focused on the study of asymptotic properties for $\varepsilon \to 0$ and $k \to \infty$, while εk remains bounded. In this chapter, we examine systems from a stability point of view. Using a two-time-scale Markov chain formulation, our main effort is directed to the long-term behavior of the systems as $\varepsilon \to 0$, $k \to \infty$, with εk being unbounded. We demonstrate that if the limit system (or the reduced system) is stable, then the original system is also stable for sufficiently small $\varepsilon > 0$. Suppose that one deals with a discrete-time system (e.g., $x_{k+1} = \rho(x_k)$ for an appropriate function $\rho(\cdot)$) directly. Even without the presence of random disturbances and the singularly perturbed Markov chain, using the Liapunov stability argument, one has to calculate $V(x_{k+1}) - V(x_k)$ as in LaSalle [101, p. 5], which is more complex than differentiation along the solution $(d/dt)V(x)$ used in a continuous-time system. In lieu of a direct approach, we use a Liapunov function of the limit system to carry out the analysis needed. An effort to reduce complexity from a different angle, this indirect approach is much simpler and more feasible. The original dynamic systems are compared with the limit systems and then perturbed Liapunov function methods are used to obtain the desired results.

The rest of the chapter is arranged as follows. Section 7.2 presents the precise formulation of the problems. Section 7.3 provides several auxiliary results that are needed in the subsequent study. The first of which is a mean squares estimate; the second concerns the weak limits of interpolated processes arising from discrete-time dynamic systems modulated by singularly perturbed Markov chains. For demonstration purposes, we also provide a couple of simple examples to illustrate the trajectory behavior of the systems. Section 7.4 gives the stability results. For ease of presentation, we first establish the results for such systems whose dominating parts have only recurrent states. Section 7.5 studies recurrence and path excursion estimates. Both mean recurrence time and probability bounds are obtained. To preserve the flow of presentation, as in most chapters, the verbatim proofs and technical details are relegated to Section 7.6. Section 7.7 gives extensions when transient states are included.

7.2   Formulation

As alluded to in the introduction, to highlight the contrasts of different transition rates, we introduce a small parameter $\varepsilon > 0$. Consider a discrete-time Markov chain α_k^ε with finite state space $\mathcal{M} = \{1, \ldots, m_0\}$ and transition matrix

$$P^\varepsilon = P + \varepsilon Q, \tag{7.1}$$

where P is a transition probability matrix of a time-homogeneous Markov chain, and $Q = (q^{ij})$ is a generator of another continuous-time Markov chain, i.e., for each $i \neq j$, $q^{ij} \geq 0$ and for each i, $\sum_j q^{ij} = 0$ or $Q\mathbb{1}_{m_0} = 0$. Recall that $\mathbb{1}_\iota \in \mathbb{R}^{\iota \times 1}$ denotes a column vector with all components being 1. We are now in a position to present the stability problems. The first one is concerned with a difference equation modulated by a singularly perturbed Markov chain, and the second one focuses on a system with an additional exogenous noise term. We are interested in whether the stability of the limit systems in the sense of Liapunov allows us to make any inference about the original systems. We answer the question affirmatively. Our result is: If the limit system is (asymptotically) stable (implied by certain bounds via a Liapunov function), then the original system also preserves stability with a Liapunov type estimate subject to $O(\varepsilon)$ perturbations.

7.2.1 Problem Setup

Difference Equations under a Markovian Regime. Let $x_k^\varepsilon \in \mathbb{R}^n$ be the state of a system at time $k \geq 0$, and $f(\cdot) : \mathbb{R}^n \times \mathcal{M} \mapsto \mathbb{R}^n$. Suppose that

$$x_{k+1}^\varepsilon = x_k^\varepsilon + \varepsilon f(x_k^\varepsilon, \alpha_k^\varepsilon),$$
$$x_0^\varepsilon = x_0, \quad \alpha_0^\varepsilon = \alpha_0. \tag{7.2}$$

We are interested in the Liapunov stability of (7.2). Instead of treating the system directly, we will examine its stability by use of the fact that as $\varepsilon \to 0$, the dynamic system is close to an averaged system or a limit system in an appropriate sense. Using the stability of the limit system, we try to figure out the large-time behavior of the dynamic system given by (7.2) as $k \to \infty$ and $\varepsilon \to 0$. Note that, effectively, rather than dealing with one vector equation, we are dealing with a system of equations in which the total number of equations is precisely $|\mathcal{M}| = m_0$, the cardinality of the state space $\mathcal{M}$.

Difference Equations under External Disturbances. Let $f(\cdot) : \mathbb{R}^n \times \mathcal{M} \mapsto \mathbb{R}^n$ and $\sigma(\cdot) : \mathbb{R}^n \times \mathcal{M} \mapsto \mathbb{R}^{n \times n}$ be appropriate functions satisfying suitable conditions (the precise conditions will be stated later), and $\{w_k\}$ be a sequence of external random noise independent of α_k^ε. Let x_k^ε be the state at time $k \geq 0$. Consider the following system:

$$x_{k+1}^\varepsilon = x_k^\varepsilon + \varepsilon f(x_k^\varepsilon, \alpha_k^\varepsilon) + \sqrt{\varepsilon}\sigma(x_k^\varepsilon, \alpha_k^\varepsilon)w_k,$$
$$x_0^\varepsilon = x_0, \quad \alpha_0^\varepsilon = \alpha_0. \tag{7.3}$$

As $\varepsilon \to 0$, the dynamic system is close to an averaged system of switching diffusions. Again, using the stability of the limit system, we make an inference about that of the dynamic system governed by (7.3).

7.2.2   Demonstrative Examples

In this section, we provide two simple examples to illustrate and reveal the stability of the systems under consideration.

Example 7.1.   Consider a Markov-modulated linear system of the form

$$x_{k+1}^\varepsilon = x_k^\varepsilon + \varepsilon A(\alpha_k^\varepsilon)x_k^\varepsilon + \sqrt{\varepsilon}\sigma(\alpha_k^\varepsilon)w_k,$$

where $\{w_k\}$ is a sequence of Gaussian random variables with mean 0 and variance 1. The transition probability matrix of α_k^ε is given by

$$P^\varepsilon = \begin{pmatrix} 0.9 & 0.1 \\ 0.15 & 0.85 \end{pmatrix} + \varepsilon \begin{pmatrix} -0.3 & 0.3 \\ 0.5 & -0.5 \end{pmatrix}. \tag{7.4}$$

Taking $A(1) = -0.3$ and $A(2) = -0.5$, we first plot the trajectory of the stochastic difference equation with $\varepsilon = 0.05$, $\sigma(1) = 0.05$, and $\sigma(2) = 0.03$. Next, keep $\varepsilon = 0.05$ and use $\sigma(1) = 0.5$ and $\sigma(2) = 0.3$. We plot the corresponding trajectories in Figure 7.1.

Example 7.2.   This example is concerned with a two-dimensional nonlinear system. Again, let P^ε be given by (7.4). The system is given by

$$x_{k+1}^\varepsilon = x_k^\varepsilon + \varepsilon f(x_k^\varepsilon, \alpha_k^\varepsilon) + \sqrt{\varepsilon}\sigma(x_k^\varepsilon, \alpha_k^\varepsilon)w_k,$$

where

$$f(x,1) = (-2x_{k,1}^\varepsilon, -3x_{k,2}^\varepsilon)',$$
$$f(x,2) = (-(x_{k,1}^\varepsilon + x_{k,2}^\varepsilon)^2, -(x_{k,2}^\varepsilon + x_{k,1}^\varepsilon x_{k,2}^\varepsilon)),$$
$$\sigma(x,1) = \text{diag}(1,1),$$
$$\sigma(x,2) = \text{diag}(0.6, 0.6).$$

Using $\varepsilon = 0.01$, we plot the trajectories of the two components in Figure 7.2.

A moment of reflection shows that the systems in both examples are stable. Of course, in examining the plots, caution must be taken since the stability refers to long-term behavior, whereas any plot can only describe finite time evolution. Nevertheless, a certain tendency can be seen from the graphs. It is clear that the dynamics of the systems depend on the size of ε; they also depend on the intensity of the noise (variance of the diffusion) in a nontrivial way.

7.3   Preliminary Results

Owing to the presence of the small parameter $\varepsilon > 0$, the matrix P has crucial influence. In the next two sections, we will study the stability of the

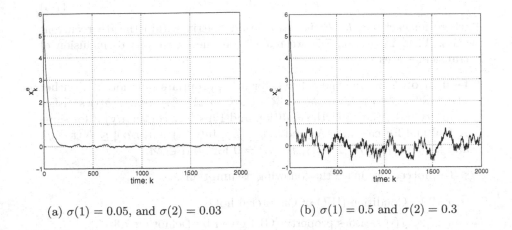

(a) $\sigma(1) = 0.05$, and $\sigma(2) = 0.03$ (b) $\sigma(1) = 0.5$ and $\sigma(2) = 0.3$

FIGURE 7.1. Trajectory of x_k^ε for one-dimensional problem: Two-state Markov chain, $\varepsilon = 0.05$

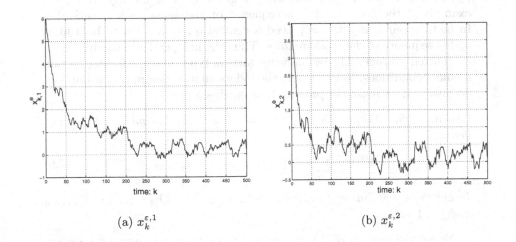

(a) $x_k^{\varepsilon,1}$ (b) $x_k^{\varepsilon,2}$

FIGURE 7.2. Trajectory of x_k^ε for two-dimensional problem: Two-state Markov chain, $\varepsilon = 0.01$, $\sigma(x,1) = \text{diag}(1,1)$, $\sigma(x,2) = \text{diag}(0.6, 0.6)$.

underlying systems when the transition matrix P has l_0 recurrent classes given by

$$P = \text{diag}(P^1, \ldots, P^{l_0}), \tag{7.5}$$

where for each $i \leq l_0$, P^i is a transition matrix within the ith recurrent class. Then in Section 7.5, we remark on the extensions to inclusion of transient states.

Definition 7.3. In what follows, for an appropriate function $h(\cdot)$ (either $h(\cdot) : \mathbb{R}^n \times \mathcal{M} \mapsto \mathbb{R}^n$ or $h(\cdot) : \mathbb{R}^n \times \mathcal{M} \mapsto \mathbb{R}^{n \times n}$), we say $h(\cdot)$ satisfies property (GL) (i.e., growth condition and Lipschitz continuity) if for some $K > 0$ and for each $\alpha \in \mathcal{M}$ and $x, y \in \mathbb{R}^n$, $|h(x, \alpha) - h(y, \alpha)| \leq K|x - y|$, and $|h_x(x, \alpha)x| \leq K(|x| + 1)$; $f(0, \alpha) = 0$ for all $\alpha \in \mathcal{M}$.

To proceed, we make the following assumptions.

(A7.1) Condition (HR) of Chapter 6 holds.

(A7.2) $f(\cdot)$ satisfies property (GL) given by Definition 7.3.

(A7.3) $\sigma(\cdot)$ satisfies property (GL).

(A7.4) $\{w_k\}$ is a sequence of independent and identically distributed random variables with zero mean and $Ew_k w_k' = I$, the identity matrix. Moreover, $\{w_k\}$ is independent of $\{\alpha_k^\varepsilon\}$.

Remark 7.4. Note that (A7.1) and (A7.2) are needed in studying (7.2), whereas (A7.1)–(A7.4) are needed in studying (7.3). Conditions (A7.2) and (A7.3) imply that $f(x, \alpha)$ and $\sigma(x, \alpha)$ grow at most linearly. A typical example of the noise $\{w_k\}$ is a sequence of Gaussian random variables. In fact, in our study, we only need a central limit theorem to hold for a scaled sequence of the random noise. Thus the independence condition can be relaxed considerably by allowing mixing-type noise satisfying certain moment conditions. Nevertheless, the independence assumption is imposed because it does make the presentation much simpler.

Lemma 7.5. *Assume* (A7.1). *For each* $i = 1, \ldots, l_0$ *and* $j = 1, \ldots, m_i$, *define*

$$\Delta^{ij} = \varepsilon \sum_{k=0}^{\infty} e^{-k\varepsilon} [I_{\{\alpha_k^\varepsilon = s_{ij}\}} - \nu^{ij} I_{\{\alpha_k^\varepsilon \in \mathcal{M}_i\}}]. \tag{7.6}$$

Under the conditions of Proposition 6.2, $E(\Delta^{ij})^2 = O(\varepsilon)$, for $i = 1, \ldots, l_0$ and $j = 1, \ldots, m_i$.

We defer the proofs of this lemma and the following results until Section 7.6. This lemma should be compared with the mean squares estimates in Chapter 4 (also cf. Proposition 6.2). It is a mean squares estimate with $\sum_{k=0}^{T/\varepsilon}$ replaced by $\sum_{k=0}^{\infty}$ and with an added "discount factor" $\exp(-k\varepsilon)$.

By taking continuous-time interpolations of the difference equations given in (7.2) and (7.3), we proceed to obtain associated limit systems. That is, suitably scaled sequences of functions converge to solutions of hybrid ordinary differential equations and hybrid stochastic differential equations, respectively. For the state x_k^ε and the modulating Markov chain α_k^ε given in (7.2) and (7.3), define continuous-time interpolations with interpolation interval of length ε as

$$x^\varepsilon(t) = x_k^\varepsilon, \quad \overline{\alpha}^\varepsilon(t) = \overline{\alpha}_k^\varepsilon, \quad \text{for } t \in [\varepsilon k, \varepsilon k + \varepsilon). \tag{7.7}$$

Lemma 7.6. *The following assertions hold:*

(a) *Under (A7.1) and (A7.2), the sequence $x^\varepsilon(\cdot)$ given in (7.2) converges weakly to $x(\cdot)$, which satisfies*

$$\frac{d}{dt}x(t) = \overline{f}(x(t), \overline{\alpha}(t)),$$
$$x(0) = x_0, \quad \overline{\alpha}(0) = \overline{\alpha}_0, \tag{7.8}$$

where

$$\overline{f}(x, i) = \sum_{j=1}^{m_i} \nu^{ij} f(x, s_{ij}), \quad \text{for } i = 1, 2, \ldots, l_0. \tag{7.9}$$

(b) *Assume (A7.1)–(A7.4). Then the sequence $(x^\varepsilon(\cdot), \overline{\alpha}^\varepsilon(\cdot))$ converges to $(x(\cdot), \overline{\alpha}(\cdot))$ weakly such that $x(\cdot)$ is the solution of*

$$dx(t) = \overline{f}(x(t), \overline{\alpha}(t))dt + \overline{\sigma}(x(t), \overline{\alpha}(t))dw, \tag{7.10}$$

where for each $i \in \overline{\mathcal{M}}$, $\overline{f}(x, i)$ is defined in (7.9), and $\overline{\sigma}(x, i)$ is defined by

$$\overline{\sigma}(x, i)\overline{\sigma}'(x, i) \stackrel{\text{def}}{=} \overline{\Xi}(x, i) = \sum_{j=1}^{m_i} \nu^{ij} \Xi(x, s_{ij}), \tag{7.11}$$

with $\Xi(x, s_{ij}) = \sigma(x, s_{ij})\sigma'(x, s_{ij})$.

7.4 Stability

For any $g(\cdot, \cdot)$ on $\mathbb{R}^n \times \mathcal{M}$ that is twice continuously differentiable with respect to the first variable, define

$$\varepsilon \mathcal{L}^\varepsilon g(x_k^\varepsilon, \overline{\alpha}_k^\varepsilon) = E_k[g(x_{k+1}^\varepsilon, \overline{\alpha}_{k+1}^\varepsilon) - g(x_k^\varepsilon, \overline{\alpha}_k^\varepsilon)], \tag{7.12}$$

where E_k denotes the conditional expectation with respect to $\mathcal{G}_k$, the σ-algebra generated by $\{x_0, \alpha_j^\varepsilon : j < k\}$ for (7.2) and the σ-algebra generated by $\{x_0, \alpha_j^\varepsilon, w_j : j < k\}$ for (7.3), respectively. For future use, define

$$\chi_k^\varepsilon = (I_{\{\alpha_k^\varepsilon = s_{ij}\}}, \ i = 1, \ldots, l_0, \ j = 1, \ldots, m_i) \in \mathbb{R}^{1 \times m_0}. \tag{7.13}$$

We are now ready to present the stability results. The first theorem deals with the system governed by (7.2), whereas the second one is concerned with that governed by (7.3). For both systems, the results indicate that if the limit system is stable in the sense $\mathcal{L}V(x,i) \leq -\lambda V(x,i)$ with an appropriate operator $\mathcal{L}$, then the singularly perturbed systems are also exponentially stable with an added error term depending on ε. We state the theorems in what follows. The proofs are in Section 7.6.

(A7.5) For each $i = 1, \ldots, l_0$ and for some positive integer d_0, there is a Liapunov function $V(\cdot, i)$ that is d_0-times continuously differentiable with respect to x, $|\partial^\iota V(x,i)||x|^\iota \leq K(V(x,j) + 1)$ for $1 \leq \iota \leq d_0 - 1$, $j = 1, \ldots, l_0$, and $|\partial^{d_0} V(x,i)| = O(1)$ (where for $1 \leq \iota \leq d_0$, $\partial^\iota V(x,i)$ denotes the ιth derivative of $V(x,i)$), $V(x,i) \to \infty$ as $|x| \to \infty$, $K_1(|x|^{d_0}+1) \leq V(x,i) \leq K_2(|x|^{d_0}+1)$.

Theorem 7.7. *Consider* (7.2). *Assume that* (A7.1), (A7.2), *and* (A7.5) *hold, that*

$$\mathcal{L}V(x,i) \leq -\gamma V(x,i) \quad \text{for some} \quad \gamma > 0, \tag{7.14}$$

with the operator defined by

$$\mathcal{L}V(x,i) = V_x'(x,i)\overline{f}(x,i) + \overline{Q}V(x,\cdot)(i), \quad i \in \overline{\mathcal{M}} = \{1, \ldots, l_0\}, \tag{7.15}$$

and that $EV(x_0, \overline{\alpha}_0) < \infty$. *Then*

$$EV(x_{k+1}^\varepsilon, \overline{\alpha}_{k+1}^\varepsilon) \leq \exp(-\varepsilon\gamma k)EV(x_0, \overline{\alpha}_0) + O(\varepsilon). \tag{7.16}$$

Example 7.8. As an illustration, let α_k^ε be a two-state Markov chain with transition matrix given by (7.1) such that P is irreducible and aperiodic. For $\iota \in \mathcal{M} = \{1, 2\}$, consider the linear system

$$x_{k+1}^\varepsilon = x_k^\varepsilon + \varepsilon A(\alpha_k^\varepsilon)x_k^\varepsilon, \quad \text{where} \quad A(\iota) = \text{diag}(a(\iota), b(\iota)), \quad x_k^\varepsilon \in \mathbb{R}^2.$$

The limit system according to (a) in Lemma 7.6 is given by

$$\dot{x} = \overline{A}x, \quad \text{where} \quad \overline{A} = \nu^1 A(1) + \nu^2 A(2).$$

Suppose that $\overline{A} < 0$. Define $V(x) = x'x$, which is a Liapunov function. Differentiation along the trajectory gives us $(d/dt)V(x) \leq -\lambda V(x)$ for some $\lambda > 0$. Thus, by Theorem 7.7 the asymptotic stability of the continuous-time system implies that of the discrete-time system.

Next, we consider the dynamic system with exogenous disturbances. The result is stated in the following theorem.

Theorem 7.9. *Consider* (7.3). *Assume that* (A7.1)–(A7.5) *hold, that* $d_0 \geq 3$ *and* $E|w_k|^{d_1} < \infty$ *for some integer* $d_1 \geq d_0$, *that*

$$\mathcal{L}V(x,i) \leq -\gamma V(x,i) \quad \text{for some} \quad \gamma > 0, \tag{7.17}$$

with the operator defined by

$$\mathcal{L}V(x,i) = V_x'(x,i)\overline{f}(x,i) + \frac{1}{2}\mathrm{tr}[V_{xx}(x,i)\overline{\Xi}(x,i)] + \overline{Q}V(x,\cdot)(i), \quad i \in \overline{\mathcal{M}},$$
(7.18)

and that $EV(x_0,\overline{\alpha}_0) < \infty$. Then

$$EV(x_{k+1}^{\varepsilon},\overline{\alpha}_{k+1}^{\varepsilon}) \leq \exp(-\varepsilon\gamma k)EV(x_0,\overline{\alpha}_0) + O(\varepsilon).$$
(7.19)

Example 7.10. Suppose that $x \in \mathbb{R}^n$, α_k^{ε} is a discrete-time Markov chain with state space $\mathcal{M} = \{1,\ldots,m_0\}$ and transition probability matrix (7.1), such that P is irreducible and aperiodic. The system modulated by the Markov chain is

$$x_{k+1}^{\varepsilon} = x_k^{\varepsilon} + \varepsilon A(\alpha_k^{\varepsilon})x_k^{\varepsilon} + \sqrt{\varepsilon}B(\alpha_k^{\varepsilon})x_k^{\varepsilon}w_k,$$
(7.20)

where for each $\iota \in \mathcal{M}$, $A(\iota)$ and $B(\iota)$ are $n \times n$ matrices such that $A(\iota)$ is Hurwitz and $x'B(\iota)B'(\iota)x + x'A(\iota)x < 0$ for $x \neq 0$, and $\{w_k\}$ is a sequence of scalar-valued random variables satisfying (A7.4). In this case, the limit system is given by

$$dx = \overline{A}x\,dt + \overline{\Xi}^{1/2}x\,dw,$$
(7.21)

where $w(\cdot)$ is a real-valued standard Brownian motion,

$$\overline{A} = \sum_{\iota=1}^{m_0} A(\iota)\nu^{\iota}, \quad \overline{\Xi} = \sum_{\iota=1}^{m_0} \nu^{\iota}B(\iota)B'(\iota),$$

and $\nu = (\nu^1,\ldots,\nu^{m_0})$ is the stationary distribution corresponding to the transition matrix P. Let $V(x) = x'x/2$. Then it is not hard to see that $\mathcal{L}V(x) \leq -\lambda V(x)$ for some $\lambda > 0$. Thus, (7.19) also holds by virtue of Theorem 7.9.

Remark 7.11. In Theorem 7.7, $V(x,i)$ is a Liapunov function for the limit hybrid ordinary differential equation, whereas $V(x,i)$ in Theorem 7.9 is a Liapunov function for the limit hybrid stochastic differential equation. The growth and smooth conditions will be satisfied if $V(x,i)$ is a polynomial of order d_0 or has polynomial growth of order d_0. It follows from this condition that

$$|\partial^{\iota}V(x,i)||f(x,i)|^{\iota} \leq K(V(x,i)+1)$$

and

$$|\partial^{\iota}V(x,i)||\sigma(x,i)|^{\iota} \leq K(V(x,i)+1).$$

Note that the notion $\partial^{\iota}V(x,i)$ can be viewed as follows: If $\iota = 1$, $\partial V(x,i)$ is the gradient; if $\iota = 2$, $\partial^2 V(x,i)$ is the Hessian; if $\iota > 2$, $\partial^{\iota}V(x,i)$ is the usual multi-index notation of mixed partial derivatives. References on matrix calculus and Kronecker products can be found, for instance, in Graham [65] among others.

As observed in Remark 7.4, if $\{w_k\}$ is a sequence of Gaussian random variables, then the moment condition $E|w_k|^{d_1}$ holds for any positive integer $d_1 < \infty$. Moreover, we can treat correlated random variables of mixing type. For notational simplicity, we confine ourselves to the current setup.

7.5   Recurrence and Path Excursion

We have established bounds on $EV(x_k^\varepsilon, \overline{\alpha}_k^\varepsilon)$ for large k and small ε. In this section, we examine the recurrence and path excursion properties of the dynamic systems associated with the singularly perturbed discrete-time Markov chains. We will use the following assumptions.

(A7.6) For each $i = 1, \ldots, l_0$, there is a twice continuously differentiable Liapunov function $V(x, i)$ such that

$$|V_x'(x, i)||x| \leq K(V(x, j) + 1), \quad \text{for each } j, \qquad (7.22)$$

that $\min_x V(x, i) = 0$, that $V_x'(x, i)f(x, i) \leq -c_0$ for some $c_0 > 0$, and that $V(x, i) \to \infty$ as $|x| \to \infty$.

For some $\lambda_0 > 0$ and $\lambda_1 > \lambda_0$, define

$$B_0 = \{x : V(x, i) \leq \lambda_0, \ i = 1, \ldots, l_0\},$$
$$B_1 = \{x : V(x, i) \leq \lambda_1, \ i = 1, \ldots, l_0\}.$$

Let τ_0 be the first exit time from B_0 of the process x_k^ε, and τ_1 be the first return time of the process after τ_0. That is,

$$\tau_0 = \min\{k : x_k^\varepsilon \notin B_0\}, \quad \tau_1 = \min\{k \geq \tau_0 : x_k^\varepsilon \in B_0\}.$$

The random time $\tau_1 - \tau_0$ is known as the recurrence time. It is the duration of the process from the first exit of B_0 to the next return to B_0.

Theorem 7.12. *Consider the system (7.2) (resp. (7.3)). Assume (A7.1), (A7.2) (resp. (A7.1)–(A7.4)), and (A7.6) are satisfied. Then for some $0 < c_1 < c_0$,*

$$E_{\tau_0}(\tau_1 - \tau_0) \leq \frac{E_{\tau_0}V(x_{\tau_0}^\varepsilon, \overline{\alpha}_{\tau_0}^\varepsilon)(1 + O(\varepsilon)) + O(\varepsilon)}{c_1}, \qquad (7.23)$$

where E_{τ_0} denotes the conditional expectation with respect to the σ-algebra $\mathcal{G}_{\tau_0}$.

Remark 7.13. The above theorem indicates that for sufficiently small $\varepsilon > 0$, x_k^ε is recurrent in the sense that if $x_{\tau_0}^\varepsilon \in B_1 - B_0$, then the conditional mean recurrence time of $\tau_1 - \tau_0$ has an upper bound $[\lambda_1(1 + O(\varepsilon)) + O(\varepsilon)]/c_1$.

In the proof of the result, we adopt the approach of Kushner [96] for the Markov chain setup. In addition to the conditional moment bound, we may also obtain the probability bound of the form

$$P\left(\sup_{\tau_0 \le k < \tau_1} V(x_k^\varepsilon, \overline{\alpha}_k^\varepsilon) \ge \kappa \Big| \mathcal{G}_{\tau_0}\right) \le \frac{E_{\tau_0} V(x_{\tau_0}^\varepsilon, \overline{\alpha}_{\tau_0}^\varepsilon)(1 + O(\varepsilon)) + O(\varepsilon)}{\kappa}.$$

In fact, it is possible to replace the first exit time τ_0 by any exit time $\widetilde{\tau}$ from the set B_0 and to replace τ_1 by the first return time after $\widetilde{\tau}$.

Compared with the conditions used in Theorems 7.7 and 7.9, the growth conditions are much more relaxed. Because we do not need the moment estimates as in Theorems 7.7 and 7.9, we can work with truncated Liapunov function.

7.6   Proofs of Results

Proof of Lemma 7.5. Expanding the summations in (7.6) leads to

$$
\begin{aligned}
E[\Delta^{ij}]^2 &= \varepsilon^2 \sum_{k=0}^{\infty} \sum_{\ell=0}^{\infty} e^{-k\varepsilon} e^{-\ell\varepsilon} E[I_{\{\alpha_k^\varepsilon = s_{ij}\}} - \nu^{ij} I_{\{\alpha_k^\varepsilon \in \mathcal{M}_i\}}] \\
&\quad \times [I_{\{\alpha_\ell^\varepsilon = s_{ij}\}} - \nu^{ij} I_{\{\alpha_\ell^\varepsilon \in \mathcal{M}_i\}}] \\
&\le 2\varepsilon^2 \sum_{k=0}^{\infty} \sum_{\ell=0}^{k} e^{-k\varepsilon} e^{-\ell\varepsilon} E[I_{\{\alpha_k^\varepsilon = s_{ij}\}} - \nu^{ij} I_{\{\alpha_k^\varepsilon \in \mathcal{M}_i\}}] \\
&\quad \times [I_{\{\alpha_\ell^\varepsilon = s_{ij}\}} - \nu^{ij} I_{\{\alpha_\ell^\varepsilon \in \mathcal{M}_i\}}].
\end{aligned}
$$

As in Chapter 4 (see also Yin, Zhang, and Badowski [165]), it can be shown that for $k \ge \ell$,

$$
\begin{aligned}
&E[I_{\{\alpha_k^\varepsilon = s_{ij}\}} - \nu^{ij} I_{\{\alpha_k^\varepsilon \in \mathcal{M}_i\}}][I_{\{\alpha_\ell^\varepsilon = s_{ij}\}} - \nu^{ij} I_{\{\alpha_\ell^\varepsilon \in \mathcal{M}_i\}}] \\
&= O\left(\varepsilon + k\varepsilon^2 + k^2 \varepsilon^3 + \lambda^\ell + \lambda^{k-\ell}\right).
\end{aligned}
$$

It follows that

$$E(\Delta^{ij})^2 \le K\varepsilon^2 \sum_{k=0}^{\infty} e^{-k\varepsilon}(1 + k\varepsilon + k^2 \varepsilon^2 + k^3 \varepsilon^3)$$

$$= O(\varepsilon).$$

The desired result thus follows.   □

Ideas of Proof of Part (b) in Lemma 7.6. Since our main efforts in this chapter focus on stability analysis, details of the weak convergence argument are omitted; related materials on the martingale problem formulation and weak convergence can be found in Chapters 4, 8, 11, and 13. We provide only a sketch of proof for Part (b) in Lemma 7.6; the proof for Part (a) in Lemma 7.6 is even simpler.

Consider the pair $\{x^\varepsilon(\cdot), \overline{\alpha}^\varepsilon(\cdot)\}$. As in the usual weak convergence analysis, first we verify the tightness of the sequence. Then we characterize the limit process.

Since it is not known *a priori* that $\{x^\varepsilon(\cdot)\}$ is bounded, we use a truncation device (see Definition 14.15). For any $0 < N < \infty$, define $q_N(\cdot)$ to be a smooth function satisfying

$$q_N(x) = \begin{cases} 1 & \text{for } x \in S_N = \{x : |x| \leq N\}, \\ 0 & \text{for } x \in \mathbb{R}^n - S_{N+1}. \end{cases}$$

Corresponding to (7.3), define the truncated difference equation as

$$\begin{aligned} x_{k+1}^{\varepsilon,N} = x_k^{\varepsilon,N} &+ \varepsilon f(x_k^{\varepsilon,N}, \alpha_k^\varepsilon) q_N(x_k^{\varepsilon,N}) \\ &+ \sqrt{\varepsilon} \sigma(x_k^{\varepsilon,N}, \alpha_k^\varepsilon) w_k q_N(x_k^{\varepsilon,N}), \end{aligned}$$

and define its interpolation by

$$x^{\varepsilon,N}(t) = x_k^{\varepsilon,N}, \quad \text{for } t \in [\varepsilon k, \varepsilon k + \varepsilon).$$

Then $x^{\varepsilon,N}(\cdot)$ is an N-truncation of $x^\varepsilon(\cdot)$; see Definition 14.15 (also Kushner and Yin [100, p. 284]). Using this truncation, the smoothness of $f(\cdot, \iota)$ and $\sigma(\cdot, \iota)$ for each ι, and the independence of w_k with $\mathcal{G}_k$, we can readily see that for sufficiently small $\delta > 0$ and $0 < s \leq \delta$,

$$E_t^\varepsilon |x^{\varepsilon,N}(t+s) - x^{\varepsilon,N}(t)|^2 \leq Ks \leq K\delta.$$

Thus, we have

$$\lim_{\delta \to 0} \limsup_{\varepsilon \to 0} E E_t^\varepsilon |x^{\varepsilon,N}(t+s) - x^{\varepsilon,N}(t)|^2 = 0.$$

By the tightness criterion in Lemma 14.12, $\{x^{\varepsilon,N}(\cdot)\}$ is tight. This and the weak convergence of $\overline{\alpha}^\varepsilon(\cdot)$ then implies that $\{x^{\varepsilon,N}(\cdot), \overline{\alpha}^\varepsilon(\cdot)\}$ is tight. By Prohorov's theorem (see Theorem 14.4), we can extract a weakly convergent subsequence. Select such a subsequence, and for simplicity, still denote it by $(x^{\varepsilon,N}(\cdot), \overline{\alpha}^\varepsilon(\cdot))$ with the limit denoted by $(x^N(\cdot), \overline{\alpha}(\cdot))$.

To characterize the limit, using t/ε and $(t+s)/\varepsilon$ as integer parts in our convention, we have

$$
x^{\varepsilon,N}(t+s) - x^{\varepsilon,N}(t)
$$

$$
= \varepsilon \sum_{k=t/\varepsilon}^{(t+s)/\varepsilon-1} \sum_{i=1}^{l_0} \sum_{j=1}^{m_i} f(x_k^{\varepsilon,N}, s_{ij}) \nu^{ij} I_{\{\overline{\alpha}_k^\varepsilon = i\}}
$$

$$
+ \sqrt{\varepsilon} \sum_{k=t/\varepsilon}^{(t+s)/\varepsilon-1} \sum_{i=1}^{l_0} \sum_{j=1}^{m_i} \sigma(x_k^{\varepsilon,N}, s_{ij}) w_k \nu^{ij} I_{\{\overline{\alpha}_k^\varepsilon = i\}} \tag{7.24}
$$

$$
+ \varepsilon \sum_{k=t/\varepsilon}^{(t+s)/\varepsilon-1} \sum_{i=1}^{l_0} \sum_{j=1}^{m_i} f(x_k^{\varepsilon,N}, s_{ij}) [I_{\{\alpha_k^\varepsilon = s_{ij}\}} - \nu^{ij} I_{\{\overline{\alpha}_k^\varepsilon = i\}}]
$$

$$
+ \sqrt{\varepsilon} \sum_{k=t/\varepsilon}^{(t+s)/\varepsilon-1} \sum_{i=1}^{l_0} \sum_{j=1}^{m_i} \sigma(x_k^{\varepsilon,N}, s_{ij}) w_k [I_{\{\alpha_k^\varepsilon = s_{ij}\}} - \nu^{ij} I_{\{\overline{\alpha}_k^\varepsilon = i\}}].
$$

Using the mean squares estimates in Theorem 4.5 (see also Proposition 6.2), it can be shown that the last two terms in (7.24) go to 0 in probability uniformly in t. Thus, we need only consider the terms on the second and the third lines of (7.24).
Define

$$
w^\varepsilon(t) = \sqrt{\varepsilon} \sum_{k=0}^{t/\varepsilon-1} w_k.
$$

Then a version of the Donsker's invariance theorem (see Billingsley [18]) yields that $w^\varepsilon(\cdot)$ converges weakly to $w(\cdot)$ a standard Brownian motion. Choose a sequence of positive integers $\{\kappa^\varepsilon\}$ satisfying $\kappa^\varepsilon \to \infty$ as $\varepsilon \to 0$, but $\delta^\varepsilon = \varepsilon \kappa^\varepsilon \to 0$. By virtue of the conditions of $f(\cdot)$, we can write

$$
\varepsilon \sum_{k=t/\varepsilon}^{(t+s)/\varepsilon-1} \sum_{i=1}^{l_0} \sum_{j=1}^{m_i} f(x_k^{\varepsilon,N}, s_{ij}) \nu^{ij} I_{\{\overline{\alpha}_k^\varepsilon = i\}}
$$

$$
= \sum_{i=1}^{l_0} \sum_{j=1}^{m_i} \sum_{\ell\delta_\varepsilon=t}^{(t+s)-\varepsilon} \delta^\varepsilon \frac{1}{\kappa^\varepsilon} \sum_{k=\ell\kappa_\varepsilon}^{\ell\kappa_\varepsilon+\kappa_\varepsilon-1} f(x_k^{\varepsilon,N}, s_{ij}) \nu^{ij} I_{\{\overline{\alpha}_k^\varepsilon = i\}}
$$

$$
= \sum_{i=1}^{l_0} \sum_{j=1}^{m_i} \sum_{\ell\delta_\varepsilon=t}^{(t+s)-\varepsilon} \delta^\varepsilon \frac{1}{\kappa^\varepsilon} \sum_{k=\ell\kappa_\varepsilon}^{\ell\kappa_\varepsilon+\kappa_\varepsilon-1} f(x_{\ell\kappa_\varepsilon}^{\varepsilon,N}, s_{ij}) \nu^{ij} I_{\{\overline{\alpha}_k^\varepsilon = i\}} + o(1),
$$

where $o(1) \to 0$ in probability as $\varepsilon \to 0$. Using the weak convergence of $(x^{\varepsilon,N}(\cdot), \overline{\alpha}^\varepsilon(\cdot))$, the Skorohod representation (without changing notation), and the structure of the limit generator of $\overline{\alpha}(\cdot)$, it can be shown that for any positive integer k_0, any bounded and continuous functions $h_\imath(\cdot)$, with

$\iota \leq k_0$, and any t_ι satisfying $0 < t_\iota \leq t \leq t + s$,

$$E \prod_{\iota=1}^{k_0} h_\iota(x^{\varepsilon,N}(t_\iota), \overline{\alpha}^\varepsilon(t_\iota))$$

$$\times \left[\sum_{i=1}^{l_0} \sum_{j=1}^{m_i} \sum_{l\delta_\varepsilon=t}^{(t+s)-\varepsilon} \delta^\varepsilon \frac{1}{\kappa^\varepsilon} \sum_{k=l\kappa_\varepsilon}^{l\kappa_\varepsilon+\kappa_\varepsilon-1} f(x_{l\kappa^\varepsilon}^{\varepsilon,N}, s_{ij}) \nu^{ij} I_{\{\overline{\alpha}_k^\varepsilon=i\}} \right]$$

$$\to E \prod_{\iota=1}^{k_0} h_\iota(x^N(t_\iota), \overline{\alpha}(t_\iota)) \left[\int_t^{t+s} f(x^N(u), \overline{\alpha}(u)) du \right] \quad \text{as } \varepsilon \to 0.$$

As for the external noise term, we find that

$$E \prod_{\iota=1}^{k_0} h_\iota(x^{\varepsilon,N}(t_\iota), \overline{\alpha}^\varepsilon(t_\iota)) \left[\sqrt{\varepsilon} \sum_{k=t/\varepsilon}^{(t+s)/\varepsilon-1} \sum_{i=1}^{l_0} \sum_{j=1}^{m_i} \sigma(x_k^{\varepsilon,N}, s_{ij}) w_k \nu^{ij} I_{\{\overline{\alpha}_k^\varepsilon=i\}} \right]$$

$$= 0,$$

and that

$$E \prod_{\iota=1}^{k_0} h_\iota(x^{\varepsilon,N}(t_\iota), \overline{\alpha}^\varepsilon(t_\iota)) \left[\sqrt{\varepsilon} \sum_{k=t/\varepsilon}^{(t+s)/\varepsilon-1} \sum_{i=1}^{l_0} \sum_{j=1}^{m_i} \sigma(x_k^{\varepsilon,N}, s_{ij}) w_k \nu^{ij} I_{\{\overline{\alpha}_k^\varepsilon=i\}} \right.$$

$$\left. \times \left[\sqrt{\varepsilon} \sum_{k=t/\varepsilon}^{(t+s)/\varepsilon-1} \sum_{i=1}^{l_0} \sum_{j=1}^{m_i} \sigma(x_k^{\varepsilon,N}, s_{ij}) w_k \nu^{ij} I_{\{\overline{\alpha}_k^\varepsilon=i\}} \right]' \right]$$

$$\to E \prod_{\iota=1}^{k_0} h_\iota(x^N(t_\iota), \overline{\alpha}(t_\iota)) \int_t^{t+s} \Xi(x^N(u), \overline{\alpha}(u)) du, \quad \text{as } \varepsilon \to 0.$$

Collecting the estimates obtained, we can show that $(x^{\varepsilon,N}(\cdot), \overline{\alpha}^\varepsilon(\cdot))$ converges to $(x^N(\cdot), \overline{\alpha}(\cdot))$ weakly. Next, using an argument similar to that of Kushner and Yin [100, Step 4, p. 285], we can show that the untruncated process is also convergent. That is, we also have $(x^\varepsilon(\cdot), \overline{\alpha}^\varepsilon(\cdot))$ converging weakly to $(x(\cdot), \overline{\alpha}(\cdot))$, which is a solution of the martingale problem with the desired generator. Moreover, the solution of the martingale problem is unique, which can be proved as in Lemma 4.9.

Proof of Theorem 7.7. The proof is divided into several steps.

Step 1: (Estimate for $V(x, i)$). Define

$$\widehat{F}(x) = \begin{pmatrix} V(x, 1) \mathbb{1}_{m_1} \\ \vdots \\ V(x, l_0) \mathbb{1}_{m_{l_0}} \end{pmatrix} \in \mathbb{R}^{m_0 \times 1}, \quad \widetilde{F}(x) = \begin{pmatrix} V(x, 1) \\ \vdots \\ V(x, l_0) \end{pmatrix} \in \mathbb{R}^{l_0 \times 1},$$

$$\widehat{V}(x, \alpha) = \sum_{i=1}^{l_0} V(x, i) I_{\{\alpha \in \mathcal{M}_i\}} = V(x, i) \quad \text{if } \alpha \in \mathcal{M}_i, \quad \text{for } i = 1, \ldots, l_0.$$

$$(7.25)$$

The function $\widehat{V}(x, \alpha)$ is a device allowing us to use α_k^ε in lieu of $\overline{\alpha}_k^\varepsilon$. In what follows, we often write $\widehat{V}(x_k^\varepsilon, \alpha_k^\varepsilon)$ and $V(x_k^\varepsilon, \overline{\alpha}_k^\varepsilon)$ interchangeably. Then we have

$$
\begin{aligned}
E_k[\widehat{V}(x_{k+1}^\varepsilon, \alpha_{k+1}^\varepsilon) &- \widehat{V}(x_k^\varepsilon, \alpha_k^\varepsilon)] \\
&= E_k[\widehat{V}(x_{k+1}^\varepsilon, \alpha_{k+1}^\varepsilon) - \widehat{V}(x_{k+1}^\varepsilon, \alpha_k^\varepsilon)] + E_k[\widehat{V}(x_{k+1}^\varepsilon, \alpha_k^\varepsilon) - \widehat{V}(x_k^\varepsilon, \alpha_k^\varepsilon)],
\end{aligned}
\tag{7.26}
$$

where E_k denotes the conditional expectation with respect to $\mathcal{G}_k = \sigma\{x_0, \alpha_j^\varepsilon : j < k\}$. Note that by use of a truncated Taylor expansion,

$$
E_k[\widehat{V}(x_{k+1}^\varepsilon, \alpha_k^\varepsilon) - \widehat{V}(x_k^\varepsilon, \alpha_k^\varepsilon)] = \varepsilon E_k \widehat{V}_x'(x_k^\varepsilon, \alpha_k^\varepsilon) f(x_k^\varepsilon, \alpha_k^\varepsilon) + \eta_{k+1}^\varepsilon, \tag{7.27}
$$

where

$$
\begin{aligned}
\eta_{k+1}^\varepsilon = {} &O\left(E_k \sum_{\iota=2}^{d_0-1} |\partial^\iota \widehat{V}(x_k^\varepsilon, \alpha_k^\varepsilon)| |x_{k+1}^\varepsilon - x_k^\varepsilon|^\iota \right) \\
&+ O\left(E_k |x_{k+1}^\varepsilon - x_k^\varepsilon|^{d_0} \int_0^1 |\partial^{d_0} \widehat{V}(x_k^\varepsilon + s(x_{k+1}^\varepsilon - x_k^\varepsilon))| ds \right).
\end{aligned}
$$

In view of the linear growth of $f(x, \alpha)$ implied by (A7.2), and the bounds on $\partial^\iota \widehat{V}(x, \alpha)$, for $2 \le \iota \le d_0$, we have

$$
\eta_{k+1}^\varepsilon = O(\varepsilon^2)(E_k \widehat{V}(x_k^\varepsilon, \alpha_k^\varepsilon) + 1). \tag{7.28}
$$

Since $P - I$ is orthogonal to $\widehat{F}(x_{k+1}^\varepsilon)$, we obtain that

$$
\begin{aligned}
E_k[\widehat{V}(x_{k+1}^\varepsilon, \alpha_{k+1}^\varepsilon) &- \widehat{V}(x_{k+1}^\varepsilon, \alpha_k^\varepsilon)] \\
&= \sum_{i_1=1}^{l_0} \sum_{j_1=1}^{m_i} E_k \sum_{i=1}^{l_0} \sum_{j=1}^{m_i} \left[\widehat{V}(x_{k+1}^\varepsilon, s_{ij}) P(\alpha_{k+1}^\varepsilon = s_{ij} | \alpha_k^\varepsilon = s_{i_1 j_1}) \right. \\
&\qquad\qquad\qquad\qquad\qquad \left. - \widehat{V}(x_{k+1}^\varepsilon, s_{i_1 j_1}) \right] I_{\{\alpha_k^\varepsilon = s_{i_1 j_1}\}} \\
&= \chi_k^\varepsilon (P^\varepsilon - I) E_k \widehat{F}(x_{k+1}^\varepsilon) \\
&= \varepsilon \chi_k^\varepsilon E_k Q \widehat{F}(x_k^\varepsilon) + O(\varepsilon^2)(E_k \widehat{V}(x_k^\varepsilon, \alpha_k^\varepsilon) + 1) \\
&= \varepsilon E_k Q \widehat{V}(x_k^\varepsilon, \cdot)(\alpha_k^\varepsilon) + O(\varepsilon^2)(E_k \widehat{V}(x_k^\varepsilon, \alpha_k^\varepsilon) + 1).
\end{aligned}
\tag{7.29}
$$

Thus, we have

$$
\begin{aligned}
E_k[\widehat{V}(x_{k+1}^\varepsilon, \alpha_{k+1}^\varepsilon) &- \widehat{V}(x_k^\varepsilon, \alpha_k^\varepsilon)] \\
&= \varepsilon E_k \widehat{V}_x'(x_k^\varepsilon, \alpha_k^\varepsilon) f(x_k^\varepsilon, \alpha_k^\varepsilon) + \varepsilon E_k Q \widehat{V}(x_k^\varepsilon, \cdot)(\alpha_k^\varepsilon) \\
&\qquad + O(\varepsilon^2)(E_k \widehat{V}(x_k^\varepsilon, \alpha_k^\varepsilon) + 1).
\end{aligned}
\tag{7.30}
$$

Step 2: (Perturbed Liapunov function). Define a perturbation of the Liapunov function by

$$V_1^\varepsilon(x,k) = \varepsilon \sum_{k_1=k}^{\infty} E_k e^{\varepsilon(k-k_1)} \widehat{V}_x'(x, \alpha_k^\varepsilon)[f(x, \alpha_{k_1}^\varepsilon) - \overline{f}(x, \overline{\alpha}_{k_1}^\varepsilon)]. \qquad (7.31)$$

Observe that for each k,

$$f(x, \alpha_k^\varepsilon) - \overline{f}(x, \overline{\alpha}_k^\varepsilon) = \sum_{i=1}^{l_0} \sum_{j=1}^{m_i} f(x, s_{ij})[I_{\{\alpha_k^\varepsilon = s_{ij}\}} - \nu^{ij} I_{\{\alpha_k^\varepsilon \in \mathcal{M}_i\}}].$$

Thus,

$$\begin{aligned}
V_1^\varepsilon(x,k) = \varepsilon \sum_{i=1}^{l_0} \sum_{j=1}^{m_i} \sum_{k_1=k}^{\infty} e^{\varepsilon(k-k_1)} E_k \widehat{V}_x'(x, \alpha_k^\varepsilon) \\
\times f(x, s_{ij})[I_{\{\alpha_{k_1}^\varepsilon = s_{ij}\}} - \nu^{ij} I_{\{\alpha_{k_1}^\varepsilon \in \mathcal{M}_i\}}].
\end{aligned}$$

Consequently, by virtue of Lemma 7.5,

$$\begin{aligned}
E|V_1^\varepsilon(x,k)| = E\Bigg|\varepsilon \sum_{i=1}^{l_0} \sum_{j=1}^{m_i} E_k \widehat{V}_x'(x, \alpha_k^\varepsilon) f(x, s_{ij}) \\
\times \sum_{k_1=k}^{\infty} e^{\varepsilon(k-k_1)} [I_{\{\alpha_{k_1}^\varepsilon = s_{ij}\}} - \nu^{ij} I_{\{\alpha_{k_1}^\varepsilon \in \mathcal{M}_i\}}]\Bigg| \\
\leq \sum_{i=1}^{l_0} \sum_{j=1}^{m_i} E^{\frac{1}{2}} [E_k \widehat{V}_x'(x, s_{ij}) f(x, s_{ij})]^2 \\
\times E^{\frac{1}{2}} \Bigg|\varepsilon \sum_{k_1=k}^{\infty} E_k e^{\varepsilon(k-k_1)} [I_{\{\alpha_{k_1}^\varepsilon = s_{ij}\}} - \nu^{ij} I_{\{\alpha_{k_1}^\varepsilon \in \mathcal{M}_i\}}]\Bigg|^2 \\
< \infty \quad \text{uniformly in } k \geq 0.
\end{aligned}$$

As a result, $\sup_{0 \leq k} E|V_1^\varepsilon(x,k)| < \infty$.

In view of (A7.2), we have for each $i = 1, \ldots, l_0$ and $j = 1, \ldots, m_i$,

$$|E_k \widehat{V}_x'(x, \alpha_k^\varepsilon) f(x, s_{ij})| \leq |E_k \widehat{V}_x(x, \alpha_k^\varepsilon)||f(x, s_{ij})| \leq K(E_k \widehat{V}(x, \alpha_k^\varepsilon) + 1),$$

for some $K > 0$, since it can be shown that for $k_1 \geq k$,

$$\begin{aligned}
E_k[I_{\{\alpha_{k_1}^\varepsilon = s_{ij}\}} - \nu^{ij} I_{\{\alpha_{k_1}^\varepsilon \in \mathcal{M}_i\}}] \\
= \sum_{i_0=1}^{l_0} \sum_{j_0=1}^{m_{i_0}} [P(\alpha_{k_1}^\varepsilon = s_{ij} | \alpha_k^\varepsilon = s_{i_0 j_0}) \\
- \nu^{ij} \sum_{j_2=1}^{m_{i_0}} P(\alpha_{k_1}^\varepsilon = s_{ij_2} | \alpha_k^\varepsilon = s_{i_0 j_0})] I_{\{\alpha_k^\varepsilon = s_{i_0 j_0}\}} \\
= O(\varepsilon + \lambda^{k_1-k}), \quad \text{for some } 0 < \lambda < 1.
\end{aligned}$$

It is easily seen that

$$\sum_{k_1=k}^{\infty} e^{\varepsilon(k-k_1)} O(\varepsilon + \lambda^{k_1-k}) = O(1).$$

Thus,

$$|V_1^\varepsilon(x, k)| \leq \varepsilon K (E_k \widehat{V}(x, \alpha_k^\varepsilon) + 1). \tag{7.32}$$

Furthermore, we obtain that

$$E_k[V_1^\varepsilon(x_{k+1}^\varepsilon, k+1) - V_1^\varepsilon(x_{k+1}^\varepsilon, k)]$$
$$= \varepsilon E_k \sum_{k_1=k+1}^{\infty} e^{\varepsilon(k+1-k_1)} [\widehat{V}_x'(x_{k+1}^\varepsilon, \alpha_{k+1}^\varepsilon) - \widehat{V}_x'(x_{k+1}^\varepsilon, \alpha_k^\varepsilon)]$$
$$\times [f(x_{k+1}^\varepsilon, \alpha_{k_1}^\varepsilon) - \overline{f}(x_{k+1}^\varepsilon, \overline{\alpha}_{k_1}^\varepsilon)]$$
$$+ \varepsilon E_k \sum_{k_1=k+1}^{\infty} e^{\varepsilon(k+1-k_1)} \widehat{V}_x'(x_{k+1}^\varepsilon, \alpha_k^\varepsilon)[f(x_{k+1}^\varepsilon, \alpha_{k_1}^\varepsilon) - \overline{f}(x_{k+1}^\varepsilon, \overline{\alpha}_{k_1}^\varepsilon)]$$
$$- \varepsilon E_k \sum_{k_1=k}^{\infty} e^{\varepsilon(k-k_1)} \widehat{V}_x'(x_{k+1}^\varepsilon, \alpha_k^\varepsilon)[f(x_{k+1}^\varepsilon, \alpha_{k_1}^\varepsilon) - \overline{f}(x_{k+1}^\varepsilon, \overline{\alpha}_{k_1}^\varepsilon)].$$

Noting that all terms but one are cancelled in the last two sums above and carrying out an estimate similar to (7.29) to the term

$$\varepsilon E_k \sum_{k_1=k+1}^{\infty} e^{\varepsilon(k+1-k_1)} [\widehat{V}_x'(x_{k+1}^\varepsilon, \alpha_{k+1}^\varepsilon) - \widehat{V}_x'(x_{k+1}^\varepsilon, \alpha_k^\varepsilon)]$$
$$\times [f(x_{k+1}^\varepsilon, \alpha_{k_1}^\varepsilon) - \overline{f}(x_{k+1}^\varepsilon, \overline{\alpha}_{k_1}^\varepsilon)],$$

we arrive at

$$E_k[V_1^\varepsilon(x_{k+1}^\varepsilon, k+1) - V_1^\varepsilon(x_{k+1}^\varepsilon, k)]$$
$$= -\varepsilon E_k \widehat{V}_x'(x_k^\varepsilon, \alpha_k^\varepsilon)[f(x_k^\varepsilon, \alpha_k^\varepsilon) - \overline{f}(x_k^\varepsilon, \alpha_k^\varepsilon)] + O(\varepsilon^2)(E_k \widehat{V}(x_k^\varepsilon, \alpha_k^\varepsilon) + 1),$$

and

$$E_k[V_1^\varepsilon(x_{k+1}^\varepsilon, k) - V_1^\varepsilon(x_k^\varepsilon, k)]$$
$$= \varepsilon E_k \sum_{k_1=k}^{\infty} e^{\varepsilon(k-k_1)} [\widehat{V}_x'(x_{k+1}^\varepsilon, \alpha_k^\varepsilon) - \widehat{V}_x'(x_k^\varepsilon, \alpha_k^\varepsilon)]$$
$$\times [f(x_{k+1}^\varepsilon, \alpha_{k_1}^\varepsilon) - \overline{f}(x_{k+1}^\varepsilon, \overline{\alpha}_{k_1}^\varepsilon)]$$
$$+ \varepsilon \sum_{k_1=k}^{\infty} e^{\varepsilon(k-k_1)} E_k \widehat{V}_x'(x_k^\varepsilon, \alpha_k^\varepsilon)[(f(x_{k+1}^\varepsilon, \alpha_{k_1}^\varepsilon) - f(x_k^\varepsilon, \alpha_{k_1}^\varepsilon))$$
$$- (\overline{f}(x_{k+1}^\varepsilon, \overline{\alpha}_{k_1}^\varepsilon) - \overline{f}(x_k^\varepsilon, \overline{\alpha}_{k_1}^\varepsilon))]$$
$$= O(\varepsilon^2)(E_k \widehat{V}(x_k^\varepsilon, \alpha_k^\varepsilon) + 1).$$

It follows that

$$
\begin{aligned}
E_k[V_1^\varepsilon(x_{k+1}^\varepsilon, k+1) &- V_1^\varepsilon(x_k^\varepsilon, k)] \\
&= -\varepsilon E_k \widehat{V}_x'(x_k^\varepsilon, \alpha_k^\varepsilon) f(x_k^\varepsilon, \alpha_k^\varepsilon) + \varepsilon E_k \widehat{V}_x'(x_k^\varepsilon, \alpha_k^\varepsilon) \overline{f}(x_k^\varepsilon, \overline{\alpha}_k^\varepsilon) \qquad (7.33) \\
&\quad + O(\varepsilon^2)(E_k \widehat{V}(x_k^\varepsilon, \alpha_k^\varepsilon) + 1).
\end{aligned}
$$

To proceed, define

$$
V^\varepsilon(x, k) = \widehat{V}(x, \alpha_k^\varepsilon) + V_1^\varepsilon(x, k). \qquad (7.34)
$$

Then, using the estimates for $V_1^\varepsilon(x, k)$,

$$
E_k[V^\varepsilon(x_k^\varepsilon, k) - \widehat{V}(x_k^\varepsilon, \alpha_k^\varepsilon)] = O(\varepsilon)(E_k \widehat{V}(x_k^\varepsilon, \alpha_k^\varepsilon) + 1),
$$

and upon cancellation, we have

$$
\begin{aligned}
E_k[V^\varepsilon(x_{k+1}^\varepsilon, k+1) &- V^\varepsilon(x_k^\varepsilon, k)] \\
&= \varepsilon E_k V_x'(x_k^\varepsilon, \overline{\alpha}_k^\varepsilon) \overline{f}(x_k^\varepsilon, \overline{\alpha}_k^\varepsilon) + \varepsilon E_k Q \widehat{V}(x_k^\varepsilon, \cdot)(\alpha_k^\varepsilon) \qquad (7.35) \\
&\quad + O(\varepsilon^2)(E_k \widehat{V}(x_k^\varepsilon, \alpha_k^\varepsilon) + 1).
\end{aligned}
$$

Step 3: (Final estimates and iteration). With the $\gamma > 0$ given in the theorem, we have

$$
\begin{aligned}
E\left[e^{\varepsilon \gamma k} V^\varepsilon(x_k^\varepsilon, k) - e^{\varepsilon \gamma(k-1)} V^\varepsilon(x_{k-1}^\varepsilon, k-1) \right] \\
= e^{\varepsilon \gamma(k-1)} [e^{\varepsilon \gamma} - 1] E V^\varepsilon(x_k^\varepsilon, k) \qquad (7.36) \\
+ e^{\varepsilon \gamma(k-1)} E[E_{k-1} V^\varepsilon(x_k^\varepsilon, k) - V^\varepsilon(x_{k-1}^\varepsilon, k-1)].
\end{aligned}
$$

Iterating on the above recursion yields

$$
\begin{aligned}
E e^{\varepsilon \gamma k} V^\varepsilon(x_k^\varepsilon, k) = EV^\varepsilon(x_0, 0) + E \sum_{k_1=0}^{k-1} e^{\varepsilon \gamma k_1} [e^{\varepsilon \gamma} - 1] E_{k_1} V^\varepsilon(x_{k_1}^\varepsilon, k_1) \\
+ E \sum_{k_1=0}^{k-1} e^{\varepsilon \gamma k_1} [E_{k_1} V^\varepsilon(x_{k_1+1}^\varepsilon, k_1+1) - V^\varepsilon(x_{k_1}^\varepsilon, k_1)].
\end{aligned}
$$

$$(7.37)$$

Note that

$$
\begin{aligned}
E \sum_{k_1=0}^{k-1} e^{\varepsilon \gamma k_1} [e^{\varepsilon \gamma} - 1] V^\varepsilon(x_{k_1}^\varepsilon, k_1) \\
\leq K\gamma\varepsilon \sum_{k_1=0}^{k-1} e^{\varepsilon \gamma k_1} EV^\varepsilon(x_{k_1}^\varepsilon, k_1),
\end{aligned}
$$

$$(7.38)$$

and

$$E \sum_{k_1=0}^{k-1} e^{\varepsilon\gamma k_1} E_{k_1}[V^\varepsilon(x^\varepsilon_{k_1+1}, k_1+1) - V^\varepsilon(x^\varepsilon_{k_1}, k_1)]$$

$$= \varepsilon E\left[E_{k_1} \sum_{k_1=0}^{k-1} e^{\varepsilon\gamma k_1} V'_x(x^\varepsilon_{k_1}, \overline{\alpha}^\varepsilon_{k_1}) \overline{f}(x^\varepsilon_{k_1}, \overline{\alpha}^\varepsilon_{k_1}) \right.$$

$$+ \sum_{k_1=0}^{k-1} e^{\varepsilon\gamma k_1} E_{k_1} \overline{Q} V(x^\varepsilon_{k_1}, \cdot)(\overline{\alpha}^\varepsilon_{k_1})$$

$$+ \sum_{k_1=0}^{k-1} e^{\varepsilon\gamma k_1} O(\varepsilon)(E_{k_1} \widehat{V}(x^\varepsilon_{k_1}, \alpha^\varepsilon_{k_1}) + 1)$$

$$\left. + \sum_{k_1=0}^{k-1} e^{\varepsilon\gamma k_1} E_{k_1} \left[\sum_{i=1}^{l_0} \sum_{j=1}^{m_i} Q\widehat{V}(x^\varepsilon_{k_1}, \cdot)(s_{ij})(I_{\{\alpha^\varepsilon_{k_1}=s_{ij}\}} - \nu^{ij} I_{\{\alpha^\varepsilon_{k_1}\in\mathcal{M}_i\}}) \right] \right].$$
$$(7.39)$$

In the last term of (7.39), replacing the term in the squared brackets by

$$QV^\varepsilon(x^\varepsilon_{k_1}, k_1) \stackrel{\text{def}}{=} Q\widehat{V}(x^\varepsilon_{k_1}, \cdot)(\alpha^\varepsilon_{k_1})$$

$$+ \varepsilon \sum_{k_2=k_1}^{\infty} E_{k_1} e^{\varepsilon(k_1-k_2)} \widehat{V}'_x(x^\varepsilon_{k_1}, \cdot)(\alpha^\varepsilon_{k_1})[f(x^\varepsilon_{k_1}, \alpha^\varepsilon_{k_2}) - \overline{f}(x^\varepsilon_{k_1}, \overline{\alpha}^\varepsilon_{k_2})]$$

yields another term of the order $O(\varepsilon)(E_{k_1} \widehat{V}(x^\varepsilon_{k_1}, \alpha^\varepsilon_{k_1}) + 1)$, which is added to the next to the last term of (7.39). Using (7.14),

$$V'_x(x^\varepsilon_{k_1}, \overline{\alpha}^\varepsilon_{k_1}) \overline{f}(x^\varepsilon_{k_1}, \overline{\alpha}^\varepsilon_{k_1}) + \overline{Q}V(x^\varepsilon_{k_1}, \cdot)(\overline{\alpha}^\varepsilon_{k_1}) \le -\gamma V(x^\varepsilon_{k_1}, \overline{\alpha}^\varepsilon_{k_1}),$$

and in addition, for sufficiently small $\varepsilon > 0$,

$$-\gamma V(x^\varepsilon_{k_1}, \overline{\alpha}^\varepsilon_{k_1}) + O(\varepsilon)V(x^\varepsilon_{k_1}, \overline{\alpha}^\varepsilon_{k_1}) \le -\gamma_1 V(x^\varepsilon_{k_1}, \overline{\alpha}^\varepsilon_{k_1})$$

for some $0 < \gamma_1 < \gamma$. As a result,

$$E\varepsilon \sum_{k_1=0}^{k-1} e^{\varepsilon\gamma k_1} E_{k_1} \left[V'_x(x^\varepsilon_{k_1}, \overline{\alpha}^\varepsilon_{k_1}) \overline{f}(x^\varepsilon_{k_1}, \overline{\alpha}^\varepsilon_{k_1}) + \overline{Q}V(x^\varepsilon_{k_1}, \cdot)(\overline{\alpha}^\varepsilon_{k_1}) \right.$$
$$(7.40)$$
$$\left. + O(\varepsilon)\widehat{V}(x^\varepsilon_{k_1}, \alpha^\varepsilon_{k_1}) \right] \le 0.$$

Using (7.38)–(7.40) in (7.37) and dividing both sides by $e^{\varepsilon\gamma k}$, we obtain

$$E \sum_{k_1=0}^{k} e^{\varepsilon\gamma(k_1-k)}[e^{\varepsilon\gamma} - 1]V^\varepsilon(x^\varepsilon_{k_1}, k_1)$$

$$\le K\gamma\varepsilon \sum_{k_1=0}^{k} e^{\varepsilon\gamma(k_1-k)} EV^\varepsilon(x^\varepsilon_{k_1}, k_1)$$

$$+ \varepsilon \sum_{k_1=0}^{k} e^{\varepsilon\gamma(k_1-k)} EQV^\varepsilon(x^\varepsilon_{k_1}, k_1) + \varepsilon \sum_{k_1=0}^{k} e^{\varepsilon\gamma(k_1-k)} O(\varepsilon).$$

It follows that

$$EV^\varepsilon(x_{k+1}^\varepsilon, k+1) \le e^{-\varepsilon\gamma k} EV^\varepsilon(x_0, 0)$$
$$+ K\gamma\varepsilon \sum_{k_1=0}^{k} e^{\varepsilon\gamma(k_1-k)} EV^\varepsilon(x_{k_1}^\varepsilon, k_1) + O(\varepsilon).$$

An application of Gronwall's inequality yields

$$EV^\varepsilon(x_{k+1}^\varepsilon, k+1) \le \left[e^{-\varepsilon\gamma k} EV^\varepsilon(x_0, 0) + O(\varepsilon) \right] \exp\left[K\gamma\varepsilon \sum_{k_1=0}^{k} e^{\varepsilon\gamma(k_1-k)} \right]$$
$$\le K e^{-\varepsilon\gamma k} EV^\varepsilon(x_0, 0) + O(\varepsilon).$$

(7.41)

Using (7.32) with $V^\varepsilon(x_k^\varepsilon, k)$ replaced by $V(x_k^\varepsilon, \overline{\alpha}_k^\varepsilon)$ in (7.41), we obtain

$$EV(x_{k+1}^\varepsilon, \overline{\alpha}_{k+1}^\varepsilon) \le K e^{-\varepsilon\gamma k} EV(x_0, \overline{\alpha}_0) + O(\varepsilon).$$

The proof of the theorem is concluded.   □

Proof of Theorem 7.9. Define $\widehat{F}(x)$, $\widetilde{F}(x)$, and $\widehat{V}(x, \iota)$ as in the proof of Theorem 7.7. Use E_k and E_k^α to denote the conditional expectations with respect to the σ-algebras

$$\mathcal{G}_k = \{x_0, \alpha_j^\varepsilon, w_j : j < k\} \quad \text{and} \quad \mathcal{G}_k^\alpha = \{x_0, \alpha_k^\varepsilon, \alpha_j^\varepsilon, w_j : j < k\},$$

respectively. Since $\widehat{V}_x(x_k^\varepsilon, \alpha_k^\varepsilon)$ and $\sigma(x_k^\varepsilon, \alpha_k^\varepsilon)$ are $\mathcal{G}_k^\alpha$-measurable and $\{w_k\}$ is independent of $\mathcal{G}_k^\alpha$,

$$E_k \widehat{V}_x'(x_k^\varepsilon, \alpha_k^\varepsilon) \sigma(x_k^\varepsilon, \alpha_k^\varepsilon) w_k = E_k E_k^\alpha \widehat{V}_x'(x_k^\varepsilon, \alpha_k^\varepsilon) \sigma(x_k^\varepsilon, \alpha_k^\varepsilon) w_k$$
$$= E_k \widehat{V}_x'(x_k^\varepsilon, \alpha_k^\varepsilon) \sigma(x_k^\varepsilon, \alpha_k^\varepsilon) E_k^\alpha w_k = 0.$$

Consequently, we have

$$E_k[\widehat{V}(x_{k+1}^\varepsilon, \alpha_{k+1}^\varepsilon) - \widehat{V}(x_k^\varepsilon, \alpha_k^\varepsilon)]$$
$$= \varepsilon E_k \widehat{V}_x'(x_k^\varepsilon, \alpha_k^\varepsilon) f(x_k^\varepsilon, \alpha_k^\varepsilon) + \varepsilon E_k Q\widehat{V}(x_k^\varepsilon, \cdot)(\alpha_k^\varepsilon)$$
$$+ \frac{1}{2} E_k \mathrm{tr}[\widehat{V}_{xx}(x_k^\varepsilon, \alpha_k^\varepsilon)(x_{k+1}^\varepsilon - x_k^\varepsilon)(x_{k+1}^\varepsilon - x_k^\varepsilon)']$$
$$+ \widetilde{\eta}_{k+1}^\varepsilon + O(\varepsilon^2)(E_k \widehat{V}(x_k^\varepsilon, \alpha_k^\varepsilon) + 1),$$

(7.42)

where

$$\widetilde{\eta}_{k+1}^\varepsilon = O\left(E_k \sum_{\iota=3}^{d_0-1} |\partial^\iota \widehat{V}(x_k^\varepsilon, \alpha_k^\varepsilon)| \|x_{k+1}^\varepsilon - x_k^\varepsilon|^\iota \right)$$
$$+ O\left(E_k |x_{k+1}^\varepsilon - x_k^\varepsilon|^{d_0} \int_0^1 |\partial^{d_0} \widehat{V}(x_k^\varepsilon + s(x_{k+1}^\varepsilon - x_k^\varepsilon))| ds \right).$$

By virtue of the independence of w_k with $\mathcal{G}_k^\alpha$ and the measurability of $\widehat{V}_{xx}(x_k^\varepsilon, \alpha_k^\varepsilon)$, $f(x_k^\varepsilon, \alpha_k^\varepsilon)$, and $\sigma(x_k^\varepsilon, \alpha_k^\varepsilon)$ with respect to $\mathcal{G}_k^\alpha$, using (A7.4), we have

$$E_k \text{tr}[\widehat{V}_{xx}(x_k^\varepsilon, \alpha_k^\varepsilon)\sigma(x_k^\varepsilon, \alpha_k^\varepsilon)w_k w_k' \sigma'(x_k^\varepsilon, \alpha_k^\varepsilon)]$$
$$= E_k \text{tr}[\widehat{V}_{xx}(x_k^\varepsilon, \alpha_k^\varepsilon)\sigma(x_k^\varepsilon, \alpha_k^\varepsilon)E_k^\alpha[w_k w_k']\sigma'(x_k^\varepsilon, \alpha_k^\varepsilon)]$$
$$= E_k \text{tr}[\widehat{V}_{xx}(x_k^\varepsilon, \alpha_k^\varepsilon)\sigma(x_k^\varepsilon, \alpha_k^\varepsilon)\sigma'(x_k^\varepsilon, \alpha_k^\varepsilon)],$$

and similarly

$$E_k \text{tr}[\widehat{V}_{xx}(x_k^\varepsilon, \alpha_k^\varepsilon)\sigma(x_k^\varepsilon, \alpha_k^\varepsilon)w_k f'(x_k^\varepsilon, \alpha_k^\varepsilon)] = 0,$$
$$\varepsilon^2 E_k \text{tr}[\widehat{V}_{xx}(x_k^\varepsilon, \alpha_k^\varepsilon)f(x_k^\varepsilon, \alpha_k^\varepsilon)f'(x_k^\varepsilon, \alpha_k^\varepsilon)] \tag{7.43}$$
$$= O(\varepsilon^2)(E_k \widehat{V}(x_k^\varepsilon, \alpha_k^\varepsilon) + 1).$$

Thus,

$$E_k \text{tr}[\widehat{V}_{xx}(x_k^\varepsilon, \alpha_k^\varepsilon)(x_{k+1}^\varepsilon - x_k^\varepsilon)(x_{k+1}^\varepsilon - x_k^\varepsilon)'] \tag{7.44}$$
$$= \varepsilon \text{tr}[E_k \widehat{V}_{xx}(x_k^\varepsilon, \alpha_k^\varepsilon)\Xi(x_k^\varepsilon, \alpha_k^\varepsilon)] + O(\varepsilon^2)(E_k \widehat{V}(x_k^\varepsilon, \alpha_k^\varepsilon) + 1).$$

As for the next to the last term in (7.42), similar to the estimates of (7.28) with the use of independence of $\{w_k\}$ of $\{\alpha_k^\varepsilon\}$, we obtain

$$\widetilde{\eta}_{k+1}^\varepsilon = O(\varepsilon^2)(E_k \widehat{V}(x_k^\varepsilon, \alpha_k^\varepsilon) + 1).$$

Therefore, we have

$$E_k[\widehat{V}(x_{k+1}^\varepsilon, \alpha_{k+1}^\varepsilon) - \widehat{V}(x_k^\varepsilon, \alpha_k^\varepsilon)]$$
$$= \varepsilon E_k \widehat{V}_x'(x_k^\varepsilon, \alpha_k^\varepsilon)f(x_k^\varepsilon, \alpha_k^\varepsilon) + \varepsilon E_k Q\widehat{V}(x_k^\varepsilon, \cdot)(\alpha_k^\varepsilon) \tag{7.45}$$
$$+ \frac{\varepsilon}{2}\text{tr}[E_k \widehat{V}_{xx}(x_k, \alpha_k^\varepsilon)\Xi(x_k^\varepsilon, \alpha_k^\varepsilon)] + O(\varepsilon^2)(E_k \widehat{V}(x_k^\varepsilon, \alpha_k^\varepsilon) + 1).$$

Define $V_1^\varepsilon(x, k)$ as in (7.31), and define

$$V_2^\varepsilon(x, k) = \frac{\varepsilon}{2}\sum_{k_1=k}^\infty E_k e^{\varepsilon(k-k_1)}\text{tr}[\widehat{V}_{xx}(x, \alpha_k^\varepsilon)[\Xi(x, \alpha_{k_1}^\varepsilon) - \overline{\Xi}(x, \overline{\alpha}_{k_1}^\varepsilon)]]. \tag{7.46}$$

Since the arguments are similar to that of Theorem 7.7, we only outline the main steps and point out the differences here. As we did in the proof of Theorem 7.7,

$$\sup_{0 \le k} E|V_\ell^\varepsilon(x_k^\varepsilon, k)| < \infty, \quad \text{for } \ell = 1, 2,$$
$$|\widehat{V}_\ell^\varepsilon(x_k, k)| = O(\varepsilon)(E_k \widehat{V}(x_k^\varepsilon, \alpha_k^\varepsilon) + 1), \quad \text{for } \ell = 1, 2.$$

Note that

$$
E_k[V_2^\varepsilon(x_{k+1}^\varepsilon, k+1) - V_2^\varepsilon(x_k^\varepsilon, k+1)]
$$

$$
= \varepsilon \sum_{k_1=k+1}^{\infty} e^{\varepsilon(k+1-k_1)} E_k \operatorname{tr}\{\widehat{V}_{xx}(x_{k+1}^\varepsilon, \alpha_{k+1}^\varepsilon)[\Xi(x_{k+1}^\varepsilon, \alpha_{k_1}^\varepsilon) - \overline{\Xi}(x_{k+1}^\varepsilon, \overline{\alpha}_{k_1}^\varepsilon)]
$$

$$
- \widehat{V}_{xx}(x_k^\varepsilon, \alpha_{k+1}^\varepsilon)[\Xi(x_k^\varepsilon, \alpha_{k_1}^\varepsilon) - \overline{\Xi}(x_k^\varepsilon, \overline{\alpha}_{k_1}^\varepsilon)]\}
$$

$$
= O(\varepsilon^2)(E_k \widehat{V}(x_k^\varepsilon, \alpha_k^\varepsilon) + 1),
$$

and that

$$
E_k[V_2^\varepsilon(x_k^\varepsilon, k+1) - V_2^\varepsilon(x_k^\varepsilon, k)]
$$

$$
= -\frac{\varepsilon}{2}\operatorname{tr}\{E_k\widehat{V}_{xx}(x_k^\varepsilon, \alpha_k^\varepsilon)\Xi(x_k^\varepsilon, \alpha_k^\varepsilon)\}
$$

$$
+ \frac{\varepsilon}{2}\operatorname{tr}\{E_k\widehat{V}_{xx}(x_k^\varepsilon, \alpha_k^\varepsilon)\overline{\Xi}(x_k^\varepsilon, \overline{\alpha}_k^\varepsilon)\} + O(\varepsilon^2)(E_k\widehat{V}(x_k^\varepsilon, \alpha_k^\varepsilon) + 1).
$$

Next define

$$
V^\varepsilon(x, k) = \widehat{V}(x, \alpha_k^\varepsilon) + V_1^\varepsilon(x, k) + V_2^\varepsilon(x, k).
$$

It follows that

$$
E_k[V^\varepsilon(x_{k+1}^\varepsilon, k+1) - V^\varepsilon(x_k^\varepsilon, k)]
$$

$$
= \varepsilon E_k V_x'(x_k^\varepsilon, \overline{\alpha}_k^\varepsilon)\overline{f}(x_k^\varepsilon, \overline{\alpha}_k^\varepsilon) + \varepsilon E_k \overline{Q}V(x_k^\varepsilon, \cdot)(\overline{\alpha}_k^\varepsilon)
$$

$$
+ \frac{\varepsilon}{2}\operatorname{tr}\{E_k V_{xx}(x_k^\varepsilon, \overline{\alpha}_k^\varepsilon)\overline{\Xi}(x_k^\varepsilon, \alpha_k^\varepsilon)\}
$$

$$
+ \varepsilon E_k[Q\widehat{V}(x_k^\varepsilon, \cdot)(\alpha_k^\varepsilon) - \overline{Q}V(x_k^\varepsilon, \cdot)(\overline{\alpha}_k^\varepsilon)]
$$

$$
+ O(\varepsilon^2)(E_k V(x_k^\varepsilon, \overline{\alpha}_k^\varepsilon) + 1).
$$

If we proceed as we did in the proof of Theorem 7.7, the desired stability follows. A few details are omitted.   □

Proof of Theorem 7.12. The proofs for the systems (7.2) and (7.3) are similar, so we shall concern ourselves with (7.2) only. Define $V_1^\varepsilon(x, k)$ as in the proof of Theorem 7.7, and define

$$
V^\varepsilon(k) = V(x_k^\varepsilon, \overline{\alpha}_k^\varepsilon) + V_1^\varepsilon(x_k^\varepsilon, k).
$$

It is easily seen that by virtue of (7.22), for each x,

$$
|V_1^\varepsilon(x, k)| \leq O(\varepsilon)(1 + E_k V(x, \overline{\alpha}_k^\varepsilon)). \tag{7.47}
$$

For any $M > 0$, define a truncated and perturbed Liapunov function as $V_M^\varepsilon(k) = V^\varepsilon(k)q_M(x)$, where $q_M(x)$ is a smooth function that is equal to 1 for $x \in S_M = \{x : |x| \leq M\}$ and equal to 0 for $x \in \mathbb{R}^n - S_{M+1}$. For any

$N > 0$, if $|x_k^\varepsilon| < N$, but $x_k^\varepsilon \notin B_0$, then for all $M > N$, $q_M(x_k^\varepsilon) = 1$ in S_N. Thus $V_M^\varepsilon(k) = V^\varepsilon(k)$ and by using the estimates similar to the proof of Theorem 7.7, we obtain for such M,

$$E_k[V_M^\varepsilon(k+1) - V_M^\varepsilon(k)] \le -c_0 + e_k^\varepsilon,$$

where $e_k^\varepsilon = o(\varepsilon)$ as $\varepsilon \to 0$. Thus, for sufficiently small $\varepsilon > 0$, we can make $-c_0 + e_k^\varepsilon \le -c_1$ for some $c_1 > 0$ with $c_1 < c_0$.

Define

$$\widetilde{\tau}_N = \inf\{k \ge \tau_0 : |x_k^\varepsilon| \ge N\}.$$

Using the recursion via telescoping,

$$E_{\tau_0}[V^\varepsilon(\tau_1 \wedge (\tau_0 + k) \wedge \widetilde{\tau}_N) - V^\varepsilon(\tau_0)]$$
$$= E_{\tau_0} \sum_{k=\tau_0}^{\tau_1 \wedge (\tau_0+k) \wedge \widetilde{\tau}_N - 1} E_k[V_M^\varepsilon(k+1) - V_M^\varepsilon(k)] \qquad (7.48)$$
$$\le -c_1 E_{\tau_0}[\tau_1 \wedge (\tau_0 + k) \wedge \widetilde{\tau}_N - \tau_0],$$

where $a \wedge b = \min(a,b)$. Replacing $V_1^\varepsilon(x,k)$ in (7.47) by $V^\varepsilon(k) - V(x_k^\varepsilon, \overline{\alpha}_k^\varepsilon)$, we obtain

$$V^\varepsilon(k) \ge (1 - O(\varepsilon))E_k V(x, \overline{\alpha}_k^\varepsilon) - O(\varepsilon).$$

We claim that $\lim_N \widetilde{\tau}_N > \tau_1$. If not, then

$$E_{\tau_0} V^\varepsilon(\tau_1 \wedge (\tau_0 + k) \wedge \widetilde{\tau}_N) \ge E_{\tau_0} V(x_k^\varepsilon, \overline{\alpha}_k^\varepsilon) - O(\varepsilon)$$
$$\to \infty \quad \text{as} \quad N \to \infty,$$

which is a contradiction.

Since $E_{\tau_0} V^\varepsilon(\tau_1 \wedge (\tau_0 + k) \wedge \widetilde{\tau}_N) \ge 0$, by (7.48), we have

$$c_1 E_{\tau_0}[\tau_1 \wedge (\tau_0 + k) \wedge \widetilde{\tau}_N - \tau_0] \le E_{\tau_0} V^\varepsilon(\tau_0). \qquad (7.49)$$

Using (7.47), we upper bound the right-hand side of (7.49) by

$$E_{\tau_0} V^\varepsilon(\tau_0) \le E_{\tau_0} V(x_{\tau_0}^\varepsilon, \overline{\alpha}_{\tau_0}^\varepsilon)(1 + O(\varepsilon)) + O(\varepsilon). \qquad (7.50)$$

Dividing both sides of (7.49) by c_1 and using the bound in (7.50), we obtain

$$E_{\tau_0}[\tau_1 \wedge (\tau_0 + k) \wedge \widetilde{\tau}_N - \tau_0] \le \frac{E_{\tau_0} V(x_{\tau_0}^\varepsilon, \overline{\alpha}_{\tau_0}^\varepsilon)(1 + O(\varepsilon)) + O(\varepsilon)}{c_1}. \qquad (7.51)$$

Letting $N \to \infty$ and $k \to \infty$ in (7.51), we obtain (7.23). $\quad\square$

7.7 Remarks

7.7.1 Extensions

The results obtained can be extended to the case that the transition matrix P in (7.1) consists of not only recurrent states, but also transient states. That is, the stability study can be extended to allow the transition matrix P to be given by

$$P = \begin{pmatrix} P^1 & & & & \\ & P^2 & & & \\ & & \ddots & & \\ & & & P^{l_0} & \\ P^{*,1} & P^{*,2} & \cdots & P^{*,l_0} & P^* \end{pmatrix}. \tag{7.52}$$

A Markov chain with transition matrix (7.52) has l_0 recurrent classes and a number of transient states. What we will need are the following conditions: Assume that P^i for $i = 1, \ldots, l_0$ satisfy (A7.1), that all the eigenvalues of P^* are inside the unit circle, and that P and P^ε are transition matrices. For ease of reference, we put this into the following condition.

(A7.1') Condition (HT) of Chapter 6 is satisfied.

To carry out the desired study, we can partition the matrix Q as $Q = \begin{pmatrix} Q^{11} & Q^{12} \\ Q^{21} & Q^{22} \end{pmatrix}$, where $Q^{11} \in \mathbb{R}^{(m-m_*) \times (m-m_*)}$, $Q^{12} \in \mathbb{R}^{(m-m_*) \times m_*}$, $Q^{21} \in \mathbb{R}^{m_* \times (m-m_*)}$, and $Q^{22} \in \mathbb{R}^{m_* \times m_*}$. Write

$$\overline{Q}_* = \mathrm{diag}(\nu^1, \ldots, \nu^{l_0})(Q^{11}\tilde{\mathbb{1}} + Q^{12}A_*), \tag{7.53}$$

where

$$A_* = (a^1, \ldots, a^{l_0}) \in \mathbb{R}^{m_* \times l_0}, \quad \text{with}$$
$$a^i = -(P^* - I)^{-1} P^{*,i} \mathbb{1}_{m_i}, \text{ for } i = 1, \ldots, l_0. \tag{7.54}$$

Now the aggregated process $\overline{\alpha}_k^\varepsilon$ is modified as follows:

$$\overline{\alpha}_k^\varepsilon = \begin{cases} i, & \text{if } \alpha_k^\varepsilon \in \mathcal{M}_i, \\ U_j, & \text{if } \alpha_k^\varepsilon = s_{*j}, \end{cases} \tag{7.55}$$

with U_j being given by

$$U_j = I_{\{0 \le U \le a^{1,j}\}} + 2I_{\{a^{1,j} < U \le a^{1,j}+a^{2,j}\}} + \ldots + l_0 I_{\{a^{1,j}+\ldots+a^{l_0-1,j} < U \le 1\}},$$

and U is a random variable uniformly distributed on $[0, 1]$, independent of α_k^ε. Define piecewise constant interpolation $\overline{\alpha}^\varepsilon(\cdot)$ of $\overline{\alpha}_k^\varepsilon$ as $\overline{\alpha}^\varepsilon(t) = \overline{\alpha}_k$ for $t \in [\varepsilon k, \varepsilon k + \varepsilon)$. Then we can show $\overline{\alpha}^\varepsilon(\cdot)$ converges weakly to $\overline{\alpha}(\cdot)$, which is a Markov chain generated by $\overline{Q}_*$. The discussion of the asymptotic expansions of the inclusion of the transient states and non-homogeneous Markov chains can be found in Chapters 3 and 4, respectively.

With the preparation above, we can carry out the stability analysis. The proofs and techniques are similar. The main idea is that the transient states do not contribute anything to the limit systems. Note that we only aggregate states in each recurrent class. The limit systems are still averages with respect to the stationary measures of the recurrent states. The results are stated as follows.

Theorem 7.14 *Assume the conditions of Theorem 7.7 (respectively, conditions of Theorem 7.9) with the replacement of (A7.1) by (A7.1'). Then the conclusions of Theorem 7.7 (respectively, Theorem 7.9) continue to hold.*

7.7.2 Further Investigation

We have studied stability problems arising from discrete-time dynamic systems. The main results indicate that under appropriate conditions, stability of the limit continuous-time dynamic systems implies that of the original problems. A number of interesting problems remain. For example, one may wish to consider the invariant sets and invariance principles associated the dynamic systems, which will be modifications of the results in LaSalle [101]. For example, associated with (7.8), we may consider

$$x_{k+1}^\varepsilon = x_k^\varepsilon + \varepsilon A(\alpha_k^\varepsilon) f(x_k^\varepsilon) x_k^\varepsilon,$$

where

$$f(x_k^\varepsilon) = \begin{pmatrix} -\gamma_1 & 1 - x_k^{\varepsilon,1} \\ 1 - x_k^{\varepsilon,2} & -\gamma_2 \end{pmatrix}, \quad x_k^\varepsilon = (x_k^{\varepsilon,1}, x_k^{\varepsilon,2})'.$$

This model (without the Markov chain) has been used as an epidemic model and/or predator-prey model (see LaSalle [101]). The $x_k^{\varepsilon,\ell}$ is the fraction of the population ℓ infected and $(1 - x_k^{\varepsilon,\ell})$ is the fraction of the population susceptible; $\gamma^\ell > 0$ and $a(\iota)\gamma^\ell$ and $b(\iota)\gamma^\ell$ are recovery coefficients. In conjunction with this formulation, asymptotic stability relative to the positive invariant set $\overline{G}$, the closure of $G = \{x : 0 < x^\ell < 1, \ell = 1, 2\}$, may be studied. This and more general invariance principles deserve to be further studied.

7.8 Notes

The results of this Chapter are based on the paper of Yin and Zhang [161]. Classical theory of stochastic stability can be found in the work of Khasminskii [83] and Kushner [95]. The main techniques used are based on the Liapunov stability. The use of perturbed Liapunov functions can be traced back to the work of Papanicolaou, Stroock, and Varadhan [122], in which they used such methods to obtain diffusion approximation results. Analysis of the stability of systems under random perturbation can be found in the work of Blankenship and Papanicolaou [22].

This chapter is concerned with systems with jumps in discrete time. The stability of systems involving jump parameters has also been studied by Badowski and Yin [9], Ji and Chizeck [74], Mao [109], and Mariton [111], among others.

Recent studies of systems with regime switching have indicated that such a formulation is more suitable for many applications; see, for example, the feedback linear system in Blair and Sworder [20], the robust control formulation of Mariton [111], Markov decision problems in Liu, Zhang, and Yin [103], and portfolio selections and near-optimal controls in Zhang and Yin [182]. For some of the recent developments on two-time-scale Markovian systems, refer to Abbad, Filar, and Bielecki [2] and Pervozvanskii and Gaitsgory [123], among others. For some recent progress in stability of hybrid systems, refer to Ji and Chizeck [74], Mao [108, 109], among others.

8
Filtering

8.1 Introduction

This chapter is concerned with hybrid filtering in discrete time. In the traditional setting of filtering problems, the coefficients of the systems are fixed parameters and are deterministic. This, however, prevents one from treating situations in which the actual systems differ from the nominal model. Therefore, efforts have been made to design more "robust" filters. Because the coefficients may be subject to random perturbations, it is particularly useful to develop filters under regime switching. In our formulation, in addition to the system and observation noises, disturbances from the random environment on the system coefficients are also considered. In particular, we incorporate random coefficients and model regime (or configuration) changes. It is natural to adopt a Markov chain model to serve as the modulating random process. In addition to the dynamics (state and observation) in the traditional setting, there is a Markovian jump process responsible for the regime changes of the underlying systems. The basic premise is as follows. Because of the regime changes caused by the random environment, in any given instance, the system chooses its configuration in accordance with the current value of the Markov chain. Corresponding to this configuration, the evolution of the systems (both state and observation) follows dynamic systems given by a pair of difference equations for a random duration until the Markov chain jumps to a new state. Then the configuration is changed, and the dynamic systems for both the state and the observation follow another pair of difference equations, and so on. Since many problems arising from applications such as target tracking, speech recognition, telecommunication, and manufacturing require solutions of filtering problems involving a regime switching, obtaining the state estimates of such systems is vital in these applications.

Note that in the filtering problem, although the state space of the Markov chain is finite, it may contain a large number of states. We introduce a small parameter $\varepsilon > 0$ into the transition probability matrix as in the previous chapters of the book. The small parameter is used to reflect the high contrast of the transition rates of the Markov chain. To carry out asymptotic analysis, it is necessary to let $\varepsilon \to 0$, which can serve as a guideline for applications, approximations, and heuristics. In real applications, however, ε might be a fixed constant and only the relative order of magnitude of this parameter matters. Following the framework used in this book, we study the asymptotic properties of the filtering problem by means of weak convergence methods.

Let α_k^ε be a discrete-time Markov chain with state space $\mathcal{M}$. Its transition matrix is of the form

$$P^\varepsilon = P + \varepsilon Q, \tag{8.1}$$

where P^ε and P are $m_0 \times m_0$ transition matrices and Q is a generator of a continuous-time Markov chain. As in Chapter 6, we consider the structure of P given by

$$P = \begin{pmatrix} P^1 & & & & \\ & P^2 & & & \\ & & \ddots & & \\ & & & P^{l_0} & \\ P^{*,1} & P^{*,2} & \cdots & P^{*,l_0} & P^* \end{pmatrix}, \tag{8.2}$$

where for each $i \leq l_0$, P^i is a transition matrix within the ith recurrent class, and the last row $(P^{*,1}, \ldots, P^{*,l_0}, P^*)$ in (8.2) corresponds to the transient states. Each of the transition matrix P^i dictates the faster transitions within the ith-ergodic class, whereas the generator Q governs the slow transitions from one ergodic class to another. Our goal is again to reduce complexity.

Based on the developments in previous chapters, we establish the natural connection between the discrete-time problem and its continuous-time limit. Under simple conditions, we show that suitably interpolated processes converge weakly to their limits, leading to continuous-time dynamic systems with regime switching. Thus, in lieu of examining the more complex original problem, we need only treat a much simplified limit system. Using the limit problem as a guide, we construct near-optimal filters. Note that the original optimal filter is infinite dimensional. The main advantage of our approach is the reduction of complexity, which makes it possible to obtain finite-dimensional near-optimal filters.

This chapter is arranged as follows. Section 8.2 presents the main results of the hybrid filtering problem. In order not to interrupt the flow of presentation, proofs and technical details are arranged in Section 8.3. Section 8.4 presents a discrete-time approximation of Wonham's filter, which is another application of the two-time-scale Markov chain and which yields an efficient numerical algorithm. Section 8.5 collects some notes and further remarks.

8.2   Main Results

Aiming at reducing the complexity of filtering problems involving large-scale hidden Markov chains, we show that a limit system can be derived in which the underlying Markov chain is replaced by an averaged chain and the system coefficients are averaged out with respect to the stationary measures of ergodic classes. Such a limit system can be used to construct approximate filtering schemes that are nearly optimal. The reduction of complexity is particularly pronounced when the transition matrix of the Markov chain consists of only one ergodic class. In this case, the limit problem becomes a standard Kalman filter free of Markovian switching processes.

8.2.1   Formulation

Following the two-time-scale formulation, let $\varepsilon > 0$ be a small parameter and $\{\alpha_k^\varepsilon\}$ be a time-homogeneous Markov chain with a finite state space $\mathcal{M}$ having m_0 elements and a transition matrix (8.1). Suppose that for some $T > 0$ and $0 \le k \le \lfloor T/\varepsilon \rfloor$ (where $\lfloor z \rfloor$ denotes the integer part of z, the largest integer that is less than or equal to z). For ease of presentation, in what follows, we suppress the floor-function notation $\lfloor \cdot \rfloor$ and write $\lfloor T/\varepsilon \rfloor$ as T/ε whenever there is no confusion.

Let $x_k^\varepsilon \in \mathbb{R}^r$ be the state to be estimated, y_k^ε be the corresponding observation, and $A(\iota)$, $C(\iota)$, $\sigma_w(\iota)$, and $\sigma_v(\iota)$ be finite for each $\iota \in \mathcal{M}$. The hybrid filtering problem is concerned with the following linear system of equations:

$$x_{k+1}^\varepsilon = x_k^\varepsilon + \varepsilon A(\alpha_k^\varepsilon)x_k^\varepsilon + \sqrt{\varepsilon}\sigma_w(\alpha_k^\varepsilon)w_k, \ x_0^\varepsilon = x,$$
$$y_{k+1}^\varepsilon = y_k^\varepsilon + \varepsilon C(\alpha_k^\varepsilon)x_k^\varepsilon + \sqrt{\varepsilon}\sigma_v(\alpha_k^\varepsilon)v_k, \ y_0^\varepsilon = 0,$$

$$(8.3)$$

where $\{w_k\}$ and $\{v_k\}$ are the system disturbance and the observation noise, respectively. The use of the $\sqrt{\varepsilon}$ in the noise terms stems from the central limit scaling. In what follows, we show that as $\varepsilon \to 0$, the above filtering problem has a limit. The limit filtering problem is still modulated by a Markov chain. However, the total number of states of the limit Markov chain is equal to l_0, the total number of ergodic classes. As mentioned before, typically $l_0 \ll m_0$, and by considering this limit filtering problem, substantial computational savings can be achieved. Although (8.3) is a discrete-time filtering problem, the limit under appropriate scaling is a continuous-time hybrid system.

In view of (8.1), the transition probabilities of α_k^ε are dominated by P, so its structure is important. Suppose that the matrix P is given by (8.2). It is clear that for sufficiently small $\varepsilon > 0$, P^ε is close to P, so P^ε is a so-called "nearly completely decomposable transition matrix" (see Courtois [41]). The corresponding state space of the Markov chain is

$$
\begin{aligned}
\mathcal{M} &= \mathcal{M}_1 \cup \mathcal{M}_2 \cup \cdots \cup \mathcal{M}_{l_0} \cup \mathcal{M}_* \\
&= \{s_{11}, \ldots, s_{1m_1}\} \cup \cdots \cup \{s_{l_0 1}, \ldots, s_{l_0 m_{l_0}}\} \cup \{s_{*1}, \ldots, s_{*m_*}\}.
\end{aligned}
\tag{8.4}
$$

For each $i = 1, \ldots, l_0$, $\mathcal{M}_i = \{s_{i1}, \ldots, s_{im_i}\}$ is the state space corresponding to the transition matrix P^i. The subspace $\mathcal{M}_* = \{s_{*1}, \ldots, s_{*m_*}\}$ collects the transient states. To proceed, we make the following assumptions about the Markov chain and the filtering system under consideration.

(A8.1) Assume Conditions (HT) in Chapter 6, i.e., P^i is irreducible and aperiodic for each $i = 1, \ldots, l_0$ and all eigenvalues of P^* are inside the unit disk.

(A8.2) For each $\iota \in \mathcal{M}$, $A(\iota), C(\iota), \sigma_w(\iota)$, and $\sigma_v(\iota)$ are finite; $\sigma_w(\iota)\sigma_w'(\iota)$ and $\sigma_v(\iota)\sigma_v'(\iota)$ are positive definite matrices.

(A8.3) The sequences $\{w_k\}$, $\{v_k\}$, and $\{\alpha_k^\varepsilon\}$ are mutually independent. The $\{w_k\}$ and $\{v_k\}$ are stationary martingale difference sequences such that

$$
E w_k w_k' = I, \ E v_k v_k' = I,
$$
$$
E|w_k|^{2+\Delta} < \infty, \ E|v_k|^{2+\Delta} < \infty \ \text{ for some } \ \Delta > 0.
$$

Owing to the assumption on the noise sequences $\{w_k\}$ and $\{v_k\}$ together with the appropriate scaling via the use of the small parameter ε, we obtain the functional central limit theorem or Donsker's invariance theorem. Its proof can be found in many references, see, for example, Ethier and Kurtz [55, Theorem 3.1, p. 351]. The following lemma is about a central-limit-theorem type approximation and it is a preparation for the limit hybrid filtering system.

Lemma 8.1. *Define*

$$
w^\varepsilon(t) = \sqrt{\varepsilon} \sum_{k=0}^{t/\varepsilon - 1} w_k \ \text{ and } \ v^\varepsilon(t) = \sqrt{\varepsilon} \sum_{k=0}^{t/\varepsilon - 1} v_k.
\tag{8.5}
$$

Under (A8.3), $w^\varepsilon(\cdot)$ *and* $v^\varepsilon(\cdot)$ *converge weakly to standard* r-*dimensional Brownian motions* $w(\cdot)$ *and* $v(\cdot)$, *respectively.*

Remark 8.2. For simplicity and ease of presentation, we assume that the noises are stationary martingale difference sequences, and that the covariance of w_k and v_k are the identity matrix. In fact, φ-mixing sequences can be treated and the corresponding functional central limit result can be obtained, but the notation will be more complex for the subsequent development. It is more informative to deal with a notationally simpler form to gain a basic understanding of the filtering problems.

Define the aggregated process $\overline{\alpha}_k^\varepsilon$ as in Section 6.4, and define its interpolation by

$$\overline{\alpha}_k^\varepsilon = \begin{cases} i, & \text{if } \alpha_k^\varepsilon \in \mathcal{M}_i, \\ U_j, & \text{if } \alpha_k^\varepsilon = s_{*j}, \end{cases} \tag{8.6}$$

$$\overline{\alpha}^\varepsilon(t) = \overline{\alpha}_k^\varepsilon \quad \text{for } t \in [\varepsilon k, \varepsilon k + \varepsilon),$$

where U_j is given by

$$U_j = I_{\{0 \le U \le a^{1,j}\}} + 2I_{\{a^{1,j} < U \le a^{1,j} + a^{2,j}\}} + \cdots + l_0 I_{\{a^{1,j} + \cdots + a^{l_0 - 1,j} < U \le 1\}},$$

and U is a random variable uniformly distributed on $[0,1]$, independent of α_k^ε. As shown in Proposition 6.5, $\overline{\alpha}^\varepsilon(\cdot)$ converges weakly to $\overline{\alpha}(\cdot)$, and the limit is a Markov chain with state space $\overline{\mathcal{M}}$.

8.2.2   Near-Optimal Filtering

In this subsection, we consider the partially observed system (8.3) with P^ε satisfying (8.1) and (8.2). In addition, we assume that $\overline{\alpha}_l^\varepsilon$ is observable. In view of the convergence results in Chapter 7, the corresponding limit system of (8.3) should have the form

$$dx = \overline{A}(\overline{\alpha}(t))x dt + \overline{\sigma}_w(\overline{\alpha}(t))dw,$$

$$dy = \overline{C}(\overline{\alpha}(t))x dt + \overline{\sigma}_v(\overline{\alpha}(t))dv,$$

with $x(0) = x$ and $y(0) = 0$, where $w(\cdot)$ and $v(\cdot)$ are independent r-dimensional, standard Brownian motions given by Lemma 8.1,

$$\overline{A}(i) = \sum_{j=1}^{m_i} \nu^{ij} A(s_{ij}), \quad \overline{C}(i) = \sum_{j=1}^{m_i} \nu^{ij} C(s_{ij}), \quad \text{for each } i \in \overline{\mathcal{M}}, \tag{8.7}$$

and for each $i \in \overline{\mathcal{M}}$, $\overline{\sigma}_w(i)$ and $\overline{\sigma}_v(i)$ satisfy

$$\Sigma_w(i) = \overline{\sigma}_w(i)\overline{\sigma}_w'(i) = \sum_{j=1}^{m_i} \nu^{ij} \sigma_w(s_{ij})\sigma_w'(s_{ij}),$$

$$\Sigma_v(i) = \overline{\sigma}_v(i)\overline{\sigma}_v'(i) = \sum_{j=1}^{m_i} \nu^{ij} \sigma_v(s_{ij})\sigma_v'(s_{ij}). \tag{8.8}$$

If both $y(t)$ and $\overline{\alpha}(t)$ are observable, then the corresponding Kalman filter is given as follows

$$d\widehat{x} = \overline{A}(\overline{\alpha}(t))\widehat{x}dt + R(t)\overline{C}'(\overline{\alpha}(t))\Sigma_v^{-1}(\overline{\alpha}(t))(dy - \overline{C}(\overline{\alpha}(t))\widehat{x}dt),$$

$$\dot{R} = \overline{A}(\overline{\alpha}(t))R(t) + R(t)\overline{A}'(\overline{\alpha}(t)) \qquad (8.9)$$

$$-R(t)\overline{C}(\overline{\alpha}(t))\Sigma_v^{-1}(\overline{\alpha}(t))\overline{C}(\overline{\alpha}(t))R(t) + \Sigma_w(\overline{\alpha}(t)),$$

with $\widehat{x}(0) = x$ and $R(0) = 0$. Here $\widehat{x}(t)$ is the best mean squares estimate of $x(t)$ given $\sigma\{y(s), \overline{\alpha}(s) : s \leq t\}$; see Fleming and Rishel [57]. Note that $y(t)$ and $\overline{\alpha}(t)$ are the weak limits of interpolations of $\{y_k^\varepsilon\}$ and $\{\overline{\alpha}_k^\varepsilon\}$, respectively. Neither of the formers are directly observable. In order to design a feasible filtering scheme, we need to replace $\{y(t)\}$ by $\{y_k^\varepsilon\}$ and $\{\overline{\alpha}(t)\}$ by $\{\overline{\alpha}_k^\varepsilon\}$, respectively. Intuitively, any filtering scheme driven by $(y_k^\varepsilon, \overline{\alpha}_k^\varepsilon)$ having a limit as in (8.9) should do. Here we adopt the following filtering scheme:

$$\widehat{x}_{k+1}^\varepsilon = \widehat{x}_k^\varepsilon + \varepsilon\overline{A}(\overline{\alpha}_k^\varepsilon)\widehat{x}_k^\varepsilon + (I + \varepsilon\overline{A}(\overline{\alpha}_k^\varepsilon))R_k^\varepsilon\overline{C}'(\overline{\alpha}_k^\varepsilon)$$
$$\times [\varepsilon\overline{C}(\overline{\alpha}_k^\varepsilon)R_k^\varepsilon\overline{C}'(\overline{\alpha}_k^\varepsilon) + \Sigma_v(\overline{\alpha}_k^\varepsilon)]^{-1} \qquad (8.10)$$
$$\times (y_{k+1}^\varepsilon - y_k^\varepsilon - \varepsilon\overline{C}(\overline{\alpha}_k^\varepsilon)\widehat{x}_k^\varepsilon),$$

and the Riccati equation

$$R_{k+1}^\varepsilon = \varepsilon\Sigma_w(\overline{\alpha}_k^\varepsilon) + (I + \varepsilon\overline{A}(\overline{\alpha}_k^\varepsilon))R_k^\varepsilon(I + \varepsilon\overline{A}'(\overline{\alpha}_k^\varepsilon))$$
$$-\varepsilon(I + \varepsilon\overline{A}(\overline{\alpha}_k^\varepsilon))R_k^\varepsilon\overline{C}'(\overline{\alpha}_k^\varepsilon)[\varepsilon\overline{C}(\overline{\alpha}_k^\varepsilon)R_k^\varepsilon\overline{C}'(\overline{\alpha}_k^\varepsilon) + \Sigma_v(\overline{\alpha}_k^\varepsilon)]^{-1}$$
$$\times \overline{C}(\overline{\alpha}_k^\varepsilon)R_k^\varepsilon(I + \varepsilon\overline{A}'(\overline{\alpha}_k^\varepsilon)),$$
$$\qquad (8.11)$$

with $\widehat{x}_0^\varepsilon = x$ and $R_0^\varepsilon = 0$.

Remark 8.3. In hybrid filtering, when $\{\alpha_k^\varepsilon\}$ jumps rapidly within a group of states, it is difficult to estimate its value over time. On the other hand, in this case, any estimated value is not useful because α_k^ε will jump to another state in a fairly short duration. Note that the aggregated process $\{\overline{\alpha}_k^\varepsilon\}$ jumps much less rapidly. Therefore it is relatively easy to estimate its value. It is reasonable to impose the observability condition in this context. To proceed, we obtain a bound on R_k^ε. Note that $R_k^\varepsilon = R_k^\varepsilon(\omega)$. Here we use ω to denote a sample point in the sample space Ω.

Lemma 8.4. *Given $T < \infty$, there exist $\varepsilon_0 > 0$ and constant K such that*

$$\sup_{\omega \in \Omega} \sup_{0 \leq k \leq T/\varepsilon} |R_k^\varepsilon| \leq K,$$

for $0 < \varepsilon < \varepsilon_0$.

Next we consider the corresponding limit filtering problem. To obtain the desired limit, we concern ourselves with suitably scaled continuous-time interpolations of piecewise constant processes. Asymptotic results regarding the hybrid filtering are provided.

To begin, for $0 \leq k \leq T/\varepsilon$, define the interpolations $x^\varepsilon(\cdot)$, $y^\varepsilon(\cdot)$, $\widehat{x}^\varepsilon(\cdot)$, $R^\varepsilon(\cdot)$, as

$$x^\varepsilon(t) = x_k^\varepsilon, \ y^\varepsilon(t) = y_k^\varepsilon, \ \widehat{x}^\varepsilon(t) = \widehat{x}_k^\varepsilon, \ R^\varepsilon(t) = R_k^\varepsilon, \quad t \in [\varepsilon k, \varepsilon k + \varepsilon), \ (8.12)$$

where x_k^ε and y_k^ε are given in (8.3) and $\widehat{x}_k^\varepsilon$ and R_k^ε given in (8.10) and (8.11), respectively. We will show that the interpolated processes converge weakly to $x(\cdot)$, $y(\cdot)$, $\widehat{x}(\cdot)$, and $R(\cdot)$. Using the martingale averaging approach (see Chapter 14 and the references therein), we show that the sequences of interests are tight, and then we characterize the limit processes by identifying the operator of the limit martingale problems. To proceed, let us first obtain the *a priori* bounds on $\{x_k^\varepsilon\}$, $\{y_k^\varepsilon\}$, and $\{\widehat{x}_k^\varepsilon\}$. The bounds are given below and their derivations are in Section 8.3.

Lemma 8.5. *Assume* (A8.1)–(A8.3). *For* $\{x_k^\varepsilon\}$ *and* $\{y_k^\varepsilon\}$ *defined in* (8.3), *and* $\{\widehat{x}_k^\varepsilon\}$ *and* $\{R_k^\varepsilon\}$ *given in* (8.10), *the following bounds hold:*

$$\sup_{0 \leq k \leq T/\varepsilon} E|x_k^\varepsilon|^4 < \infty,$$

$$\sup_{0 \leq k \leq T/\varepsilon} E|y_k^\varepsilon|^4 < \infty, \qquad (8.13)$$

$$\sup_{0 \leq k \leq T/\varepsilon} E|\widehat{x}_k^\varepsilon|^4 < \infty.$$

Tightness. Let $\mathcal{F}_k$ and $\mathcal{F}_t^\varepsilon$ be the σ-algebras generated by $\{\alpha_{k_1}^\varepsilon, w_{k_1}, v_{k_1} : k_1 \leq k\}$ and $\{\alpha^\varepsilon(s), w^\varepsilon(s), v^\varepsilon(s) : s \leq t\}$, and E_k and E_t^ε be the corresponding conditional expectations with respect to $\mathcal{F}_k$ and $\mathcal{F}_t^\varepsilon$, respectively. We are in a position to derive the tightness of $\{(x^\varepsilon(\cdot), y^\varepsilon(\cdot), \widehat{x}^\varepsilon(\cdot), R^\varepsilon(\cdot))\}$. The tightness is really a compactness result; see Chapter 14 for further reading and references.

Recall that $D([0,T]; \mathbb{S})$ is the space of $\mathbb{S}$-valued functions that are right continuous, have left-hand limits, endowed with the Skorohod topology. To prove the desired assertion, we verify a tightness criterion.

Lemma 8.6. *Under* (A8.1)–(A8.3), $\{x^\varepsilon(\cdot), y^\varepsilon(\cdot), \widehat{x}^\varepsilon(\cdot), R^\varepsilon(\cdot), \overline{\alpha}^\varepsilon(\cdot)\}$ *is tight in* $D([0,T]; \mathbb{R}^r \times \mathbb{R}^r \times \mathbb{R}^r \times \mathbb{R}^{r \times r} \times \overline{\mathcal{M}})$.

Weak Convergence. We proceed to obtain the weak convergence of $(x^\varepsilon(\cdot), y^\varepsilon(\cdot), \widehat{x}^\varepsilon(\cdot), R^\varepsilon(\cdot))$ using a martingale problem formulation. Since the sequence is tight, by using Prohorov's theorem, we can extract a convergent subsequence. For notational simplicity and without loss of generality, still denote the subsequence by $(x^\varepsilon(\cdot), y^\varepsilon(\cdot), \widehat{x}^\varepsilon(\cdot), R^\varepsilon(\cdot))$ with limit

$(x(\cdot), y(\cdot), \widehat{x}(\cdot), R(\cdot))$. Thus, our task is reduced to finding the limit by characterizing the operator of the limit martingale problem. The technique used is essentially an averaging approach. Different from the diffusion approximation for wideband noise systems (cf. Kushner [96]), where the drift and diffusion coefficients are deterministic functions, the Markov chain $\overline{\alpha}(\cdot)$ is in both the limit drift and diffusion coefficients.

Theorem 8.7. *Assume Conditions* (A8.1)–(A8.3) *hold. Then*

$$(x^{\varepsilon}(\cdot), y^{\varepsilon}(\cdot), \widehat{x}^{\varepsilon}(\cdot), R^{\varepsilon}(\cdot), \overline{\alpha}^{\varepsilon}(\cdot)) \text{ converges to } (x(\cdot), y(\cdot), \widehat{x}(\cdot), R(\cdot), \overline{\alpha}(\cdot))$$

weakly such that $\overline{\alpha}(\cdot)$ *is a Markov chain generated by* $\overline{Q}_{*}$ *defined in* (6.16) *and* $(x(\cdot), y(\cdot), \widehat{x}(\cdot), R(\cdot))$ *is the limit satisfies the system of equations*

$$dx = \overline{A}(\overline{\alpha}(t))x dt + \overline{\sigma}_w(\overline{\alpha}(t)) dw,$$

$$dy = \overline{C}(\overline{\alpha}(t))x dt + \overline{\sigma}_v(\overline{\alpha}(t)) dv,$$

$$d\widehat{x} = \overline{A}(\overline{\alpha}(t))\widehat{x} dt + R(t)\overline{C}'(\overline{\alpha}(t))\Sigma_v^{-1}(\overline{\alpha}(t))(dy - \overline{C}(\overline{\alpha}(t))\widehat{x} dt), \quad (8.14)$$

$$\dot{R} = \overline{A}(\overline{\alpha}(t))R(t) + R(t)\overline{A}'(\overline{\alpha}(t))$$

$$\qquad - R(t)\overline{C}(\overline{\alpha}(t))\Sigma_v^{-1}(\overline{\alpha}(t))\overline{C}(\overline{\alpha}(t))R(t) + \Sigma_w(\overline{\alpha}(t)),$$

with $x(0) = x$, $y(0) = 0$, $\widehat{x}(0) = x$, *and* $R(0) = 0$, *where* $w(\cdot)$ *and* $v(\cdot)$ *are independent* r-*dimensional, standard Brownian motions given by Lemma 8.1,* $\overline{A}(i)$ *and* $\overline{C}(i)$ *are given in* (8.7), *and* $\overline{\sigma}_w(i)$ *and* $\overline{\sigma}_v(i)$ *in* (8.8), *respectively.*

Remark 8.8. Note that $x(\cdot)$ being a solution of the first equation in (8.14) is equivalent to $(x(\cdot), \overline{\alpha}(\cdot))$ being a solution of the martingale problem with operator

$$\mathcal{L}f(x, i) = f_x'(x, i)\overline{A}(i)x + \frac{1}{2}\text{tr}[f_{xx}(x, i)\overline{\sigma}_w(i)\overline{\sigma}_w'(i)] + \overline{Q}_* f(x, \cdot)(i), \quad i \in \overline{\mathcal{M}},$$
$$(8.15)$$

where $\text{tr}(A)$ is the trace of the matrix A,

$$\overline{Q}_* f(x, \cdot)(i) = \sum_{j \in \overline{\mathcal{M}}} \overline{q}^{ij} f(x, j) = \sum_{j \in \overline{\mathcal{M}}, \, j \neq i} \overline{q}^{ij}(f(x, j) - f(x, i)),$$

for each $i \in \overline{\mathcal{M}}$, and $f(\cdot, i) \in C_0^2$ (the class of twice continuously differentiable function with compact support), and that the martingale problem has a unique solution. Such a representation will be used in the proof of results in Section 8.3. A similar comment holds for $y(\cdot)$, $\widehat{x}(\cdot)$, and $R(\cdot)$.

Recall that $\widehat{x}(t)$ is the best mean squares estimate of $x(t)$ given $\mathcal{F}_t = \sigma\{y(s), \overline{\alpha}(s) : s \leq t\}$. That is, for each given t,

$$E|x(t) - \widehat{x}(t)|^2 = \min_{\forall z \text{ adapted to } \mathcal{F}_t} E|x(t) - z|^2.$$

Therefore, $E \int_0^T |x(t) - \widehat{x}(t)|^2 dt$ is a minimizer of $E \int_0^T |x(t) - z(t)| dt$ over all $\mathcal{F}_t$-adapted $z(t)$.

Theorem 8.9. *Assume (A8.1)–(A8.3). The following hold:*

(a) $\{\widehat{x}_k^\varepsilon\}$ *is nearly optimal in the sense that*

$$E \sum_{k=0}^{T/\varepsilon} \varepsilon |x_k^\varepsilon - \widehat{x}_k^\varepsilon|^2 \to E \int_0^T |x(t) - \widehat{x}(t)|^2 dt = \min. \qquad (8.16)$$

(b) *Assume, in addition, that* $\overline{\sigma}_v(i) = \delta I$ *and that* $\overline{C}(i)\overline{C}'(i)$ *is positive definite. Then,*

$$\lim_{\delta \to 0} \lim_{\varepsilon \to 0} E \sum_{k=0}^{T/\varepsilon} \varepsilon |x_k^\varepsilon - \widehat{x}_k^\varepsilon|^2 = 0.$$

Remark 8.10. The condition $\overline{\sigma}_v(i) = \delta I$ in (b) corresponds to a partially observed system with small observation noise. Related literature in filtering with small observation noise can be found in Fleming and Zhang [58] and the references therein.

Remark 8.11. In this chapter, we assume that the aggregated process $\overline{\alpha}_k^\varepsilon$ is observable. When it is not directly observable, one may estimate the values of $\overline{\alpha}_k^\varepsilon$ over time. To carry out such an estimation procedure, either a quadratic variation test or a maximum likelihood approach can be used to estimate the value of $\overline{\alpha}_k^\varepsilon$; we refer the reader to Haussmann and Zhang [68], and Zhang [174] for details along this direction.

Example 8.12. In this example, we consider a one-block transition matrix case. That is, the transition matrix P in (8.1) is irreducible and aperiodic. Let $\nu = (\nu^1, \ldots, \nu^{m_0})$ denote the stationary distribution of P and let

$$\overline{A}^0 = \sum_{j=1}^{m_0} A(j)\nu^j, \quad \text{and} \quad \Sigma_w^0 = \overline{\sigma}_w^0 (\overline{\sigma}_w^0)' = \sum_{j=1}^{m_0} \nu^j \sigma_w(j)\sigma_w'(j),$$

$$\overline{C}^0 = \sum_{j=1}^{m_0} C(j)\nu^j, \quad \text{and} \quad \Sigma_v^0 = \overline{\sigma}_v^0 (\overline{\sigma}_v^0)' = \sum_{j=1}^{m_0} \nu^j \sigma_v(j)\sigma_v'(j).$$

Then the filtering equations in (8.10) and (8.11) become

$$\widehat{x}_{k+1}^\varepsilon = \widehat{x}_k^\varepsilon + \varepsilon \overline{A}^0 \widehat{x}_k^\varepsilon + (I + \varepsilon \overline{A}^0) R_k^\varepsilon (\overline{C}^0)'$$
$$\times [\varepsilon \overline{C}^0 R_k^\varepsilon (\overline{C}^0)' + \Sigma_v^0]^{-1} (y_{k+1}^\varepsilon - y_k^\varepsilon - \varepsilon \overline{C}^0 \widehat{x}_k^\varepsilon),$$

and

$$R_{k+1}^\varepsilon = \varepsilon \Sigma_w^0 + (I + \varepsilon \overline{A}^0) R_k^\varepsilon (I + \varepsilon (\overline{A}^0)')$$
$$- \varepsilon (I + \varepsilon \overline{A}^0) R_k^\varepsilon (\overline{C}^0)' [\varepsilon \overline{C}^0 R_k^\varepsilon (\overline{C}^0)' + \Sigma_v^0]^{-1} \times \overline{C}^0 R_k^\varepsilon (I + \varepsilon (\overline{A}^0)'),$$

with $\widehat{x}_0^\varepsilon = x$ and $R_0^\varepsilon = 0$.

Moreover, the limit of $(x_k^\varepsilon, y_k^\varepsilon, \widehat{x}_k^\varepsilon, R_k^\varepsilon)$ satisfies

$$dx = \overline{A}^0 x dt + \overline{\sigma}_w^0 dw,$$
$$dy = \overline{C}^0 x dt + \overline{\sigma}_v^0 dv,$$
$$d\widehat{x} = \overline{A}^0 \widehat{x} dt + R(t)(\overline{C}^0)'(\Sigma_v^0)^{-1}(dy - \overline{C}^0 \widehat{x} dt),$$
$$\dot{R} = \overline{A}^0 R(t) + R(t)(\overline{A}^0)' - R(t)\overline{C}^0(\Sigma_v^0)^{-1}\overline{C}^0 R(t) + \Sigma_w^0,$$

with $x(0) = x$, $y(0) = 0$, $\widehat{x}(0) = x$, and $R(0) = 0$. Then Theorem 8.9 implies that as $\varepsilon \to 0$,

$$E \sum_{k=0}^{T/\varepsilon} \varepsilon |x_k^\varepsilon - \widehat{x}_k^\varepsilon|^2 \to E \int_0^T |x(t) - \widehat{x}(t)|^2 dt = \min,$$

and if $\overline{\sigma}_v^{\,0} = \delta I$ and $\overline{C}^0(\overline{C}^0)'$ is positive definite, then,

$$\lim_{\delta \to 0} \lim_{\varepsilon \to 0} E \sum_{k=0}^{T/\varepsilon} \varepsilon |x_k^\varepsilon - \widehat{x}_k^\varepsilon|^2 = 0.$$

As was emphasized, our main purpose is to reduce the complexity of the underlying filtering problem. The foregoing shows that the reduction of complexity is particularly pronounced if the transition matrix (8.2) consists of only one aperiodic ergodic class.

Remark 8.13. In this chapter, we considered discrete-time filters. For the continuous-time counterpart, one may consider the following system

$$\begin{aligned} dx^\varepsilon(t) &= A(\alpha^\varepsilon(t))x^\varepsilon(t)dt + \sigma_w(\alpha^\varepsilon(t))dw, \quad x^\varepsilon(0) = x \\ dy^\varepsilon(t) &= C(\alpha^\varepsilon(t))x^\varepsilon(t)dt + \sigma_v(\alpha^\varepsilon(t))dv, \quad y^\varepsilon(0) = 0, \end{aligned} \tag{8.17}$$

where $w(\cdot)$ and $v(\cdot)$ are independent standard Brownian motions, and $\alpha^\varepsilon(\cdot)$ is a continuous-time singularly perturbed Markov chain with finite state space $\mathcal{M}$ and generator

$$Q^\varepsilon(t) = \frac{\widetilde{Q}(t)}{\varepsilon} + \widehat{Q}(t).$$

Assuming that $\overline{\alpha}(t)$ is observable, one may design the corresponding near-optimal filter. We refer the reader to Wang, Zhang, and Yin [147] for details.

8.3   Proofs of Results

Proof of Lemma 8.4. First, it is easy to see that R_k^ε is symmetric and non-negative definite because it can be represented as the conditional covariance matrix of $x_k^\varepsilon - \widehat{x}_k^\varepsilon$ given $\mathcal{Y}_{k-1}^\varepsilon$, where $\mathcal{Y}_k^\varepsilon = \sigma\{(y_l^\varepsilon, \overline{\alpha}_l^\varepsilon) : l \leq k\}$. It remains to show its boundedness. Here we use the matrix norm $|A| = \sqrt{\mathrm{tr}(AA')}$. It follows that $\mathrm{tr}(AB) \leq |A| \cdot |B|$ for any square matrices A and B of the same dimension. Under this norm, for any symmetric nonnegative definite matrix R, we have

$$|R| = (\mathrm{tr}(R^2))^{\frac{1}{2}}$$
$$= (\lambda_1^2 + \cdots + \lambda_n^2)^{\frac{1}{2}}$$
$$\leq (\lambda_1 + \cdots + \lambda_n) = \mathrm{tr}(R),$$

where $\lambda_1 \geq 0, \ldots, \lambda_n \geq 0$ are the eigenvalues of R. Therefore, it suffices to show that $\mathrm{tr}(R_k^\varepsilon)$ is bounded. Take trace on both sides of the Riccati equation in (8.11) to obtain

$$\mathrm{tr}(R_{k+1}^\varepsilon) \leq K_0 \varepsilon + \mathrm{tr}[(I + \varepsilon\overline{A}(\overline{\alpha}_k^\varepsilon))R_k^\varepsilon(I + \varepsilon\overline{A}'(\overline{\alpha}_k^\varepsilon))],$$

for some $K_0 > 0$. Note also that

$$\mathrm{tr}[(I + \varepsilon\overline{A}(\overline{\alpha}_k^\varepsilon))R_k^\varepsilon(I + \varepsilon\overline{A}'(\overline{\alpha}_k^\varepsilon))]$$
$$= \mathrm{tr}[R_k^\varepsilon + \varepsilon(R_k^\varepsilon\overline{A}'(\overline{\alpha}_k^\varepsilon) + \overline{A}(\overline{\alpha}_k^\varepsilon)R_k^\varepsilon) + \varepsilon^2\overline{A}(\overline{\alpha}_k^\varepsilon)R_k^\varepsilon\overline{A}'(\overline{\alpha}_k^\varepsilon)]$$
$$= \mathrm{tr}(R_k^\varepsilon) + \varepsilon\mathrm{tr}(R_k^\varepsilon\overline{A}'(\overline{\alpha}_k^\varepsilon) + \overline{A}(\overline{\alpha}_k^\varepsilon)R_k^\varepsilon)) + \varepsilon^2\mathrm{tr}(\overline{A}(\overline{\alpha}_k^\varepsilon)R_k^\varepsilon\overline{A}'(\overline{\alpha}_k^\varepsilon))$$
$$\leq \mathrm{tr}(R_k^\varepsilon) + 2\varepsilon|\overline{A}(\overline{\alpha}_k^\varepsilon)| \cdot |R_k^\varepsilon| + \varepsilon^2|A(\overline{\alpha}_k^\varepsilon)|^2 \cdot |R_k^\varepsilon|$$
$$\leq \mathrm{tr}(R_k^\varepsilon)(1 + 2\varepsilon|\overline{A}(\overline{\alpha}_k^\varepsilon)| + \varepsilon^2|A(\overline{\alpha}_k^\varepsilon|^2)$$
$$\leq \mathrm{tr}(R_k^\varepsilon)(1 + \varepsilon K_1 + \varepsilon^2 K_2),$$

for some constants K_1 and K_2. Moreover, the above bounds hold uniformly in both ω and $0 \leq k \leq T/\varepsilon$. Let $\phi_k = \mathrm{tr}(R_k^\varepsilon)$. Then

$$\phi_{k+1} \leq K_0\varepsilon + (1 + \varepsilon K_1 + \varepsilon^2 K_2)\phi_k, \quad \phi_0 = 0.$$

This implies

$$\phi_{k+1} \leq K_0\varepsilon\left(1 + (1 + \varepsilon K_1 + \varepsilon^2 K_2) + \cdots + (1 + \varepsilon K_1 + \varepsilon^2 K_2)^k\right)$$
$$= K_0\varepsilon\left(\frac{(1 + \varepsilon K_1 + \varepsilon^2 K_2)^k - 1}{\varepsilon K_1 + \varepsilon^2 K_2}\right)$$
$$\leq \frac{K_0}{K_1 + \varepsilon K_2}[(1 + \varepsilon K_1 + \varepsilon^2 K_2)^{\frac{T}{\varepsilon}} - 1] \leq K,$$

for some $K < \infty$ uniformly in $\omega \in \Omega$ and $0 \leq k \leq T/\varepsilon$. $\square$

Proof of Lemma 8.5. Let us first work with x_k^ε. For $0 \leq k \leq T/\varepsilon$, iterating on the first equation in (8.3) gives us

$$x_{k+1}^\varepsilon = x_0^\varepsilon + \varepsilon \sum_{k_1=0}^{k} A(\alpha_{k_1}^\varepsilon)x_{k_1}^\varepsilon + \sqrt{\varepsilon} \sum_{k_1=0}^{k} \sigma_w(\alpha_{k_1}^\varepsilon)w_{k_1}. \tag{8.18}$$

For any $z \in \mathbb{R}^r$, we use $|z|^2 = \text{tr}(zz')$. Recall that K is a generic positive constant. Thus

$$E|x_{k+1}^\varepsilon|^2 \leq K\left[E|x_0^\varepsilon|^2 + \varepsilon^2 E\left|\sum_{k_1=0}^{k} A(\alpha_{k_1}^\varepsilon)x_{k_1}^\varepsilon\right|^2 + \varepsilon E\left|\sum_{k_1=0}^{k} \sigma_w(\alpha_{k_1}^\varepsilon)w_{k_1}\right|^2\right]$$

$$\leq KE|x_0^\varepsilon|^2 + K\varepsilon \sum_{k_1=0}^{k} E|x_{k_1}^\varepsilon|^2$$

$$+\varepsilon K \sum_{k_1=0}^{k}\sum_{k_2=0}^{k} E[\text{tr}(\sigma_w(\alpha_{k_1}^\varepsilon)w_{k_1}w_{k_2}'\sigma_w'(\alpha_{k_2}^\varepsilon))].$$

$$(8.19)$$

By the independence of $\{\alpha_k^\varepsilon\}$ and $\{w_k\}$, the boundedness of $\sigma_w(\iota)$ for each $\iota \in \mathcal{M}$, noting

$$Ew_{k_1}w_k' = 0 \quad \text{if} \quad k_1 \neq k,$$

we have

$$\varepsilon \sum_{k_1=0}^{k}\sum_{k_2=0}^{k} \text{tr}\left(E\sigma_w(\alpha_{k_1}^\varepsilon)w_{k_1}w_{k_2}'\sigma_w'(\alpha_{k_2}^\varepsilon)\right)$$

$$\leq \varepsilon K\left|\sum_{k_1=0}^{k}\sum_{k_2=0}^{k} E\{\sigma_w(\alpha_{k_1}^\varepsilon)[Ew_{k_1}w_{k_2}']\sigma_w'(\alpha_{k_2}^\varepsilon)\}\right|$$

$$\leq \varepsilon K \sum_{k_1=0}^{k} |Ew_{k_1}w_{k_1}'|$$

$$\leq \varepsilon K\frac{T}{\varepsilon} = KT < \infty.$$

$$(8.20)$$

Using (8.20) in (8.19), and applying Gronwall's inequality (Lemma 14.41), we have

$$E|x_{k+1}^\varepsilon|^2 \leq K + K\varepsilon \sum_{k_1=0}^{k} E|x_{k_1}^\varepsilon|^2$$

$$\leq K\exp(K\varepsilon k)$$

$$\leq K < \infty,$$

since $k \leq T/\varepsilon$. Moreover, the bound holds uniformly in k for $0 \leq k \leq T/\varepsilon$. As for y_k^ε, using the bounds on $\sup_{0 \leq k \leq T/\varepsilon} E|x_k^\varepsilon|^2$, we have

$$\sup_{0 \leq k \leq T/\varepsilon} E|y_{k+1}^\varepsilon|^2$$

$$\leq K \sup_{0 \leq k \leq T/\varepsilon}\left[E|y_0^\varepsilon|^2 + \varepsilon^2 E\left|\sum_{k_1=0}^{k} C(\alpha_{k_1}^\varepsilon)x_{k_1}^\varepsilon\right|^2 + \varepsilon E\left|\sum_{k_1=0}^{k} \sigma_v(\alpha_{k_1}^\varepsilon)v_{k_1}\right|^2\right]$$

$$\leq KE|y_0^\varepsilon|^2 + K\varepsilon \sup_{0 \leq k \leq T/\varepsilon} \sum_{k_1=0}^{k} E|x_{k_1}^\varepsilon|^2$$

$$+\varepsilon K \sup_{0\le k\le T/\varepsilon} \sum_{k_1=0}^{k}\sum_{k_2=0}^{k} E\big[\mathrm{tr}(\sigma_v(\alpha_{k_1}^\varepsilon)v_{k_1}v_{k_2}'\sigma_v'(\alpha_{k_2}^\varepsilon))\big]$$
$$\le K < \infty.$$

The uniform bound for y_k^ε is obtained for $0\le k\le T/\varepsilon$.

Continuing along this line, to obtain the bounds for the fourth moments, by repeated applications of the familiar inequality $|a+b|^l \le 2^{l-1}[|a|^l + |b|^l]$ (for $a,b\in\mathbb{R}$ and l a positive integer) in (8.18),

$$E|x_{k+1}^\varepsilon|^4 \le KE\Big[|x_0^\varepsilon|^4 + \varepsilon^4\Big|\sum_{k_1=0}^{k} A(\alpha_{k_1}^\varepsilon)x_{k_1}^\varepsilon\Big|^4 + \varepsilon^2\Big|\sum_{k_1=0}^{k}\sigma_w(\alpha_{k_1}^\varepsilon)w_{k_1}\Big|^4\Big].$$

Just as in the derivation of $E|x_k^\varepsilon|^2$, we obtain

$$E|x_{k+1}^\varepsilon|^4 \le K + K\varepsilon\sum_{k_1=0}^{k} E|x_{k_1}^\varepsilon|^4.$$

Gronwall's inequality leads to the desired estimate. Similarly, it can be shown that $E|y_{k+1}^\varepsilon|^4 < \infty$ uniformly in $0\le k\le T/\varepsilon$.

Since $\{R_k^\varepsilon\}$ is uniformly bounded, $E|R_k^\varepsilon|^4 < \infty$. To obtain the bound for $E|\widehat{x}_k^\varepsilon|^4$, define

$$H_k^\varepsilon = (I + \varepsilon\overline{A}(\overline{\alpha}_k^\varepsilon))R_k^\varepsilon \overline{C}'(\overline{\alpha}_k^\varepsilon)[\varepsilon\overline{C}(\overline{\alpha}_k^\varepsilon)R_k^\varepsilon\overline{C}'(\overline{\alpha}_k^\varepsilon) + \Sigma_v(\overline{\alpha}_k^\varepsilon)]^{-1}. \qquad (8.21)$$

Note that $\{H_k^\varepsilon\}$ is bounded uniformly in $0\le k\le T/\varepsilon$ and $\omega\in\Omega$.

Expanding (8.10) gives us

$$\begin{aligned}\widehat{x}_{k+1}^\varepsilon = \widehat{x}_k^\varepsilon &+\varepsilon\overline{A}(\overline{\alpha}_k^\varepsilon)\widehat{x}_k^\varepsilon - \varepsilon H_k^\varepsilon\overline{C}(\overline{\alpha}_k^\varepsilon)\widehat{x}_k^\varepsilon \\ &+\varepsilon H_k^\varepsilon C(\alpha_k^\varepsilon)x_k^\varepsilon + \sqrt{\varepsilon}H_k^\varepsilon\sigma_v(\alpha_k^\varepsilon)v_k.\end{aligned} \qquad (8.22)$$

Iterating on (8.22) yields

$$\begin{aligned}\widehat{x}_{k+1}^\varepsilon = x_0^\varepsilon &+\varepsilon\sum_{k_1=0}^{k}\overline{A}(\overline{\alpha}_{k_1})\widehat{x}_{k_1}^\varepsilon - \varepsilon\sum_{k_1=0}^{k} H_{k_1}^\varepsilon\overline{C}(\overline{\alpha}_k^\varepsilon)\widehat{x}_{k_1}^\varepsilon \\ &+\varepsilon\sum_{k_1=0}^{k} H_{k_1}^\varepsilon C(\alpha_{k_1}^\varepsilon)x_{k_1}^\varepsilon + \sqrt{\varepsilon}\sum_{k_1=0}^{k} H_{k_1}^\varepsilon\sigma_v(\alpha_{k_1}^\varepsilon)v_{k_1}.\end{aligned}$$

Using the already derived moment bounds for $E|x_k^\varepsilon|^4$ and the independence of $\{v_k\}$ with α_k^ε, we obtain

$$E|\widehat{x}_{k+1}^\varepsilon|^4 \le K + K\varepsilon\sum_{k_1=0}^{k} E|\widehat{x}_{k_1}^\varepsilon|^4.$$

The desired bounds then follow from Gronwall's inequality.   □

Proof of Lemma 8.6. Since $\{\overline{\alpha}^\varepsilon(\cdot)\}$ is tight, we need only show the tightness of $\{x^\varepsilon(\cdot)\}$, $\{y^\varepsilon(\cdot)\}$, $\{R^\varepsilon(\cdot)\}$, and $\{\widehat{x}^\varepsilon(\cdot)\}$ separately. Again, we first consider the sequence $\{x^\varepsilon(\cdot)\}$. For any $\delta > 0$, $t > 0$, and $s > 0$ with $s \leq \delta$, consider

$$
E_t^\varepsilon |x^\varepsilon(t+s) - x^\varepsilon(t)|^2
$$
$$
= E_t^\varepsilon \left| \varepsilon \sum_{k=t/\varepsilon}^{(t+s)/\varepsilon-1} A(\alpha_k^\varepsilon) x_k^\varepsilon + \sqrt{\varepsilon} \sum_{k=t/\varepsilon}^{(t+s)/\varepsilon-1} \sigma_w(\alpha_k^\varepsilon) w_k \right|^2 \qquad (8.23)
$$
$$
\stackrel{\text{def}}{=} e_1^\varepsilon(t,s) + e_2^\varepsilon(t,s) + e_3^\varepsilon(t,s),
$$

where

$$
e_1^\varepsilon(t,s) = \varepsilon^2 \sum_{k_1=t/\varepsilon}^{(t+s)/\varepsilon-1} \sum_{k=t/\varepsilon}^{(t+s)/\varepsilon-1} E_t^\varepsilon \left[\mathrm{tr}[A(\alpha_{k_1}^\varepsilon) x_{k_1}^\varepsilon x_k^{\varepsilon\prime} A'(\alpha_k^\varepsilon)] \right],
$$
$$
e_2^\varepsilon(t,s) = 2\sqrt{\varepsilon^3} \sum_{k_1=t/\varepsilon}^{(t+s)/\varepsilon-1} \sum_{k=t/\varepsilon}^{(t+s)/\varepsilon-1} E_t^\varepsilon \left[\mathrm{tr}[A(\alpha_{k_1}^\varepsilon) x_{k_1}^\varepsilon w_k' \sigma_w'(\alpha_k^\varepsilon)] \right],
$$
$$
e_3^\varepsilon(t,s) = \varepsilon \sum_{k_1=t/\varepsilon}^{(t+s)/\varepsilon-1} \sum_{k=t/\varepsilon}^{(t+s)/\varepsilon-1} E_t^\varepsilon \left[\mathrm{tr}[\sigma_w(\alpha_{k_1}^\varepsilon) w_{k_1} w_k' \sigma_w'(\alpha_k^\varepsilon)] \right].
$$

Consider each of the terms on the right-hand side of the equations above separately. First, by the finiteness of $A(\iota)$ for each $\iota \in \mathcal{M}$,

$$
e_1^\varepsilon(t,s) \leq K\varepsilon^2 \sum_{k_1=t/\varepsilon}^{(t+s)/\varepsilon-1} \sum_{k=t/\varepsilon}^{(t+s)/\varepsilon-1} E_t^\varepsilon |x_{k_1}^\varepsilon| |x_k^\varepsilon|.
$$

By virtue of Lemma 8.5, an application of the Cauchy–Schwarz inequality leads to

$$
E e_1^\varepsilon(t,s) \leq K\varepsilon^2 \sum_{k_1=t/\varepsilon}^{(t+s)/\varepsilon-1} \sum_{k=t/\varepsilon}^{(t+s)/\varepsilon-1} E^{1/2} |x_{k_1}^\varepsilon|^2 E^{1/2} |x_k^\varepsilon|^2
$$
$$
\leq K\varepsilon^2 \left(\frac{t+s}{\varepsilon} - \frac{t}{\varepsilon} \right)^2 \leq Ks^2 = O(\delta^2).
$$

Thus

$$
\lim_{\delta \to 0} \limsup_{\varepsilon \to 0} E e_1^\varepsilon(t,s) = \lim_{\delta \to 0} O(\delta^2) = 0. \qquad (8.24)
$$

As for $e_2^\varepsilon(t,s)$, note that $x_{k_1}^\varepsilon$ and $A(\alpha_{k_1}^\varepsilon)$ are $\mathcal{F}_{k_1}$-measurable. Since for $k_1 < k$, $E_{k_1} w_k = 0$, the independence of $\{\alpha_k^\varepsilon\}$ and $\{w_k\}$ in (A8.3), and the finiteness of $A(\iota)$ and $\sigma_w(\iota)$ for each $\iota \in \mathcal{M}$ and the Cauchy–Schwarz inequality yield

$$
\begin{aligned}
e_2^\varepsilon(t,s) &\leq K\sqrt{\varepsilon^3} \sum_{k_1=t/\varepsilon}^{(t+s)/\varepsilon-1} \sum_{k\geq k_1} \left| \mathrm{tr}\left[E_t^\varepsilon A(\alpha_{k_1}^\varepsilon) x_{k_1}^\varepsilon (E_{k_1} w_k')(E_{k_1}\sigma_w'(\alpha_k^\varepsilon)) \right] \right| \\
&\leq K\sqrt{\varepsilon^3} \sum_{k=t/\varepsilon}^{(t+s)/\varepsilon-1} \sqrt{E_t^\varepsilon |x_k^\varepsilon|^2} \sqrt{E_t^\varepsilon |w_k|^2}.
\end{aligned}
$$

Therefore,

$$
\lim_{\delta\to 0}\limsup_{\varepsilon\to 0} E e_2^\varepsilon(t,s) = \lim_{\delta\to 0}\limsup_{\varepsilon\to 0} O(\sqrt{\varepsilon}) = 0. \tag{8.25}
$$

Next, we consider the last term of (8.23). Using the martingale difference property, the independence of $\{\alpha_k^\varepsilon\}$ and $\{w_k\}$, and $E_{k_1} w_k = 0$ for $k_1 < k$ and $E_k w_{k_1} = 0$ for $k < k_1$, we obtain

$$
e_3^\varepsilon(t,s) = \varepsilon \sum_{k=t/\varepsilon}^{(t+s)/\varepsilon-1} \left| \mathrm{tr}[E_t^\varepsilon \sigma_w(\alpha_k^\varepsilon) w_k w_k' \sigma_w'(\alpha_k^\varepsilon)] \right|.
$$

Noting $s \leq \delta$,

$$
E e_3^\varepsilon(t,s) \leq K\varepsilon \left(\frac{t+s}{\varepsilon} - \frac{t}{\varepsilon} \right) = O(\delta),
$$

and hence

$$
\lim_{\delta\to 0}\limsup_{\varepsilon\to 0} E e_3^\varepsilon(t,s) = \lim_{\delta\to 0} O(\delta) = 0. \tag{8.26}
$$

Combining (8.24), (8.25), and (8.26), we obtain

$$
\lim_{\delta\to 0}\limsup_{\varepsilon\to 0} E|x^\varepsilon(t+s) - x^\varepsilon(t)|^2 = 0.
$$

The tightness criterion in Lemma 14.12 then yields that $\{x^\varepsilon(\cdot)\}$ is tight in $D([0,T];\mathbb{R}^r)$.

As for the estimates of $y^\varepsilon(\cdot)$, we merely note that

$$
\begin{aligned}
&E_t^\varepsilon |y^\varepsilon(t+s) - y^\varepsilon(t)|^2 \\
&= E_t^\varepsilon \left| \varepsilon \sum_{k=t/\varepsilon}^{(t+s)/\varepsilon-1} C(\alpha_k^\varepsilon) x_k^\varepsilon + \sqrt{\varepsilon} \sum_{k=t/\varepsilon}^{(t+s)/\varepsilon-1} \sigma_v(\alpha_k^\varepsilon) v_k \right|^2 \\
&\leq K E_t^\varepsilon \left| \varepsilon \sum_{k=t/\varepsilon}^{(t+s)/\varepsilon-1} C(\alpha_k^\varepsilon) x_k^\varepsilon \right|^2 + K E_t^\varepsilon \left| \sqrt{\varepsilon} \sum_{k=t/\varepsilon}^{(t+s)/\varepsilon-1} \sigma_v(\alpha_k^\varepsilon) v_k \right|^2.
\end{aligned}
$$

The rest of the estimates are similar to that of $x^\varepsilon(\cdot)$. Thus we also have that $\{y^\varepsilon(\cdot)\}$ is tight in $D([0,T];\mathbb{R}^r)$.

Rewrite (8.11) as

$$
\begin{aligned}
R_{k+1}^\varepsilon = R_k^\varepsilon &+ \varepsilon R_k^\varepsilon \overline{A}'(\overline{\alpha}_k^\varepsilon) + \varepsilon \overline{A}(\overline{\alpha}_k^\varepsilon) R_k^\varepsilon \\
&- \varepsilon R_k^\varepsilon \overline{C}'(\overline{\alpha}_k^\varepsilon)(M_k^\varepsilon)^{-1}\overline{C}(\overline{\alpha}_k^\varepsilon) R_k^\varepsilon + \varepsilon \Sigma_w(\overline{\alpha}_k^\varepsilon) + \varepsilon^2 N_k^\varepsilon,
\end{aligned}
\tag{8.27}
$$

where

$$
\begin{aligned}
M_k^\varepsilon &= \varepsilon \overline{C}(\overline{\alpha}_k^\varepsilon) R_k^\varepsilon \overline{C}(\overline{\alpha}_k^\varepsilon) + \Sigma_v(\overline{\alpha}_k^\varepsilon), \\
N_k^\varepsilon &= \overline{A}(\overline{\alpha}_k^\varepsilon) R_k^\varepsilon \overline{A}'(\overline{\alpha}_k^\varepsilon) - R_k^\varepsilon \overline{C}'(\overline{\alpha}_k^\varepsilon)(M_k^\varepsilon)^{-1}\overline{C}(\overline{\alpha}_k^\varepsilon) R_k^\varepsilon \overline{A}'(\overline{\alpha}_k^\varepsilon) \\
&\quad - \overline{A}(\overline{\alpha}_k^\varepsilon) R_k^\varepsilon \overline{C}'(\overline{\alpha}_k^\varepsilon)(M_k^\varepsilon)^{-1}\overline{C}(\overline{\alpha}_k^\varepsilon) R_k^\varepsilon \\
&\quad - \varepsilon \overline{A}(\overline{\alpha}_k^\varepsilon) R_k^\varepsilon \overline{C}'(\overline{\alpha}_k)(M_k^\varepsilon)^{-1}\overline{C}(\overline{\alpha}_k^\varepsilon) R_k^\varepsilon \overline{A}'(\overline{\alpha}_k^\varepsilon).
\end{aligned}
$$

Owing to the boundedness of R_k^ε, it is readily seen that N_k^ε is bounded uniformly in ω and $0 \le k \le T/\varepsilon$. Calculation similar to that for $x^\varepsilon(\cdot)$ and $y^\varepsilon(\cdot)$ implies that for any $t, s \in [0, T]$ with $t + s \le T$ and $s \le \delta$,

$$
\lim_{\delta \to 0} \limsup_{\varepsilon \to 0} E|R^\varepsilon(t + s) - R^\varepsilon(t)|^2 = 0.
$$

The tightness of $\{R^\varepsilon(\cdot)\}$ then holds. Finally, using (8.22) together with the already established estimates thus far, we also have

$$
\lim_{\delta \to 0} \limsup_{\varepsilon \to 0} E|\widehat{x}^\varepsilon(t + s) - \widehat{x}^\varepsilon(t)|^2 = 0.
$$

The lemma is proved. $\square$

Proof of Theorem 8.7. Consider $\{x^\varepsilon(\cdot)\}$ first. In fact, we work with the pair $(x^\varepsilon(\cdot), \overline{\alpha}^\varepsilon(\cdot))$. Owing to the tightness of $\{x^\varepsilon(\cdot)\}$ and the weak convergence of $\{\overline{\alpha}^\varepsilon(\cdot)\}$, $\{(x^\varepsilon(\cdot), \overline{\alpha}^\varepsilon(\cdot))\}$ is tight. By virtue of Prohorov's theorem (see Theorem 14.4), we can extract a weakly convergent subsequence. Select such a subsequence and denote it by $\{(x^\varepsilon(\cdot), \overline{\alpha}^\varepsilon(\cdot))\}$ for simplicity. Denote the limit of the sequence by $(x(\cdot), \overline{\alpha}(\cdot))$. By the Skorohod representation (see Theorem 14.5), we may assume without loss of generality that $(x^\varepsilon(\cdot), \overline{\alpha}^\varepsilon(\cdot))$ converges to $(x(\cdot), \overline{\alpha}(\cdot))$ w.p.1. In addition, the convergence is uniform on each bounded time interval. We proceed to use martingale averaging techniques to figure out the limit.

To obtain the desired limit, it suffices to show that the limit $(x(\cdot), \overline{\alpha}(\cdot))$ is the unique solution of a martingale problem with operator $\mathcal{L}$ given by (8.15). Noting the system equation being linear in the state variable and using an argument similar to that used in the proof of Lemma 4.9, it can be shown that the corresponding martingale problem with operator $\mathcal{L}$ given in (8.15) has a unique solution.

To obtain the desired results, it suffices to show (see Theorem 14.7) that for each $i \in \overline{\mathcal{M}}$ and any $f(\cdot, i) \in C_0^2$ (the class of C^2 functions with compact support),

$$f(x(t), \overline{\alpha}(t)) - f(x(0), \overline{\alpha}(0)) - \int_0^t \mathcal{L}f(x(u), \bar{\alpha}(u))du \text{ is a martingale.}$$

To verify this martingale property, we need only show that for any positive integer k_0, any bounded and continuous function $h_{k_2}(\cdot)$ with $k_2 \leq k_0$, any $t, s > 0$, and $t_{k_2} \leq t \leq t + s$, the following equation holds:

$$E \prod_{k_2=1}^{k_0} h_{k_2}(x(t_{k_2}), \overline{\alpha}(t_{k_2}))\Big(f(x(t+s), \overline{\alpha}(t+s)) - f(x(t), \overline{\alpha}(t)) \tag{8.28}$$
$$- \int_t^{t+s} \mathcal{L}f(x(u), \overline{\alpha}(u))du \Big) = 0.$$

To obtain (8.28), we begin with the pair $(x^\varepsilon(\cdot), \overline{\alpha}^\varepsilon(\cdot))$. For each x and $\alpha \in \mathcal{M}$, we define $\widehat{f}(x, \alpha)$ by

$$\widehat{f}(x, \alpha) = \sum_{i=1}^{l_0} f(x, i) I_{\{\alpha \in \mathcal{M}_i\}} \text{ for each } \alpha \in \mathcal{M}. \tag{8.29}$$

Note that for each $\alpha = s_{ij} \in \mathcal{M}_i$, $\widehat{f}(x, \alpha)$ takes a constant value $f(x, i)$. Note also that at any time instant t, $\alpha^\varepsilon(t) = \alpha^\varepsilon_{t/\varepsilon}$ takes on one of the m_0 possible values from $\mathcal{M}$. Moreover, $\widehat{f}(x_k^\varepsilon, \alpha_k^\varepsilon) = f(x_k^\varepsilon, \overline{\alpha}_k^\varepsilon)$ for each k. Then

$$\widehat{f}(x^\varepsilon(t+s), \alpha^\varepsilon(t+s)) - \widehat{f}(x^\varepsilon(t), \alpha^\varepsilon(t)) - \int_0^t \mathcal{L}\widehat{f}(x^\varepsilon(u), \alpha^\varepsilon(u))du$$

is a martingale.

Choose a sequence of positive integers $\{\kappa_\varepsilon\}$ such that $\kappa_\varepsilon \to \infty$ but $\delta_\varepsilon = \varepsilon \kappa_\varepsilon \to 0$ as $\varepsilon \to 0$. The piecewise constant interpolation implies that

$$\widehat{f}(x^\varepsilon(t+s), \alpha^\varepsilon(t+s)) - \widehat{f}(x^\varepsilon(t), \alpha^\varepsilon(t))$$
$$= \sum_{l\kappa_\varepsilon=t/\varepsilon}^{(t+s)/\varepsilon-1} [\widehat{f}(x^\varepsilon_{l\kappa_\varepsilon+\kappa_\varepsilon}, \alpha^\varepsilon_{l\kappa_\varepsilon+\kappa_\varepsilon}) - \widehat{f}(x^\varepsilon_{l\kappa_\varepsilon+\kappa_\varepsilon}, \alpha^\varepsilon_{l\kappa_\varepsilon})] \tag{8.30}$$
$$+ \sum_{l\kappa_\varepsilon=t/\varepsilon}^{(t+s)/\varepsilon-1} [\widehat{f}(x^\varepsilon_{l\kappa_\varepsilon+\kappa_\varepsilon}, \alpha^\varepsilon_{l\kappa_\varepsilon}) - \widehat{f}(x^\varepsilon_{l\kappa_\varepsilon}, \alpha^\varepsilon_{l\kappa_\varepsilon})],$$

and hence

$$\lim_{\varepsilon \to 0} E \prod_{k_2=1}^{k_0} h_{k_2}(x^\varepsilon(t_{k_2}), \overline{\alpha}^\varepsilon(t_{k_2}))[\widehat{f}(x^\varepsilon(t+s), \alpha^\varepsilon(t+s)) - \widehat{f}(x^\varepsilon(t), \alpha^\varepsilon(t))]$$
$$\overset{\text{def}}{=} \lim_{\varepsilon \to 0} E \prod_{k_2=1}^{k_0} h_{k_2}(x^\varepsilon(t_{k_2}), \overline{\alpha}^\varepsilon(t_{k_2}))[g_1^\varepsilon + g_2^\varepsilon],$$

$$\tag{8.31}$$

where

$$g_1^\varepsilon = \sum_{l\kappa_\varepsilon=t/\varepsilon}^{(t+s)/\varepsilon-1} [\widehat{f}(x_{l\kappa_\varepsilon+\kappa_\varepsilon}^\varepsilon, \alpha_{l\kappa_\varepsilon+\kappa_\varepsilon}^\varepsilon) - \widehat{f}(x_{l\kappa_\varepsilon+\kappa_\varepsilon}^\varepsilon, \alpha_{l\kappa_\varepsilon}^\varepsilon)],$$

$$g_2^\varepsilon = \sum_{l\kappa_\varepsilon=t/\varepsilon}^{(t+s)/\varepsilon-1} [\widehat{f}(x_{l\kappa_\varepsilon+\kappa_\varepsilon}^\varepsilon, \alpha_{l\kappa_\varepsilon}^\varepsilon) - \widehat{f}(x_{l\kappa_\varepsilon}^\varepsilon, \alpha_{l\kappa_\varepsilon}^\varepsilon)].$$

We proceed to obtain the desired limit by examining g_i^ε for $i = 1, 2$ given above.

By using a Taylor expansion, rewrite g_2^ε as

$$
\begin{aligned}
g_2^\varepsilon &= \sum_{l\kappa_\varepsilon=t/\varepsilon}^{(t+s)/\varepsilon-1} \widehat{f}_x'(x_{l\kappa_\varepsilon}^\varepsilon, \alpha_{l\kappa_\varepsilon}^\varepsilon)[x_{l\kappa_\varepsilon+\kappa_\varepsilon}^\varepsilon - x_{l\kappa_\varepsilon}^\varepsilon] \\
&\quad + \frac{1}{2} \sum_{l\kappa_\varepsilon=t/\varepsilon}^{(t+s)/\varepsilon-1} [x_{l\kappa_\varepsilon+\kappa_\varepsilon}^\varepsilon - x_{l\kappa_\varepsilon}^\varepsilon]' \widehat{f}_{xx}(x_{l\kappa_\varepsilon}^+, \alpha_{l\kappa_\varepsilon}^\varepsilon)[x_{l\kappa_\varepsilon+\kappa_\varepsilon}^\varepsilon - x_{l\kappa_\varepsilon}^\varepsilon] \\
&\overset{\text{def}}{=} [g_{2,1}^\varepsilon + \frac{1}{2} g_{2,2}^\varepsilon],
\end{aligned}
\tag{8.32}
$$

where

$$g_{2,1}^\varepsilon = \sum_{l\kappa_\varepsilon=t/\varepsilon}^{(t+s)/\varepsilon-1} \sum_{k=l\kappa_\varepsilon}^{l\kappa_\varepsilon+\kappa_\varepsilon-1} \widehat{f}_x'(x_{l\kappa_\varepsilon}^\varepsilon, \alpha_{l\kappa_\varepsilon}^\varepsilon)[\varepsilon A(\alpha_k^\varepsilon)x_k^\varepsilon + \sqrt{\varepsilon}\sigma_w(\alpha_k^\varepsilon)w_k],$$

$$
\begin{aligned}
g_{2,2}^\varepsilon = \frac{1}{2} \sum_{l\kappa_\varepsilon=t/\varepsilon}^{(t+s)/\varepsilon-1} \sum_{k=l\kappa_\varepsilon}^{l\kappa_\varepsilon+\kappa_\varepsilon-1} & [\varepsilon A(\alpha_k^\varepsilon)x_k^\varepsilon + \sqrt{\varepsilon}\sigma_w(\alpha_k^\varepsilon)w_k]' \widehat{f}_{xx}(x_{l\kappa_\varepsilon}^+, \alpha_{l\kappa_\varepsilon}^\varepsilon) \\
& \times \sum_{k_1=l\kappa_\varepsilon}^{l\kappa_\varepsilon+\kappa_\varepsilon-1} [\varepsilon A(\alpha_{k_1}^\varepsilon)x_{k_1}^\varepsilon + \sqrt{\varepsilon}\sigma_w(\alpha_{k_1}^\varepsilon)w_{k_1}],
\end{aligned}
$$

for some $x_{l\kappa_\varepsilon}^+$ being on the line segment joining $x_{l\kappa_\varepsilon}^\varepsilon$ and $x_{l\kappa_\varepsilon+\kappa_\varepsilon}^\varepsilon$.

It follows that

$$
\begin{aligned}
E \prod_{k_2=1}^{k_0} & h_{k_2}(x^\varepsilon(t_{k_2}), \overline{\alpha}^\varepsilon(t_{k_2})) \Big[\sqrt{\varepsilon} \sum_{l\kappa_\varepsilon=t/\varepsilon}^{(t+s)/\varepsilon-1} \widehat{f}_x'(x_{l\kappa_\varepsilon}^\varepsilon, \alpha_{l\kappa_\varepsilon}^\varepsilon) \sum_{k=l\kappa_\varepsilon}^{l\kappa_\varepsilon+\kappa_\varepsilon-1} \sigma_w(\alpha_k^\varepsilon)w_k \Big] \\
&= E \prod_{k_2=1}^{k_0} h_{k_2}(x^\varepsilon(t_{k_2}), \overline{\alpha}^\varepsilon(t_{k_2})) \\
&\quad \times \Big[\sqrt{\varepsilon} \sum_{l\kappa_\varepsilon=t/\varepsilon}^{(t+s)/\varepsilon-1} \widehat{f}_x'(x_{l\kappa_\varepsilon}^\varepsilon, \alpha_{l\kappa_\varepsilon}^\varepsilon) \sum_{k=l\kappa_\varepsilon}^{l\kappa_\varepsilon+\kappa_\varepsilon-1} E_{l\kappa_\varepsilon}\sigma_w(\alpha_k^\varepsilon)E_{l\kappa_\varepsilon}w_k \Big].
\end{aligned}
$$

The above estimate is obtained by noting the independence of $\{\alpha_k^\varepsilon\}$ and $\{w_k\}$ and the measurability of $x_{l\kappa_\varepsilon}^\varepsilon$ and $\alpha_{l\kappa_\varepsilon}^\varepsilon$ w.r.t. $\mathcal{F}_{l\kappa_\varepsilon}$ Since $t_{k_2} \leq t$,

$\prod_{k_2=1}^{k_0} h_{k_2}(x^\varepsilon(t_{k_2}), \overline{\alpha}^\varepsilon(t_{k_2}))$ is $\mathcal{F}_{l\delta_\varepsilon}^\varepsilon$-measurable. By the finiteness of $\sigma_w(\alpha_k^\varepsilon)$, we obtain

$$
E \prod_{k_2=1}^{k_0} h_{k_2}(x^\varepsilon(t_{k_2}), \overline{\alpha}^\varepsilon(t_{k_2}))
$$
$$
\times \left[\sqrt{\varepsilon} \sum_{l\kappa_\varepsilon=t/\varepsilon}^{(t+s)/\varepsilon-1} \sum_{k=l\kappa_\varepsilon}^{l\kappa_\varepsilon+\kappa_\varepsilon-1} \widehat{f}_x'(x_{l\kappa_\varepsilon}^\varepsilon, \alpha_{l\kappa_\varepsilon}^\varepsilon) E_{l\kappa_\varepsilon} \sigma_w(\alpha_k^\varepsilon) E_{l\kappa_\varepsilon} w_k \right] \tag{8.33}
$$
$$
\to 0 \quad \text{as} \quad \varepsilon \to 0.
$$

Let us consider the term involving $A(\alpha_k^\varepsilon)x_k^\varepsilon$ in $g_{2,1}^\varepsilon$. Note that

$$
\varepsilon \sum_{l\kappa_\varepsilon=t/\varepsilon}^{(t+s)/\varepsilon-1} \widehat{f}_x'(x_{l\kappa_\varepsilon}^\varepsilon, \alpha_{l\kappa_\varepsilon}^\varepsilon) \sum_{k=l\kappa_\varepsilon}^{l\kappa_\varepsilon+\kappa_\varepsilon-1} A(\alpha_k^\varepsilon)x_k^\varepsilon
$$
$$
= \varepsilon \sum_{i=1}^{l_0} \sum_{j=1}^{m_i} \sum_{l\kappa_\varepsilon=t/\varepsilon}^{(t+s)/\varepsilon-1} \widehat{f}_x'(x_{l\kappa_\varepsilon}^\varepsilon, \alpha_{l\kappa_\varepsilon}^\varepsilon) \sum_{k=l\kappa_\varepsilon}^{l\kappa_\varepsilon+\kappa_\varepsilon-1} A(s_{ij})x_k^\varepsilon \nu^{ij} I_{\{\alpha_k^\varepsilon \in \mathcal{M}_i\}}
$$
$$
+ \varepsilon \sum_{i=1}^{l_0} \sum_{j=1}^{m_i} \sum_{l\kappa_\varepsilon=t/\varepsilon}^{(t+s)/\varepsilon-1} \widehat{f}_x'(x_{l\kappa_\varepsilon}^\varepsilon, \alpha_{l\kappa_\varepsilon}^\varepsilon) \sum_{k=l\kappa_\varepsilon}^{l\kappa_\varepsilon+\kappa_\varepsilon-1} A(s_{ij})x_k^\varepsilon
$$
$$
\times [I_{\{\alpha_k^\varepsilon=s_{ij}\}} - \nu^{ij} I_{\{\alpha_k^\varepsilon \in \mathcal{M}_i\}}]. \tag{8.34}
$$

Thus we need only examine the terms with fixed indices i and j.
For $l\kappa_\varepsilon \le k \le l\kappa_\varepsilon + \kappa_\varepsilon - 1$,

$$
\gamma_k = \widehat{f}'(x_{l\kappa_\varepsilon}^\varepsilon, \alpha_{l\kappa_\varepsilon}^\varepsilon) A(s_{ij})[I_{\{\alpha_k^\varepsilon=s_{ij}\}} - \nu^{ij} I_{\{\alpha_k^\varepsilon \in \mathcal{M}_i\}}].
$$

By a partial summation together with (8.3), the last term in (8.34) with fixed i and j and without $\sum_{i=1}^{l_0} \sum_{j=1}^{m_i}$ becomes

$$
\varepsilon \sum_{k=t/\varepsilon}^{(t+s)/\varepsilon-1} \gamma_k x_k^\varepsilon = \left[\varepsilon \sum_{k=0}^{(t+s)/\varepsilon-1} \gamma_k \right] x_{(t+s)/\varepsilon-1}^\varepsilon - \left[\varepsilon \sum_{k=0}^{t/\varepsilon-1} \gamma_k \right] x_{t/\varepsilon-1}^\varepsilon
$$
$$
+ \varepsilon \sum_{k=t/\varepsilon}^{(t+s)/\varepsilon-2} \left[\sum_{k_1=0}^{k} \gamma_{k_1} \right] (x_{k+1}^\varepsilon - x_k^\varepsilon)
$$
$$
= \left[\varepsilon \sum_{k=0}^{(t+s)/\varepsilon-1} \gamma_k \right] x_{(t+s)/\varepsilon-1}^\varepsilon - \left[\varepsilon \sum_{k=0}^{t/\varepsilon-1} \gamma_k \right] x_{t/\varepsilon-1}^\varepsilon
$$
$$
+ \varepsilon^2 \sum_{k=t/\varepsilon}^{(t+s)/\varepsilon-2} \left[\sum_{k_1=0}^{k} \gamma_{k_1} \right] A(\alpha_k^\varepsilon) x_k^\varepsilon
$$
$$
+ \varepsilon^{\frac{3}{2}} \sum_{k=t/\varepsilon}^{(t+s)/\varepsilon-2} \left[\sum_{k_1=0}^{k} \gamma_{k_1} \right] A(\alpha_k^\varepsilon) \sigma_w(\alpha_k^\varepsilon) w_k.
$$

Similar to (8.33), it can be shown that

$$E \prod_{k_2=1}^{k_0} h_{k_2}(x^\varepsilon(t_{k_2}), \overline{\alpha}^\varepsilon(t_{k_2})) \left[\varepsilon^{\frac{3}{2}} \sum_{k=t/\varepsilon}^{(t+s)/\varepsilon-2} \left[\sum_{k_1=0}^{k} \gamma_{k_1} \right] E_k A(\alpha_k^\varepsilon) \sigma_w(\alpha_k^\varepsilon) E_k w_k \right]$$

$$\to 0 \quad \text{as} \quad \varepsilon \to 0.$$

(8.35)

We claim that

$$\sup_{0 \le n \le T/\varepsilon} E \left| \varepsilon \sum_{k=0}^{n} \gamma_k \right|^2 \to 0 \quad \text{as} \quad \varepsilon \to 0,$$

(8.36)

which, in fact, is a version of Part (c) in Proposition 6.2 (see also Yin and Zhang [158, Lemma 7.14, p. 189] for a continuous-time counter part).

Note that $\sup_{0 \le k \le T/\varepsilon} E|x_k^\varepsilon|^2 < \infty$ by virtue of Lemma 8.5. This together with (8.36) and the Cauchy–Schwarz inequality yields

$$E \left| \left[\varepsilon \sum_{k=0}^{(t+s)/\varepsilon-1} \gamma_k \right] x_{(t+s)/\varepsilon-1}^\varepsilon \right|$$

$$\le E^{1/2} \left| \varepsilon \sum_{k=0}^{(t+s)/\varepsilon-1} \gamma_k x_{(t+s)/\varepsilon-1}^\varepsilon \right|^2 E^{1/2} |x_{l\kappa_\varepsilon+\kappa_\varepsilon-1}^\varepsilon|^2$$

(8.37)

$$\to 0 \quad \text{as} \quad \varepsilon \to 0.$$

Similarly,

$$E \left| \left[\varepsilon \sum_{k=0}^{t/\varepsilon-1} \gamma_k \right] x_{t/\varepsilon-1}^\varepsilon \right| \to 0 \quad \text{as} \quad \varepsilon \to 0.$$

Next, note that

$$E \left| \varepsilon^2 \sum_{k=t/\varepsilon}^{(t+s)/\varepsilon-2} \left[\sum_{k_1=0}^{k} \gamma_{k_1} \right] A(\alpha_k^\varepsilon) x_k^\varepsilon \right|$$

$$\le \varepsilon \sum_{k=t/\varepsilon}^{(t+s)/\varepsilon-2} E \left| \varepsilon \sum_{k_1=0}^{k} \gamma_{k_1} \right| |A(\alpha_k^\varepsilon) x_k^\varepsilon|$$

(8.38)

$$\le K\varepsilon \sum_{k=t/\varepsilon}^{(t+s)/\varepsilon-2} E^{1/2} \left| \varepsilon \sum_{k_1=0}^{k} \gamma_{k_1} \right|^2 E^{1/2} |x_k^\varepsilon|^2$$

$$\to 0 \quad \text{as} \quad \varepsilon \to 0.$$

Using estimates (8.33)–(8.38) and the continuity of $\widehat{f}_x(\cdot, \alpha)$ for each $\alpha \in$

$\mathcal{M}$,

$$
E \prod_{k_2=1}^{k_0} h_{k_2}(x^\varepsilon(t_{k_2}), \overline{\alpha}^\varepsilon(t_{k_2})) \left[\varepsilon \sum_{l\kappa_\varepsilon=t/\varepsilon}^{(t+s)/\varepsilon-1} \widehat{f}_x'(x_{l\kappa_\varepsilon}^\varepsilon, \alpha_{l\kappa_\varepsilon}^\varepsilon) \sum_{k=l\kappa_\varepsilon}^{l\kappa_\varepsilon+\kappa_\varepsilon-1} A(\alpha_k^\varepsilon)x_k^\varepsilon \right]
$$

$$
= E \prod_{k_2=1}^{k_0} h_{k_2}(x^\varepsilon(t_{k_2}), \overline{\alpha}^\varepsilon(t_{k_2})) \left[\sum_{l\kappa_\varepsilon=t/\varepsilon}^{(t+s)/\varepsilon-1} \sum_{j=1}^{m_i} \nu^{ij} \widehat{f}_x'(x_{l\kappa_\varepsilon}^\varepsilon, \alpha_{l\kappa_\varepsilon}^\varepsilon) \right.
$$

$$
\left. \times \frac{\delta_\varepsilon}{\kappa_\varepsilon} \sum_{k=l\kappa_\varepsilon}^{l\kappa_\varepsilon+\kappa_\varepsilon-1} A(s_{ij})x_{l\kappa_\varepsilon}^\varepsilon \nu^{ij} I_{\{\alpha_k^\varepsilon \in \mathcal{M}_i\}} \right] + o(1),
$$

(8.39)

where $o(1) \to 0$ as $\varepsilon \to 0$. Then as $\varepsilon \to 0$, letting $\varepsilon l\kappa_\varepsilon \to u$, and using the techniques of weak convergence, (8.39) together with (8.33) leads to

$$
E \prod_{k_2=1}^{k_0} h_{k_2}(x^\varepsilon(t_{k_2}), \overline{\alpha}^\varepsilon(t_{k_2})) g_{2,1}^\varepsilon
$$

$$
\to E \prod_{k_2=1}^{k_0} h_{k_2}(x(t_{k_2}), \overline{\alpha}(t_{k_2})) \left(\int_t^{t+s} f_x'(x(u), \overline{\alpha}(u)) A(\overline{\alpha}(u))x(u)du \right)
$$

(8.40)

as $\varepsilon \to 0$. As for $g_{2,2}^\varepsilon$, we have by the continuity of $f_{xx}(\cdot, \alpha)$ for each $\alpha \in \mathcal{M}$, $x_{l\kappa_\varepsilon}^+ - x_{l\kappa_\varepsilon}^\varepsilon \to 0$ in probability as $\varepsilon \to 0$. Consequently,

$$
E \prod_{k_2=1}^{k_0} h_{k_2}(x^\varepsilon(t_{k_2}), \overline{\alpha}^\varepsilon(t_{k_2})) g_{2,2}^\varepsilon
$$

$$
\stackrel{\text{def}}{=} E \prod_{k_2=1}^{k_0} h_{k_2}(x^\varepsilon(t_{k_2}), \overline{\alpha}^\varepsilon(t_{k_2})) \widetilde{g}_{2,2}^\varepsilon + o(1),
$$

where $o(1) \to 0$ as $\varepsilon \to 0$ uniformly in t, and

$$
\widetilde{g}_{2,2}^\varepsilon = \left[\sum_{l\kappa_\varepsilon=t/\varepsilon}^{(t+s)/\varepsilon-1} \sum_{k=l\kappa_\varepsilon}^{l\kappa_\varepsilon+\kappa_\varepsilon-1} [\varepsilon A(\alpha_k^\varepsilon)x_k^\varepsilon + \sqrt{\varepsilon}\sigma_w(\alpha_k^\varepsilon)w_k]' \widehat{f}_{xx}(x_{l\kappa_\varepsilon}^\varepsilon, \alpha_{l\kappa_\varepsilon}^\varepsilon) \right.
$$

$$
\left. \times \sum_{k_1=l\kappa_\varepsilon}^{l\kappa_\varepsilon+\kappa_\varepsilon-1} [\varepsilon A(\alpha_{k_1}^\varepsilon)x_{k_1}^\varepsilon + \sqrt{\varepsilon}\sigma_w(\alpha_{k_1}^\varepsilon)w_{k_1}] \right].
$$

It then follows that

$$
\tilde{g}_{2,2}^{\varepsilon} = \Bigg[\varepsilon^2 \sum_{l\kappa_\varepsilon = t/\varepsilon}^{(t+s)/\varepsilon - 1} \sum_{k=l\kappa_\varepsilon}^{l\kappa_\varepsilon + \kappa_\varepsilon - 1} (A(\alpha_k^\varepsilon) x_k^\varepsilon)' \widehat{f}_{xx}(x_{l\kappa_\varepsilon}^\varepsilon, \alpha_{l\kappa_\varepsilon}^\varepsilon) \sum_{k_1 = l\kappa_\varepsilon}^{l\kappa_\varepsilon + \kappa_\varepsilon - 1} A(\alpha_{k_1}^\varepsilon) x_{k_1}^\varepsilon
$$

$$
+ \sqrt{\varepsilon^3} \sum_{l\kappa_\varepsilon = t/\varepsilon}^{(t+s)/\varepsilon - 1} \sum_{k=l\kappa_\varepsilon}^{l\kappa_\varepsilon + \kappa_\varepsilon - 1} (A(\alpha_k^\varepsilon) x_k^\varepsilon)' \widehat{f}_{xx}(x_{l\kappa_\varepsilon}^\varepsilon, \alpha_{l\kappa_\varepsilon}^\varepsilon) \sum_{k_1 = l\kappa_\varepsilon}^{l\kappa_\varepsilon + \kappa_\varepsilon - 1} \sigma_w(\alpha_{k_1}^\varepsilon) w_{k_1}
$$

$$
+ \sqrt{\varepsilon^3} \sum_{l\kappa_\varepsilon = t/\varepsilon}^{(t+s)/\varepsilon - 1} \sum_{k=l\kappa_\varepsilon}^{l\kappa_\varepsilon + \kappa_\varepsilon - 1} (\sigma_w(\alpha_k^\varepsilon) w_k)' \widehat{f}_{xx}(x_{l\kappa_\varepsilon}^\varepsilon, \alpha_{l\kappa_\varepsilon}^\varepsilon) \sum_{k_1 = l\kappa_\varepsilon}^{l\kappa_\varepsilon + \kappa_\varepsilon - 1} A(\alpha_{k_1}^\varepsilon) x_{k_1}^\varepsilon
$$

$$
+ \varepsilon \sum_{l\kappa_\varepsilon = t/\varepsilon}^{(t+s)/\varepsilon - 1} \sum_{k=l\kappa_\varepsilon}^{l\kappa_\varepsilon + \kappa_\varepsilon - 1} (\sigma_w(\alpha_k^\varepsilon) w_k)' \widehat{f}_{xx}(x_{l\kappa_\varepsilon}^\varepsilon, \alpha_{l\kappa_\varepsilon}^\varepsilon) \sum_{k_1 = l\kappa_\varepsilon}^{l\kappa_\varepsilon + \kappa_\varepsilon - 1} \sigma_w(\alpha_{k_1}^\varepsilon) w_{k_1} \Bigg]
$$

$$
= \varepsilon \sum_{l\kappa_\varepsilon = t/\varepsilon}^{(t+s)/\varepsilon - 1} \sum_{k=l\kappa_\varepsilon}^{l\kappa_\varepsilon + \kappa_\varepsilon - 1} (\sigma_w(\alpha_k^\varepsilon) w_k)' \widehat{f}_{xx}(x_{l\kappa_\varepsilon}^\varepsilon, \alpha_{l\kappa_\varepsilon}^\varepsilon) \sigma_w(\alpha_k^\varepsilon) w_k + o(1),
$$

where $o(1) \to 0$ in probability as $\varepsilon \to 0$. Furthermore, using the idea of the estimates leading to (8.39) and the mean squares estimates in Part (c) Proposition 6.2, it can be shown that

$$
\varepsilon \sum_{l\kappa_\varepsilon = t/\varepsilon}^{(t+s)/\varepsilon - 1} \sum_{k=l\kappa_\varepsilon}^{l\kappa_\varepsilon + \kappa_\varepsilon - 1} (\sigma_w(\alpha_k^\varepsilon) w_k)' \widehat{f}_{xx}(x_{l\kappa_\varepsilon}^\varepsilon, \alpha_{l\kappa_\varepsilon}^\varepsilon) \sigma_w(\alpha_k^\varepsilon) w_k
$$

$$
= \sum_{l\kappa_\varepsilon = t/\varepsilon}^{(t+s)/\varepsilon - 1} \sum_{j=1}^{m_i} \frac{\delta_\varepsilon}{\kappa_\varepsilon} \sum_{k=l\kappa_\varepsilon}^{l\kappa_\varepsilon + \kappa_\varepsilon - 1} \mathrm{tr}[\widehat{f}_{xx}(x_{l\kappa_\varepsilon}^\varepsilon, \alpha_{l\kappa_\varepsilon}^\varepsilon) \sigma_w(s_{ij}) w_k w_k' \sigma_w'(s_{ij})]
$$

$$
\times I_{\{\alpha_k^\varepsilon = s_{ij}\}} + o(1)
$$

$$
= \sum_{l\kappa_\varepsilon = t/\varepsilon}^{(t+s)/\varepsilon - 1} \sum_{j=1}^{m_i} \frac{\delta_\varepsilon}{\kappa_\varepsilon} \sum_{k=l\kappa_\varepsilon}^{l\kappa_\varepsilon + \kappa_\varepsilon - 1} \mathrm{tr}[\widehat{f}_{xx}(x_{l\kappa_\varepsilon}^\varepsilon, \alpha_{l\kappa_\varepsilon}^\varepsilon) \sigma_w(s_{ij}) w_k w_k' \sigma_w'(s_{ij})]
$$

$$
\times \nu^{ij} I_{\{\alpha_k^\varepsilon \in \mathcal{M}_i\}} + o(1),
$$

where $o(1) \to 0$ in probability as $\varepsilon \to 0$ uniformly in t. It then follows that

$$
\lim_{\varepsilon \to 0} E \prod_{k_2 = 1}^{k_0} h_{k_2}(x^\varepsilon(t_{k_2}), \overline{\alpha}^\varepsilon(t_{k_2}))
$$

$$
\times \Bigg[\varepsilon \sum_{l\kappa_\varepsilon = t/\varepsilon}^{(t+s)/\varepsilon - 1} \sum_{k=l\kappa_\varepsilon}^{l\kappa_\varepsilon + \kappa_\varepsilon - 1} (\sigma_w(\alpha_k^\varepsilon) w_k)' \widehat{f}_{xx}(x_{l\kappa_\varepsilon}^\varepsilon, \alpha_{l\kappa_\varepsilon}^\varepsilon) \sigma_w(\alpha_k^\varepsilon) w_k \Bigg]
$$
(8.41)

$$
= E \prod_{k_2 = 1}^{k_0} h_{k_2}(x(t_{k_2}), \overline{\alpha}(t_{k_2}))
$$

$$
\times \Bigg[\int_t^{t+s} \mathrm{tr}[f_{xx}(x(u), \overline{\alpha}(u)) \sigma_w(\overline{\alpha}(u)) \sigma_w'(\overline{\alpha}(u))] du \Bigg].
$$

Next, we consider the term g_1^ε. Using the continuity of $\widehat{f}(\cdot, \alpha)$ for each $\alpha \in \mathcal{M}$, the Markov property of α_k^ε, the mean squares estimate of the occupation measures in Part (c) Proposition 6.2, (8.1), and Proposition 6.2, we have $g_1^\varepsilon = \widetilde{g}_1^\varepsilon + \widetilde{g}_{1,1}^\varepsilon$ such that $E \prod_{k_2=1}^{k_0} h_{k_2}(x^\varepsilon(t_{k_2}), \overline{\alpha}^\varepsilon(t_{k_2})) \widetilde{g}_{1,1}^\varepsilon \to 0$, and

$$E \prod_{k_2=1}^{k_0} h_{k_2}(x^\varepsilon(t_{k_2}), \overline{\alpha}^\varepsilon(t_{k_2})) \widetilde{g}_1^\varepsilon$$

$$= E \prod_{k_2=1}^{k_0} h_{k_2}(x^\varepsilon(t_{k_2}), \overline{\alpha}^\varepsilon(t_{k_2})) \left[\sum_{l\kappa_\varepsilon=t/\varepsilon}^{(t+s)/\varepsilon-1} [\widehat{f}(x_{l\kappa_\varepsilon}^\varepsilon, \alpha_{l\kappa_\varepsilon+\kappa_\varepsilon}^\varepsilon) - \widehat{f}(x_{l\kappa_\varepsilon}^\varepsilon, \alpha_{l\kappa_\varepsilon}^\varepsilon)] \right]$$

$$= E \prod_{k_2=1}^{k_0} h_{k_2}(x^\varepsilon(t_{k_2}), \overline{\alpha}^\varepsilon(t_{k_2}))$$

$$\times \left[\sum_{l\kappa_\varepsilon=t/\varepsilon}^{(t+s)/\varepsilon-1} \sum_{k=l\kappa_\varepsilon}^{l\kappa_\varepsilon+\kappa_\varepsilon-1} \sum_{i_1=1}^{l_0} \sum_{j_1=1}^{m_{i_1}} \left[\sum_{i=1}^{l_0} \sum_{j=1}^{m_i} \widehat{f}(x_{l\kappa_\varepsilon}^\varepsilon, s_{ij}) \right. \right.$$

$$\left. \left. \times P(\alpha_{k+1}^\varepsilon = s_{ij} | \alpha_k^\varepsilon = s_{i_1 j_1}) - \widehat{f}(x_{l\kappa_\varepsilon}^\varepsilon, s_{i_1 j_1}) \right] I_{\{\alpha_k^\varepsilon = s_{i_1 j_1}\}} \right]$$

$$= E \prod_{k_2=1}^{k_0} h_{k_2}(x^\varepsilon(t_{k_2}), \overline{\alpha}^\varepsilon(t_{k_2}))$$

$$\times \left[\varepsilon \sum_{l\kappa_\varepsilon=t/\varepsilon}^{(t+s)/\varepsilon-1} \sum_{k=l\kappa_\varepsilon}^{l\kappa_\varepsilon+\kappa_\varepsilon-1} ((P-I)+\varepsilon Q)\widehat{f}(x_{l\kappa_\varepsilon}^\varepsilon, \cdot)(\alpha_k^\varepsilon) \right]$$

$$= E \prod_{k_2=1}^{k_0} h_{k_2}(x^\varepsilon(t_{k_2}), \overline{\alpha}^\varepsilon(t_{k_2})) \left[\varepsilon \sum_{l\kappa_\varepsilon=t/\varepsilon}^{(t+s)/\varepsilon-1} \sum_{k=l\kappa_\varepsilon}^{l\kappa_\varepsilon+\kappa_\varepsilon-1} Q\widehat{f}(x_{l\kappa_\varepsilon}^\varepsilon, \cdot)(\alpha_k^\varepsilon) \right].$$

Thus,

$$\lim_{\varepsilon \to 0} E \prod_{k_2=1}^{k_0} h_{k_2}(x^\varepsilon(t_{k_2}), \overline{\alpha}^\varepsilon(t_{k_2})) g_1^\varepsilon$$

$$= E \prod_{k_2=1}^{k_0} h_{k_2}(x(t_{k_2}), \overline{\alpha}(t_{k_2})) \left[\int_t^{t+s} \overline{Q} f(x(u), \overline{\alpha}(u)) du \right]. \quad (8.42)$$

Combining (8.40), (8.41), and (8.42),

$$\lim_{\varepsilon \to 0} E \prod_{k_2=1}^{k_0} h_{k_2}(x^\varepsilon(t_{k_2}), \overline{\alpha}^\varepsilon(t_{k_2})) \left[\widehat{f}(x^\varepsilon(t+s), \alpha^\varepsilon(t+s)) - \widehat{f}(x^\varepsilon(t), \alpha^\varepsilon(t)) \right]$$

$$= E \prod_{k_2=1}^{k_0} h_{k_2}(x(t_{k_2}), \overline{\alpha}(t_{k_2})) \left[\int_t^{t+s} \mathcal{L}f(x(u), \overline{\alpha}(u)) du \right]$$

$$(8.43)$$

as desired. On the other hand, by the weak convergence of $(x^\varepsilon(\cdot), \overline{\alpha}^\varepsilon(\cdot))$ to $(x(\cdot), \overline{\alpha}(\cdot))$, the Skorohod representation (without changing notation), and the definition of $\widehat{f}(\cdot)$, we have

$$\lim_{\varepsilon \to 0} E \prod_{k_2=1}^{k_0} h_{k_2}(x^\varepsilon(t_{k_2}), \overline{\alpha}^\varepsilon(t_{k_2}))[\widehat{f}(x^\varepsilon(t+s), \alpha^\varepsilon(t+s)) - \widehat{f}(x^\varepsilon(t), \alpha^\varepsilon(t))]$$

$$= E \prod_{k_2=1}^{k_0} h_{k_2}(x(t_{k_2}), \overline{\alpha}(t_{k_2}))[f(x(t+s), \overline{\alpha}(t+s)) - f(x(t), \overline{\alpha}(t))].$$

$$(8.44)$$

By (8.43) and (8.44), (8.28) holds. Thus the limit of $x^\varepsilon(\cdot)$ (the first equation in (8.14)) is obtained.

In exactly the same way, we prove that $(y(\cdot), \overline{\alpha}(\cdot))$, the weak limit of $(y^\varepsilon(\cdot), \overline{\alpha}^\varepsilon(\cdot))$, is the unique solution of the martingale problem with operator

$$\mathcal{L}_1 f(y, i) = f_y'(y, i)\overline{C}(i)y + \frac{1}{2}\mathrm{tr}[f_{yy}(y, i)\overline{\sigma}_v(i)\overline{\sigma}_v'(i)] + \overline{Q}_* f(y, \cdot)(i), \quad i \in \mathcal{M}.$$

The limit equation for $y(\cdot)$ in (8.14) is established.

Next, we consider $(R^\varepsilon(\cdot), \overline{\alpha}^\varepsilon(\cdot))$. Using (8.27) and the boundedness of R_k^ε, we obtain that $(R(\cdot), \overline{\alpha}(\cdot))$ is the solution of the degenerate martingale problem with operator

$$\mathcal{L}_2 \widetilde{f}(R, i) = \widetilde{f}_R'(R, i)[\overline{A}(i)R + R\overline{A}'(i)$$
$$- R\overline{C}(i)\Sigma_v^{-1}(i)\overline{C}(i)R + \Sigma_w(i)] + \overline{Q}_* \widetilde{f}(R, \cdot)(i), \quad i \in \overline{M},$$

where $\widetilde{f}(\cdot, i) : \mathbb{R}^{r \times r} \mapsto \mathbb{R}$ is twice continuously differentiable with compact support. Thus the equation for $R(\cdot)$ in (8.14) is obtained.

To complete the proof, we examine the limit of $\widehat{x}^\varepsilon(\cdot)$. By virtue of the weak convergence and the Skorohod representation (without changing notation), we may assume that $(x^\varepsilon(\cdot), y^\varepsilon(\cdot), R^\varepsilon(\cdot), \overline{\alpha}^\varepsilon(\cdot))$ converges to $(x(\cdot), y(\cdot), R(\cdot), \overline{\alpha}(\cdot))$ w.p.1 and the convergence is uniform on any bounded subinterval of $[0, T]$. In view of (8.22),

$$\widehat{x}^\varepsilon(t+s) - \widehat{x}^\varepsilon(t) = \varepsilon \sum_{k=t/\varepsilon}^{(t+s)/\varepsilon-1} \overline{A}(\overline{\alpha}_k^\varepsilon)\widehat{x}_k^\varepsilon - \varepsilon \sum_{k=t/\varepsilon}^{(t+s)/\varepsilon-1} H_k^\varepsilon \overline{C}(\overline{\alpha}_k^\varepsilon)\widehat{x}_k^\varepsilon$$
$$+ \varepsilon \sum_{k=t/\varepsilon}^{(t+s)/\varepsilon-1} H_k^\varepsilon C(\alpha_k^\varepsilon)x_k^\varepsilon + \sqrt{\varepsilon} \sum_{k=t/\varepsilon}^{(t+s)/\varepsilon-1} H_k^\varepsilon \sigma_v(\alpha_k^\varepsilon)v_k.$$

$$(8.45)$$

Using the same idea for evaluating the limits of $x^\varepsilon(\cdot)$ and $y^\varepsilon(\cdot)$, it can be shown that

$$\varepsilon \sum_{k=t/\varepsilon}^{(t+s)/\varepsilon-1} \overline{A}(\overline{\alpha}_k^\varepsilon)\widehat{x}_k^\varepsilon \to \int_t^{t+s} \overline{A}(\overline{\alpha}(u))\widehat{x}(u)du.$$

Note that by virtue of (8.21), $H_k^\varepsilon = H(R_k^\varepsilon, \overline{\alpha}_k^\varepsilon)$ and $H(\cdot, i)$ is a continuous function for each $i \in \overline{\mathcal{M}}$. Owing to the linearity in $\widehat{x}_k^\varepsilon$ and the continuity of $H(\cdot, i)$, the limit of $\varepsilon \sum_{k=t/\varepsilon}^{(t+s)/\varepsilon - 1} H_k^\varepsilon \overline{C}(\overline{\alpha}_k^\varepsilon) \widehat{x}_k^\varepsilon$ is the same as that of

$$\varepsilon \sum_{l\kappa_\varepsilon = t/\varepsilon}^{(t+s)/\varepsilon - 1} \delta_\varepsilon \frac{1}{\kappa_\varepsilon} \sum_{k=l\kappa_\varepsilon}^{l\kappa_\varepsilon + \kappa_\varepsilon - 1} H(R_{l\kappa_\varepsilon}^\varepsilon, \overline{\alpha}_k^\varepsilon) \overline{C}(\overline{\alpha}_k^\varepsilon) \widehat{x}_{l\kappa_\varepsilon}^\varepsilon$$

$$= \sum_{l\kappa_\varepsilon = t/\varepsilon}^{(t+s)/\varepsilon - 1} \delta_\varepsilon \frac{1}{\kappa_\varepsilon} \sum_{k=l\kappa_\varepsilon}^{l\kappa_\varepsilon + \kappa_\varepsilon - 1} H(R^\varepsilon(\varepsilon l\kappa_\varepsilon), \overline{\alpha}_k^\varepsilon) \overline{C}(\overline{\alpha}_k^\varepsilon) \widehat{x}^\varepsilon(\varepsilon l\kappa_\varepsilon)$$

$$= \sum_{i=1}^{l_0} \sum_{l\kappa_\varepsilon = t/\varepsilon}^{(t+s)/\varepsilon - 1} \delta_\varepsilon \frac{1}{\kappa_\varepsilon} \sum_{k=l\kappa_\varepsilon}^{l\kappa_\varepsilon + \kappa_\varepsilon - 1} H(R^\varepsilon(\varepsilon l\kappa_\varepsilon), i) \overline{C}(i) \widehat{x}^\varepsilon(\varepsilon l\kappa_\varepsilon) I_{\{\overline{\alpha}^\varepsilon(\varepsilon k) = i\}}.$$

$$(8.46)$$

Sending $\varepsilon l\kappa_\varepsilon \to u$, we have for any k satisfying $l\kappa_\varepsilon \leq k \leq l\kappa_\varepsilon + \kappa_\varepsilon$, $\varepsilon k \to u$ as well. Thus using the defining relation (8.21), the weak limit in (8.46) is

$$\sum_{i=1}^{l_0} \int_t^{t+s} R(u) \overline{C}'(i) \Sigma_v^{-1}(i) \overline{C}(i) \widehat{x}(u) I_{\{\overline{\alpha}(u) = i\}} du$$

$$= \int_t^{t+s} R(u) \overline{C}'(\overline{\alpha}(u)) \Sigma_v^{-1}(\overline{\alpha}(u)) \overline{C}(\overline{\alpha}(u)) \widehat{x}(u) du.$$

Likewise, detailed arguments lead to the weak convergence of

$$\varepsilon \sum_{k=t/\varepsilon}^{(t+s)/\varepsilon - 1} H_k^\varepsilon C(\alpha_k^\varepsilon) x_k^\varepsilon + \sqrt{\varepsilon} \sum_{k=t/\varepsilon}^{(t+s)/\varepsilon - 1} H_k^\varepsilon \sigma_v(\alpha_k^\varepsilon) v_k$$

$$\to \int_t^{t+s} [R(u) \overline{C}'(\overline{\alpha}(u)) \Sigma_v^{-1}(\overline{\alpha}(u)) [dy(u) - \overline{C}(\overline{\alpha}(u)) \widehat{x}(u) du].$$

The limit equation for $\widehat{x}(\cdot)$ in (8.14) is obtained. Therefore, the proof of the theorem is completed. $\square$

Proof of Theorem 8.9. First of all, in view of the definition of $x^\varepsilon(\cdot)$ and $\widehat{x}^\varepsilon(\cdot)$ and their convergence in Theorem 8.7, we have, as $\varepsilon \to 0$,

$$\sum_{k=0}^{T/\varepsilon} \varepsilon |x_k^\varepsilon - \widehat{x}_k^\varepsilon|^2 = \int_0^T |x^\varepsilon(t) - \widehat{x}^\varepsilon(t)|^2 dt \to \int_0^T |x(t) - \widehat{x}(t)|^2 dt,$$

in distribution. In order to show

$$E \sum_{k=0}^{T/\varepsilon} \varepsilon |x_k^\varepsilon - \widehat{x}_k^\varepsilon|^2 \to E \int_0^T |x(t) - \widehat{x}(t)|^2 dt,$$

it suffices that

$$\left\{ \sum_{k=0}^{T/\varepsilon} \varepsilon |x_k^\varepsilon - \widehat{x}_k^\varepsilon|^2 \right\} \text{ is uniformly integrable.}$$

A sufficient condition is

$$E\left(\sum_{k=0}^{T/\varepsilon}\varepsilon|x_k^\varepsilon - \widehat{x}_k^\varepsilon|^2\right)^2 \leq K < \infty.$$

In fact,

$$E\left(\sum_{k=0}^{T/\varepsilon}\varepsilon|x_k^\varepsilon - \widehat{x}_k^\varepsilon|^2\right)^2 \leq TE\sum_{k=0}^{T/\varepsilon}\varepsilon|x_k^\varepsilon - \widehat{x}_k^\varepsilon|^4$$

$$\leq 8T\sum_{k=0}^{T/\varepsilon}\varepsilon\left(E|x_k^\varepsilon|^4 + E|\widehat{x}_k^\varepsilon|^4\right) \leq K.$$

We now show Part (b). In view of Lemma 14.39, we have

$$E\int_0^T |x(t) - \widehat{x}(t)|^2 dt = O(\delta).$$

Therefore,

$$\lim_{\delta\to0}\lim_{\varepsilon\to0} E\sum_{k=0}^{T/\varepsilon}\varepsilon|x_k^\varepsilon - \widehat{x}_k^\varepsilon|^2 = \lim_{\delta\to0}E\int_0^T |x(t) - \widehat{x}(t)|^2 dt = 0.$$

The proof is thus concluded.   □

8.4   Discrete-Time Approximation of Wonham Filter

In hybrid filtering, it is useful to provide conditional probabilities of the underlying switching process. This can be accomplished by introducing an additional observation equation for the Markov chain. Such a model gives rise to the Wonham filter; see Wonham [150]. In this section, we consider a discrete-time approximation of the Wonham filter. This is another application of the two-time-scale Markov chain. The algorithms developed provide a simple method for needed calculation.

Working with a finite time horizon $t \in [0,T]$ for some $T > 0$, suppose that $\alpha(t)$ is a finite-state Markov chain with state space $\mathcal{M} = \{z^1, \ldots, z^{m_0}\}$ and generator $Q = (q^{ij}) \in \mathbb{R}^{m_0 \times m_0}$, $w(\cdot)$ is a standard one-dimensional Brownian motion that is independent of $\alpha(t)$, and $\sigma(\cdot) : \mathbb{R} \mapsto \mathbb{R}$ with $\sigma(t) \geq c > 0$ is an appropriate function satisfying suitable conditions. Consider the following observation process

$$dy(t) = \alpha(t)dt + \sigma(t)dw(t), \ y(0) = 0. \tag{8.47}$$

Note that due to the jump processes $\alpha(t)$, the distribution of $y(t)$ is non-Gaussian, but a Gaussian mixture. Let $p(t) = (p^1(t), \ldots, p^{m_0}(t)) \in \mathbb{R}^{1 \times m_0}$, with $p^i(t) = P(\alpha(t) = z^i | y(s), \ 0 \leq s \leq t)$, for $i = 1, \ldots, m_0$, $p^i(0) = p_0^i$. It was proved in Wonham [150] that this conditional density satisfies the following system of stochastic differential equations

$$
\begin{cases}
dp^i(t) = \left[q^{ii} p^i(t) + \sum_{j \neq i} q^{ji} p^j(t) - \sigma^{-2}(t)(z^i - \widehat{\alpha}(t))\widehat{\alpha}(t)p^i(t) \right] dt \\
\qquad\qquad + \sigma^{-2}(t)(z^i - \widehat{\alpha}(t))p^i(t)dy(t), \quad \text{for } i = 1, \ldots, m_0, \\
p^i(0) = p_0^i.
\end{cases}
$$

$$(8.48)$$

Although the filter provides precise results on the posterior probabilities, the system often has to be solved numerically because it is nonlinear and because observations are frequently collected in discrete moments.

To construct approximation algorithms, one may wish to discretize the stochastic differential equations (8.48) directly. Nevertheless, numerical experiments and simulations show that such a procedure is numerically unstable due to the white noise perturbations. It may produce a non-probability vector (e.g., some components might be less than 0 or the sum of the components might not be 1). To overcome the difficulty, we first transform the stochastic differential equations and then design a numerical procedure for the transformed system.

Define $v^i(t)$ to be the natural logarithm of $p^i(t)$, i.e.,

$$
v^i(t) = \log p^i(t), \quad \text{for } t \geq 0 \text{ and } i = 1, \ldots, m_0.
$$

It follows that $p^i(t) = \exp(v^i(t))$. A straightforward application of the Ito's rule leads to the following. For each $i = 1, \ldots, m_0$,

$$
\begin{cases}
dv^i(t) = \left[q^{ii} + \sum_{j \neq i} q^{ji} \dfrac{p^j(t)}{p^i(t)} - \sigma^{-2}(t)(z^i - \widehat{\alpha}(t))\widehat{\alpha}(t) \right. \\
\qquad\qquad \left. - \dfrac{1}{2}\sigma^{-2}(t)(z^i - \widehat{\alpha}(t))^2 \right] dt + \sigma^{-2}(t)(z^i - \widehat{\alpha}(t))dy(t) \\
v^i(0) = \log p_0^i.
\end{cases}
$$

$$(8.49)$$

Let $\varepsilon > 0$ be the step size, and let the (actual physical) observation be given by

$$
y_{k+1} = y_k + \varepsilon \alpha_k^\varepsilon + \sqrt{\varepsilon}\sigma_k \xi_k,
$$

$$
y_0 = 0 \text{ w.p.1,}
$$

$$(8.50)$$

where $\{\xi_k\}$ is a sequence of i.i.d. zero mean random variables, $\{\alpha_k^\varepsilon\}$ is a Markov chain with state space $\mathcal{M}$, and the one-step transition matrix is given by

$$
P^\varepsilon = I + \varepsilon Q,
$$

$$(8.51)$$

where I is the m_0-dimensional identity matrix.

In what follows, we show that under suitable scaling and interpolation, the recursive algorithm given above leads to the limit process, which is the posterior probability vector satisfying (8.48). Note that in the above, to reinforce that $\{p_k^i\}$ verifying $0 \le p_k^i \le 1$, we may use

$$p_{k+1}^i = \frac{\exp(v_{k+1}^i)}{\sum\limits_{j=1}^{m_0} \exp(v_{k+1}^j)} \tag{8.52}$$

in (8.53). Our numerical experiments show that even without using this renormalization, the approximation to probability vector is rather robust, however. Also in the above, we have assumed implicitly that $\{p_k^i\}$ is bounded below from zero. We will comment on the relaxation of this condition at the end of this section.

Remark 8.14. Note that (8.50) mimics the dynamics given in (8.47). If one discretized the observed signal $y(t)$ in (8.47) by using a constant step ε with

$$\alpha_k^\varepsilon = \alpha(\varepsilon k), \ \ \sigma_k = \sigma(\varepsilon k), \ \ \Delta w_k = w(\varepsilon(k+1)) - w(\varepsilon k),$$

and (8.50) with
$$\xi_k = [w(\varepsilon(k+1)) - w(\varepsilon k)]/\sqrt{\varepsilon},$$

then one would obtain an Euler–Maruyama type approximation of (8.47) (see Kloeden and Platen [88]).

Note that in such a case, α_k^ε is a discrete-time Markov chain with a one-step transition matrix given by $\exp(\varepsilon Q)$. That is, α_k^ε is an ε-skeleton of the continuous-time Markov chain $\alpha(t)$; see Chung [38, p.132]. Our approximation (8.50) uses another approximation by replacing the one-step transition probability matrix $\exp(\varepsilon Q)$ with the first few terms in the Taylor expansions, namely, $I + \varepsilon Q$. For simplicity, we assume that $\{\xi_k\}$ is a sequence of independent and identically distributed random variables with mean 0 and unit variance. More general correlated noise may be considered. What is essential is that the limit of a suitably scaled sequence is a "white noise" in continuous time.

Discretizing the transformed system (8.49) and using the notation $\Delta y_k = y_{k+1} - y_k$ yield the following algorithm

$$\begin{cases} v_{k+1}^i = v_k^i + \varepsilon \tilde{r}_k^i + \sqrt{\varepsilon}\sigma_k^{-2}(z^i - \widehat{\alpha}_k^\varepsilon)\Delta y_k, \quad v_0^i = \log p_0^i, \\[2mm] \tilde{r}_k^i = q^{ii} + \sum\limits_{j \ne i} q^{ji}\frac{p_k^j}{p_k^i} - \sigma_k^{-2}(z^i - \widehat{\alpha}_k^\varepsilon)\widehat{\alpha}_k^\varepsilon - \frac{1}{2}\sigma_k^{-2}(z^i - \widehat{\alpha}_k^\varepsilon)^2, \\[2mm] p_{k+1}^i = \exp(v_{k+1}^i), \quad p_0^i = p^i(0). \end{cases}$$

Equivalently, we can write the above equations in terms of the white noise ξ_k as follows:

$$\begin{cases} v_{k+1}^i = v_k^i + \varepsilon r_k^i + \sqrt{\varepsilon}\sigma_k^{-1}(z^i - \widehat{\alpha}_k^\varepsilon)\xi_k, \quad v_0^i = \log p_0^i, \\ r_k^i = q^{ii} + \sum_{j \neq i} q^{ji}\frac{p_k^j}{p_k^i} - \sigma_k^{-2}(z^i - \widehat{\alpha}_k^\varepsilon)\widehat{\alpha}_k^\varepsilon + \sigma_k^{-2}(z^i - \widehat{\alpha}_k^\varepsilon)\alpha_k^\varepsilon \\ \qquad - \frac{1}{2}\sigma_k^{-2}(z^i - \widehat{\alpha}_k^\varepsilon)^2, \\ p_{k+1}^i = \exp(v_{k+1}^i), \quad p_0^i = p^i(0). \end{cases} \tag{8.53}$$

This expression is convenient for convergence verifications. To proceed, let us first present the conditions needed.

(A8.4) The following conditions hold:

(1) α_k^ε is a discrete-time Markov chain with a one-step transition probability matrix (8.51).

(2) $\{\sigma_k\}$ is a sequence of real numbers such that $\sigma_k = \sigma(\varepsilon k)$, where $\sigma(\cdot)$ is as given in (8.48), a bounded and continuously differentiable function satisfying $\sigma(t) \geq c > 0$ for some constant c.

(3) $\{\xi_k\}$ in (8.50) is a sequence of independent and identically distributed random variables satisfying $E\xi_k = 0$, $E\xi_k^2 = 1$, and $E|\xi_k|^{2+\widehat{\gamma}} < \infty$ for some $\widehat{\gamma} > 0$.

(4) The sequences $\{\alpha_k^\varepsilon\}$ and $\{\xi_k\}$ are independent.

To obtain the desired result, we take continuous-time interpolations. Define

$$\left.\begin{aligned} &\alpha^\varepsilon(t) = \alpha_k^\varepsilon, \ \sigma^\varepsilon(t) = \sigma_k, \ y^\varepsilon(t) = y_k, \ p^{\varepsilon,i}(t) = p_k^i, \\ &\widehat{\alpha}^\varepsilon(t) = \sum_{j=1}^{m_0} z^j p_k^j, \ v^{\varepsilon,i}(t) = v_k^i, \ i = 1,\ldots,m_0, \end{aligned}\right\} \quad \text{for } t \in [k\varepsilon, k\varepsilon+\varepsilon).$$

$$\tag{8.54}$$

Then for some $T > 0$, $\alpha^\varepsilon(\cdot) \in D([0,T] : \mathcal{M})$, the space of functions that are defined on $[0,T]$, that take values in $\mathcal{M}$, and that are right continuous and have left-hand limits endowed with the Skorohod topology.

Theorem 8.15. *Assume* (A8.4). *Then* $(v^{\varepsilon,i}(\cdot), \alpha^\varepsilon(\cdot))$ *converges weakly to* $(v^i(\cdot), \alpha(\cdot))$. *As a result,* $(p^{\varepsilon,i}(\cdot), \alpha^\varepsilon(\cdot))$ *converges weakly to* $(p^i(\cdot), \alpha(\cdot))$ *such that* $\{p^i(\cdot)\}$ *satisfies the system of Wonham filter equations* (8.48).

We omit the proof of this theorem but provide the remark below. The proof consists of several steps. As a preliminary, we first show that under (A8.4), $(y^\varepsilon(\cdot), \alpha^\varepsilon(\cdot))$ converges weakly to $(y(\cdot), \alpha(\cdot))$ such that $y(\cdot)$ is a solution of (8.47) and $\alpha(\cdot)$ is a continuous-time Markov chain whose generator is given by Q. In the second step, we prove the tightness of the sequences $\{v^{\varepsilon,i}(\cdot), \alpha^{\varepsilon}(\cdot)\}$. Using Prohorov's theorem, we then extract weakly convergent subsequences of $(v^{\varepsilon,i}(\cdot), \alpha^{\varepsilon}(\cdot))$ with the limit denoted by $(v^i(\cdot), \alpha(\cdot))$. By the continuous mapping theorem (see Theorem 14.21), $p^{\varepsilon,i}(\cdot)$ converges weakly to $p^i(\cdot)$ and by virtue of the Skorohod representation (without changing notation), we may assume that the convergence is w.p.1 and it takes place uniformly on any bounded interval. Then we characterize the limit process by showing that the limit $(v^i(\cdot), \alpha(\cdot))$ is the solution of the martingale problem with operator $\mathcal{L}^i$ defined as follows. For each $z^{\ell} \in \mathcal{M}$,

$$
\begin{aligned}
\mathcal{L}^i f(t, y, z^{\ell}) &= \frac{1}{2}[f_{yy}(t, y, z^{\ell})\sigma^{-2}(z^i - \widehat{a})^2] \\
&\quad + [q^{ii} + \sum_{j \neq i} q^{ji}\frac{p^j}{p^i} - \sigma^{-2}(z^i - \widehat{a})\widehat{a} \\
&\quad + \sigma^{-2}(z^i - \widehat{a})z^{\ell}]f_y(t, y, z^{\ell}) + Qf(t, y, \cdot)(z^{\ell}),
\end{aligned}
$$
(8.55)

for a smooth function $f(\cdot, \cdot, z^i)$ ($i = 1, \ldots, m_0$), and f_y and f_{yy} denote the first and the second derivatives of f w.r.t. the variable y. In the last step, using Ito's rule, $p^i(t)$ is a solution of the Wonham filter (8.48).

As a demonstration of the numerical algorithm, we provide an example below. Suppose that the generator of the Markov chain is

$$
Q = \begin{pmatrix}
-1.5 & 0.3 & 0.7 & 0.5 \\
0.5 & -1.5 & 1.0 & 0 \\
3.0 & 4.0 & -8.0 & 1.0 \\
1.0 & 2.0 & 3.0 & -6.0
\end{pmatrix},
$$

and the initial probability distribution is given by $p_0 = (0.4, 0.3, 0.2, 0.1)$. In what follows, to display the sample paths of the approximation sequences, we take a single simulation run (a single sample path), and present the evolution of the sample paths as functions of elapsed time. For the finite-time horizon $[0, T]$, we delineate the sample paths of the approximation sequence for the iteration number k range from $0 \leq k \leq T/\varepsilon$. The corresponding graphs are given in Figure 8.1.

To obtain the related frequency distributions, we fix the time T at $T = 10$, compute $p_{T/\varepsilon}^{\varepsilon,i}$ with $\varepsilon = 0.001$, and repeat the simulation $N = 1,000$ times. The corresponding frequency distributions of the four components are displayed in Figure 8.2.

Remark 8.16. For simplicity, only scalar problems were considered. Extensions to vector-valued problems are straightforward. For example, we may consider the observation of the form

$$
dy(t) = g(\alpha(t))dt + \sigma(t)dw(t), \quad y(0) = 0,
$$

where $y(t) \in \mathbb{R}^d$, $g(\cdot) : \mathcal{M} \mapsto \mathbb{R}^d$, $\sigma(\cdot) : \mathbb{R}^{d \times d} \mapsto \mathbb{R}^d$, $w(\cdot)$ is standard d-dimensional Brownian motion and $\alpha(t)$ is a continuous-time Markov chain

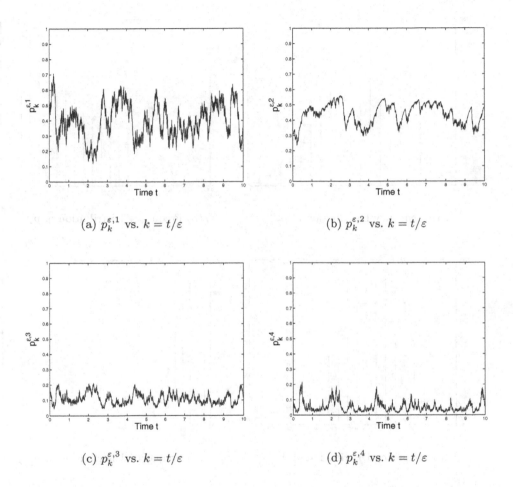

(a) $p_k^{\varepsilon,1}$ vs. $k = t/\varepsilon$

(b) $p_k^{\varepsilon,2}$ vs. $k = t/\varepsilon$

(c) $p_k^{\varepsilon,3}$ vs. $k = t/\varepsilon$

(d) $p_k^{\varepsilon,4}$ vs. $k = t/\varepsilon$

FIGURE 8.1. Trajectories of p_k^ε for a four-state Markov chain, $\varepsilon = 0.001$; t is the real time elapsed with $0 \leq t \leq T = 10$; k is the iteration number given by $k = t/\varepsilon$; CPU time is 7.86 seconds.

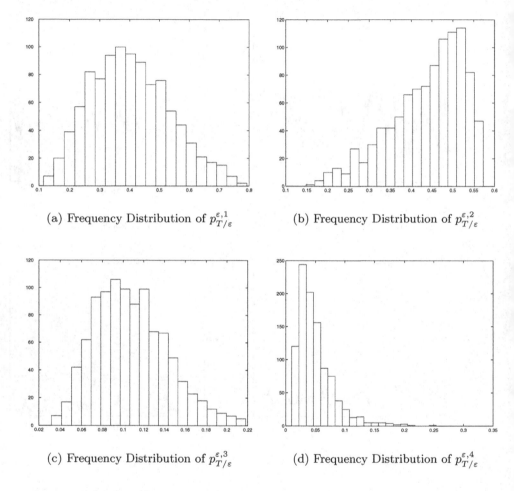

(a) Frequency Distribution of $p_{T/\varepsilon}^{\varepsilon,1}$

(b) Frequency Distribution of $p_{T/\varepsilon}^{\varepsilon,2}$

(c) Frequency Distribution of $p_{T/\varepsilon}^{\varepsilon,3}$

(d) Frequency Distribution of $p_{T/\varepsilon}^{\varepsilon,4}$

FIGURE 8.2. Frequency distribution (histogram) of $p_{T/\varepsilon}^{\varepsilon,i}$, for $i = 1, 2, 3, 4$ with $T = 10$, $\varepsilon = 0.001$, and $N = 1,000$ simulation runs. Horizontal axis: Values of $p_{T/\varepsilon}^{\varepsilon,i}$. Vertical axis: Frequency count.

taking values in $\mathcal{M}$. Then we can proceed with the corresponding Wonham filter, the approximation when the state values are observed with noise, and the discrete-time approximation. The results carry over.

As was mentioned earlier, we have concentrated on the case p_k^i being bounded away from 0. A modification can be made to take into consideration the case of $p_k^i = 0$. This is done as follows. In lieu of (8.53), let $N > 0$ be a fixed but otherwise arbitrarily large real number. Construct the approximation according to

$$
\begin{cases}
v_{k+1}^i = v_k^i + \varepsilon r_k^i + \sqrt{\varepsilon}\sigma_k^{-1}(z^i - \widehat{\alpha}_k^\varepsilon)\xi_k, \quad v_0^i = 0, \\
r_k^i = \left[q^{ii} + \sum_{j \neq i} q^{ji}\frac{p_k^j}{p_k^i} - \sigma_k^{-2}(z^i - \widehat{\alpha}_k^\varepsilon)\widehat{\alpha}_k^\varepsilon + \sigma_k^{-2}(z^i - \widehat{\alpha}_k^\varepsilon)\alpha_k^\varepsilon \right] \\
\qquad \times I_{\{p_k^i \geq e^{-N}\}} - N I_{\{p_k^i < e^{-N}\}}, \\
p_{k+1}^i = \exp(v_{k+1}^i), \quad p_0^i = p^i(0).
\end{cases}
$$

$$(8.56)$$

For such fixed N, we can derive the tightness and weak convergence just as in the previous cases. Then we let $N \to \infty$ to complete the proof. The steps are essentially the same as in the previous case only modifications are needed as outlined above.

8.5 Notes

The first rigorous development of nonlinear filters for diffusion-type processes was given by Kushner [94]. The first finite-dimensional filter for a jump Markovian system was developed by Wonham in [150], which is referred to as Wonham filter nowadays.

The main motivation of our study is to reduce complexity. Considering (8.3) for the time horizon $0 \leq k \leq \lfloor T/\varepsilon \rfloor$, as pointed out in Runggaldier and Visentin [132], if we treat the discrete-time case directly, it can be reduced to an $(m_0)^{\lfloor T/\varepsilon \rfloor}$-dimensional recursive system of equations. In our case, m_0 is a fairly large number, which renders the computation needed intensive. Virtually, one cannot complete the computation in polynomial time. Using weak convergence methods, we have obtained a reduced system of filtering equations, which provides us with a way to find nearly optimal filters with less complexity. In particular, if the transition matrix P given in (8.1) is irreducible, the limit becomes a Kalman filter.

As recognized in Björk [19], Dufour and Bertrand [52], Miller and Runggaldier [116], continuous-time Kalman filter problems with Markovian switching are generally of infinite dimension, just like the nonlinear filter cases in Liptser and Shiryayev [105]. Nevertheless, Björk [19] proved that a finite-dimensional filter exists for a linear hybrid system only if the observation is independent of the Markov chain, which corresponds to

$$x_{k+1}^\varepsilon = x_k^\varepsilon + \varepsilon A(\alpha_k^\varepsilon)x_k^\varepsilon + \sqrt{\varepsilon}\sigma_w(\alpha_k^\varepsilon)w_k, \ x^\varepsilon(0) = x,$$
$$y_{k+1}^\varepsilon = y_k^\varepsilon + \varepsilon C(\alpha_k^\varepsilon) + \sqrt{\varepsilon}\sigma_v w_k, \ y^\varepsilon(0) = 0,$$

(8.57)

in our problem. Similar to the derivation of Theorem 8.7, we obtain the limit filtering equations

$$dx = \overline{A}(\overline{\alpha}(t))x\,dt + \overline{\sigma}_w(\overline{\alpha}(t))dw, \ x(0) = x,$$
$$dy = \overline{C}(\overline{\alpha}(t))dt + \sigma_v dw, \ y(0) = 0.$$

(8.58)

The calculation of (8.57) leads to recursive filters of dimension $(m_0)^{\lfloor T/\varepsilon \rfloor}$, whereas (8.58) yields the solution to a finite-dimensional filtering problem.

Related results on hybrid filtering in discrete time can be found in Bar-Shalom and Li [11], Costa [40], Doucet, Gordon and Krishnamurthy [50], Yang, Bar-Shalom, Lin [152] and others. Marcus and Westwood [110] considered a discrete-time problem involving a Markov chain with transition probability matrix $\exp(\varepsilon Q)$, with Q being a generator. Recently, Dey derived reduced-complexity filtering results for hidden Markov models, in which the underlying Markov chains are nearly completely decomposable [46]. The work of Zhang [175] dealt with hybrid filters in continuous time and treated problems involving non-Gaussian noise. The paper by Yin and Dey [155] contains some numerical experiments and simulation data that demonstrate the relationship between the original system and a reduced system.

For solution of continuous-time hybrid filtering problems involving jump Markov processes, see Björk [19], Dufour and P. Bertrand [52], Dufour and Elliott [53], and Miller and Runggaldier [116].

Related to the problem treated in Section 8.4, in a recent paper [107], Malcome, Elliott, and van der Hoek used Clark's transformation and then discretized the resulting equations, which may be viewed as a robustness treatment. In [23], Blom reported a scheme using logarithm transformation of nonlinear filtering for continuous-time systems. One of the main features behind our approach is that we do not discretize the stochastic differential equation (8.48) directly, but rather work with a discrete system. Using a simple transformation via Ito lemma, we show that another computable scheme can be obtained, which yields good computational results. Further numerical results, including sample means and variances, assessment of approximation errors, can be found in Yin, Zhang, and Liu [167].

Finally, we emphasize that the results presented in this chapter are all based on the assumption that the Markov chain is time-homogeneous, which is extendible to nonstationary cases. The work of Yin and Dey [155] demonstrates that in lieu of the transition probability matrix (8.1), we may consider filtering problems with a time-dependent transition matrix of the form

$$P^\varepsilon(\varepsilon k) = P(\varepsilon k) + \varepsilon Q(\varepsilon k).$$

Results similar to those presented in this chapter can be obtained.

9

Markov Decision Processes

9.1 Introduction

This chapter presents a two-time-scale approach for discrete-time Markov
decision processes (MDP) with finite state spaces. The primary motivation
for our study stems from many applications in resource allocation, queueing
networks, machine replacement, and command control. Markov decision
processes are convenient in modeling these systems, since no difference or
differential equations are needed.

A common practice in dealing with MDPs is the method of dynamic
programming (DP). Using such an approach to find the optimal solutions
requires solving a set of associated DP equations. Since the total number
of equations to be solved is exactly the same as the total number of states
of the Markov chain, the DP approach is computationally feasible only if
the dimension of the underlying system is not too large. For large-scale
systems, one has to resort to approximately optimal schemes.

Due to the complex nature of the real-world problems, we often face large-
dimensional systems with uncertainty. Using Markov chains to model the
uncertain events, the states of the Markov chain often evolve at different
rates. Some of them vary fast, whereas others change slowly. To reduce
complexity, one of the possible solutions is to utilize the idea of a two-time-
scale approach so that in lieu of a large-dimensional system, one deals with
a much simpler system with less complexity. Such an approach provides a
powerful and efficient tool in treating large and complex systems to find
approximately optimal solutions.

In this chapter, we illustrate how to deal with the aforementioned problems for MDPs in discrete time by using their weak and strong interactions. Both discounted costs and long-run average costs are considered. The basic idea of our approach is to aggregate the states in each recurrent class into a single state and to derive a limit problem with lower dimensions. Using the optimal solution of the limit problem, we then construct a solution of the original problem, which will be shown to be asymptotically optimal. One of the interesting aspects is that the limit problem derived is a continuous-time Markov decision process, although the original one is in discrete time. Thus this chapter explores the interface between continuous-time and discrete-time problems. For MDPs with long-run average cost, we show that the generator of the limit problem is irreducible, derive the limit value function, and obtain error bounds on the constructed control via the limit problem. As one of the results, we establish the asymptotic optimality.

The rest of the chapter is arranged as follows. Section 9.2 presents the formulation and results of an MDP under discounted cost. The transition probability matrix of the underlying MDP consists of a rapid-changing part and a slow-varying part. The rapid-changing part either consists of several classes of recurrent states or includes transient states as well. Section 9.3 is devoted to long-run average cost problems. Section 9.4 presents an example for illustration. Given in Section 9.5, the proofs in this chapter are made for cases that transient states are included; those cases that include only recurrent states can be handled similarly. Finally, some notes are given in Section 9.6.

9.2   Discounted Cost

We consider a discrete-time Markov chain α_k^ε, $k = 0, 1, \ldots$, with finite state space $\mathcal{M} = \{1, \ldots, m_0\}$, where $\varepsilon > 0$ is a small parameter. Let the control space Γ be a compact subset of an Euclidean space. We consider feedback control $u_k = u(\alpha_k^\varepsilon)$ such that $u_k \in \Gamma$, $k = 0, 1, \ldots$ Let $P^\varepsilon(u) = (p^{\varepsilon, ij}(u))_{m_0 \times m_0}$ denote the transition probability matrix of α_k^ε given by

$$P^\varepsilon(u) = P(u) + \varepsilon Q(u) \text{ for } u \in \Gamma, \tag{9.1}$$

where $P(u) = (p^{ij}(u))_{m_0 \times m_0}$ is a transition probability matrix, i.e., $0 \leq p^{ij}(u) \leq 1$, $\sum_j p^{ij}(u) = 1$, and $Q(u) = (q^{ij}(u))_{m_0 \times m_0}$ is a generator of a continuous-time Markov chain, i.e., $q^{ij}(u) \geq 0$ for $j \neq i$ and $\sum_j q^{ij}(u) = 0$ for each i. In view of (9.1), it is clear that the dominating factor is given by the transition matrix $P(u)$. Note that both $P(u)$ and $Q(u)$ are control dependent.

Suppose that the Markov chain has transition matrix $P(u)$ of the form

$$
P(u) = \begin{pmatrix}
P^1(u) & & & \\
& \ddots & & \\
& & P^{l_0}(u) & \\
P^{*,1}(u) & \cdots & P^{*,l_0}(u) & P^*(u)
\end{pmatrix}. \tag{9.2}
$$

Then the Markov chain corresponding to the transition matrix $P(u)$ consists of a number of recurrent classes and some transient states. In this case, denote the subspace of recurrent states by $\mathcal{M}_i = \{s_{i1}, \ldots, s_{im_i}\}$, $i = 1, \ldots, l_0$, and the collection of transient states by $\mathcal{M}_* = \{s_{*1}, \ldots, s_{*m_*}\}$. Then the state space can be written as

$$
\begin{aligned}
\mathcal{M} &= \mathcal{M}_1 \cup \cdots \cup \mathcal{M}_{l_0} \cup \mathcal{M}_* \\
&= \{s_{11}, \ldots, s_{1m_1}\} \cup \cdots \cup \{s_{l_01}, \ldots, s_{l_0 m_{l_0}}\} \cup \{s_{*1}, \ldots, s_{*m_*}\},
\end{aligned}
$$

with $m_1 + \cdots + m_{l_0} + m_* = m_0$.

Definition 9.1. Let $u(\cdot) = \{u(\alpha) : \alpha \in \mathcal{M}\}$ be a function such that $u(\alpha) \in \Gamma$ for all $\alpha \in \mathcal{M}$. Then $u(\cdot)$ is called an admissible control and the collection of all such functions is denoted by $\mathcal{A}^\varepsilon$.

Consider the following cost function $J^\varepsilon(\alpha, u(\cdot))$ defined on $\mathcal{M} \times \mathcal{A}^\varepsilon$:

$$
J^\varepsilon(\alpha, u(\cdot)) = E\left(\varepsilon \sum_{k=0}^\infty (1 - \beta\varepsilon)^k g(\alpha_k^\varepsilon, u(\alpha_k^\varepsilon)) \right), \tag{9.3}
$$

where $\alpha = \alpha_0^\varepsilon$ is the initial state of the chain, $g(\alpha, u)$ is the cost-to-go function, and $\beta > 0$ is a given constant. Note that $(1 - \beta\varepsilon)$ is the discount factor. Its use is based on the rationale of giving more weight on the current state values and putting relatively less emphasis on the future.

Our objective is to find a function $u(\cdot) \in \mathcal{A}^\varepsilon$ that minimizes $J^\varepsilon(\alpha, u(\cdot))$. The original MDP problem, termed $\mathcal{P}^\varepsilon$, takes the form

$$
\mathcal{P}^\varepsilon : \begin{cases}
\text{minimize: } J^\varepsilon(\alpha, u(\cdot)) = E\left(\varepsilon \sum_{k=0}^\infty (1 - \beta\varepsilon)^k g(\alpha_k^\varepsilon, u(\alpha_k^\varepsilon)) \right), \\
\text{subject to: } \alpha_k^\varepsilon \sim P^\varepsilon(u(\alpha_k^\varepsilon)), \ k = 0, 1, \ldots, \ \alpha_0^\varepsilon = \alpha, \ u(\cdot) \in \mathcal{A}^\varepsilon, \\
\text{value function: } v^\varepsilon(\alpha) = \inf_{u(\cdot) \in \mathcal{A}^\varepsilon} J^\varepsilon(\alpha, u(\cdot)),
\end{cases}
$$

where by $\alpha_k^\varepsilon \sim P^\varepsilon(u(\alpha_k^\varepsilon))$ we mean that α_k^ε is a discrete-time Markov chain whose transition probability matrix is $P^\varepsilon(u(\alpha_k^\varepsilon))$.

We can use the DP approach to solve the problem $\mathcal{P}^\varepsilon$. For each $\alpha \in \mathcal{M}$, the associated discrete-time DP equation is

$$v^\varepsilon(\alpha) = \min_{u \in \Gamma}\left\{ \varepsilon g(\alpha, u) + (1 - \beta\varepsilon) \sum_\varpi p^{\varepsilon,\alpha\varpi}(u)v^\varepsilon(\varpi) \right\}. \qquad (9.4)$$

Subtracting $(1 - \beta\varepsilon)v^\varepsilon(\alpha)$ from both sides leads to

$$\beta\varepsilon v^\varepsilon(\alpha) = \min_{u \in \Gamma}\left\{ \varepsilon g(\alpha, u) + (1 - \beta\varepsilon) \sum_\varpi (p^{\varepsilon,\alpha\varpi}(u) - \delta^{\alpha\varpi})v^\varepsilon(\varpi) \right\}, \qquad (9.5)$$

where

$$\delta^{\alpha\varpi} = \begin{cases} 1, & \text{if } \varpi = \alpha, \\ 0, & \text{if } \varpi \neq \alpha. \end{cases}$$

Dividing both sides of (9.5) by ε yields

$$\beta v^\varepsilon(\alpha) = \min_{u \in \Gamma}\left\{ g(\alpha, u) + (1 - \beta\varepsilon) \sum_\varpi b^{\varepsilon,\alpha\varpi}(u)v^\varepsilon(\varpi) \right\}, \qquad (9.6)$$

where

$$(b^{\varepsilon,\alpha\varpi}(u))_{m_0 \times m_0} = \frac{1}{\varepsilon}(P(u) - I) + Q(u) \qquad (9.7)$$

and I denotes the corresponding identity matrix.

It can be shown that $v^\varepsilon(\alpha)$ is the unique solution to the DP equation (9.6). Moreover, for each $\alpha \in \mathcal{M}$, let $u^{\varepsilon,\mathrm{o}}(\alpha)$ denote the minimizer of

$$g(\alpha, u) + (1 - \beta\varepsilon) \sum_\varpi b^{\varepsilon,\alpha\varpi}(u)v^\varepsilon(\varpi).$$

Then $u^{\varepsilon,\mathrm{o}}(\cdot) = \{u^{\varepsilon,\mathrm{o}}(\alpha_k^\varepsilon)\}$ is optimal. That is, $J^\varepsilon(\alpha, u^{\varepsilon,\mathrm{o}}(\cdot)) = v^\varepsilon(\alpha)$.

It is clear that in order to get the optimal solution of $\mathcal{P}^\varepsilon$, one has to solve m_0 DP equations of the form (9.6). If m_0 is very large, this approach is not computationally feasible. To resolve this problem, we use a two-time-scale approach to reduce the dimensionality of the problem under consideration. We will show that as $\varepsilon \to 0$, the original problem can be approximated by a limit problem. This section is devoted to the derivation of the corresponding limit control problem. Let $p^{i,jj_1}(u)$ denote the jj_1th component of $P^i(u)$ i.e., $P^i(u) = (p^{i,jj_1}(u))_{m_i \times m_i}$. Similarly, define $p^{*,jj_1}(u)$ and $p^{*i,jj_1}(u)$ such that $P^*(u) = (p^{*,jj_1}(u))_{m_* \times m_*}$ and $P^{*,i}(u) = (p^{*i,jj_1}(u))_{m_* \times m_i}$, respectively. Next, we define the control sets for the limit problem. For each $i = 1, \ldots, l_0, *$, define

$$\Gamma^i = \{U^i := (u^{i1}, \ldots, u^{im_i}) : u^{ij} \in \Gamma, \ j = 1, \ldots, m_i\},$$

$$\overline{\Gamma} = \Gamma^1 \times \cdots \times \Gamma^{l_0} = \{U = (U^1, \ldots, U^{l_0}) : U^i \in \Gamma^i, \ i = 1, \ldots, l_0\},$$

$$\widetilde{\Gamma} = \overline{\Gamma} \times \Gamma^* = \{\widetilde{U} = (U, U^*) : U = (U^1, \ldots, U^{l_0}) \in \overline{\Gamma}, \ U^* \in \Gamma^*\}.$$

For each $\widetilde{U} = (U, U^*) = (U^1, \ldots, U^{l_0}, U^*) \in \widetilde{\Gamma}$, let

$$P^i(U^i) = (p^{i, jj_1}(u^{ij}))_{m_i \times m_i}, \text{ for } i = 1, \ldots, l_0,$$

$$P^*(U^*) = (p^{*, jj_1}(u^{*j}))_{m_* \times m_*},$$

$$P^{*,i}(U^*) = (p^{*i, jj_1}(u^{*j}))_{m_* \times m_i}, \text{ for } i = 1, \ldots, l_0,$$

and $Q(\widetilde{U}) = (q^{ij}(\widetilde{U}))_{m_0 \times m_0}$, where

$$q^{ij}(\widetilde{U}) = \begin{cases} q^{ij}(u^{1i}), & \text{if } 1 \leq i \leq m_1, \\ q^{ij}(u^{2(i-m_1)}), & \text{if } m_1 < i \leq m_1 + m_2, \\ \ldots & \ldots \\ q^{ij}(u^{l_0(i - m_0 + m_{l_0} + m_*)}), & \text{if } m_0 - m_{l_0} - m_* < i \leq m_0 - m_*, \\ q^{ij}(u^{*(i - m_0 + m_*)}), & \text{if } m_0 - m_* < i \leq m_0. \end{cases}$$

Define

$$\overline{P}^\varepsilon(\widetilde{U}) = P(\widetilde{U}) + \varepsilon Q(\widetilde{U}), \tag{9.8}$$

where

$$P(\widetilde{U}) = \begin{pmatrix} P^1(U^1) & & & \\ & \ddots & & \\ & & P^{l_0}(U^{l_0}) & \\ P^{*,1}(U^*) & \cdots & P^{*,l_0}(U^*) & P^*(U^*) \end{pmatrix}.$$

Here $P(u)$ and $P(\widetilde{U})$ are defined differently depending on context. (Note the difference between u and $\widetilde{U}$ used above.) If we denote $\widetilde{U} \in \widetilde{\Gamma}$ as an m_0-vector $\widetilde{U} = (u^1, \ldots, u^{m_0})$, then all these newly defined matrices above are obtained from $P^\varepsilon(u)$ by replacing the control variable u in the ith row with u^i. Such a definition reveals the dependence of the matrices on the controls.

For each $u(\cdot) \in \mathcal{A}^\varepsilon$, if we define

$$\widetilde{U} = (u^{11}, \ldots, u^{1m_1}, \ldots, u^{l_0 1}, \ldots, u^{l_0 m_{l_0}}, u^{*1}, \ldots, u^{*m_*}), \text{ with } u^{ij} = u(s_{ij}),$$

then the Markov chain determined by $P^\varepsilon(u(\alpha_k^\varepsilon))$ and the one determined by $\overline{P}^\varepsilon(\widetilde{U})$ have the same probability distribution. The following assumptions are made on the transition matrix $P^\varepsilon(u)$ and the cost-to-go function $g(\alpha, u)$.

(A9.1) For each $\alpha \in \mathcal{M}$, $g(\alpha, \cdot)$ is a continuous function on Γ.

(A9.2) For each $\varepsilon > 0$, $P^\varepsilon(u)$ is a continuous function of u. Moreover, for each $U^i \in \Gamma^i$, $i = 1, \ldots, l_0$, $P^i(U^i)$ is irreducible and aperiodic.

For each $U^i \in \Gamma^i$, $i = 1, \ldots, l_0$, let $\nu^i(U^i) = (\nu^{i1}(U^i), \ldots, \nu^{im_i}(U^i))$ be the stationary distribution of $P^i(U^i)$. That is, $\nu^i(U^i)$ is the unique solution of the following system of equations

$$\nu^i(U^i)P^i(U^i) = \nu^i(U^i), \qquad \sum_{j=1}^{m_i} \nu^{ij}(U^i) = 1.$$

To proceed, we impose the following additional conditions on the matrices $P^*(U^*)$ and $P^{*,i}(U^*)$ for $i = 1, \ldots, l_0$.

(A9.3) For each $U^* \in \Gamma^*, P^*(U^*)$ has all of its eigenvalues inside the unit circle. Moreover, for $i = 1, \ldots, l_0$, $P^{*,i}(U^*) = B(U^*)B_0^i$, and $P^*(U^*) - I = B(U^*)B_0$, where $B(U^*) \in \mathbb{R}^{m_* \times m_*}$ is a Lipschitz continuous function, $B_0^i \in \mathbb{R}^{m_* \times m_i}$ and $B_0 \in \mathbb{R}^{m_* \times m_*}$ are constant matrices.

Note that the last condition in (A9.3) stipulates that the states corresponding to $P^*(U^*)$ are transient for all U^*. For each $U^* \in \Gamma^*$, by (A9.3) $(P^*(U^*) - I)$ is invertible. Define

$$a^i(U^*) = -(P^*(U^*) - I)^{-1}P^{*,i}(U^*)\mathbb{1}_{m_i}, \quad \text{for } i = 1, \ldots, l_0, \qquad (9.9)$$

where $a^i = (a^{i,1}, \ldots, a^{i,m_*})' \in \mathbb{R}^{m_* \times 1}$ and $\mathbb{1}_{m_i} = (1, \ldots, 1)' \in \mathbb{R}^{m_i \times 1}$.

Remark 9.2. Under (A9.3), it is readily seen that $B(U^*)$ is nonsingular for each $U^* \in \Gamma^*$. Then, $a^i(U^*) = a^i$ (i.e., they are independent of the control U^*). Moreover, it can be shown that $a^{i,j} \geq 0$ and $\sum_{i=1}^{l_0} a^{i,j} = 1$ for each $j = 1, \ldots, m_*$. That is, $(a^{1,j}, \ldots, a^{l_0,j})$ can be viewed as a probability row vector.

Remark 9.3. One of the main focuses of this chapter is the limit behavior of the system as $\varepsilon \to 0$. Intuitively, as ε gets smaller and smaller, from any transient state s_{*j}, the Markov chain will jump to the set of recurrent states with a small probability of return. The role of control in the set of transient states is not as important as that of the recurrent states.

Given $\widetilde{U}(U, U^*)$, partition $Q(\widetilde{U})$ as

$$Q(\widetilde{U}) = \begin{pmatrix} Q^{11}(U) & Q^{12}(U) \\ Q^{21}(U^*) & Q^{22}(U^*) \end{pmatrix},$$

where

$$Q^{11}(U) \in \mathbb{R}^{(m_0 - m_*) \times (m_0 - m_*)},$$
$$Q^{12}(U) \in \mathbb{R}^{(m_0 - m_*) \times m_*},$$
$$Q^{21}(U^*) \in \mathbb{R}^{m_* \times (m_0 - m_*)}, \quad \text{and}$$
$$Q^{22}(U^*) \in \mathbb{R}^{m_* \times m_*}.$$

For $U = (U^1, \ldots, U^{l_0}) \in \overline{\Gamma}$, let

$$\overline{Q}_*(U) = \text{diag}(\nu^1(U^1), \ldots, \nu^{l_0}(U^{l_0})) \left(Q^{11}(U)\tilde{\mathbb{1}} + Q^{12}(U)(a^1, \ldots, a^{l_0}) \right), \tag{9.10}$$

where

$$\tilde{\mathbb{1}} = \text{diag}(\mathbb{1}_{m_1}, \ldots, \mathbb{1}_{m_{l_0}}) \in \mathbb{R}^{(m_0 - m_*) \times l_0}.$$

It is easy to check that $\overline{Q}_*(U)$ is a generator. Let $\tilde{x}(\cdot) = \{\tilde{x}(t) : t \geq 0\}$ be the continuous-time Markov chain generated by $\overline{Q}_*(U)$. The state space of $\tilde{x}(\cdot)$ is denoted by $\overline{\mathcal{M}} = \{1, \ldots, l_0\}$. Note that for $i = 1, \ldots, l_0$, the ith row of $\overline{Q}_*(U)$ depends only on U^i. That is, $\overline{Q}_*(U) = (\overline{q}^{jj_1}(U^i))_{l_0 \times l_0}$. For a function $f(\cdot)$ defined on $\overline{\mathcal{M}}$, with a slight abuse of notation, write $\overline{Q}_*(U^i)f(\cdot)(i_1)$ instead of $\overline{Q}_*(U)f(\cdot)(i_1)$, where

$$\overline{Q}_*(U^i)f(\cdot)(i_1) = \sum_{i_2 \neq i_1} \overline{q}^{i_1 i_2}(U^i)(f(i_2) - f(i_1)).$$

Thus, the process $\tilde{x}(\cdot)$ generated by $\overline{Q}_*(U)$ can be viewed as a Markov chain generated by $\overline{Q}_*(U(\tilde{x}(t)))$, $t \geq 0$, with the understanding that $U(\tilde{x}(t)) = U^i$ if $\tilde{x}(t) = i$.

We aim to show that as $\varepsilon \to 0$, there is a limit problem in an appropriate sense. Our task is to characterize the limit and to prove the convergence. To proceed, define

$$\overline{g}(i, U^i) = \sum_{j=1}^{m_i} \nu^{ij}(U^i)g(s_{ij}, u^{ij}), \quad i = 1, \ldots, l_0. \tag{9.11}$$

Let $\mathcal{A}^0$ denote a class of functions $U(\cdot) = \{U(i) : i \in \overline{\mathcal{M}}\}$ such that $U(i) \in \Gamma^i$ for $i = 1, \ldots, l_0$. For convenience, call $U = (U(1), \ldots, U(l_0)) \in \mathcal{A}^0$ an admissible control for the limit problem, termed $\mathcal{P}^0$. We use $\tilde{x}(t) \sim \overline{Q}_*(U(\tilde{x}(t)))$ to denote that $\tilde{x}(t)$ is a Markov chain generated by $\overline{Q}_*(U(\tilde{x}(t)))$.

Now we have an auxiliary Markov decision problem:

$$\mathcal{P}^0 : \begin{cases} \text{minimize: } J^0(i, U) = E \int_0^\infty e^{-\beta t}\overline{g}(\tilde{x}(t), U(\tilde{x}(t)))dt, \\ \text{subject to: } \tilde{x}(t) \sim \overline{Q}_*(U(\tilde{x}(t))), \ t \geq 0, \ \tilde{x}_0 = i, \ U \in \mathcal{A}^0, \\ \text{value function: } v(i) = \inf_{U \in \mathcal{A}^0} J^0(i, U). \end{cases}$$

The system of dynamic programming equations for the limit problem $\mathcal{P}^0$ is given by

$$\beta v(i) = \min_{U^i \in \Gamma^i} \left\{ \overline{g}(i, U^i) + \overline{Q}_*(U^i)v(\cdot)(i) \right\}, \quad \text{for } i = 1, \ldots, l_0. \tag{9.12}$$

Let $U^\circ = (U^{1,\circ}, \ldots, U^{l_0,\circ}) \in \overline{\Gamma}$ denote a minimizer of the right-hand side of (9.12). Then $U^\circ \in \mathcal{A}^0$ is optimal for $\mathcal{P}^0$.

Remark 9.4. Note that the limit problem is a continuous-time Markov decision process, although the original problem $\mathcal{P}^\varepsilon$ is a discrete-time MDP. Effectively, we are using solutions of a continuous-time MDP to approximate that of the corresponding discrete-time MDP.

We next show that $\mathcal{P}^0$ is indeed the limit problem by proving the convergence of $v^\varepsilon(\cdot)$ to $v(\cdot)$. We then derive a strategy based on the optimal control of the limit problem $\mathcal{P}^0$ and prove the asymptotic optimality of constructed control for $\mathcal{P}^\varepsilon$.

Lemma 9.5. *If there exists a subsequence of $\varepsilon \to 0$ (still denoted by ε for simplicity) such that $v^\varepsilon(\alpha) \to v^0(\alpha)$ for $\alpha \in \mathcal{M}$, the following assertions hold:*

(a) *For $\alpha \in \mathcal{M}_i$, the limit function $v^0(\alpha)$ depends only on i, i.e., $v^0(\alpha) = v(i)$ for some function $v(i)$.*

(b) *For $j = 1, \ldots, m_*$, denote the limit of $v^\varepsilon(s_{*j})$ by $v(*j) = v^0(s_{*j})$ and write $v(*) = (v(*1), \ldots, v(*m_*))'$. Then*

$$v(*) = a^1 v(1) + \cdots + a^{l_o} v(l_0), \tag{9.13}$$

where $v(i)$ is given in Part (a).

Remark 9.6. This lemma indicates that if there is a convergent subsequence of the value functions of the original problem, the limit value function depends only on i, the index of $\mathcal{M}_i$ if it is in one of the recurrent classes, and the limit is an average (with respect to the probabilities $a^{i,j}$) of $v(\alpha)$, if it is in one of the transient states.

Theorem 9.7. *Assume (A9.1)–(A9.3). For each $\alpha \in \mathcal{M}_i$, $i = 1, \ldots, l_0$,*

$$\lim_{\varepsilon \to 0} v^\varepsilon(\alpha) = v(i), \tag{9.14}$$

where $v(i)$ is the value function of the limit problem $\mathcal{P}^0$; for each $\alpha = s_{*j} \in \mathcal{M}_*$, $j = 1, \ldots, m_*$,

$$\lim_{\varepsilon \to 0} v^\varepsilon(s_{*j}) = v(*j), \tag{9.15}$$

where $v(*j)$ is determined by (9.13).

Next we construct an asymptotic optimal control policy for the original problem $\mathcal{P}^\varepsilon$ from the optimal decision of the limit problem $\mathcal{P}^0$. Let $U^o = (U^{1,o}, \ldots, U^{l_0,o}) \in \mathcal{A}^0$ be an optimal control for the limit problem $\mathcal{P}^0$, which is obtained by minimizing the right-hand side of (9.12). Pick out any vector $U^* = (u^{*1}, \ldots, u^{*m_*}) \in \Gamma^*$. For this $\widetilde{U}^o = (U^o, U^*) \in \widetilde{\Gamma}$, define a control $u^\varepsilon(\cdot) = \{u^\varepsilon(\alpha) : \alpha \in \mathcal{M}\}$ for the original problem $\mathcal{P}^\varepsilon$:

$$u^\varepsilon(\alpha) = \sum_{i=1}^{l_0} \sum_{j=1}^{m_i} I_{\{\alpha = s_{ij}\}} u^{ij,o} + \sum_{j=1}^{m_*} I_{\{\alpha = s_{*j}\}} u^{*j}. \tag{9.16}$$

It is clear that $u^\varepsilon(\cdot) \in \mathcal{A}^\varepsilon$. In what follows, we will show that the constructed control $u^\varepsilon(\cdot)$ is asymptotically optimal.

Remark 9.8. Note that in the above construction, the control corresponding to the transient state $\alpha \in \mathcal{M}_*$ can be taken to be any value. This indicates that for the purpose of establishing nearly optimal policies, the control on the transition probabilities of transient states has little impact on the overall system performance.

Theorem 9.9. *The control* $u^\varepsilon(\cdot) = \{u^\varepsilon(\alpha)\}$ *constructed in* (9.16) *is asymptotically optimal in that*

$$\lim_{\varepsilon \to 0} \left| J^\varepsilon(\alpha, u^\varepsilon(\cdot)) - v^\varepsilon(\alpha) \right| = 0, \text{ for } \alpha \in \mathcal{M}.$$

It is interesting from a computational point of view to estimate the convergence rate of $v^\varepsilon(\cdot)$ to $v(\cdot)$ and to obtain the error bound of the control $u^\varepsilon(\cdot)$ constructed in (9.16). The following theorem demonstrates that such a convergence rate and error bound are of the order ε.

Theorem 9.10. *Assume that* (A9.1)–(A9.3) *hold and that the control set* Γ *contains finitely many elements. Then,*

$$v^\varepsilon(\alpha) - v(i) = O(\varepsilon), \text{ for } \alpha \in \mathcal{M}_i, \ i = 1, \ldots, l_0,$$

$$v^\varepsilon(s_{*j}) - v(*j) = O(\varepsilon), \text{ for } \alpha = s_{*j} \in \mathcal{M}_*, \ j = 1, \ldots, m_*.$$

Moreover,

$$J^\varepsilon(\alpha, u^\varepsilon(\cdot)) - v^\varepsilon(\alpha) = O(\varepsilon), \text{ for } \alpha \in \mathcal{M}.$$

Remark 9.11. Let us comment on the choice of the discount factor $(1 - \beta\varepsilon)$. To study a discounted-cost MDP, perhaps the first choice coming to mind would be

$$J = E\left(\sum_{k=0}^{\infty} \rho^k g(\alpha_k^\varepsilon, u(\alpha_k^\varepsilon)) \right), \tag{9.17}$$

for a constant $\rho > 0$ that is independent of ε. This turns out to be a completely different problem than ours, which is not as interesting as (9.3). Mainly, the formulation will not lead to the reduction of complexity in any way. To see this, let v^ε denote the corresponding value function. Then the DP equation becomes

$$v^\varepsilon(\alpha) = \min_{u \in \Gamma} \left\{ g(\alpha, u) + \rho \sum_j p^{\varepsilon,ij}(u) v^\varepsilon(j) \right\}. \tag{9.18}$$

For simplicity, let us consider only the case of one irreducible class, i.e., $l_0 = 1$. Noting that $p^{\varepsilon,ij}(u) = p^{ij}(u) + \varepsilon q^{ij}(u)$, and sending $\varepsilon \to 0$, the term Q will vanish in (9.18). The system reduces to an MDP with the Markov chain $\alpha_k^\varepsilon = \alpha_k$, independent of ε and governed by the transition matrix P. Since the matrix Q plays no role in the limit system, we will not be able to take advantage of the weak and strong interactions, and will not be able to get averages w.r.t. the stationary distribution. As a result, the formulation of such a regular perturbation does not reduce the dimensionality. In lieu of such an obvious choice, we should choose a discount factor that does not discount too much of the future and that should be ε-dependent. Therefore, $\rho = 1 - \beta\varepsilon$ appears to be a suitable choice. This leads to cancellation of ε in the associated DP equations and is necessary for the development of the results in this chapter.

9.3   Long-Run Average Cost

Up to now, we have discussed discrete-time MDPs with discounted cost criteria. In this section we deal with MDPs with a long-run average cost. In what follows, we need to revise the assumptions (A9.1) and (A9.2), and we require a stronger version of irreducibility.

(A9.1') Γ contains finitely many elements.

(A9.2') For each $U^i \in \Gamma^i$, $i = 1, \ldots, l_0$, $P^i(U^i)$ is irreducible and aperiodic. Moreover, for sufficiently small $\varepsilon > 0$, $\overline{P}^\varepsilon(\tilde{U})$ defined by (9.8) is irreducible and aperiodic.

Consider the following problem:

$$
\mathcal{P}_{av}^\varepsilon : \begin{cases}
\text{minimize: } J^\varepsilon(u(\cdot)) = \limsup_{N\to\infty} \frac{1}{N} E \sum_{k=0}^{N} g(\alpha_k^\varepsilon, u(\alpha_k^\varepsilon)), \\[2mm]
\text{subject to: } \alpha_k^\varepsilon \sim P^\varepsilon(u(\alpha_k^\varepsilon)), \ k = 0, 1, \ldots, \ \alpha_0^\varepsilon = \alpha \in \mathcal{M}, \ u(\cdot) \in \mathcal{A}^\varepsilon, \\[2mm]
\text{value function: } \lambda^\varepsilon = \inf_{u(\cdot)\in\mathcal{A}^\varepsilon} J^\varepsilon(u(\cdot)).
\end{cases}
$$

Note that for any given $\tilde{U} \in \tilde{\Gamma}$, $\overline{P}^\varepsilon(\tilde{U})$ is irreducible and aperiodic. Thus the associated Markov chain α_k^ε has a stationary distribution. Consequently, the average cost function $J^\varepsilon(u(\cdot))$ does not depend on the initial state $\alpha_0^\varepsilon = \alpha$, nor does the value function λ^ε.

Using dynamic programming approach, one can write the associated DP equation of $\mathcal{P}_{av}^\varepsilon$ as follows:

$$
\lambda^\varepsilon + h^\varepsilon(\alpha) = \min_{u\in\Gamma}\left\{ g(\alpha, u) + \sum_\varpi p^{\varepsilon,\alpha\varpi}(u) h^\varepsilon(\varpi) \right\}, \tag{9.19}
$$

where $h^\varepsilon(\alpha)$ is a function to be determined later. Subtracting $h^\varepsilon(\alpha)$ from both sides of (9.19) leads to

$$
\lambda^\varepsilon = \min_{u\in\Gamma}\left\{ g(\alpha, u) + \sum_\varpi (p^{\varepsilon,\alpha\varpi}(u) - \delta^{\alpha\varpi}) h^\varepsilon(\varpi) \right\}. \tag{9.20}
$$

Note that the irreducibility of $\overline{P}^\varepsilon(\widetilde{U})$ implies that $\overline{P}^\varepsilon(\widetilde{U}) - I$ is an irreducible generator for a continuous-time Markov chain. Using standard techniques in Ross [131], one obtains the following theorem that reveals the optimality of problem $\mathcal{P}^\varepsilon_{\mathrm{av}}$.

Theorem 9.12. *Under (A9.1') and (A9.2'), the following assertions hold:*

(a) *For each fixed $\varepsilon > 0$, there exists a pair $(\lambda^\varepsilon, h^\varepsilon(\cdot))$ that satisfies the DP equation (9.20).*

(b) *The DP equation (9.20) has a unique solution up to an additive constant. That is, if $(\tilde{\lambda}^\varepsilon, \tilde{h}^\varepsilon(\cdot))$ is another solution to (9.20), then $\tilde{\lambda}^\varepsilon = \lambda^\varepsilon$ and for some constant K_0, $\tilde{h}^\varepsilon(\alpha) = h^\varepsilon(\alpha) + K_0$, for $\alpha \in \mathcal{M}$.*

(c) *Let $u^{\varepsilon,\circ}(\cdot) = \{u^{\varepsilon,\circ}(\alpha) \in \Gamma, \ \alpha \in \mathcal{M}\}$ denote a minimizer of the right-hand side of (9.20). Then $u^{\varepsilon,\circ}(\cdot) \in \mathcal{A}^\varepsilon$ is optimal and $J^\varepsilon(u^{\varepsilon,\circ}(\cdot)) = \lambda^\varepsilon$.*

Analogous to the discounted cost case, the limit problem with the long-run average cost is given as follows. Again we obtain a continuous-time MDP limit problem:

$$
\mathcal{P}^0_{\mathrm{av}} : \begin{cases}
\text{minimize: } J^0(U) = \limsup_{T\to\infty} \frac{1}{T} E \int_0^T \overline{g}(\tilde{x}(t), U(\tilde{x}(t)))dt, \\[2mm]
\text{subject to: } \tilde{x}(t) \sim \overline{Q}_*(U(\tilde{x}(t))), \ t \geq 0, \ \tilde{x}_0 = i, \ U \in \mathcal{A}^0, \\[2mm]
\text{value function: } \lambda^0 = \inf_{U \in \mathcal{A}^0} J^0(U).
\end{cases}
$$

Remark 9.13. To relate the two-time-scale problem and the limit problem, it is essential that there exists a stationary distribution of $\tilde{x}(\cdot)$. A sufficient condition that guarantees the existence of this stationary distribution is the irreducibility of the generator $\overline{Q}_*(U)$. Nevertheless, the irreducibility of $\overline{P}^\varepsilon(U)$ does not imply that of $\overline{Q}_*(U)$. To illustrate, consider the Markov chain with transition matrix given by

$$
\overline{P}^\varepsilon = \begin{pmatrix} 1 & 0 & 0 & 0 & 0 \\ 0 & 1 & 0 & 0 & 0 \\ 0 & 0 & 1 & 0 & 0 \\ \frac{1}{2} & 0 & 0 & \frac{1}{2} & 0 \\ 0 & \frac{1}{2} & 0 & 0 & \frac{1}{2} \end{pmatrix} + \varepsilon \begin{pmatrix} -1 & 0 & 0 & 1 & 0 \\ 0 & -1 & 0 & 0 & 1 \\ 0 & 1 & -1 & 0 & 0 \\ 1 & 1 & 1 & -4 & 1 \\ 1 & 1 & 1 & 1 & -4 \end{pmatrix},
$$

where $\varepsilon < 1/8$. In this example, we have

$$
P^1 = P^2 = P^3 = 1, \quad P^* = \begin{pmatrix} \frac{1}{2} & 0 \\ 0 & \frac{1}{2} \end{pmatrix}.
$$

It follows that

$$\nu^1 = \nu^2 = \nu^3 = 1, \; a^1 = (1,0)', \; a^2 = (0,1)', \; a^3 = (0,0)'.$$

Thus, by (9.10) one can find that

$$\overline{Q}_* = \mathrm{diag}(\nu^1, \nu^2, \nu^3)\left(Q^{11}\tilde{1} + Q^{12}(a^1, a^2, a^3)\right) = \begin{pmatrix} 0 & 0 & 0 \\ 0 & 0 & 0 \\ 0 & 1 & -1 \end{pmatrix},$$

which is not irreducible since the rank of $\overline{Q}_*$ is only 1. On the other hand, it is easy to see that $\overline{P}^\varepsilon$ is irreducible for sufficiently small $\varepsilon > 0$. In fact, the unique ε-dependent stationary distribution is given by

$$\left(\frac{2(5\varepsilon + 1)}{5(10\varepsilon + 1)}, \frac{20\varepsilon + 3}{5(10\varepsilon + 1)}, \frac{10\varepsilon}{5(10\varepsilon + 1)}, \frac{4\varepsilon}{5(10\varepsilon + 1)}, \frac{6\varepsilon}{5(10\varepsilon + 1)}\right).$$

To guarantee the desired irreducibility, we need a condition on the vectors a^i for $i = 1, \ldots, l_0$. Based on the condition, we establish the irreducibility of the generator $\overline{Q}_*(U)$.

(A9.4) for $i = 1, \ldots, l_0$, $j = 1, \ldots, m_*$, $0 < a^{i,j} < 1$.

Lemma 9.14. *Assume* (A9.1'), (A9.2'), (A9.3), *and* (A9.4). *The generator $\overline{Q}_*(U)$ is irreducible for each $U \in \overline{\Gamma}$.*

Next, let us consider the DP equation for the limit problem $\mathcal{P}^0_{\mathrm{av}}$,

$$\lambda^0 = \min_{U^i \in \Gamma^i} \left\{\overline{g}(i, U^i) + \overline{Q}_*(U^i)h^0(\cdot)(i)\right\}, \qquad (9.21)$$

for some function $h^0(\cdot)$. The verification theorem on the limit problem is given below; its proof is standard (see Ross [131]).

Theorem 9.15. *Assume* (A9.1'), (A9.2'), (A9.3), *and* (A9.4). *Then the following assertions hold:*

(a) *There exists a pair $(\lambda^0, h^0(\cdot))$ that satisfies the DP equation (9.21).*

(b) *The DP equation (9.21) has a unique solution up to an additive constant. That is, if $(\tilde{\lambda}^0, \tilde{h}^0(\cdot))$ is another solution to (9.21), then $\tilde{\lambda}^0 = \lambda^0$ and for some constant K_0, $\tilde{h}^0(i) = h^0(i) + K_0$, for $i = 1, \ldots, l_0$.*

(c) *Let $U^\circ = (U^{1,\circ}, \ldots, U^{l_0,\circ}) \in \overline{\Gamma}$ denote a minimizer of the right-hand side of (9.21). Then $U^\circ \in \mathcal{A}^0$ is optimal and $J^0(U^\circ) = \lambda^0$.*

The following lemma establishes that the stationary probability distribution associated with $\overline{P}^\varepsilon(\tilde{U})$ can be approximated by the quasi-stationary distribution of a continuous-time MDP corresponding to $\overline{Q}_*(U)$ and the stationary probability distributions of $P^i(U^i)$, $i = 1, \ldots, l_0$.

Lemma 9.16. *For any* $\widetilde{U} = (U, U^*) = (U^1, \ldots, U^{l_0}, U^*) \in \widetilde{\Gamma}$, *let* $\nu^\varepsilon(\widetilde{U})$ *denote the stationary distribution of* $\overline{P}^\varepsilon(\widetilde{U})$. *Then*

$$\nu^\varepsilon(\widetilde{U}) = \nu^0(U) + O(\varepsilon),$$

where

$$\nu^0(U) = (\overline{\nu}^1(U)\nu^1(U^1), \ldots, \overline{\nu}^{l_0}(U)\nu^{l_0}(U^{l_0}), 0_{1 \times m_*}),$$

$\nu^i(U^i)$ *is the stationary distribution of* $P^i(U^i)$ *for* $i = 1, \ldots, l_0$, *and moreover* $(\overline{\nu}^1(U), \ldots, \overline{\nu}^{l_0}(U))$ *is the quasi-stationary distribution of a continuous-time MDP corresponding to* $\overline{Q}_*(U)$.

This lemma can be proved using the irreducibility condition on $\overline{P}^\varepsilon$. Next, we present the main theorem of this section.

Theorem 9.17. *Let* $U^\circ \in \mathcal{A}^0$ *be an optimal control for* $\mathcal{P}^0_{\mathrm{av}}$ *and construct* $u^\varepsilon(\cdot)$ *as in (9.16). Then* $u^\varepsilon(\cdot)$ *is asymptotically optimal with an error bound of the order* ε, *i.e.,*

$$J^\varepsilon(u^\varepsilon(\cdot)) - \lambda^\varepsilon = O(\varepsilon).$$

9.4 An Example

This section presents a numerical example of a four-state MDP problem. Consider a manufacturing system consisting of two machines that are subject to breakdown and repair. Each machine has two states, up and down, denoted by 1 and 0, respectively. Let $\alpha_k^\varepsilon = (\alpha_k^{\varepsilon,1}, \alpha_k^{\varepsilon,2})$ denote the machine states and let $\mathcal{M} = \{s_{11}, s_{12}, s_{21}, s_{22}\}$ be the state space of α_k^ε, where

$$s_{11} = (1,1), s_{12} = (0,1), s_{21} = (1,0), s_{22} = (0,0).$$

We assume that α_k^ε is an MDP whose control variable u represents a service rate such as rate of preventive maintenance and repair. The problem is to choose u to keep the average machine capacity at a reasonable level and to avoid an excessive service rate, i.e., to choose u over time to minimize

$$J^\varepsilon(\alpha, u(\cdot)) = E\varepsilon \sum_{k=0}^{\infty} (1 - \beta\varepsilon)^k g(\alpha_k^\varepsilon, u_k).$$

In this example, we choose $u \in \Gamma = \{1, 2\}$, with 1 and 2 representing a low service rate and high service rate, respectively, and

$$g(\alpha, u) = c_1|\alpha^1 - \gamma^1| + |\alpha^2 - \gamma^2| + c_2 u,$$

where γ^i is the expected machine capacity level and c_i, $i = 1, 2$ are the rates of service cost.

Suppose that the state of the first machine changes more rapidly than that of the second one. Then the transition matrix of α_k^ε can be given as follows:

$$P^\varepsilon(u) = \begin{pmatrix} 1-p^1(u) & p^1(u) & 0 & 0 \\ p^2(u) & 1-p^2(u) & 0 & 0 \\ 0 & 0 & 1-p^1(u) & p^1(u) \\ 0 & 0 & p^2(u) & 1-p^2(u) \end{pmatrix}$$

$$+\varepsilon \begin{pmatrix} -p^3(u) & 0 & p^3(u) & 0 \\ 0 & -p^3(u) & 0 & p^3(u) \\ p^4(u) & 0 & -p^4(u) & 0 \\ 0 & p^4(u) & 0 & -p^4(u) \end{pmatrix},$$

where $p^1(u)$ and $p^2(u)$ are the breakdown and repair probabilities, respectively, and $\varepsilon p^3(u)$ and $\varepsilon p^4(u)$ are those for the second machine.

In this example, $\mathcal{M}_1 = \{s_{11}, s_{12}\}$ corresponds to the states when second machine is up, and $\mathcal{M}_2 = \{s_{21}, s_{22}\}$ corresponds to these states when the second machine is down. The corresponding DP equations are given by

$$v^\varepsilon(\alpha) = \min_{u \in \Gamma} \left\{ \varepsilon g(\alpha, u) + (1-\beta\varepsilon) \sum_\varpi p^{\varepsilon, \alpha\varpi}(u) v^\varepsilon(\varpi) \right\}.$$

Let

$$\Gamma^1 = \left\{ U^1 := (u^{11}, u^{12}) \text{ such that } u^{11}, u^{12} \in \Gamma \right\},$$
$$\Gamma^2 = \left\{ U^2 := (u^{21}, u^{22}) \text{ such that } u^{21}, u^{22} \in \Gamma \right\}.$$

The control set is

$$\tilde{\Gamma} = \overline{\Gamma} = \Gamma^1 \times \Gamma^2$$
$$= \left\{ U = (U^1, U^2) = (u^{11}, u^{12}, u^{21}, u^{22}) \right\}.$$

For $\tilde{U} \in \tilde{\Gamma}$, let

$$\overline{P}^\varepsilon(U) = P(U) + \varepsilon Q(U) = \begin{pmatrix} P^1(U^1) & \\ & P^2(U^2) \end{pmatrix} + \varepsilon Q(U),$$

where

$$P^1(U^1) = \begin{pmatrix} 1-p^1(u^{11}) & p^1(u^{11}) \\ p^2(u^{12}) & 1-p^2(u^{12}), \end{pmatrix},$$

$$P^2(U^2) = \begin{pmatrix} 1-p^1(u^{21}) & p^1(u^{21}) \\ p^2(u^{22}) & 1-p^2(u^{22}) \end{pmatrix},$$

and

$$Q(U) = \begin{pmatrix} -p^3(u^{11}) & 0 & p^3(u^{11}) & 0 \\ 0 & -p^3(u^{12}) & 0 & p^3(u^{12}) \\ p^4(u^{21}) & 0 & -p^4(u^{21}) & 0 \\ 0 & p^4(u^{22}) & 0 & -p^4(u^{22}) \end{pmatrix}.$$

The corresponding stationary distributions are given by

$$\nu^1(U^1) = \left(\frac{p^2(u^{12})}{p^1(u^{11}) + p^2(u^{12})}, \frac{p^1(u^{11})}{p^1(u^{11}) + p^2(u^{12})} \right)$$

and

$$\nu^2(U^2) = \left(\frac{p^2(u^{22})}{p^1(u^{21}) + p^2(u^{22})}, \frac{p^1(u^{21})}{p^1(u^{21}) + p^2(u^{22})} \right).$$

The limit generator is

$$\begin{aligned} \overline{Q}_*(U) &= \mathrm{diag}(\nu^1(U^1), \nu^2(U^2))Q(U)\mathrm{diag}(\mathbb{1}_{m_1}, \mathbb{1}_{m_2}) \\ &= \begin{pmatrix} -\eta_1(U) & \eta_1(U) \\ \eta_2(U) & -\eta_2(U) \end{pmatrix}, \end{aligned}$$

with

$$\eta_1(U) = \frac{p^2(u^{12})p^3(u^{11}) + p^1(u^{11})p^3(u^{12})}{p^1(u^{11}) + p^2(u^{12})}$$

and

$$\eta_2(U) = \frac{p^2(u^{22})p^4(u^{21}) + p^1(u^{21})p^4(u^{22})}{p^1(u^{21}) + p^2(u^{22})}.$$

Let $\overline{\alpha}(\cdot) \in \{1, 2\}$ be a MDP generated by $\overline{Q}_*(U)$. Define

$$\overline{g}(1, U^1) = \nu^{11}(U^1)g(s_{11}, u^{11}) + \nu^{12}(U^1)g(s_{12}, u^{12}),$$
$$\overline{g}(2, U^2) = \nu^{21}(U^2)g(s_{21}, u^{21}) + \nu^{22}(U^2)g(s_{22}, u^{22}).$$

Note that the number of the DP equations for $\mathcal{P}^\varepsilon$ is equal to 4, while the number of that for $\mathcal{P}^0$ is only 2.

In our numerical experiments, we take

$$\beta = 0.9, \ \gamma^1 = \gamma^2 = 1, \ c_1 = 2, \ c_2 = 0.7.$$

In addition, we take

$$p^1(1) = 0.5, \ p^1(2) = 0.4, \ p^2(1) = 0.2, \ p^2(2) = 0.5,$$
$$p^3(1) = 0.5, \ p^3(2) = 0.4, \ p^4(1) = 0.2, \ p^4(2) = 0.5.$$

We plot the value functions v^ε for $0 < \varepsilon < 1$ in Figure 9.1 and λ^ε in Figure 9.2, where the horizontal axes give values of ε. The convergence of v^ε and λ^ε can be seen clearly from the first column in these figures as $\varepsilon \to 0$. In the second columns of these pictures, we plot the difference between the cost under constructed control u^ε and value functions. In this example, $u^\varepsilon = U_0$ for smaller ε under the discounted cost, while in the average cost case, u^ε is identical to U_0 for all ε. Therefore, the resulting difference between J^ε and λ^ε is zero for all ε. This is mainly because in the average cost case, the error $J^\varepsilon - \lambda^\varepsilon$ depends on $\nu^\varepsilon - \nu^0$, which is very small in this example.

The optimal controls for the limit problems with both the discounted and long-run average costs are identical in this case and are given by

$$U^\circ = (u^{11,\circ}, u^{12,\circ}, u^{21,\circ}, u^{22,\circ}) = (1, 2, 1, 2).$$

This policy indicates that a higher service rate is only needed whenever the first machine is down.

9.5   Proofs of Results

This section presents the proofs of the results. Technical details and complements of this chapter are provided in sequential order.

Proof of Lemma 9.5. Let

$$v^\varepsilon = (v^\varepsilon(s_{ij}), \ i = 1, \ldots, l_0, *, j = 1, \ldots, m_i)' \in \mathbb{R}^{m_0 \times 1}$$
$$= (v^{\varepsilon,i}(s_{ij}), \ i = 1, \ldots, l_0, *, j = 1, \ldots, m_i),$$

where

$$v^{\varepsilon,i} = (v^\varepsilon(s_{i1}), \ldots, v^\varepsilon(s_{im_i}))', \ \text{for } i = 1, \ldots, l_0,$$

and

$$v^{\varepsilon,*} = (v^\varepsilon(s_{*1}), \ldots, v^\varepsilon(s_{*m_*}))'.$$

Define

$$g = (g(s_{ij}, u), i = 1, \ldots, l_0, *, j = 1, \ldots, m_i)' \in \mathbb{R}^{m_0 \times 1}.$$

Similarly, define their corresponding limits v^0, $v^{0,i}$, $v^{0,*}$. Given $\widetilde{U}$, the DP equation (9.6) implies

$$\beta v^\varepsilon \leq g + (1 - \beta\varepsilon)\left(\frac{1}{\varepsilon}\left(P(\widetilde{U}) - I\right) + Q(\widetilde{U})\right)v^\varepsilon. \tag{9.22}$$

Using the hypothesis $v^\varepsilon(\alpha) \to v^0(\alpha)$, multiplying both sides of (9.22) by ε, and sending $\varepsilon \to 0$ lead to

$$(P(\widetilde{U}) - I)v^0 \geq 0.$$

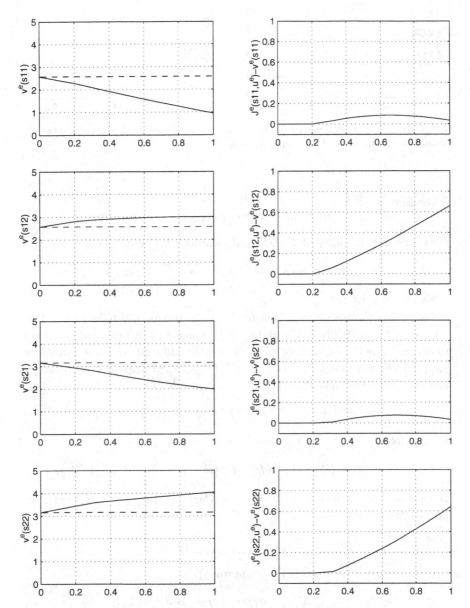

FIGURE 9.1. Convergence of v^ε and near optimality of u^ε. Horizontal axes give values of ε. The dashed lines represent v^0.

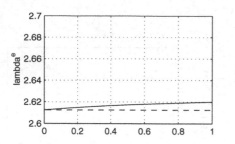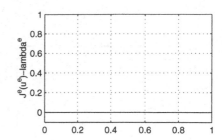

FIGURE 9.2. Convergence of λ^ε and near optimality of u^ε. Horizontal axes give values of ε. The dashed line represents λ^0.

Therefore,
$$P^i(U^i)v^{0,i} \geq v^{0,i}, \text{ for } i = 1, \dots, l_0.$$

Now, the irreducibility of $P^i(U^i)$ and Lemma 14.37 imply that
$$v(i) := v^{0,i}(s_{i1}) = v^{0,i}(s_{i2}) = \cdots = v^{0,i}(s_{im_i}).$$

This proves Part (a).

Next we establish Part (b). Let $\widetilde{U}^{\varepsilon,\mathrm{o}} \in \widetilde{\Gamma}$ denote an optimal control. Then the equality in (9.22) holds under $\widetilde{U}^{\varepsilon,\mathrm{o}}$. Since $\widetilde{\Gamma}$ is a bounded set, there exists a further subsequence of $\varepsilon \to 0$ (still indexed by ε for notational simplicity) such that $\widetilde{U}^{\varepsilon,\mathrm{o}} \to \widetilde{U}^{\mathrm{o}} = (U^{\mathrm{o}}, U^*) \in \widetilde{\Gamma}$. Multiplying both sides of the resulting equation by ε and sending $\varepsilon \to 0$ lead to
$$P^{*,1}(U^*)v^{0,1} + \cdots + P^{*,l_0}(U^*)v^{0,l_0} + (P^*(U^*) - I)v^{0,*} = 0,$$

which yields
$$v^{0,*} = -(P^*(U^*) - I)^{-1}(P^{*,1}(U^*)v^{0,1} + \cdots + P^{*,l_0}(U^*)v^{0,l_0}).$$

In view of the definition of a^i by (9.9), Part (a), i.e., $v^{0,i} = \mathbb{1}_{m_i}v(i)$, and (A9.3), we obtain

$$
\begin{aligned}
v^{0,*} &= (v^0(s_{*1}), \dots, v^0(s_{*m_*}))' \\
&= (v(*1), \dots, v(*m_*))' \\
&= \sum_{i=1}^{l_0} -(P^*(U^*) - I)^{-1}P^{*,i}(U^*)v(i) \\
&= \sum_{i=1}^{l_0} a^i v(i).
\end{aligned}
$$

This proves Part (b). $\square$

Proof of Theorem 9.7. Note that $|g| \leq K$ for some constant $K > 0$. This implies that

$$|v^\varepsilon(\alpha)| \leq K\varepsilon \sum_{k=0}^{\infty} (1 - \beta\varepsilon)^k = \frac{K}{\beta} < \infty.$$

That is, $v^\varepsilon(\alpha)$ is uniformly bounded. Thus, for each subsequence of $\varepsilon \to 0$, there exists a further subsequence (for notational simplicity, still index both subsequences by ε) and a $v^0(\alpha)$, such that $v^\varepsilon(\alpha) \to v^0(\alpha)$. In view of Lemma 9.5, if $\alpha \in \mathcal{M}_i$ then $v^0(\alpha) = v(i)$, and $v(*j)$ can be determined by $v(i)$ for $i = 1, \ldots, l_0$, through the relation (9.13). In what follows, we will show that such a limit $v(i)$, $i = 1, \ldots, l_0$ is a solution to the DP equation (9.12). Since this DP equation has a unique solution, we conclude that (9.14) holds, which leads to (9.15). As in (9.22), the inequality holds for any control $\widetilde{U}$. Multiplying both sides of (9.22) by

$$\text{diag}(\nu^1(U^1), \ldots, \nu^{l_0}(U^{l_0}), 0_{m_* \times m_*}),$$

and letting $\varepsilon \to 0$, we have, in view of the definition of $\overline{Q}_*(U)$ in (9.10) and Lemma 9.5,

$$\beta v(i) \leq \overline{g}(k, U^i) + \overline{Q}_*(U^i)v(\cdot)(i).$$

Since this inequality holds for every $U^i \in \Gamma^i$, we have

$$\beta v(i) \leq \min_{U^i \in \Gamma^i} \left\{ \overline{g}(i, U^i) + \overline{Q}_*(U^i)v(\cdot)(i) \right\}.$$

To derive the reverse inequality, let

$$\widetilde{U}^{\varepsilon,o} = (U^{\varepsilon,1,o}, \ldots, U^{\varepsilon,l_0,o}, U^{\varepsilon,*}) \in \overline{\Gamma}$$

be a minimizer of the right-hand side of (9.6) corresponding to the recurrent states. Then,

$$\beta v^\varepsilon = g + (1 - \beta\varepsilon) \left(\frac{1}{\varepsilon}(P(\widetilde{U}^{\varepsilon,o}) - I) + Q(\widetilde{U}^{\varepsilon,o}) \right) v^\varepsilon. \tag{9.23}$$

Recall that $\widetilde{\Gamma}$ is a bounded set. There exists a further subsequence of $\varepsilon \to 0$ (still indexed by ε) such that $\widetilde{U}^{\varepsilon,o} \to \widetilde{U}^o \in \widetilde{\Gamma}$. Multiplying both sides of (9.23) by

$$\text{diag}(\nu^1(U^{\varepsilon,1,o}), \ldots, \nu^{l_0}(U^{\varepsilon,l_0,o}), 0_{m_* \times m_*}),$$

and letting $\varepsilon \to 0$ lead to

$$\beta v(i) = \overline{g}(i, U^{i,o}) + \overline{Q}_*(U^{i,o})v(\cdot)(i)$$

$$\geq \min_{U^i \in \Gamma^i} \left\{ \overline{g}(i, U^i) + \overline{Q}_*(U^i)v(\cdot)(i) \right\}.$$

This completes the proof. $\square$

We give a lemma which will be used in proof of Theorem 9.9. The lemma can be derived similarly as in Proposition 6.6.

Lemma 9.18. *Given* $\tilde{U} = (U, U^*) \in \tilde{\Gamma}$, *let* $\alpha^\varepsilon(\cdot)$ *denote the discrete-time Markov chain with transition probability matrix* $\overline{P}^\varepsilon(\tilde{U}) = P(\tilde{U}) + \varepsilon Q(\tilde{U})$. *defined by (9.8). Then there exist positive constants* K *and* $0 < \lambda < 1$ *(both being independent of* ε *and* k*) such that for* $j = 1, \ldots, m_i$, $i = 1, \ldots, l_0$,

$$\left| P(\alpha_k^\varepsilon = s_{ij}) - \nu^{ij}(U^i)\theta^i(\varepsilon k) \right| \le K(k\varepsilon^2 + \varepsilon + \lambda^k), \qquad (9.24)$$

and for $j = 1, \ldots, m_*$,

$$P(\alpha_k^\varepsilon = s_{*j}) \le K(k\varepsilon^2 + \varepsilon + \lambda^k), \qquad (9.25)$$

where $\theta^i(\varepsilon k)$, $i = 1, \ldots, l_0$, *can be obtained from the solution* $\theta^i(t)$ *of the following differential equation*

$$\begin{cases} \dfrac{d}{dt}\left(\theta^1(t), \ldots, \theta^{l_0}(t)\right) = \left(\theta^1(t), \ldots, \theta^{l_0}(t)\right) \overline{Q}_*(U), \\ \theta^i(0) = p_0^{\varepsilon,i}\mathbb{1}_{m_i} + p_0^{\varepsilon,*}a^i, \end{cases} \qquad (9.26)$$

where

$$p_0^\varepsilon = (p_0^{\varepsilon,1}, \ldots, p_0^{\varepsilon,l_0}, p_0^{\varepsilon,*}) = (P(\alpha_0^\varepsilon = 1), \ldots, P(\alpha_0^\varepsilon = m_0))$$

with $p_0^{\varepsilon,i} \in \mathbb{R}^{1 \times m_i}$ *and* $p_0^{\varepsilon,*} \in \mathbb{R}^{1 \times m_*}$. *Moreover, let* $\tilde{x}(\cdot)$ *denote the Markov chain generated by* $\overline{Q}_*(U)$. *Then*

$$\theta^i(t) = P(\tilde{x}(t) = i). \qquad (9.27)$$

Proof of Theorem 9.9. In view of the convergence of $v^\varepsilon(\alpha)$, it suffices to show that for $\alpha_0^\varepsilon = \alpha \in \mathcal{M}_i$, $i = 1, \ldots, l_0$,

$$\lim_{\varepsilon \to 0} J^\varepsilon(\alpha, u^\varepsilon(\cdot)) = v(i), \qquad (9.28)$$

and for $\alpha_0^\varepsilon = s_{*j} \in \mathcal{M}_*$, $j = 1, \ldots, m_*$,

$$\lim_{\varepsilon \to 0} J^\varepsilon(s_{*j}, u^\varepsilon(\cdot)) = v(*j). \qquad (9.29)$$

Let α_k^ε be the Markov chain with transition matrix $P^\varepsilon(u^\varepsilon(\alpha_k^\varepsilon))$, $k = 0, 1, \ldots$, with $u^\varepsilon(\alpha)$ given in (9.16). Then $P^\varepsilon(u^\varepsilon(\alpha_k^\varepsilon))$ and $\overline{P}^\varepsilon(\tilde{U}^\circ)$, defined in (9.8), determine chains with identical probability distribution.

Using the definition of $u^\varepsilon(\alpha)$, we have that for any $\alpha \in \mathcal{M}$,

$$
\begin{aligned}
J^\varepsilon(\alpha, u^\varepsilon(\cdot)) &= E\left(\varepsilon \sum_{k=0}^\infty (1 - \beta\varepsilon)^k g(\alpha_k^\varepsilon, u(\alpha_k^\varepsilon)) \right) \\
&= E\left(\sum_{i=1}^{l_0} \sum_{j=1}^{m_i} \sum_{k=0}^\infty \varepsilon(1 - \beta\varepsilon)^k g(s_{ij}, u^{ij,o}) I_{\{\alpha_k^\varepsilon = s_{ij}\}} \right) \\
&\quad + E\left(\sum_{j=1}^{m_*} \sum_{k=0}^\infty \varepsilon(1 - \beta\varepsilon)^k g(s_{*j}, u^{*j}) I_{\{\alpha_k^\varepsilon = s_{*j}\}} \right) \\
&= \sum_{i=1}^{l_0} \sum_{j=1}^{m_i} g(s_{ij}, u^{ij,o}) \left(\sum_{k=0}^\infty \varepsilon(1 - \beta\varepsilon)^k P(\alpha_k^\varepsilon = s_{ij}) \right) \\
&\quad + \sum_{j=1}^{m_*} g(s_{*j}, u^{*j}) \left(\sum_{k=0}^\infty \varepsilon(1 - \beta\varepsilon)^k P(\alpha_k^\varepsilon = s_{*j}) \right).
\end{aligned}
$$

Now, we consider the limit cost function $J^0(i, U^o)$. For $i = 1, \ldots, l_0$, in view of the definition of $\overline{g}(i, U^i)$ by (9.11), we have

$$
\begin{aligned}
J^0(i, U^o) &= E \int_0^\infty e^{-\beta t} \overline{g}(\tilde{x}(t), U^o(\tilde{x}(t))) dt \\
&= E \sum_{i_1=1}^{l_0} \int_0^\infty e^{-\beta t} \overline{g}(i_1, U^{i_1,o}) I_{\{\tilde{x}(t)=i_1\}} dt \\
&= \sum_{i_1=1}^{l_0} \int_0^\infty e^{-\beta t} \overline{g}(i_1, U^{i_1,o}) P(\tilde{x}(t) = i_1) dt \\
&= \sum_{i_1=1}^{l_0} \sum_{j=1}^{m_{i_1}} g(s_{i_1 j}, u^{i_1 j,o}) \int_0^\infty e^{-\beta t} \nu^{i_1,j}(U^{i_1,o}) P(\tilde{x}(t) = i_1) dt.
\end{aligned}
$$

Then we have for $\alpha \in \mathcal{M}_i$,

$$
\begin{aligned}
\left| J^\varepsilon(\alpha, u^\varepsilon(\cdot)) - v(i) \right| &= \left| J^\varepsilon(\alpha, u^\varepsilon(\cdot)) - J^0(i, U^o) \right| \\
&\leq \sum_{i=1}^{l_0} \sum_{j=1}^{m_i} |g(s_{ij}, u^{ij,o})| \left| \sum_{k=0}^\infty \varepsilon(1 - \beta\varepsilon)^k P(\alpha_k^\varepsilon = s_{ij}) \right. \\
&\quad \left. - \int_0^\infty e^{-\beta t} \nu^{ij}(U^{i,o}) P(\tilde{x}(t) = i) dt \right| \\
&\quad + \sum_{j=1}^{m_*} |g(s_{*j}, u^{*j})| \sum_{k=0}^\infty \varepsilon(1 - \beta\varepsilon)^k P(\alpha_k^\varepsilon = s_{*j}).
\end{aligned}
$$

(9.30)

Recall the uniform boundedness of g. For the term on the last line of (9.30), in view of (6.21), we obtain

$$\sum_{j=1}^{m_*} |g(s_{*j}, u^{*j})| \sum_{k=0}^{\infty} \varepsilon(1 - \beta\varepsilon)^k P(\alpha_k^{\varepsilon} = s_{*j})$$

$$\leq \sum_{j=1}^{m_*} K \sum_{k=0}^{\infty} \varepsilon(1 - \beta\varepsilon)^k K(k\varepsilon^2 + \varepsilon + \lambda^k) \leq K\varepsilon. \tag{9.31}$$

To estimate the terms in the first summation in (9.30), note that

$$\left| \sum_{k=0}^{\infty} \varepsilon(1 - \beta\varepsilon)^k P(\alpha_k^{\varepsilon} = s_{ij}) - \int_0^{\infty} e^{-\beta t} \nu^{ij}(U^{i,o}) P(\tilde{x}(t) = i) dt \right|$$

$$\leq \sum_{k=0}^{\infty} \varepsilon(1 - \beta\varepsilon)^k \left| P(\alpha_k^{\varepsilon} = s_{ij}) - \nu^{ij}(U^{i,o}) P(\tilde{x}(\varepsilon k) = i) \right|$$

$$+ \nu^{ij}(U^{i,o}) \left| \sum_{k=0}^{\infty} \varepsilon(1 - \beta\varepsilon)^k P(\tilde{x}(\varepsilon k) = i) - \sum_{k=0}^{\infty} \varepsilon e^{-\beta\varepsilon k} P(\tilde{x}(\varepsilon k) = i) \right|$$

$$+ \nu^{ij}(U^{i,o}) \left| \sum_{k=0}^{\infty} \varepsilon e^{-\beta\varepsilon k} P(\tilde{x}(\varepsilon k) = i) - \int_0^{\infty} e^{-\beta t} P(\tilde{x}(t) = i) dt \right|. \tag{9.32}$$

In view of Lemma 9.18, we have

$$\sum_{k=0}^{\infty} \varepsilon(1 - \beta\varepsilon)^k \left| P(\alpha_k^{\varepsilon} = s_{ij}) - \nu^{ij}(U^{i,o}) P(\tilde{x}(\varepsilon k) = i) \right|$$

$$\leq \sum_{k=0}^{\infty} \varepsilon(1 - \beta\varepsilon)^k K(k\varepsilon^2 + \varepsilon + \lambda^k) \leq K\varepsilon. \tag{9.33}$$

Note that $e^{-\beta\varepsilon} \geq 1 - \beta\varepsilon$ and both $\sum_{k=0}^{\infty} \varepsilon(1 - \beta\varepsilon)^k$ and $\sum_{k=0}^{\infty} \varepsilon e^{-\beta\varepsilon k}$ are uniformly bounded. Using these facts and $P(\tilde{x}(\varepsilon k) = i) \leq 1$, it is easy to show that

$$\left| \sum_{k=0}^{\infty} \varepsilon(1 - \beta\varepsilon)^k P(\tilde{x}(\varepsilon k) = i) - \sum_{k=0}^{\infty} \varepsilon e^{-\beta\varepsilon k} P(\tilde{x}(\varepsilon k) = i) \right|$$

$$\leq \sum_{k=0}^{\infty} \varepsilon(e^{-\beta\varepsilon k} - (1 - \beta\varepsilon)^k) \leq K\varepsilon. \tag{9.34}$$

It remains to show that the last line of (9.32) goes to 0 as $\varepsilon \to 0$. Moreover, we have

$$\left| P(\tilde{x}(\varepsilon k) = i) - P(\tilde{x}(t) = i) \right| \leq K\varepsilon \text{ for } \varepsilon k \leq t \leq \varepsilon(k+1),$$

which implies that the convergence is of the order ε. To this end,

$$\left| \sum_{k=0}^{\infty} \varepsilon e^{-\beta \varepsilon k} P(\tilde{x}(\varepsilon k) = i) - \int_0^{\infty} e^{-\beta t} P(\tilde{x}(t) = i) dt \right|$$

$$\leq \sum_{k=0}^{\infty} \int_{\varepsilon k}^{\varepsilon(k+1)} \left| e^{-\beta \varepsilon k} P(\tilde{x}(\varepsilon k) = i) - e^{-\beta t} P(\tilde{x}(t) = i) \right| dt$$

$$\leq \sum_{k=0}^{\infty} \left(\int_{\varepsilon k}^{\varepsilon(k+1)} \left| e^{-\beta \varepsilon k} - e^{-\beta t} \right| P(\tilde{x}(\varepsilon k) = i) dt \right. \tag{9.35}$$

$$\left. + \int_{\varepsilon k}^{\varepsilon(k+1)} e^{-\beta t} \left| P(\tilde{x}(\varepsilon k) = i) - P(\tilde{x}(t) = i) \right| dt \right)$$

$$\leq \sum_{k=0}^{\infty} \left(\int_{\varepsilon k}^{\varepsilon(k+1)} (e^{-\beta \varepsilon k} - e^{-\beta t}) dt + \int_{\varepsilon k}^{\varepsilon(k+1)} e^{-\beta t} K \varepsilon dt \right)$$

$$= \left(\frac{\varepsilon}{1 - e^{-\beta \varepsilon}} - \frac{1}{\beta} \right) + \frac{K \varepsilon}{\beta} = O(\varepsilon).$$

Combining (9.30)–(9.35), we have

$$\left| J^{\varepsilon}(\alpha, u^{\varepsilon}(\cdot)) - v(i) \right| = O(\varepsilon), \quad \text{for } \alpha \in \mathcal{M}_i, \ i = 1, \ldots, l_0. \tag{9.36}$$

For $s_{*j} \in \mathcal{M}_*$, in view of Remark 9.2, (9.13), and (9.36), we have

$$J^{\varepsilon}(s_{*j_0}, u^{\varepsilon}(\cdot))$$

$$= \sum_{i=0}^{l_0} \sum_{j=1}^{m_i} g(s_{ij}, u^{ij,o}) \sum_{k=0}^{\infty} \sum_{i_1=1}^{l_0} a^{i_1,j_0} \theta_{i_1 i}(\varepsilon k) \nu^{ij}(U^{i,o}) + O(\varepsilon) \tag{9.37}$$

$$= \sum_{i_1=1}^{l_0} a^{i_1,j_0} v(i_1) + O(\varepsilon)$$

$$= v(*j_0) + O(\varepsilon). \quad \square$$

Proof of Theorem 9.10. In view of (9.36), (9.37), and the triangle inequalities, we have

$$\left| J^{\varepsilon}(\alpha, u^{\varepsilon}(\cdot)) - v^{\varepsilon}(\alpha) \right| \leq \left| J^{\varepsilon}(\alpha, u^{\varepsilon}(\cdot)) - v(i) \right| + \left| v^{\varepsilon}(\alpha) - v(i) \right|, \quad \alpha \in \mathcal{M}_i,$$

$$\left| J^{\varepsilon}(s_{*j}, u^{\varepsilon}(\cdot)) - v^{\varepsilon}(s_{*j}) \right| \leq \left| J^{\varepsilon}(s_{*j}, u^{\varepsilon}(\cdot)) - v(*j) \right| + \left| v^{\varepsilon}(s_{*j}) - v(*j) \right|,$$

for $j = 1, \ldots, m_*$. Recall Lemma 9.5 Part (b). It suffices to show that $v^{\varepsilon}(\alpha) - v(i) = O(\varepsilon)$, for $\alpha \in \mathcal{M}_i$. Note that the inequalities (9.36) and (9.37) imply that

$$v^{\varepsilon}(\alpha) - v(i) \leq J^{\varepsilon}(\alpha, u^{\varepsilon}(\cdot)) - v(i) \leq O(\varepsilon), \quad \text{for } \alpha \in \mathcal{M}_i.$$

To derive the reverse inequalities, let $u^{\varepsilon,\mathrm{o}}(\cdot) = \{u^{\varepsilon,\mathrm{o}}(\alpha) : \alpha \in \mathcal{M}\}$ be an optimal control for $\mathcal{P}^{\varepsilon}$ and let

$$\widetilde{U}^{\varepsilon,\mathrm{o}} = (U^{\varepsilon,\mathrm{o}}, U^{\varepsilon,*,\mathrm{o}}) = (U^{\varepsilon,1,\mathrm{o}}, \ldots, U^{\varepsilon,l_0,\mathrm{o}}, U^{\varepsilon,*,\mathrm{o}}) \in \widetilde{\Gamma},$$

where

$$U^{\varepsilon,i,\mathrm{o}} = (u^{\varepsilon,i1,\mathrm{o}}, \ldots, u^{\varepsilon,im_i,\mathrm{o}}) := (u^{\varepsilon,\mathrm{o}}(s_{i1}), \ldots, u^{\varepsilon,\mathrm{o}}(s_{im_i})), \ i = 1, \ldots, l_0, *.$$

The same arguments as in the proof of (9.36) yield

$$\left| J^{\varepsilon}(\alpha, u^{\varepsilon,\mathrm{o}}(\cdot)) - J^0(i, U^{\varepsilon,\mathrm{o}}) \right| \leq O(\varepsilon), \ \text{for } \alpha \in \mathcal{M}, \ i = 1, \ldots, l_0. \quad (9.38)$$

The control set Γ contains finitely many elements by the hypothesis, so does $\overline{\Gamma}$, the control set for the limit problem $\mathcal{P}^0$. Suppose $\overline{\Gamma} = \{\gamma_1, \ldots, \gamma_{l_1}\}$ for some positive integer l_1. Define

$$\mathcal{E}_j = \left\{ \varepsilon : \varepsilon \in (0,1), \ \widetilde{U}^{\varepsilon,\mathrm{o}} = (\gamma_j, U^{\varepsilon,*,\mathrm{o}}) \text{ for an } U^{\varepsilon,*,\mathrm{o}} \in \Gamma^* \right\}, \ j = 1, \ldots, l_1.$$

Then $\{\mathcal{E}_j\}$ consists of a finite number of sets such that $(0,1) = \mathcal{E}_1 \cup \ldots \cup \mathcal{E}_{l_1}$. For fixed j and any $\varepsilon \in \mathcal{E}_j$, consider α_k^{ε} with transition matrix $\overline{P}^{\varepsilon}(\widetilde{U}^{\varepsilon,\mathrm{o}})$. Then in view of (9.38), the optimality of $u^{\varepsilon,\mathrm{o}}(\cdot)$, and (9.13), we have, for $\varepsilon \in \mathcal{E}_j$,

$$v^{\varepsilon}(\alpha) = J^{\varepsilon}(\alpha, u^{\varepsilon,\mathrm{o}}(\cdot)) = J^0(i, \gamma_j) + O(\varepsilon) \geq v(i) + O(\varepsilon),$$

for $\alpha \in \mathcal{M}_i$, $i = 1, \ldots, l_0$. Thus, for $0 < \varepsilon < 1$ and for $\alpha \in \mathcal{M}_i$, $v^{\varepsilon}(\alpha) - v(i) \geq O(\varepsilon)$. $\square$

Proof of Lemma 9.14. The following proof is along the lines of a Gaussian elimination procedure in which elementary row operations do not alter the rank of a matrix. It proceeds in two steps. The first step derives the weak irreducibility of $\overline{Q}_*(\widetilde{U})$, and the second step shows that it is also irreducible. For notational simplicity, the control variable $\widetilde{U}$ will be suppressed in the proof whenever no confusion arises.

Step 1: We first show the weak irreducibility. In view of Lemma 14.38, it suffices to show $\mathrm{rank}(\overline{Q}_*) = l_0 - 1$. To this end, write $Q^{11} = (Q^{11,ij})$ as the blocks of sub-matrices such that $Q^{11,ij}$ has dimension $m_i \times m_j$, $i, j = 1, \ldots, l_0$; $Q^{12} = (Q^{12,i*})$ such that $Q^{12,i*}$ has dimension $m_i \times m_*$, $i = 1, \ldots, l_0$. Then in view of the definition of $\overline{Q}_*$, we have $\overline{Q}_* = (\overline{q}^{ij})_{l_0 \times l_0}$ with $\overline{q}^{ij} = \nu^i (Q^{11,ij} \mathbb{1}_{m_j} + Q^{12,i*} a^j)$. Since $\nu^i > 0$, $\mathbb{1}_{m_j} > 0$, and $a^j > 0$, it follows that for $j \neq i$, if $\overline{q}^{ij} = 0$, then

$$Q^{11,ij} = 0 \text{ and } Q^{12,i*} = 0. \quad (9.39)$$

Note that every entry of $Q^{11,ij}$ (resp. $Q^{12,i*}$) is nonnegative for $j \neq i$.

Next we show that the irreducibility of $\overline{P}^{\varepsilon}(\widetilde{U})$ implies $\overline{q}^{ii} < 0$, for $i = 1, \ldots, l_0$. In fact, if $\overline{q}^{ii} = 0$ for some $1 \leq i \leq l_0$, then since $\overline{Q}_*$ is a generator, we must have $\overline{q}^{ij} = 0$ for $j \neq i$. In view of (9.39), $Q^{11,ij} = 0$ for $j \neq i$, and $Q^{12,i*} = 0$. But this implies that $\overline{P}^{\varepsilon}(\widetilde{U})$ cannot be irreducible since a state outside $\mathcal{M}_i$ is not accessible from a state in $\mathcal{M}_i$. Therefore, $\overline{q}^{ii} < 0$ for $i = 1, \ldots, l_0$.

We now apply the Gaussian elimination procedure on $\overline{Q}_* = (\overline{q}^{ij})$. Multiply the first row of $\overline{Q}_*$ by $-\overline{q}^{i1}/\overline{q}^{11}$ and add to the ith row, $i = 2, \ldots, l_0$ such that the first component of that row equals zero. Denote the resulting matrix by $\overline{Q}^{(1),*} = (\overline{q}^{(1),ij})$. Then we have

$$\overline{q}^{(1),1j} = \overline{q}^{1j}, \ j = 1, \ldots, l_0,$$

$$\overline{q}^{(1),i1} = 0, \ i = 2, \ldots, l_0,$$

$$\overline{q}^{(1),ij} = \overline{q}^{ij} - \overline{q}^{1j}\frac{\overline{q}^{i1}}{\overline{q}^{11}}, \ i, j = 2, \ldots, l_0,$$

with $\overline{q}^{(1),ii} \leq 0$, $\overline{q}^{(1),ij} \geq 0$, for $j \neq i$, and $\sum_{j=2}^{l_0} \overline{q}^{(1),ij} = 0$.

We claim that $\overline{q}^{(1),ii} < 0$ for $i = 2, \ldots, l_0$. For $i = 2$, if $\overline{q}^{(1),22} = 0$, then $\overline{q}^{(1),2j} = 0$, $j = 3, \ldots, l_0$, or, equivalently,

$$(\overline{q}^{23}, \ldots, \overline{q}^{2l_0}) + \left(-\frac{\overline{q}^{21}}{\overline{q}^{11}}\right)(\overline{q}^{13}, \ldots, \overline{q}^{1l_0}) = 0. \tag{9.40}$$

Recall that $\overline{q}^{22} < 0$. We must have $\overline{q}^{21} > 0$, since $\overline{q}^{21} = 0$ implies that $\overline{q}^{22} = \overline{q}^{(1),22} = 0$, which contradicts the fact that $\overline{q}^{ii} < 0$ for $i = 1, \ldots, l_0$. Therefore, $-\overline{q}^{21}/\overline{q}^{11} > 0$. It follows that both vectors in (9.40) equal zero. That is, $(\overline{q}^{23}, \ldots, \overline{q}^{2l_0}) = 0$ and $(\overline{q}^{13}, \ldots, \overline{q}^{1l_0}) = 0$. As a result, one has

$$Q^{11,1j} = 0, \ Q^{11,2j} = 0, \ \text{for } j = 3, \ldots, l_0, \ \text{and } Q^{12,1*} = 0, \ Q^{12,2*} = 0.$$

This again implies that $\overline{P}^{\varepsilon}(\widetilde{U})$ cannot be irreducible since a state outside $\mathcal{M}_1 \cup \mathcal{M}_2$ cannot be reached from a state in $\mathcal{M}_1 \cup \mathcal{M}_2$. By this contradiction we obtain that $\overline{q}^{(1),22} < 0$. Similarly, we can show that $\overline{q}^{(1),ii} < 0$ for $i > 2$.

Repeat this procedure. Multiply the second row of $\overline{Q}^{(1),*}$ by $-\overline{q}^{(1),i2}/\overline{q}^{(1),22}$ and add to the ith row, for $i = 3, \ldots, l_0$. Let $\overline{Q}^{(2),*} = (\overline{q}_{(2),ij})$ denote the resulting matrix. Then

$$\overline{q}^{(2),ij} = \overline{q}^{(1),ij}, \ i = 1, 2, \ j = 1, \ldots, l_0,$$

$$\overline{q}^{(2),ij} = 0, \ i = 3, \ldots, l_0, \ j = 1, 2,$$

$$\overline{q}^{(2),ij} = \overline{q}^{(1),ij} - \overline{q}^{(1),2j}\frac{\overline{q}^{(1),i2}}{\overline{q}^{(1),22}}, \ i, j = 3, \ldots, l_0,$$

with $\overline{q}^{(2),ii} \leq 0$, $\overline{q}^{(2),ij} \geq 0$, for $j \neq i$, and $\sum_{j=3}^{l_0} \overline{q}^{(2),ij} = 0$. Similarly, we can prove that $\overline{q}^{(2),ii} < 0$ for $i \geq 3$.

Continue this transformation process. We obtain a sequence of matrices

$$\overline{Q}_* \to \overline{Q}^{(1),*} \to \cdots \to \overline{Q}^{(l_0-1),*}$$

with the last one $\overline{Q}^{(l_0-1),*} = (\overline{q}^{(l_0-1),ij})$ such that

$$\overline{q}^{(l_0-1),ij} = 0, \; i > j$$

$$\overline{q}^{(l_0-1),ii} < 0, \; i = 1, \ldots, l_0 - 1,$$

$$\sum_{j=1}^{l_0} \overline{q}^{(l_0-1),ij} = 0, \; i = 1, \ldots, l_0, \; \text{and} \; \overline{q}^{(l_0-1),l_0 l_0} = 0.$$

Since these transformations do not change the rank of the original matrix,

$$\text{rank}\left(\overline{Q}_*\right) = \text{rank}\left(\overline{Q}^{(1),*}\right) = \cdots = \text{rank}\left(\overline{Q}^{(l_0-1),*}\right) = l_0 - 1.$$

In view of Lemma 14.38, $\overline{Q}_*$ is weakly irreducible.

Step 2: Show that $\overline{Q}_*$ is irreducible. That is, let $(\overline{\nu}^1, \ldots, \overline{\nu}^{l_0})$ be the quasi-stationary distribution corresponding to $\overline{Q}_*$, then $(\overline{\nu}^1, \ldots, \overline{\nu}^{l_0}) > 0$. Suppose that this is not true. Without loss of generality, we may assume for some $1 \leq i_0 < l_0$, $\overline{\nu}^1 > 0, \ldots, \overline{\nu}^{i_0} > 0$, and $\overline{\nu}^{i_0+1} = 0, \ldots, \overline{\nu}^{l_0} = 0$. Then the fact that $(\overline{\nu}^1, \ldots, \overline{\nu}^{l_0})\overline{Q}_* = 0$ would imply that $\overline{q}^{ij} = 0$ for $i = 1, \ldots, i_0$ and $j = i_0 + 1, \ldots, l_0$, which in turn implies that $Q_{ij}^{11} = 0$ for $i = 1, \ldots, i_0, j = i_0 + 1, \ldots, l_0$, and $\overline{Q}_{i*}^{12} = 0$ for $i = 1, \ldots, i_0$. Again, $\overline{P}^\varepsilon(\widetilde{U})$ is not irreducible since the chain $\alpha^\varepsilon(\cdot)$ cannot jump from a state in $\mathcal{M}_1 \cup \mathcal{M}_2 \cup \cdots \cup \mathcal{M}_{i_0}$ to a state in $\mathcal{M}_{i_0+1} \cup \mathcal{M}_{i_0+2} \cup \cdots \cup \mathcal{M}_{l_0}$. Therefore, $\overline{Q}_*$ is irreducible.   □

Proof of Theorem 9.17. Let α_k^ε denote the Markov chain with transition probability matrix $\overline{P}^\varepsilon(\widetilde{U}^\circ)$ and $\tilde{x}(t)$ denote the continuous-time Markov chain generated by $\overline{Q}_*(U^\circ)$. In view of the irreducibility of $\overline{P}^\varepsilon(\widetilde{U}^\circ)$ and $\overline{Q}_*(U^\circ)$, we have

$$\lim_{k \to \infty} P(\alpha_k^\varepsilon = s_{ij}) = \nu^{\varepsilon,ij}(\widetilde{U}^\circ), \; j = 1, \ldots, m_i, \; i = 1, \ldots, l_0, *,$$

$$\lim_{t \to \infty} P(\tilde{x}(t) = i) = \overline{\nu}^i(U^\circ), \; i = 1, \ldots, l_0,$$

where $\nu^{\varepsilon,ij}(\widetilde{U}^\circ)$ denotes the jth component of $\nu^{\varepsilon,i}(\widetilde{U}^\circ)$ for each $i = 1, \ldots, l_0, *,$

and $\nu^{\varepsilon,i}(\widetilde{U}^o) = (\nu^{\varepsilon,i1}(\widetilde{U}^o), \ldots, \nu^{\varepsilon,il_0}(\widetilde{U}^o), \nu^{\varepsilon,i*}(\widetilde{U}^o))$. Therefore, we have

$$
\begin{aligned}
J^\varepsilon(u^\varepsilon(\cdot)) &= \limsup_{N\to\infty} \frac{1}{N} E \sum_{k=0}^{N} g(\alpha_k^\varepsilon, u(\alpha_k^\varepsilon)) \\
&= \limsup_{N\to\infty} \frac{1}{N} \sum_{k=0}^{N} \Bigg[\sum_{i=1}^{l_0} \sum_{j=1}^{m_i} g(s_{ij}, u^{ij,o}) P(\alpha_k^\varepsilon = s_{ij}) \\
&\qquad\qquad\qquad + \sum_{j=1}^{m_*} g(s_{*j}, u^{*j}) P(\alpha_k^\varepsilon = s_{*j}) \Bigg] \\
&= \sum_{i=1}^{l_0} \sum_{j=1}^{m_i} g(s_{ij}, u^{ij,o}) \nu^{\varepsilon,ij}(\widetilde{U}^o) + \sum_{j=1}^{m_*} g(s_{*j}, u^{*j}) \nu^{\varepsilon,*j}(\widetilde{U}^o),
\end{aligned}
\tag{9.41}
$$

and

$$
\begin{aligned}
\lambda^0 = J^0(U^o) &= \limsup_{T\to\infty} \frac{1}{T} E \int_0^T \overline{g}(\tilde{x}(t), U^o(\tilde{x}(t))) dt \\
&= \limsup_{T\to\infty} \frac{1}{T} \int_0^T \sum_{i=1}^{l_0} \overline{g}(i, U^{i,o}) P(\tilde{x}(t) = i) dt \\
&= \sum_{i=1}^{l_0} \overline{g}(i, U^{i,o}) \overline{\nu}^i(U^o) \\
&= \sum_{i=1}^{l_0} \sum_{j=1}^{m_i} g(s_{ij}, u^{ij,o}) \nu^{ij}(U^{i,o}) \overline{\nu}^i(U^o).
\end{aligned}
\tag{9.42}
$$

By virtue of Lemma 9.16, we have

$$
\left| \nu^{\varepsilon,ij}(\widetilde{U}^o) - \nu^{ij}(U^{i,o}) \overline{\nu}^k(U^o) \right| = O(\varepsilon), \ j = 1, \ldots, m_i, \ i = 1, \ldots, l_0, \text{ and}
$$
$$
\nu^{\varepsilon,*j}(\widetilde{U}^o) = O(\varepsilon), \ j = 1, \ldots, m_*.
$$

It follows that
$$
J^\varepsilon(u^\varepsilon(\cdot)) - \lambda^0 = O(\varepsilon).
\tag{9.43}
$$

As a result,
$$
\lambda^\varepsilon \le J^\varepsilon(u^\varepsilon(\cdot)) \le \lambda^0 + O(\varepsilon).
\tag{9.44}
$$

Next, let $u^{\varepsilon,o}(\cdot) = \{u^{\varepsilon,o}(\alpha) : \alpha \in \mathcal{M}\} \in \mathcal{A}^\varepsilon$ denote an optimal control for $\mathcal{P}_{av}^\varepsilon$ and let

$$
\widetilde{U}^{\varepsilon,o} = (U^{\varepsilon,o}, U^{\varepsilon,*,o}) = (U^{\varepsilon,1,o}, \ldots, U^{\varepsilon,l_0,o}, U^{\varepsilon,*,o}) \in \widetilde{\Gamma},
$$

where

$$
U^{\varepsilon,i,o} = (u^{\varepsilon,i1,o}, \ldots, u^{\varepsilon,im_i,o}) := (u^{\varepsilon,o}(s_{i1}), \ldots, u^{\varepsilon,o}(s_{im_i})), \ i = 1, \ldots, l_0, *.
$$

Using the same arguments as in the proof of (9.43), we obtain

$$J^\varepsilon(u^{\varepsilon,o}(\cdot)) - J^0(U^{\varepsilon,o}) = O(\varepsilon). \tag{9.45}$$

The control set Γ contains finitely many elements, so does $\widetilde{\Gamma}$, the control set for the limit problem $\mathcal{P}_{av}^0$. As in the proof of Theorem 9.10, suppose $\widetilde{\Gamma} = \{\gamma_1, \ldots, \gamma_{l_1}\}$ for some positive integer l_1. For $j = 1, \ldots, l_1$, define

$$\mathcal{E}_j = \{\varepsilon : \varepsilon \in (0,1), \ \widetilde{U}^{\varepsilon,o} = (\gamma_j, U^{\varepsilon,*,o}) \text{ for an } U^{\varepsilon,*,o} \in \Gamma^*\}.$$

Then $\{\mathcal{E}_j\}$ consists of finitely many sets such that $(0,1) = \mathcal{E}_1 \cup \ldots \cup \mathcal{E}_{l_1}$. For fixed j and any $\varepsilon \in \mathcal{E}_j$, in view of the optimality of $u^{\varepsilon,o}(\cdot)$ and (9.45), we have

$$\lambda^\varepsilon = J^\varepsilon(u^{\varepsilon,o}(\cdot)) = J^0(U^{\varepsilon,o}) + O(\varepsilon) \geq \lambda^0 + O(\varepsilon). \tag{9.46}$$

Thus for $0 < \varepsilon < 1$,

$$\lambda^\varepsilon \geq \lambda^0 + O(\varepsilon). \tag{9.47}$$

Combining (9.44) and (9.47), we obtain

$$\lambda^\varepsilon = \lambda^0 + O(\varepsilon). \tag{9.48}$$

Finally, in view of (9.43) and (9.48),

$$J^\varepsilon(u^\varepsilon(\cdot)) - \lambda^\varepsilon = (J^\varepsilon(u^\varepsilon(\cdot)) - \lambda^0) + (\lambda^0 - \lambda^\varepsilon) = O(\varepsilon). \quad \square$$

9.6   Notes

This chapter is concerned with two-time-scale Markov decision processes involving weak and strong interactions. We have developed asymptotically optimal strategies for two-time-scale Markov decision processes. The hierarchical control approach developed in this chapter is useful to reduce dimensionality for a wide variety of stochastic systems of practical concerns. For solving DP equations, where the number of equations is the dominant factor that affects the computational effort, our results can substantially reduce complexity and consequently, the computational requirement. This chapter is based on the work of Liu, Zhang and Yin [103]. Classical treatments of discrete-time-MDP models can be found in Derman [45], Ross [131], and White [149] among others.

Singularly perturbed MDPs have been studied by many researchers in the past decades. In Delebecque and Quadrat [44], a class of DP equations arising from continuous-time discounted MDPs was considered. The underlying Markov chain is assumed to have weak and strong interactions. Under such a probabilistic structure, asymptotic expansions of solutions to these DP equations were derived. The results in [44] were extended in Delebecque [43] and Quadrat [126] to incorporate Markov chains with multiple-time scales. The corresponding probabilistic interpretation of the reduction of perturbed Markov chains was discussed in depth in [43]. The main focus of these papers is on asymptotic expansions of DP equations rather than the construction of near-optimal policies via limit control problems.

In Abbad, Filar, and Bielecki [1] and Bielecki and Filar [15], a discrete-time singularly perturbed MDP with average cost was considered. The basic idea is to let the time go to infinity and then study the dependence of the system on the small parameter ε. By sending $\varepsilon \to 0$, a limit problem is obtained. Using this limit problem, a δ-optimal control policy was derived for any given $\delta > 0$. In [1], it was shown that there exists a deterministic control policy that is uniformly optimal for ε sufficiently small, while the emphasis of [15] was on the relaxed control policies. These results were further extended to general state spaces in Bielecki and Stettner [17]. Additional results along this line can be found in Abbad, Filar, and Bielecki [2], which contains several algorithms for solving the corresponding limit problems. For a more detailed discussion on singularly perturbed Markov systems and a review of the literature, we refer the reader to the book of Pervozvanskii and Gaitsgory [123]; see also the recent survey in Avrachenkov, Filar, and Haviv [7], and the references therein.

The long-run average results in this section are similar in spirit to those of Bielecki and Filar [15] and Bielecki and Stettner [17]. The main differences are that we consider classical feedback control policies and obtain asymptotic optimality of these policies as $\varepsilon \to 0$, while in [15] and [17] relaxed controls are considered and only δ-optimal policies are obtained for any given $\delta > 0$.

In studying MDP problems, it is interesting to examine system sensitivity in terms of ε, i.e., how the corresponding control and the value function depend on ε. We refer the reader to a series of papers by Cao [30, 31, 32], in the context of long-run average-cost MDP and connections to potential functions and perturbation analysis.

10
LQ Controls

10.1 Introduction

This chapter is concerned with near-optimal controls of a class of hybrid discrete-time linear quadratic (LQ) regulator problems. The LQ models are advantageous since the resulting control laws are linear with respect to the state variable and are therefore easy to compute. However, the control policies based on traditional LQ models are often unable to capture the system structural changes because the classical design is based on a plant with fixed-deterministic parameters. To address this issue, much effort has been directed to designing the so-called "robust" controls that can adapt to varying system environment. One of the approaches uses the so-called hybrid systems. This approach is applicable to process control, speech and/or pattern recognition, signal processing, telecommunications, and manufacturing, among others. Hybrid LQ systems belong to the class of hybrid systems, in which dynamics are governed by a number of linear subsystems coupled by a jump process (often assumed to be a Markov chain) that dictates structural changes.

In this chapter, we focus on a hybrid linear system consisting of a large number of configurations modulated by a finite-state Markov chain. In any given instance, the system takes one of the possible configurations, in which the coefficients depend on the state of the underlying Markov chain. Such a hybrid model is frequently referred to as a system with regime switching. Clearly, this model has a greater capability to account for disturbances in a realistic random environment. The inclusion of additional variables and the consideration of multiple factors make the system larger and more complex, however. We use a Markov chain to model the factor process responsible for the regime switching. Using the dynamic programming (DP) approach to the underlying hybrid system results in a system of Riccati equations leading to optimal control laws. Hybrid optimal control problems, where the modulating Markov chain has a large state space, resulting in a large number of Riccati equations to be solved, are our main concern here. To overcome computational difficulties, using time-scale separation, we introduce a small parameter in the underlying Markov chains to reflect the different rates of change among different states, yielding a two-time-scale (a fast time scale and a slowly varying one) formulation. This leads to singularly perturbed Markovian models with weak and strong interactions.

In this chapter, we first decompose the state space of the underlying Markov chain into a number of recurrent classes and a group of transient states according to the different jump rates. We aggregate the states in each recurrent class and replace the original system with its "average." Subsequently, under suitable scaling, we obtain a limit control system that requires solving fewer Riccati equations. One interesting aspect is that the corresponding limit problem is a continuous-time one. Using the optimal control law of the limit system, we construct controls for the original system, which leads to a feasible approximation scheme. We demonstrate that controls so constructed are asymptotically optimal. Since the LQ problem is completely determined by the solutions of Riccati equations, the decomposition and aggregation can substantially reduce complexity of the problem.

The rest of the chapter is arranged as follows. The formulation of the optimal control problem is presented in Section 10.2; the optimal control law and the associated Riccati equations are obtained by using the DP method. In Section 10.3, we analyze the convergence of the original system as the small parameter goes to 0, derive the limit control system, and construct controls of the actual system based on the optimal control law of the limit system. We also derive the asymptotic optimality of control policies so constructed. Section 10.4 presents a couple of numerical examples to demonstrate the effectiveness of our approximation scheme; proofs of results are provided in Section 10.5.

10.2   Problem Formulation

Let $\{\alpha_k^\varepsilon : \ k \geq 0\}$ be a finite-state Markov chain with state space $\mathcal{M} = \{1, 2, \ldots, m_0\}$ and transition matrix $P^\varepsilon = (p^{\varepsilon, ij})_{m_0 \times m_0}$, where ε is a small parameter. For a given $0 < T < \infty$, the discrete-time control system is governed by

$$\begin{cases} x_{k+1} = x_k + \varepsilon[A(\alpha_k^\varepsilon)x_k + B(\alpha_k^\varepsilon)u_k] + \sqrt{\varepsilon}\xi_k, \\ x_0 = x, \ \alpha_0^\varepsilon = \alpha, \ 0 \le k \le \lfloor T/\varepsilon \rfloor, \end{cases} \tag{10.1}$$

where $x_k \in \mathbb{R}^{n_1}$ represents the state, $\xi_k \in \mathbb{R}^{n_1}$ denotes the system disturbance, $u_k \in \mathbb{R}^{n_2}$ is the control, $A(\alpha) \in \mathbb{R}^{n_1 \times n_1}$, $B(\alpha) \in \mathbb{R}^{n_1 \times n_2}$, and for any $z \in \mathbb{R}$, $\lfloor z \rfloor$ denotes the integer part of z. Let $u = \{u_0, u_1, \ldots, u_{\lfloor T/\varepsilon \rfloor - 1}\}$ denote the control sequence. We consider the quadratic cost function

$$\begin{aligned} &J^\varepsilon(x, \alpha, u) \\ &= E\left[\varepsilon \sum_{k=0}^{\lfloor T/\varepsilon \rfloor - 1} [x_k' M(\alpha_k^\varepsilon)x_k + u_k' N(\alpha_k^\varepsilon)u_k] + x_{\lfloor T/\varepsilon \rfloor}' D x_{\lfloor T/\varepsilon \rfloor} \right], \end{aligned} \tag{10.2}$$

where for each $\alpha \in \mathcal{M}$, $M(\alpha) \in \mathbb{R}^{n_1 \times n_1}$, $N(\alpha) \in \mathbb{R}^{n_2 \times n_2}$, and $D \in \mathbb{R}^{n_1 \times n_1}$.

Remark 10.1. The control problem (10.1) and (10.2) may be obtained from a continuous-time problem via discretization. To see this, consider a continuous-time hybrid control problem:

$$\text{Minimize } J^\varepsilon = E\left[\int_0^T \left(x'(t)M(\alpha^\varepsilon(t))x(t) \right. \right.$$
$$\left. \left. + u'(t)N(\alpha^\varepsilon(t))u(t) \right)dt + x'(T)Dx(T) \right]$$
$$\text{subject to } dx(t) = (A(\alpha^\varepsilon(t))x(t) + B(\alpha^\varepsilon(t))u(t))dt + \sigma dw(t), \tag{10.3}$$

where $w(\cdot)$ is an $\mathbb{R}^{n_1}$-valued standard Brownian motion and $\alpha^\varepsilon(t)$ is a continuous-time Markov chain generated by $\widetilde{Q}/\varepsilon + \widehat{Q}$, such that both $\widetilde{Q}$ and $\widehat{Q}$ are generators and $\widetilde{Q}$ has block-diagonal form (see Yin and Zhang [158, Chapter 9] for more detail). In (10.3), making a time change $\tau = t/\varepsilon$ and then discretizing the resulting system using a step size ε lead to a control problem of the form given by (10.1) and (10.2). It should also be mentioned that discrete-time control problems also come from applications directly because in many cases measurements of state variables can only be obtained in discrete time.

We make the following assumptions for the system (10.1).

(A10.1) (a) For each $\alpha \in \mathcal{M}$, $M(\alpha)$ is symmetric and nonnegative definite, and $N(\alpha)$ and D are symmetric positive definite.

(b) $\{\xi_k\}$ is a sequence of independent and identically distributed (i.i.d.) Gaussian random variables with mean 0 and variance Σ.

(c) $\{\alpha_k^\varepsilon\}$ and $\{\xi_k\}$ are independent.

Suppose that the transition probability matrix, P^ε, of the Markov chain α_k^ε has the form

$$P^\varepsilon = P + \varepsilon Q, \tag{10.4}$$

where ε is a small parameter, $P = (p^{ij})_{m_0 \times m_0}$ is a probability transition matrix (i.e., $p^{ij} \geq 0$ and, for each i, $\sum_j p^{ij} = 1$), and $Q = (q^{ij})_{m_0 \times m_0}$ is a generator of a continuous-time Markov chain (i.e., $q^{ij} \geq 0$ for $i \neq j$ and, for each $i \in \mathcal{M}$, $\sum_j q^{ij} = 0$). Clearly, P is the dominating part. Its structure is of crucial importance to the system's behavior. As in Chapter 6, suppose that P has a partitioned block form

$$P = \begin{pmatrix} P^1 & & & \\ & \ddots & & \\ & & P^{l_0} & \\ P^{*,1} & \cdots & P^{*,l_0} & P^* \end{pmatrix}, \tag{10.5}$$

where the matrices are such that $P^i \in \mathbb{R}^{m_i \times m_i}$, $P^{*,i} \in \mathbb{R}^{m_* \times m_i}$, for $i = 1, \ldots, l_0$, $P^* \in \mathbb{R}^{m_* \times m_*}$. For the transition probability matrices, we assume that the following conditions hold:

(A10.2) (a) P^i is a transition probability matrix and is irreducible and aperiodic, for each $i = 1, \ldots, l_0$;

(b) P^* has all of its eigenvalues inside the unit circle.

Under Assumption (A10.2), there exists a stationary distribution of the Markov chain corresponding to P^i for each $i = 1, \ldots, l_0$. Denote the stationary distribution by $\nu^i = (\nu^{i1}, \ldots, \nu^{im_i})$. Then, ν^i is the unique solution of the following system of equations

$$\nu^i P^i = \nu^i, \quad \sum_{j=1}^{m_i} \nu^{ij} = 1. \tag{10.6}$$

For each fixed $\varepsilon > 0$, the problem of interest is to choose $u(\cdot)$ to minimize J^ε. We use the DP approach to solve the problem. Let

$$\lambda_k^\varepsilon(x_k, \alpha_k) =$$

$$\min_{u_k, \ldots, u_{\lfloor T/\varepsilon \rfloor - 1}} E\left[\varepsilon \sum_{k_1 = k}^{\lfloor T/\varepsilon \rfloor - 1} [x_{k_1}' M(\alpha_{k_1}^\varepsilon) x_{k_1} + u_{k_1}' N(\alpha_{k_1}^\varepsilon) u_{k_1}] + x_{\lfloor T/\varepsilon \rfloor}' D x_{\lfloor T/\varepsilon \rfloor}\right]$$

with $\alpha_k^\varepsilon = \alpha_k \in \mathcal{M}$. In particular, for $k = 0$,

$$\lambda_0^\varepsilon(x_0, \alpha_0) = \min_u J^\varepsilon(x, \alpha, u) \text{ with } x_0 = x, \ \alpha_0 = \alpha, \tag{10.7}$$

where $u = \{u_{k_1} : k_1 \le k \le \lfloor T/\varepsilon \rfloor - 1\}$. The associated DP equations (see Bertsekas [13]) are given by

$$
\begin{cases}
\lambda^\varepsilon_{\lfloor T/\varepsilon \rfloor}(x_{\lfloor T/\varepsilon \rfloor}, \alpha_{\lfloor T/\varepsilon \rfloor}) = x'_{\lfloor T/\varepsilon \rfloor} D x_{\lfloor T/\varepsilon \rfloor}, \\
\lambda^\varepsilon_k(x_k, \alpha_k) = \min_{u_k} E\Big[\varepsilon[x'_k M(\alpha_k)x_k + u'_k N(\alpha_k)u_k] \\
\qquad\qquad + \lambda^\varepsilon_{k+1}(x_{k+1}, \alpha_{k+1})\Big], \ 0 \le k < \lfloor T/\varepsilon \rfloor.
\end{cases} \tag{10.8}
$$

Thus, in view of the independence of $\{\alpha^\varepsilon_k\}$ and $\{\xi_k\}$, for any given x_k and $\alpha_k = \alpha \in \mathcal{M}$, we have

$$
\lambda^\varepsilon_k(x_k, \alpha) = \min_{u_k} E\Big\{\varepsilon[x'_k M(\alpha)x_k + u'_k N(\alpha)u_k] \\
+ \sum_{j \in \mathcal{M}} p^\varepsilon_{\alpha j} \lambda^\varepsilon_{k+1}(x_{k+1}, j)\Big\}. \tag{10.9}
$$

To solve (10.9), we suppose that $\lambda^\varepsilon_k(x, \alpha)$ has the following quadratic form

$$
\lambda^\varepsilon_k(x, \alpha) = x' R^\varepsilon_k(\alpha)x + q^\varepsilon_k(\alpha), \tag{10.10}
$$

where $R^\varepsilon_k(\alpha) \in \mathbb{R}^{n_1 \times n_1}$ and $q^\varepsilon_k(\alpha) \in \mathbb{R}^1$ are functions to be determined. We will show in what follows that both $R^\varepsilon_k(\alpha)$ and $q^\varepsilon_k(\alpha)$ exist uniquely and $\lambda^\varepsilon_k(x, \alpha)$ is indeed a solution to (10.9). Then using the verification theorem given in Fleming and Rishel [57], it is easy to show that $\lambda^\varepsilon_k(x, \alpha)$ is the only solution to (10.9) and the corresponding minimizer u^*_k of the right-hand side of (10.9) is an optimal control.

To proceed, we use (10.1) and (10.9) to obtain

$$
\lambda^\varepsilon_k(x_k, \alpha) = \min_{u_k}\Big\{\varepsilon[x'_k M(\alpha)x_k + u'_k N(\alpha)u_k] \\
+ \sum_{j \in \mathcal{M}} p^{\varepsilon,\alpha j}\Big[x'_k(I + \varepsilon A(\alpha))' R^\varepsilon_{k+1}(j)(I + \varepsilon A(\alpha))x_k \\
+ \varepsilon^2 u'_k B'(\alpha) R^\varepsilon_{k+1}(j)B(\alpha)u_k + \varepsilon\mathrm{tr}\left(R^\varepsilon_{k+1}(j)\Sigma\right) \\
+ 2\varepsilon x'_k(I + \varepsilon A(\alpha))' R^\varepsilon_{k+1}(j)B(\alpha)u_k + q^\varepsilon_{k+1}(j)\Big]\Big\}. \tag{10.11}
$$

For any given function $f(\cdot)$ and for $\alpha \in \mathcal{M}$, define

$$
\widehat{f}(\alpha) = \sum_{j \in \mathcal{M}} p^{\varepsilon,\alpha j} f(j).
$$

Using this notation, we write (10.11) as follows:

$$
\lambda_k^\varepsilon(x_k, \alpha) = \min_{u_k} \Big\{ \varepsilon x_k' M(\alpha) x_k + \varepsilon u_k' N(\alpha) u_k
$$
$$
+ x_k'(I + \varepsilon A(\alpha))' \widehat{R}_{k+1}^\varepsilon(\alpha)(I + \varepsilon A(\alpha)) x_k
$$
$$
+ \varepsilon^2 u_k' B'(\alpha) \widehat{R}_{k+1}^\varepsilon(\alpha) B(\alpha) u_k + \varepsilon \mathrm{tr}\left(\widehat{R}_{k+1}^\varepsilon(\alpha) \Sigma \right)
$$
$$
+ 2\varepsilon x_k'(I + \varepsilon A(\alpha))' \widehat{R}_{k+1}^\varepsilon(\alpha) B(\alpha) u_k + \widehat{q}_{k+1}^\varepsilon(\alpha) \Big\}.
$$
$$
(10.12)
$$

Differentiating the right-hand side of (10.12) w.r.t. u_k and setting the derivative to 0, we obtain the corresponding minimizer

$$
u_k^{\varepsilon, o}(x_k, \alpha) = -\Big(N(\alpha) + \varepsilon B'(\alpha) \widehat{R}_{k+1}^\varepsilon(\alpha) B(\alpha) \Big)^{-1}
$$
$$
\times B'(\alpha) \widehat{R}_{k+1}^\varepsilon(\alpha)(I + \varepsilon A(\alpha)) x_k \qquad (10.13)
$$
$$
\overset{\text{def}}{=} -\Phi_k(\alpha) x_k.
$$

Using (10.10)–(10.13) with $\alpha \in \mathcal{M}$, we obtain a system of Riccati equations for $R_k^\varepsilon(\alpha)$,

$$
\begin{cases}
R_k^\varepsilon(\alpha) = \varepsilon M(\alpha) + (I + \varepsilon A(\alpha))' \widehat{R}_{k+1}^\varepsilon(\alpha)(I + \varepsilon A(\alpha)) \\
\qquad - \varepsilon(I + \varepsilon A(\alpha))' \widehat{R}_{k+1}^\varepsilon(\alpha) B(\alpha) \\
\qquad \times \Big(N(\alpha) + \varepsilon B'(\alpha) \widehat{R}_{k+1}^\varepsilon(\alpha) B(\alpha) \Big)^{-1} \qquad (10.14) \\
\qquad \times B'(\alpha) \widehat{R}_{k+1}^\varepsilon(\alpha)(I + \varepsilon A(\alpha)), \\
R_{\lfloor T/\varepsilon \rfloor}^\varepsilon(\alpha) = D, \ \alpha \in \mathcal{M},
\end{cases}
$$

and a system of equations for $q_k^\varepsilon(\cdot)$,

$$
\begin{cases}
q_k^\varepsilon(\alpha) = \varepsilon \mathrm{tr}\left(\widehat{R}_{k+1}^\varepsilon(\alpha) \Sigma \right) + \widehat{q}_{k+1}^\varepsilon(\alpha), \\
q_{\lfloor T/\varepsilon \rfloor}^\varepsilon(\alpha) = 0, \ \alpha \in \mathcal{M}.
\end{cases}
\qquad (10.15)
$$

Lemma 10.2. *Under Assumption (A10.1), the following assertions hold:*

(a) *For small ε, $R_k^\varepsilon(\alpha)$ are positive definite for each $k \leq \lfloor T/\varepsilon \rfloor$.*

(b) *There exists a constant K_T depending only on T such that for each $k \leq \lfloor T/\varepsilon \rfloor$ and $\alpha \in \mathcal{M}$,*

$$
|R_k^\varepsilon(\alpha)| \leq K_T \ \text{and} \ |q_k^\varepsilon(\alpha)| \leq K_T.
$$

The proof of Lemma 10.2 is postponed until Section 10.6. Note that (10.14) and (10.15) both consist of m_0 equations coupled by P^ε. To find the optimal control $\{u_k^o\}$, we have to solve these equations. The difficulty typically arises when the state space of α_k^ε is large. One has to resort to approximation schemes. In the next section, we present an approach based on the aggregation of the underlying Markov chain.

10.3   Main Results

In this section, we show that there exists a limit problem as $\varepsilon \to 0$, which is a continuous-time LQ control problem. The limit problem is simpler to solve than that of the original one. We then use the optimal controls of the limit problem to construct controls for the original problem and to show that the controls so constructed are asymptotically optimal.

To proceed, for each $k \in \mathcal{M}$, define the piecewise constant interpolations $R^\varepsilon(\cdot, \alpha)$ and $q^\varepsilon(\cdot, \alpha)$ as follows:

$$R^\varepsilon(t, \alpha) = R_k^\varepsilon(\alpha), \ q^\varepsilon(t, \alpha) = q_k^\varepsilon(\alpha), \text{ for } t \in [k\varepsilon, \ k\varepsilon + \varepsilon). \qquad (10.16)$$

It follows from Lemma 10.2 that $R^\varepsilon(t, \alpha)$ are positive definite for each $t \in [0, T]$, $|R^\varepsilon(t, \alpha)| \le K_T$, and $|q^\varepsilon(t, \alpha)| \le K_T$.

We next demonstrate the convergence of $R^\varepsilon(\cdot, \alpha)$ and $q^\varepsilon(\cdot, \alpha)$ as $\varepsilon \to 0$. Let

$$Q^\varepsilon = \frac{1}{\varepsilon}\widetilde{Q} + \widehat{Q}, \qquad (10.17)$$

where $\widetilde{Q} = P - I$ and $\widehat{Q} = Q$. In view of (10.5), we write

$$\widetilde{Q} = \begin{pmatrix} \widetilde{Q}^1 & & & \\ & \ddots & & \\ & & \widetilde{Q}^{l_0} & \\ \widetilde{Q}^{*,1} & \cdots & \widetilde{Q}^{*,l_0} & \widetilde{Q}^* \end{pmatrix} = \begin{pmatrix} P^1 - I & & & \\ & \ddots & & \\ & & P^{l_0} - I & \\ P^{*,1} & \cdots & P^{*,l_0} & P^* - I \end{pmatrix}.$$

For each $\alpha = s_{ij} \in \mathcal{M}$ with $i = 1, \ldots, l_0$ and $j = 1, \ldots, m_i$, and a given function $H(\cdot, \alpha)$ (real valued, or vector valued, or matrix valued), define

$$\overline{H}(t, i) = \sum_{j=1}^{m_i} \nu^{ij} H(t, s_{ij}), \ i = 1, \ldots, l_0.$$

Theorem 10.3. *Under Assumptions* (A10.1) *and* (A10.2). *As* $\varepsilon \to 0$, *we have*

$$R^\varepsilon(t, s_{ij}) \to \overline{R}(t, i)$$

and

$$q^\varepsilon(t, s_{ij}) \to \overline{q}(t, i),$$

for $i = 1, \ldots, l_0$ and $j = 1, \ldots, m_i$;

$$R^\varepsilon(t, s_{*j}) \to \overline{R}(t, *j)$$

and

$$q^\varepsilon(t, s_{*j}) \to \overline{q}(t, *j)$$

for $j = 1, \ldots, m_*$, uniformly on $[0, T]$ where

$$
\begin{aligned}
\overline{R}(t, *j) &= a^{1,j}\overline{R}(t, 1) + \cdots + a^{l_0,j}\overline{R}(t, l_0), \\
\overline{q}(t, *j) &= a^{1,j}\overline{q}(t, 1) + \cdots + a^{l_0,j}\overline{q}(t, l_0),
\end{aligned}
\tag{10.18}
$$

$a^i = (a^{i,1}, \ldots, a^{i,m_*})$ *are defined in* (6.13). *Moreover, for $i = 1, \ldots, l_0$, $\overline{R}(t, i)$ and $\overline{q}(t, i)$ are the unique solutions to the following differential equations*

$$
\begin{cases}
\dot{\overline{R}}(t, i) = -\overline{R}(t, i)\overline{A}(i) - \overline{A}'(i)\overline{R}(t, i) - \overline{M}(i) \\
\qquad\qquad + \overline{R}(t, i)\overline{BN^{-1}B'}(i)\overline{R}(t, i) - \overline{Q}_*\overline{R}(t, \cdot)(i), \\
\overline{R}(T, i) = D,
\end{cases}
\tag{10.19}
$$

with $\overline{Q}_*$ defined in (6.16), *and*

$$
\begin{cases}
\dot{\overline{q}}(t, i) = -\mathrm{tr}\left(\overline{R}(t, i)\Sigma\right) - \overline{Q}_*\overline{q}(t, \cdot)(i), \\
\overline{q}(T, i) = 0,
\end{cases}
\tag{10.20}
$$

respectively.

To proceed, define the piecewise constant interpolation $v^\varepsilon(\cdot, x, \alpha)$ as

$$v^\varepsilon(t, x, \alpha) = \lambda_k^\varepsilon(x, \alpha) \text{ for } t \in [k\varepsilon, k\varepsilon + \varepsilon). \tag{10.21}$$

Corollary 10.4. *As $\varepsilon \to 0$, $v^\varepsilon(t, x, s_{ij}) \to v(t, x, i)$ for $i = 1, \ldots, l_0$, $j = 1, \ldots, m_i$, and $v^\varepsilon(t, x, s_{*j}) \to v(t, x, *j)$ for $j = 1, \ldots, m_*$, where*

$$
\begin{aligned}
v(t, x, i) &= x'\overline{R}(t, i)x + \overline{q}(t, i), & \text{for } i = 1, \ldots, l_0, \\
v(t, x, *j) &= a^{1,j}v(t, x, 1) + \cdots + a^{l_0,j}v(t, x, l_0), & \text{for } j = 1, \ldots, m_*.
\end{aligned}
\tag{10.22}
$$

Next, we introduce a limit control problem in which the value functions $v(t, x, i)$ are the limits of $v^\varepsilon(t, x, s_{ij})$ as in Corollary 10.4 for $i = 1, \ldots, l_0$. First, let $\mathcal{U}$ denote the control set for the limit control system

$$\mathcal{U} = \left\{ U = (U^1, \ldots, U^{l_0}) : U^i = (u^{i1}, \ldots, u^{im_i}), u^{ij} \in \mathbb{R}^{n_2} \right\}. \tag{10.23}$$

Define

$$f(i, x, U) = \overline{A}(i)x + \sum_{j=1}^{m_i} \nu^{ij} B(s_{ij}) u^{ij},$$

$$\widetilde{N}(i, U) = \sum_{j=1}^{m_i} \nu^{ij} \left(u^{ij,\prime} N(s_{ij}) u^{ij} \right),$$

for $i = 1, \ldots, l_0$. Given the limit equations in (10.19) and (10.20), the HJB equations for $v(t, x, i)$ with $i = 1, \ldots, l_0$ are

$$0 = \frac{\partial v(t, x, i)}{\partial t} + \min_{U \in \mathcal{U}} \left\{ f(i, x, U) \frac{\partial v(t, x, i)}{\partial x} + x' \overline{M}(i) x \right.$$
$$\left. + \widetilde{N}(i, U) + \frac{1}{2} \mathrm{tr} \left(\frac{\partial^2 v(t, x, i)}{\partial x^2} \Sigma \right) + \overline{Q}_* v(t, x, \cdot)(i) \right\} \tag{10.24}$$

with $v(T, x, i) = x' D x$. Corresponding to the value functions $v(t, x, i)$ for $i = 1, \ldots, l_0$, there is an associate control problem. We can it the limit control problem, which is given by

$$\begin{cases} \text{Minimize: } J(s, x, i, U(\cdot)) = \\ \qquad E \left\{ \int_s^T \left[\overline{x}'(t) \overline{M}(\overline{\alpha}(t)) \overline{x}(t) + \widetilde{N}(\overline{\alpha}(t), U(t)) \right] dt + \overline{x}'(T) D \overline{x}(T) \right\}, \\ \text{subject to: } d\overline{x}(t) = f(\overline{\alpha}(t), \overline{x}(t), U(t)) dt + \sigma dw(t), \ \overline{x}(s) = x, \\ \text{value function: } v(s, x, i) = \min_{U(\cdot)} J(s, x, i, U(\cdot)), \end{cases}$$

where $\sigma \sigma' = \Sigma$ and $w(t)$ is a standard Brownian motion. Note that the limit problem is a continuous-time quadratic regulator problem. Denote the optimal control for this limit control problem by

$$U^{\circ}(t, x) = \left(U^{1,\circ}(t, x), \ldots, U^{l_0,\circ}(t, x) \right), \tag{10.25}$$

where for $i = 1, \ldots, l_0$, $U^{i,\circ}(t, x) = \left(u^{i1,\circ}(t, x), \ldots, u^{im_i,\circ}(t, x) \right)$ and

$$u^{ij,\circ}(t, x) = -N^{-1}(s_{ij}) B'(s_{ij}) \overline{R}(t, i) x, \ j = 1, \ldots, m_i.$$

Now we construct a control for the original discrete-time control system as

$$u_k^{\varepsilon} = u_k^{\varepsilon}(x_k, \alpha) = u^{\varepsilon}(k\varepsilon, x_k, \alpha), \tag{10.26}$$

where

$$u^{\varepsilon}(t, x, \alpha) = \sum_{i=1}^{l_0} \sum_{j=1}^{m_i} I_{\{\alpha = s_{ij}\}} u^{ij,\circ}(t, x) + \sum_{j=1}^{m_*} I_{\{\alpha = s_{*j}\}} u^{*j,\circ}(t, x),$$

and $u^{ij,\circ}(t, x)$ is given above and

$$u^{*j,\circ}(t, x) = -N^{-1}(s_{*j}) B'(s_{*j}) \overline{R}(t, *j) x, \tag{10.27}$$

where $\overline{R}(t, *j)$ is defined in (10.18). Equivalently,

$$
u_k^\varepsilon(x_k, \alpha) = \begin{cases} -N^{-1}(s_{ij})B'(s_{ij})\overline{R}(k\varepsilon, i)x_k, & \text{if } \alpha = s_{ij} \in \mathcal{M}_i, \\ -N^{-1}(s_{*j})B'(s_{*j})\overline{R}(k\varepsilon, *j)x_k, & \text{if } \alpha = s_{*j} \in \mathcal{M}_*. \end{cases}
$$

(10.28)

The next theorem gives the asymptotic optimality resulting from this approximation scheme.

Theorem 10.5. *Under Assumptions* (A10.1) *and* (A10.2), *the control* $u^\varepsilon(\cdot)$ *defined in* (10.28) *is asymptotically optimal for the original control system* (10.1) *and* (10.2) *in the sense that*

$$
\lim_{\varepsilon \to 0} \left| J^\varepsilon(x, \alpha, u^\varepsilon(\cdot)) - \lambda_0^\varepsilon(x, \alpha) \right| = 0,
$$

(10.29)

where $\lambda_0^\varepsilon(x, \alpha)$ *is the value function defined in* (10.7).

Remark 10.6. It would be interesting to come up with an error bound for $J^\varepsilon(x, \alpha, u^\varepsilon(\cdot)) - \lambda_0^\varepsilon(x, \alpha)$ in terms of ε. However, such a bound is usually difficult to obtain since the near optimality proof is based on weak convergence ideas yielding no error estimates. Nevertheless, in certain cases, error bounds can be obtained, for example, in Chapter 9 in connection with a Markov decision problem. In addition, the numerical example in the next section suggests that the error bound is of the order $\sqrt{\varepsilon}$.

10.4   Numerical Example

This section provides two numerical examples. The first example is a scalar dynamic system modulated by a discrete-time Markov chain. By constructing the sample paths of the Markov chain, the original dynamic system, the limit system, and the corresponding Riccati equations, we demonstrate that the limit systems closely approximate the original one. Different values of ε are used for demonstration purposes. The second example is taken from Yang, Yin, Yin, and Zhang [153], in which a vector-valued dynamic system is treated.

10.4.1   A Scalar LQ System

Consider a Markov chain $\alpha_k^\varepsilon \in \mathcal{M} = \{1, 2, 3\}$ with transition matrix

$$
P^\varepsilon = \begin{pmatrix} 0.8 & 0.2 & 0 \\ 0.1 & 0.9 & 0 \\ 0.3 & 0.2 & 0.5 \end{pmatrix} + \varepsilon \begin{pmatrix} -0.5 & 0.3 & 0.2 \\ 0.2 & -0.5 & 0.3 \\ 0.3 & 0.2 & -0.5 \end{pmatrix}.
$$

ε	$|R^\varepsilon - \overline{R}|$	$|x^\varepsilon - \overline{x}^\varepsilon|$	$|\lambda^\varepsilon - v|$	$|J^\varepsilon - \lambda^\varepsilon|$
0.1	7.54ε	0.42ε	18.78ε	$4.99\sqrt{\varepsilon}$
0.01	11.63ε	0.10ε	23.85ε	$7.52\sqrt{\varepsilon}$
0.001	12.20ε	0.069ε	24.58ε	$15.30\sqrt{\varepsilon}$
0.0001	12.26ε	0.11ε	26.59ε	$26.07\sqrt{\varepsilon}$

TABLE 10.1. Error bounds for various values of ε

Note that the state space $\mathcal{M}$ consists of a pair of recurrent states $\{1,2\}$ and a transient state $\{3\}$.

Consider the following one-dimensional discrete-time dynamic system model

$$x_{k+1} = x_k + \varepsilon[A(\alpha_k^\varepsilon)x_k + B(\alpha_k^\varepsilon)u_k] + \sqrt{\varepsilon}\xi_k,$$

and the objective function (10.2) with the following specifications:

$$A(1) = 0.5, \ A(2) = -0.3, \ A(3) = 0.2, \ B(1) = 1, \ B(2) = 2, \ B(3) = 1.5,$$

$$M(1) = M(2) = M(3) = N(1) = N(2) = N(3) = D = 1.$$

To obtain the optimal control law, one needs to solve a system of three Riccati equations in (10.14). Nevertheless, one need only solve a single scalar equation (10.19) for the limit problem to get the nearly optimal control policy. This leads to considerable reduction of the needed computation.

Take $T = 10$, $x_0 = x = 0$, $\alpha_0^\varepsilon = \alpha = 1$. Let $\lambda^\varepsilon = \lambda_0^\varepsilon(0,1)$, $v = v(0,0,1)$, and $J^\varepsilon = J^\varepsilon(0,1,u^\varepsilon(\cdot))$.

Define the norms

$$|R^\varepsilon - \overline{R}| = \frac{1}{\lfloor T/\varepsilon \rfloor} \sum_{k=0}^{\lfloor T/\varepsilon \rfloor - 1} \left(|R_k^\varepsilon(1) - \overline{R}(k\varepsilon)| + |R_k^\varepsilon(2) - \overline{R}(k\varepsilon)| \right.$$

$$\left. + |R_k^\varepsilon(3) - \overline{R}(k\varepsilon)| \right),$$

$$|x^\varepsilon - \overline{x}^\varepsilon| = \frac{1}{\lfloor T/\varepsilon \rfloor} \sum_{k=0}^{\lfloor T/\varepsilon \rfloor - 1} |x_k - \overline{x}(\varepsilon k)|,$$

where $R_k^\varepsilon(i)$, $i = 1,2,3$ are the solutions of (10.14), $\overline{R}(t)$ denotes the solution of the limit Riccati equation (10.19), x_k is the optimal trajectory under the optimal control law (10.13), and $\overline{x}(\varepsilon k)$ is the near optimal trajectory under the asymptotic control (10.28). Table 10.1 gives error bounds for various ε values, which are based on computations using 100 sample paths.

Sample paths of various trajectories of α_k^ε, x_k, $\overline{x}(\varepsilon k)$, and $|x_k - \overline{x}(\varepsilon k)|$ are plotted in the left column in Figure 10.1 for $\varepsilon = 0.01$. The corresponding

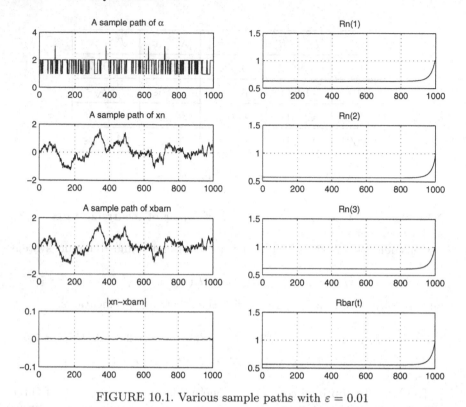

FIGURE 10.1. Various sample paths with $\varepsilon = 0.01$

$R_k^\varepsilon(1)$, $R_k^\varepsilon(2)$, $R_k^\varepsilon(3)$, and $\overline{R}(k\varepsilon)$ are given in the right column. Note that with the total iterations fixed at 10,000 ($\varepsilon = 0.001$), the average CPU time for solving the original Riccati equations is 42.1 seconds, whereas the average CPU time for solving the limit Riccati equation is only 15.7 seconds. Thus, compared with the solution of the original system, only a little more than one third of the computational effort is used to find the near-optimal control policy. The numerical bounds provided in Table 10.1. It is clear that the approximation scheme performs quite well.

10.4.2   A Vector-Valued LQ System

In this section, we consider a multidimensional problem. Let α_k^ε be a four-state Markov chain with state space $\mathcal{M} = \{1, 2, 3, 4\}$, and transition probability matrix

$$P^\varepsilon = P + \varepsilon Q,$$

where

$$P = \begin{pmatrix} 0.50 & 0.50 & 0 & 0 \\ 0.55 & 0.45 & 0 & 0 \\ 0 & 0 & 0.4 & 0.6 \\ 0 & 0 & 0.5 & 0.5 \end{pmatrix},$$

$$Q = \begin{pmatrix} -0.6 & 0 & 0.3 & 0.3 \\ 0 & -0.3 & 0.1 & 0.2 \\ 0.2 & 0.3 & -0.5 & 0 \\ 0.1 & 0.3 & 0 & -0.4 \end{pmatrix}.$$

For a two-dimensional dynamic system (10.1) and the cost function (10.2), let

$$x_0 = \begin{pmatrix} 0 \\ 1 \end{pmatrix}, \quad \Sigma = \begin{pmatrix} 1.5 & 0.5 \\ 0.5 & 2.0 \end{pmatrix}, \quad D = \begin{pmatrix} 2 & 1 \\ 1 & 2 \end{pmatrix},$$

$$A(1) = \begin{pmatrix} -1 & 0 \\ 0 & 2 \end{pmatrix}, \quad A(2) = \begin{pmatrix} -2 & -1 \\ -1 & 1 \end{pmatrix},$$

$$A(3) = \begin{pmatrix} -3 & -2 \\ -2 & 0 \end{pmatrix}, \quad A(4) = \begin{pmatrix} -4 & -3 \\ -3 & -1 \end{pmatrix},$$

$$B(1) = \begin{pmatrix} 1 & 2 \\ 2 & 4 \end{pmatrix}, \quad B(2) = \begin{pmatrix} 2 & 3 \\ 3 & 5 \end{pmatrix},$$

$$B(3) = \begin{pmatrix} 3 & 4 \\ 4 & 6 \end{pmatrix}, \quad B(4) = \begin{pmatrix} 4 & 5 \\ 5 & 7 \end{pmatrix},$$

$$M(1) = \begin{pmatrix} 5 & 3 \\ 3 & 7 \end{pmatrix}, \quad M(2) = \begin{pmatrix} 4 & 3/2 \\ 3/2 & 5 \end{pmatrix},$$

$$M(3) = \begin{pmatrix} 11/3 & 1 \\ 1 & 13/3 \end{pmatrix}, \quad M(4) = \begin{pmatrix} 7/2 & 3/4 \\ 3/4 & 4 \end{pmatrix},$$

$$N(1) = \begin{pmatrix} 8 & 3 \\ 3 & 10 \end{pmatrix}, \quad N(2) = \begin{pmatrix} 10 & 6 \\ 6 & 14 \end{pmatrix},$$

$$N(3) = \begin{pmatrix} 12 & 9 \\ 9 & 18 \end{pmatrix}, \quad N(4) = \begin{pmatrix} 14 & 12 \\ 12 & 22 \end{pmatrix}.$$

The time horizon for this discrete-time model is $0 \le k \le \lfloor T/\varepsilon \rfloor$ with $T = 5$. We use step size $h = 0.01$ to discretize the limit system of Riccati equations.

Take $\alpha_0^\varepsilon = 1$. The trajectories of x_k vs. $\bar{x}(t)$, $R_k^\varepsilon(i)$ vs. $\overline{R}(t, \cdot)$, and $v_k^\varepsilon(x, i)$ vs. $v(\cdot)$ are given in Figure 10.2 for $\varepsilon = 0.01$. The simulation results show that the discrete-time linear quadratic regulator problem is closely approximated by the corresponding continuous-time hybrid LQG problem, which allows us further to construct nearly optimal controls for the original system.

10.4.3 Remarks

This chapter focuses on approximation schemes for a class of discrete-time hybrid systems. It provides a systematic approach to reduce the complexity of the underlying systems. To find the approximate control, one need only solve l_0 Riccati equations. The computation load is reduced considerably

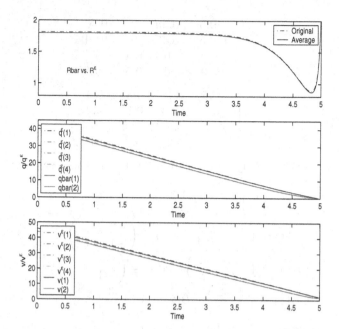

FIGURE 10.2. Discrete-Time LQG (Vector-Valued)

compared with the optimal solution to the original problem that requires solving m_0 Riccati equations. For example, if $m_0 = 120$, $l_0 = 10$, $m_1 = m_2 = \cdots = m_{10} = 10$, and $m_* = 20$, then instead of solving 120 Riccati equations jointly, one only has to deal with 10 Riccati equations. This is the most attractive feature of our approach. Furthermore, the asymptotic optimality ensures that such an approximation is almost as good as the optimal one for sufficiently small ε.

Note that in order to implement this procedure in application, the value of ε need not be too small. For example, if all the coefficients are of order 1, then $\varepsilon = 0.1$ should yield a decent approximation.

10.5   Proofs of Results

Proof of Lemma 10.2. First of all, (a) follows directly from the system of equations in (10.14) for small ε and the positive definiteness of D. To verify (b), note that by setting $u_k = 0$ in (10.1), we can show that

$$E|x_k|^2 \leq K_T(|x_0|^2 + 1), \tag{10.30}$$

for some K_T. It follows that for all x,

$$\lambda_k(x, \alpha) \leq J^\varepsilon(x, \alpha, 0) \leq K_{0,T}(|x|^2 + 1),$$

for some $K_{0,T}$. This implies in view of the quadratic representation of λ_k,

$$x' R_k^\varepsilon(\alpha) x \leq 2K_{0,T}$$

for any x with $|x| = 1$. The uniform boundedness of $R_k^\varepsilon(\alpha)$ follows.   □

Proof of Theorem 10.3. For any $t, s \in [0, T]$ satisfying $t + s \leq T$, we use $(t+s)/\varepsilon$ and t/ε in lieu of $\lfloor (t+s)/\varepsilon \rfloor$ and $\lfloor t/\varepsilon \rfloor$ for simplicity. We first show that $R^\varepsilon(\cdot, \alpha)$ is equicontinuous in the extended sense (see Definition 14.34). Note that $v^\varepsilon(t + s, x, \alpha)$ and $v^\varepsilon(t, x, \alpha)$ are the minimal costs defined in (10.21). Using (10.13), (10.30), and the boundedness of $M(\cdot)$, $N(\cdot)$ and $R_n^\varepsilon(\cdot)$, we have

$$
\begin{aligned}
|v^\varepsilon(t + s, x, \alpha) - v^\varepsilon(t, x, \alpha)| &= \left| \lambda_{(t+s)/\varepsilon}(x, \alpha) - \lambda_{t/\varepsilon}(x, \alpha) \right| \\
&\leq Ks(1 + |x|^2) + O(\varepsilon).
\end{aligned}
$$

Given x, for any $\eta > 0$, there is a $\delta > 0$ such that

$$\limsup_{\varepsilon \to 0} \sup_{\substack{0 \leq s \leq \delta \\ 0 \leq t+s \leq T}} |v^\varepsilon(t + s, x, \alpha) - v^\varepsilon(t, x, \alpha)| \leq \eta.$$

Thus, for each (x, α), $v^\varepsilon(\cdot, x, \alpha)$ is equicontinuous in the extended sense; see Definition 14.34. Note that

$$v^\varepsilon(t, x, \alpha) = x' R^\varepsilon(t, \alpha) x + q^\varepsilon(t, \alpha), \tag{10.31}$$

taking $x = 0$ in the above yields the equicontinuity of $q^\varepsilon(\cdot, x, \alpha)$ in the extended sense. Since $x' R^\varepsilon(t, \alpha) x = v^\varepsilon(t, x, \alpha) - q^\varepsilon(t, \alpha)$, the quadratic form is equicontinuous in the extended sense. By repeatedly choosing the appropriate vector x, we can show that all the entries of $R^\varepsilon(t, \alpha)$ are equicontinuous in the extended sense. This implies that, for each $\alpha \in \mathcal{M}$, $\{R^\varepsilon(\cdot, \alpha)\}$ is uniformly bounded. In view of Theorem 14.35, there is a subsequence that converges uniformly on $[0, T]$ to a continuous limit $R^0(\cdot, \alpha)$. We next characterize the limit. First rewrite (10.14) as follows:

$$
\begin{aligned}
R_k^\varepsilon(\alpha) &= \widehat{R}_{k+1}^\varepsilon(\alpha) + \varepsilon A'(\alpha) \widehat{R}_{k+1}^\varepsilon(\alpha) + \varepsilon \widehat{R}_{k+1}^\varepsilon(\alpha) A(\alpha) + \varepsilon M(\alpha) \\
&\quad - \varepsilon \widehat{R}_{k+1}^\varepsilon(\alpha) B(\alpha) \left(N(\alpha) + \varepsilon B'(\alpha) \widehat{R}_{k+1}^\varepsilon(\alpha) B(\alpha) \right)^{-1} \\
&\quad \times B'(\alpha) \widehat{R}_{k+1}^\varepsilon(\alpha) + O(\varepsilon^2).
\end{aligned}
\tag{10.32}
$$

Moreover, using $Q^\varepsilon = (q^{\varepsilon, ij})$ given in (10.17) instead of $p^{\varepsilon, ij}$, we have

$$\widehat{R}_{k+1}^\varepsilon(\alpha) = R_{k+1}^\varepsilon(\alpha) + \varepsilon \sum_{j \in \mathcal{M}} q^{\varepsilon, \alpha j} R_{k+1}^\varepsilon(j), \tag{10.33}$$

Using (10.33), we write

$$\varepsilon A'(\alpha) \widehat{R}_{k+1}^\varepsilon(\alpha) = \varepsilon A'(\alpha) R_{k+1}^\varepsilon(\alpha) + g_k^\varepsilon,$$

where
$$g_k^\varepsilon = \varepsilon^2 A'(\alpha) \sum_{j \in \mathcal{M}} q^{\varepsilon,\alpha j} R_{k+1}^\varepsilon(j).$$

Similarly, we can treat the other terms involving $R_{k+1}^\varepsilon(\alpha)$. We write (10.32) as

$$
\begin{aligned}
R_{k+1}^\varepsilon(\alpha) = \ & R_k^\varepsilon(\alpha) - \varepsilon A'(\alpha) R_{k+1}^\varepsilon(\alpha) - \varepsilon R_{k+1}^\varepsilon(\alpha) A(\alpha) - \varepsilon M(\alpha) \\
& + \varepsilon R_{k+1}^\varepsilon(\alpha) B(\alpha) \left(N(\alpha) + O(\varepsilon)\right)^{-1} B'(\alpha) R_{k+1}^\varepsilon(\alpha) \\
& - \varepsilon Q^\varepsilon R_{k+1}^\varepsilon(\cdot)(\alpha) + \overline{g}_k^\varepsilon + O(\varepsilon^2),
\end{aligned}
$$

where $\overline{g}_k^\varepsilon$ denotes a collection of the terms involves g_k^ε. Summing both sides from $\lfloor t/\varepsilon \rfloor$ to $\lfloor T/\varepsilon \rfloor - 1$ gives

$$
\begin{aligned}
R^\varepsilon(t,\alpha) = \ & D + \varepsilon \sum_{k=\lfloor t/\varepsilon \rfloor}^{\lfloor T/\varepsilon \rfloor - 1} \left[A'(\alpha) R_{k+1}^\varepsilon(\alpha) + R_{k+1}^\varepsilon(\alpha) A(\alpha) + M(\alpha) \right] \\
& - \varepsilon \sum_{k=\lfloor t/\varepsilon \rfloor}^{\lfloor T/\varepsilon \rfloor - 1} R_{k+1}^\varepsilon(\alpha) B(\alpha) (N(\alpha) + O(\varepsilon))^{-1} B'(\alpha) R_{k+1}^\varepsilon(\alpha) \\
& + \varepsilon \sum_{k=\lfloor t/\varepsilon \rfloor}^{\lfloor T/\varepsilon \rfloor - 1} Q^\varepsilon R_{k+1}^\varepsilon(\cdot)(\alpha) + G^\varepsilon(t) + O(\varepsilon),
\end{aligned}
$$

$$(10.34)$$

where
$$G^\varepsilon(t) = \sum_{k=\lfloor t/\varepsilon \rfloor}^{\lfloor T/\varepsilon \rfloor - 1} \overline{g}_k^\varepsilon.$$

Multiplying both sides of (10.34) by ε and send $\varepsilon \to 0$, we have

$$\varepsilon^2 \sum_{k=\lfloor t/\varepsilon \rfloor}^{\lfloor T/\varepsilon \rfloor - 1} Q^\varepsilon R_{k+1}^\varepsilon(\cdot)(\alpha) \to 0.$$

This implies
$$G^\varepsilon(t) \to 0 \text{ for all } t \in [0, T].$$

Moreover, we have

$$\varepsilon \sum_{k=\lfloor t/\varepsilon \rfloor}^{\lfloor T/\varepsilon \rfloor - 1} \widetilde{Q} R_{k+1}^\varepsilon(\cdot)(\alpha) \to 0.$$

Recall that $R^0(t,\alpha)$ is the limit of $R^\varepsilon(t,\alpha)$. It follows that for all t,

$$\int_t^T \widetilde{Q} R^0(s,\cdot)(\alpha) ds = 0.$$

Thus, in view of the continuity of $R^0(t, \alpha)$, we obtain that for all $t \in [0, T]$,

$$\widetilde{Q} R^0(t, \cdot)(\alpha) = 0. \qquad (10.35)$$

Following the irreducibility of $\widetilde{Q}^i$, we have $R^0(t, s_{ij}) = R^0(t, s_{ij_1})$ for any $j_1 = 1, \ldots, m_i$. That is, the limit is independent of j. With a slight abuse of notation, we denote $R^0(t, s_{ij}) = R^0(t, i)$ for $i = 1, \ldots, l_0$. Using this notation, it follows that the last m_* rows in (10.35) are equivalent to

$$R^0(t, s_{*j}) = a^{1,j} R^0(t, 1) + \cdots + a^{l_0, j} R^0(t, l_0). \qquad (10.36)$$

It remains to show that $R^\varepsilon(t, s_{ij}) \to R^0(t, i) = \overline{R}(t, i)$. For each $i = 1, \ldots, l_0$ and $j = 1, \ldots, m_i$, multiplying both sides of the equation corresponding to $\alpha = s_{ij}$ in (10.34) by ν^{ij} and then taking summation over the index j, we have

$$\sum_{j=1}^{m_i} \nu^{ij} R^\varepsilon(t, s_{ij})$$

$$= D + \varepsilon \sum_{t/\varepsilon}^{\lfloor T/\varepsilon \rfloor - 1} \left[\sum_{j=1}^{m_i} \nu^{ij} A'(s_{ij}) R^\varepsilon_{k+1}(s_{ij}) \right.$$

$$+ \sum_{j=1}^{m_i} \nu^{ij} R^\varepsilon_{k+1}(s_{ij}) A(s_{ij}) + \left. \sum_{j=1}^{m_i} \nu^{ij} M(s_{ij}) \right]$$

$$- \varepsilon \sum_{t/\varepsilon}^{\lfloor T/\varepsilon \rfloor - 1} \sum_{j=1}^{m_i} \nu^{ij} R^\varepsilon_{k+1}(s_{ij}) B(s_{ij}) (N(s_{ij}))^{-1} B'(s_{ij}) R^\varepsilon_{k+1}(s_{ij})$$

$$+ \varepsilon \sum_{t/\varepsilon}^{\lfloor T/\varepsilon \rfloor - 1} \sum_{j=1}^{m_i} \nu^{ij} Q^\varepsilon R^\varepsilon_{k+1}(\cdot)(s_{ij}) + o_\varepsilon(1),$$

where $o_\varepsilon(1) \to 0$ as $\varepsilon \to 0$. Note that

$$\mathrm{diag}(\nu^1, \ldots, \nu^{l_0}, 0_{m_* \times m_*}) \widetilde{Q} = 0.$$

This together with (10.36) implies that

$$\mathrm{diag}(\nu^1, \ldots, \nu^{l_0}, 0_{m_* \times m_*}) Q^\varepsilon \widetilde{\mathbb{1}}_* = \mathrm{diag}(\overline{Q}_*, 0_{m_* \times m_*}).$$

Noting the uniform convergence of $R^\varepsilon(t, s_{ij}) \to R^0(t, i)$ and $\sum_{j=1}^{m_i} \nu^{ij} = 1$, we obtain (10.19). Thus, the uniqueness of the Riccati equation implies that $R^0(t, i) = \overline{R}(t, i)$. As a result, $R^\varepsilon(t, s_{ij}) \to \overline{R}(t, i)$. $\square$

Proof of Corollary 10.4. Following from (10.16), (10.21), and (10.31), the convergence of $R^\varepsilon(\cdot, \alpha)$ and $q^\varepsilon(\cdot, \alpha)$, we obtain the convergence of the interpolated sequence of value functions $v^\varepsilon(\cdot, x, \alpha)$. $\square$

Proof of Theorem 10.5. First, we need to derive some convergence results. Let

$$F(t, \alpha) = \begin{cases} A(s_{ij}) - B(s_{ij})N^{-1}(s_{ij})B'(s_{ij})\overline{R}(t, i), & \text{if } \alpha = s_{ij}, \\ A(s_{*j}) - B(s_{*j})N^{-1}(s_{*j})B'(s_{*j})\overline{R}(t, *j), & \text{if } \alpha = s_{*j}. \end{cases}$$
(10.37)

Then under the optimal control $U^\circ(t, x)$ in (10.25), the limit control system can be written as

$$d\overline{x}(t) = \sum_{i=1}^{l_0} \sum_{j=1}^{m_i} \nu^{ij} I_{\{\overline{\alpha}(t)=i\}} F(t, s_{ij})\overline{x}(t)dt + \sigma dw(t), \quad \overline{x}(0) = x. \quad (10.38)$$

Let

$$\widetilde{F}(t, i) = \sum_{j=1}^{m_i} \nu^{ij} F(t, s_{ij}), \text{ for } i = 1, \dots, l_0.$$

We rewrite (10.38) as

$$d\overline{x}(t) = \widetilde{F}(t, \overline{\alpha}(t))\overline{x}(t)dt + \sigma dw(t). \quad (10.39)$$

Under feedback control (10.28), the system equation of the original discrete-time problem (10.1) becomes

$$x_{k+1} = x_k + \varepsilon F(k\varepsilon, \alpha_k^\varepsilon)x_k + \sqrt{\varepsilon}\xi_k, \quad x_0 = x. \quad (10.40)$$

Equivalently, by separating the recurrent and transient terms, we have

$$\begin{aligned} x_{k+1} = x_k &+ \sum_{i=1}^{l_0} \sum_{j=1}^{m_i} I_{\{\alpha_k^\varepsilon = s_{ij}\}} F(k\varepsilon, s_{ij})x_k \\ &+ \sum_{j=1}^{m_*} I_{\{\alpha_k^\varepsilon = s_{*j}\}} F(k\varepsilon, s_{*j})x_k + \sqrt{\varepsilon}\xi_k. \end{aligned} \quad (10.41)$$

Define the aggregated process $\overline{\alpha}_k^\varepsilon$ as in (6.18). Define also the piecewise constant interpolated processes $\alpha^\varepsilon(\cdot)$, $\overline{\alpha}^\varepsilon(\cdot)$, and $x^\varepsilon(\cdot)$ as

$$\alpha^\varepsilon(t) = \alpha_k^\varepsilon, \quad \overline{\alpha}^\varepsilon(t) = \overline{\alpha}_k^\varepsilon, \text{ and } x^\varepsilon(t) = x_k, \text{ for } t \in [k\varepsilon, k\varepsilon + \varepsilon). \quad (10.42)$$

To proceed, we introduce an intermediate auxiliary system defined by

$$d\overline{x}^\varepsilon(t) = \sum_{i=1}^{l_0} \sum_{j=1}^{m_i} \nu^{ij} I_{\{\alpha^\varepsilon(t) \in \mathcal{M}_i\}} F(t, s_{ij})\overline{x}^\varepsilon(t)dt + \sigma dw(t), \quad \overline{x}^\varepsilon(0) = x,$$
(10.43)

or,

$$d\overline{x}^\varepsilon(t) = \widetilde{F}(t, \overline{\alpha}^\varepsilon(t))\overline{x}^\varepsilon(t)dt + \sigma dw(t), \quad \overline{x}^\varepsilon(0) = x. \quad (10.44)$$

By a direct calculation, it can be shown that

$$\sup_{0 \le t \le T} E|x^\varepsilon(t)|^4 < \infty,$$

$$\sup_{0 \le t \le T} E|\overline{x}^\varepsilon(t)|^4 < \infty, \tag{10.45}$$

$$\sup_{0 \le t \le T} E|\overline{x}(t)|^4 < \infty.$$

Next, let us establish an estimate of $E|x^\varepsilon(t) - \overline{x}^\varepsilon(t)|$. For $0 \le t \le T$, in view of (10.41) and (10.42), we have

$$
\begin{aligned}
x^\varepsilon(t) &= x + \sum_{i=1}^{l_0} \sum_{j=1}^{m_i} \int_0^t I_{\{\alpha^\varepsilon(s)=s_{ij}\}} F(s, s_{ij}) x^\varepsilon(s) ds \\
&\quad - \sum_{i=1}^{l_0} \sum_{j=1}^{m_i} \int_{\lfloor t/\varepsilon \rfloor \varepsilon}^t I_{\{\alpha^\varepsilon(s)=s_{ij}\}} F(s, s_{ij}) x^\varepsilon(s) ds \\
&\quad + \sum_{i=1}^{l_0} \sum_{j=1}^{m_i} \sum_{k=0}^{\lfloor t/\varepsilon \rfloor - 1} \int_{k\varepsilon}^{(k+1)\varepsilon} I_{\{\alpha_k^\varepsilon=s_{ij}\}} \Big[F(k\varepsilon, s_{ij}) - F(s, s_{ij}) \Big] x_k ds \\
&\quad + \sum_{j=1}^{m_*} \varepsilon \left(\sum_{k=0}^{\lfloor t/\varepsilon \rfloor - 1} I_{\{\alpha_k^\varepsilon=s_{*j}\}} F(k\varepsilon, s_{*j}) x_k \right) \\
&\quad + \sqrt{\varepsilon} \sum_{k=0}^{\lfloor t/\varepsilon \rfloor - 1} \xi_k.
\end{aligned}
$$

On the other hand, we rewrite (10.43) in terms of the corresponding integrals

$$
\begin{aligned}
\overline{x}^\varepsilon(t) &= x + \sum_{i=1}^{l_0} \sum_{j=1}^{m_i} \int_0^t \nu^{ij} I_{\{\alpha^\varepsilon(s) \in \mathcal{M}_i\}} F(s, s_{ij}) \overline{x}^\varepsilon(s) ds \\
&\quad + \sigma \int_0^t dw(s).
\end{aligned}
$$

It is easy to see from definition (10.37) and Lemma 10.2 that $|F(t, \alpha)| \le K$. In view of the Skorohod representation (see Theorem 14.5), we can choose a probability space such that

$$\left(\sqrt{\varepsilon} \sum_{k=0}^{\lfloor t/\varepsilon \rfloor - 1} \xi_k, \overline{\alpha}^\varepsilon(\cdot) \right) \to \left(\sigma \int_0^t dw(s), \overline{\alpha}(\cdot) \right) \text{ w.p.1.} \tag{10.46}$$

We have

$$E|x^\varepsilon(t) - \overline{x}^\varepsilon(t)| \le$$

$$K \int_0^t E|x^\varepsilon(s) - \overline{x}^\varepsilon(s)|ds$$

$$+K \sum_{i=1}^{l_0} \sum_{j=1}^{m_i} E \left| \int_0^t \left(I_{\{\alpha^\varepsilon(s)=s_{ij}\}} - \nu^{ij} I_{\{\overline{\alpha}^\varepsilon(s)\in\mathcal{M}_i\}} \right)\overline{x}^\varepsilon(s)ds \right|$$

$$+K \int_{\lfloor t/\varepsilon\rfloor\varepsilon}^t E|x^\varepsilon(s)|ds$$

$$+K \sum_{i=1}^{l_0} \sum_{j=1}^{m_i} \sum_{k=0}^{\lfloor t/\varepsilon\rfloor-1} \int_{k\varepsilon}^{(k+1)\varepsilon} |F(k\varepsilon, s_{ij}) - F(s, s_{ij})| E|x_k|ds \tag{10.47}$$

$$+K \sum_{j=1}^{m_*} \varepsilon \left(\sum_{k=0}^{\lfloor t/\varepsilon\rfloor-1} E|I_{\{\alpha_k^\varepsilon=s_{*j}\}} x_k| \right)$$

$$+E \left| \sqrt{\varepsilon} \sum_{k=0}^{\lfloor t/\varepsilon\rfloor-1} \xi_k - \sigma \int_0^t dw(s) \right|.$$

Integrating the term in the second line above by parts yields

$$E \left| \int_0^t \left(I_{\{\alpha^\varepsilon(s)=s_{ij}\}} - \nu^{ij} I_{\{\overline{\alpha}^\varepsilon(s)\in\mathcal{M}_i\}} \right) \overline{x}^\varepsilon(s)ds \right|$$

$$\le E \left| \overline{x}^\varepsilon(t) \int_0^t \left(I_{\{\alpha^\varepsilon(s)=s_{ij}\}} - \nu^{ij} I_{\{\overline{\alpha}^\varepsilon(s)\in\mathcal{M}_i\}} \right) ds \right|$$

$$+E \left| \int_0^t \left(\int_0^s \left(I_{\{\alpha^\varepsilon(\tau)=s_{ij}\}} - \nu^{ij} I_{\{\overline{\alpha}^\varepsilon(\tau)\in\mathcal{M}_i\}} \right) d\tau \right) d\overline{x}^\varepsilon(s) \right|.$$

By virtue of the Cauchy–Schwarz inequality, (10.45), and Proposition 6.5, we have that as $\varepsilon \to 0$,

$$E \left| \overline{x}^\varepsilon(t) \int_0^t \left(I_{\{\alpha^\varepsilon(s)=s_{ij}\}} - \nu^{ij} I_{\{\overline{\alpha}^\varepsilon(s)\in\mathcal{M}_i\}} \right) ds \right|$$

$$\le \left(E|\overline{x}^\varepsilon(t)|^2 \right)^{\frac{1}{2}} \left(E \left(\int_0^t \left(I_{\{\alpha^\varepsilon(s)=s_{ij}\}} - \nu^{ij} I_{\{\overline{\alpha}^\varepsilon(s)\in\mathcal{M}_i\}} \right) ds \right)^2 \right)^{\frac{1}{2}} \to 0.$$

Similarly, in view of (10.43), as $\varepsilon \to 0$,

$$E \left| \int_0^t \left(\int_0^s \left(I_{\{\alpha^\varepsilon(\tau)=s_{ij}\}} - \nu^{ij} I_{\{\overline{\alpha}^\varepsilon(\tau)\in\mathcal{M}_i\}} \right) d\tau \right) d\overline{x}^\varepsilon(s) \right|$$

$$\le K \int_0^t E \left| \left(\int_0^s \left(I_{\{\alpha^\varepsilon(\tau)=s_{ij}\}} - \nu^{ij} I_{\{\overline{\alpha}^\varepsilon(\tau)\in\mathcal{M}_i\}} \right) d\tau \right) \overline{x}^\varepsilon(s) \right| ds$$

$$+K \left(\int_0^t E \left(\int_0^s \left(I_{\{\alpha^\varepsilon(\tau)=s_{ij}\}} - \nu^{ij} I_{\{\overline{\alpha}^\varepsilon(\tau)\in\mathcal{M}_i\}} \right) d\tau \right)^2 ds \right)^{\frac{1}{2}} \to 0.$$

Thus, we obtain

$$E\left|\int_0^t \left(I_{\{\alpha^\varepsilon(s)=s_{ij}\}} - \nu^{ij}I_{\{\overline{\alpha}^\varepsilon(s)\in\mathcal{M}_i\}}\right)\overline{x}^\varepsilon(s)ds\right| \to 0 \quad \text{as } \varepsilon \to 0. \quad (10.48)$$

We claim that the rest terms in (10.47) all go to zero as $\varepsilon \to 0$. In fact, for the third line, in view of (10.45),

$$\int_{\lfloor t/\varepsilon\rfloor\varepsilon}^t E|x^\varepsilon(s)|ds \le K\varepsilon(t/\varepsilon - \lfloor t/\varepsilon\rfloor) \le K\varepsilon \to 0.$$

For the fourth line, in view of (10.19) and (10.37), it is readily seen that $F(t,i)$ is Lipschitz in t, i.e., $|F(t,s_{ij}) - F(s,s_{ij})| \le K|t-s|$. Then

$$\sum_{k=0}^{\lfloor t/\varepsilon\rfloor-1} \int_{k\varepsilon}^{(k+1)\varepsilon} |F(k\varepsilon,s_{ij}) - F(s,s_{ij})|E|x_k|ds$$

$$\le K \sum_{k=0}^{\lfloor t/\varepsilon\rfloor-1} \int_{k\varepsilon}^{(k+1)\varepsilon} (s-k\varepsilon)ds \le K \sum_{k=0}^{\lfloor t/\varepsilon\rfloor-1} \varepsilon^2 \le K\lfloor t/\varepsilon\rfloor\varepsilon^2 \to 0.$$

For the fifth line of (10.47), by virtue of Proposition 6.5, we have

$$P(\alpha_k^\varepsilon = s_{*j}) \le K(\varepsilon + \lambda^k), \quad (10.49)$$

for some constants K and $0 < \lambda < 1$ that are independent of ε and k. Then, using Cauchy–Schwarz inequality, we have

$$\varepsilon \sum_{k=0}^{\lfloor t/\varepsilon\rfloor-1} E|I_{\{\alpha_k^\varepsilon=s_{*j}\}}x_k| \le \varepsilon \sum_{k=0}^{\lfloor t/\varepsilon\rfloor-1} (P(\alpha_k^\varepsilon = s_{*j}))^{\frac{1}{2}} (E|x_k|^2)^{\frac{1}{2}}$$

$$\le K\varepsilon \sum_{k=0}^{\lfloor T/\varepsilon\rfloor-1} (\sqrt{\varepsilon} + (\sqrt{\lambda})^k)$$

$$\le K\left(\sqrt{\varepsilon}T + \frac{\varepsilon}{1-\sqrt{\lambda}}\right) \to 0.$$

The last line of (10.47) goes to zero due to the convergence in (10.46). Now, applying Gronwall's inequality to (10.47), and using the estimates above, we obtain

$$E|x^\varepsilon(t) - \overline{x}^\varepsilon(t)|^2 \to 0 \quad \text{as } \varepsilon \to 0. \quad (10.50)$$

In view of (10.41) and (10.42), $\{x^\varepsilon(\cdot)\}$ is tight. Then (10.50) implies that $\{\overline{x}^\varepsilon(\cdot)\}$ is also tight. By virtue of this tightness and the weak convergence of $\overline{\alpha}^\varepsilon(\cdot)$ to $\overline{\alpha}(\cdot)$, we can show that $(\overline{x}^\varepsilon(\cdot), \overline{\alpha}^\varepsilon(\cdot))$ converges weakly to $(\overline{x}(\cdot), \overline{\alpha}(\cdot))$. Furthermore, Skorohod representation (without changing notation) can be used.

To complete the proof of the theorem, recall that

$$\lambda_0^\varepsilon(x,\alpha) = v^\varepsilon(0,x,\alpha) = v^\varepsilon(s,x,\alpha)|_{s=0}.$$

For any $\alpha \in \mathcal{M}_i$, we have

$$|J^\varepsilon(x, \alpha, u^\varepsilon(\cdot)) - \lambda_0^\varepsilon(x, \alpha)| \leq |J^\varepsilon(x, \alpha, u^\varepsilon(\cdot)) - v(0, x, i)|$$
$$+ |v^\varepsilon(0, x, \alpha) - v(0, x, \alpha)|;$$

for any $\alpha \in \mathcal{M}_*$ for some $j \in \{s_{*1}, \ldots, s_{*m_*}\}$, and we have

$$|J^\varepsilon(x, s_{*j}, u^\varepsilon(\cdot)) - \lambda_0^\varepsilon(x, s_{*j})| \leq |J^\varepsilon(x, s_{*j}, u^\varepsilon(\cdot)) - v(0, x, *j)|$$
$$+ |v^\varepsilon(0, x, s_{*j}) - v(0, x, *j)|.$$

In view of the Corollary 10.4, as $\varepsilon \to 0$,

$$|v^\varepsilon(0, x, \alpha) - v(0, x, i)| \to 0 \text{ for } \alpha \in \mathcal{M}_i, \; i = 1, \ldots, l_0,$$
$$|v^\varepsilon(0, x, s_{*j}) - v(0, x, *j)| \to 0 \text{ for } j = 1, \ldots, m_*.$$

Thus, to prove (10.29), it suffices to show

$$|J^\varepsilon(x, \alpha, u^\varepsilon(\cdot)) - v(0, x, i)| \to 0 \text{ for } \alpha \in \mathcal{M}_i, \; i = 1, \ldots, l_0, \text{ and}$$
$$|J^\varepsilon(x, s_{*j}, u^\varepsilon(\cdot)) - v(0, x, *j)| \to 0, \text{ for } j = 1, \ldots, m_*.$$

Let

$$S(t, \alpha) = \begin{cases} M(s_{ij}) + \overline{R}(t, i)B(s_{ij})N^{-1}(s_{ij})B'(s_{ij})\overline{R}(t, i), & \text{if } \alpha = s_{ij}, \\ M(s_{*j}) + \overline{R}(t, *j)B(s_{*j})N^{-1}(s_{*j})B'(s_{*j})\overline{R}(t, *j), & \text{if } \alpha = s_{*j}. \end{cases}$$
$$(10.51)$$

Then under the constructed control $u_k^\varepsilon(x_k, \alpha)$ given in (10.28), we have that for any $\alpha \in \mathcal{M}$,

$$J^\varepsilon(x, \alpha, u^\varepsilon(\cdot))$$
$$= \sum_{i=1}^{l_0} \sum_{j=1}^{m_i} E\left(\sum_{k=0}^{\lfloor T/\varepsilon \rfloor - 1} \int_{k\varepsilon}^{(k+1)\varepsilon} I_{\{\alpha^\varepsilon(t)=s_{ij}\}} x^{\varepsilon,'}(t)S(t, s_{ij})x^\varepsilon(t)dt \right)$$
$$+ \sum_{i=1}^{l_0} \sum_{j=1}^{m_i} E\left[\sum_{k=0}^{\lfloor T/\varepsilon \rfloor - 1} \int_{k\varepsilon}^{(k+1)\varepsilon} I_{\{\alpha_k^\varepsilon=s_{ij}\}} x_k' [S(k\varepsilon, s_{ij}) - S(t, s_{ij})]x_k dt \right]$$
$$+ \sum_{j=1}^{m_*} E\left(\varepsilon \sum_{k=0}^{\lfloor T/\varepsilon \rfloor - 1} I_{\{\alpha_k^\varepsilon=s_{*j}\}} x_k' S(k\varepsilon, s_{*j})x_k \right) + E x_{\lfloor T/\varepsilon \rfloor}' D x_{\lfloor T/\varepsilon \rfloor}.$$

Recall that $U^\circ(\cdot)$ is the optimal control of the limit system; we have

$$
\begin{aligned}
&v(0, x, i) \\
&= J(0, x, i, U^\circ(\cdot)) \\
&= E\left[\int_0^T \left[\overline{x}'(t)\overline{M}(\overline{\alpha}(t))\overline{x}(t) + \tilde{N}(\overline{\alpha}(t), U^\circ(t))\right]dt + \overline{x}'(T)D\overline{x}(T)\right] \\
&= \sum_{i=1}^{l_0}\sum_{j=1}^{m_i} E\int_0^T \nu^{ij} I_{\{\overline{\alpha}(t)=i\}}\overline{x}'(t)S(t, s_{ij})\overline{x}(t)dt + E\left(\overline{x}'(T)D\overline{x}(T)\right).
\end{aligned}
$$

Define $\overline{v}^\varepsilon(0, x, i)$ by replacing $(\overline{x}(\cdot), \overline{\alpha}(\cdot))$ by $(\overline{x}^\varepsilon(\cdot), \overline{\alpha}^\varepsilon(\cdot))$ in $v(0, x, i)$, i.e.,

$$
\begin{aligned}
&\overline{v}^\varepsilon(0, x, i) \\
&= \sum_{i=1}^{l_0}\sum_{j=1}^{m_i} E\int_0^T \nu^{ij} I_{\{\overline{\alpha}^\varepsilon(t)=i\}}\overline{x}^{\varepsilon,'}(t)S(t, s_{ij})\overline{x}^\varepsilon(t)dt + E\left(\overline{x}^{\varepsilon,'}(T)D\overline{x}^\varepsilon(T)\right).
\end{aligned}
$$

Using the weak convergence of $(\overline{x}^\varepsilon(\cdot), \overline{\alpha}^\varepsilon(\cdot)) \to (\overline{x}(\cdot), \overline{\alpha}(\cdot))$, we can show that

$$
\overline{v}^\varepsilon(0, x, i) \to v(0, x, i),
$$

as $\varepsilon \to 0$. It remains to show that $|J^\varepsilon(x, \alpha, u^\varepsilon(\cdot)) - \overline{v}^\varepsilon(0, x, i)| \to 0$. Note that for $\alpha \in \mathcal{M}_i$,

$$
\begin{aligned}
&|J^\varepsilon(x, \alpha, u^\varepsilon(\cdot)) - \overline{v}^\varepsilon(0, x, i)| \\
&\leq \left|E\left(x'_{\lfloor T/\varepsilon\rfloor}Dx_{\lfloor T/\varepsilon\rfloor} - \overline{x}^{\varepsilon,'}(T)D\overline{x}^\varepsilon(T)\right)\right| \\
&\quad + \sum_{j=1}^{m_*}\varepsilon \sum_{k=0}^{\lfloor T/\varepsilon\rfloor-1}\left|E\left(I_{\{\alpha_k^\varepsilon=s_{*j}\}}x_k'S(k\varepsilon, s_{*j})x_k\right)\right| \\
&\quad + \sum_{i=1}^{l_0}\sum_{j=1}^{m_i}\sum_{k=0}^{\lfloor T/\varepsilon\rfloor-1}\int_{k\varepsilon}^{(k+1)\varepsilon} E\left|I_{\{\alpha_k^\varepsilon=s_{ij}\}}x_k'\left[S(k\varepsilon, s_{ij}) - S(t, s_{ij})\right]x_k\right|dt \\
&\quad + \sum_{i=1}^{l_0}\sum_{j=1}^{m_i}\left|E\left(\int_0^{\lfloor T/\varepsilon\rfloor\varepsilon} I_{\{\alpha^\varepsilon(t)=s_{ij}\}}\overline{x}^{\varepsilon,'}(t)S(t, s_{ij})x^\varepsilon(t)dt\right.\right. \\
&\qquad\qquad\qquad \left.\left. - \int_0^T \nu^{ij}I_{\{\overline{\alpha}^\varepsilon(t)=i\}}\overline{x}^{\varepsilon,'}(t)S(t, s_{ij})\overline{x}^\varepsilon(t)dt\right)\right|.
\end{aligned}
\tag{10.52}
$$

Note that for $i \in \mathcal{M}$,

$$
|S(t, i)| \leq K,
$$

$$
|S(t, i) - S(s, i)| \leq K|t - s|.
$$

In view of (10.45) and (10.49), it is easy to show that as $\varepsilon \to 0$,

$$
\left|E\left(x'_{\lfloor T/\varepsilon\rfloor}Dx_{\lfloor T/\varepsilon\rfloor} - \overline{x}^{\varepsilon,'}(T)D\overline{x}^\varepsilon(T)\right)\right| \to 0,
$$

and

$$\varepsilon \sum_{k=0}^{\lfloor T/\varepsilon \rfloor -1} \left| E\left(I_{\{\alpha_k^\varepsilon = s_{*j}\}} x_k' S(k\varepsilon, s_{*j}) x_k \right) \right|$$

$$\leq K\varepsilon \sum_{k=0}^{\lfloor T/\varepsilon \rfloor -1} \left(\sqrt{\varepsilon} + (\sqrt{\lambda})^k \right) \to 0.$$

In addition,

$$\sum_{k=0}^{\lfloor T/\varepsilon \rfloor -1} \int_{k\varepsilon}^{(k+1)\varepsilon} E \left| I_{\{\alpha_k^\varepsilon = s_{ij}\}} x_k' \left[S(k\varepsilon, s_{ij}) - S(t, s_{ij}) \right] x_k \right| dt$$

$$\leq \sum_{k=0}^{\lfloor T/\varepsilon \rfloor -1} \int_{k\varepsilon}^{(k+1)\varepsilon} (t - k\varepsilon) dt$$

$$= \sum_{k=0}^{\lfloor T/\varepsilon \rfloor -1} O(\varepsilon^2) \to 0.$$

For the last line of (10.52), using the triangular inequality, we obtain

$$\left| E\left(\int_0^{\lfloor T/\varepsilon \rfloor \varepsilon} I_{\{\alpha^\varepsilon(t)=s_{ij}\}} x^{\varepsilon,'}(t) S(t, s_{ij}) x^\varepsilon(t) dt \right.\right.$$

$$\left.\left. - \int_0^T \nu^{ij} I_{\{\overline{\alpha}^\varepsilon(t)=i\}} \overline{x}^{\varepsilon,'}(t) S(t, s_{ij}) \overline{x}^\varepsilon(t) dt \right) \right|$$

$$\leq \left| E \int_0^{\lfloor T/\varepsilon \rfloor \varepsilon} I_{\{\alpha^\varepsilon(t)=s_{ij}\}} \left[x^{\varepsilon,'}(t) S(t, s_{ij}) x^\varepsilon(t) - \overline{x}^{\varepsilon,'}(t) S(t, s_{ij}) \overline{x}^\varepsilon(t) \right] dt \right|$$

$$+ \left| E \int_0^{\lfloor T/\varepsilon \rfloor \varepsilon} \left[I_{\{\alpha^\varepsilon(t)=s_{ij}\}} - \nu^{ij} I_{\{\overline{\alpha}^\varepsilon(t)=i\}} \right] \overline{x}^{\varepsilon,'}(t) S(t, s_{ij}) \overline{x}^\varepsilon(t) dt \right|$$

$$+ \left| E \int_{\lfloor T/\varepsilon \rfloor \varepsilon}^T \nu^{ij} I_{\{\overline{\alpha}^\varepsilon(t)=i\}} \overline{x}^{\varepsilon,'}(t) S(t, s_{ij}) \overline{x}^\varepsilon(t) dt \right|.$$

$$(10.53)$$

In view of (10.45) and (10.46), we can show that the second and the fourth to the sixth lines go to zero as $\varepsilon \to 0$. It can be proved (via integration by parts) as in the proof of (10.48) that the third line also converges to zero. As a result,

$$\left| J^\varepsilon(x, \alpha, u^\varepsilon(\cdot)) - \overline{v}^\varepsilon(0, x, i) \right| \to 0 \text{ for any } \alpha \in \mathcal{M}_i, \ i = 1, \ldots, l_0. \quad (10.54)$$

For $\alpha = s_{*j} \in \mathcal{M}_*$, we write

$$J^\varepsilon(x, s_{*j}, u^\varepsilon(\cdot)) = \varepsilon E[g_0 | \alpha_0^\varepsilon = s_{*j}]$$

$$+ E\left[\varepsilon \sum_{k=1}^{\lfloor T/\varepsilon \rfloor -1} g_k + x'_{\lfloor T/\varepsilon \rfloor} D x_{\lfloor T/\varepsilon \rfloor} \Big| \alpha_0^\varepsilon = s_{*j} \right]$$

$$:= O(\varepsilon) + E[G | \alpha_0^\varepsilon = s_{*j}],$$

where $g_k = x'_k M(\alpha_k^\varepsilon) x_k + u'_k N(\alpha_k^\varepsilon) u_k$ and G is the corresponding sum plus the terminal cost. By conditioning on α_1^ε, we have

$$E[G|\alpha_0^\varepsilon = s_{*j}] = \sum_{\alpha \in \mathcal{M}} E[G|\alpha_1^\varepsilon = \alpha]P(\alpha_1^\varepsilon = \alpha|\alpha_0^\varepsilon = s_{*j}].$$

It can be shown that

$$E[G|\alpha_1^\varepsilon = s_{ij}] = v(0, x, i) + O(\varepsilon),$$
$$E[G|\alpha_1^\varepsilon = s_{*j_1}] = J^\varepsilon(x, s_{*j_1}, u^\varepsilon(\cdot)) + O(\varepsilon).$$

Let

$$J^\varepsilon(*) = \Big(J^\varepsilon(x, s_{*1}, u^\varepsilon(\cdot)), \ldots, J^\varepsilon(x, s_{*m_*}, u^\varepsilon(\cdot)) \Big)'.$$

It follows that

$$J^\varepsilon(*) = (P^{*,1}, \ldots, P^{*,l_0}, P^*) \begin{pmatrix} v(0, x, 1)\mathbb{1}_{m_1} \\ \vdots \\ v(0, x, l_0)\mathbb{1}_{m_{l_0}} \\ J^\varepsilon(*) \end{pmatrix} + O(\varepsilon).$$

Using the definition of $a^{i,j}$, we have

$$J^\varepsilon(x, s_{*j}, u^\varepsilon(\cdot)) = \sum_{i=1}^{l_0} a^{i,j} v(0, x, i) + O(\varepsilon),$$

and therefore,

$$|J^\varepsilon(x, s_{*j}, u^\varepsilon(\cdot)) - v(0, x, *j)| \to 0.$$

This completes the proof. □

10.6 Notes

This chapter is based on the work of Liu, Zhang, and Yin [104]. In Yang et al. [153], weak convergence methods were used to establish the desired limit system for the numerical approximation. For classical design of linear feedback controls, see Fleming and Rishel [57] and Bertsekas [13], among others. For results on optimal control and related issues such as system stability, we refer the reader to Blair and Sworder [20], Caines and Chen [29], Mariton and Bertrand [111], Rishel [128], Ji and Chizeck [74, 75], and Yin and Zhang [158]; see also the related work of Altman and Gaitsgory [4] and the references therein. For results on singularly perturbed Markovian models with weak and strong interactions, see, for example, Abbad,

Filar, and Bielecki [2], Pan and Basar [120], and Pervozvanskii and Gaits-gory [123], among others. A continuous-time hybrid LQG problem with singularly perturbed Markov chain was considered in Zhang and Yin [181] in which a limit control problem is obtained and nearly optimal control policy is constructed using the solution of the limit problem. Additional applications to control and filtering can be found in Zhang [175].

Note that the limit Riccati equation in (10.19) is identical to the limit Riccati equation corresponding to the continuous-time LQG problem; see Zhang and Yin [181]. The LQ formulation can be extended in several directions. The diffusion coefficients σ can be replaced by $\sigma(x, \alpha)$ (i.e., state and Markov dependent). The system (10.1) can be replaced by

$$\begin{cases} x_{k+1} = x_k + \varepsilon[A(\alpha_k^\varepsilon)x_k + B(\alpha_k^\varepsilon)u_k] + \sqrt{\varepsilon}\sigma(x_k, \alpha_k^\varepsilon)\xi_k, \\ x_0 = x, \ \alpha_0^\varepsilon = \alpha, \ 0 \le k \le \lfloor T/\varepsilon \rfloor, \end{cases}$$

The main steps remain the same, whereas the state and Markov dependent diffusion can be dealt with using weak convergence methods; some of these ideas can be found in the treatments discussed in Chapter 11.

Another direction is to consider the indefinite control problems. It enables us to relax the conditions on the nonnegative definiteness and positive definiteness assumptions on (A10.1) (a). This line of work uses ideas from the work of Chen, Li, and Zhou [35].

11
Mean-Variance Controls

11.1 Introduction

This chapter is concerned with hybrid mean-variance control problems arising from portfolio selections in financial engineering. It belongs to the class of hybrid LQ control problems with indefinite control weights. The hybrid feature is represented by a Markov chain leading to a regime-switching model.

Markowitz's single period mean-variance portfolio selection model (see Markowitz [112, 113]) laid the foundation for modern finance theory. It aims to maximize terminal wealth, in the mean time to minimize risk, using variance as a criterion. It enables investors to seek the highest return upon specifying their acceptable risk levels. Markowitz's Nobel-prize-winning work has inspired numerous extensions. The original single-stage model has been extended to continuous-time formulations, in which stochastic differential equations and geometric Brownian motion models are used. Despite their extensive use, the geometric Brownian motion models have certain limitations, since the parameters, including the interest rates of bonds and the appreciation and volatility rates of stocks, are all deterministic. As a result, they are insensitive to drastic changes in the market. Financial markets tend to have trends; the simplest categorization includes up's and down's. These trends can be regarded as market "modes." In addition to trends, there are certain factor processes that reflect economic factors and that dictate the configuration or mode (regime) changes of the market. Corresponding to different regimes, the coefficients (the appreciation and volatility rates) are markedly different. The mode of the market may reflect the state of the underlying economy, the general mood of the investors, business cycles, and other economic factors. For example, a coarse division of the market modes leads to the characterization of "bullish" and "bearish;" a refinement adds more intermediate states between these two states. To take the factor process into consideration, we model the appreciation and volatility rates of the market as functions of the modes. This leads to a switching model formulated as a system of stochastic difference or differential equations whose coefficients are modulated by a discrete-time or a continuous-time Markov chain.

Motivated by these developments, we consider a class of discrete-time mean-variance portfolio selection models and consider portfolio selection policy that minimizes the overall risk with a given expected return. In addition, we explore their relationship to the continuous-time counterparts. One of the main features is that all the market coefficients are modulated by a discrete-time Markov chain that has a finite state space. Taking into account various economic factors, the state space of the Markov chain, which represents the market modes, is often large. To reduce complexity, we make use of the hierarchy of the market by lumping many states at the same hierarchical level together to form an aggregated "state." Such an aggregation substantially reduces the size of the problem to be solved. To highlight the different rates of changes, by introducing a small parameter $\varepsilon > 0$ into the transition matrix, we formulate it as a control problem with two time scales.

The rest of the chapter is arranged as follows. We begin with the formulation of the discrete-time mean-variance portfolio selection problem in Section 11.2. Section 11.3 presents the main results. We derive *a priori* bounds of certain processes, obtain the limit system using weak convergence method, and establish natural connections between the discrete-time models and continuous-time systems. To reduce complexity of the underlying systems, based on the limit system, we construct policies leading to near optimality. Section 11.4 collects the proofs of results. Additional notes and remarks are provided in Section 11.5.

11.2 Formulation

Suppose that $T > 0$ and that $\varepsilon > 0$ is a small parameter. Working with discrete time k, we consider $0 \leq k \leq \lfloor T/\varepsilon \rfloor$, where for a real number z, $\lfloor z \rfloor$ denotes the integer part of z. Let α_k^ε, for $0 \leq k \leq \lfloor T/\varepsilon \rfloor$, be a discrete-time Markov chain, which is parameterized by ε, with a finite state space $\mathcal{M} = \{1, 2, \ldots, m_0\}$. Consider a wealth model as follows. Suppose that there are $d + 1$ assets in the underlying market. One of them is a bond (or risk-free asset) and the rest of them are stocks (risky assets). Let $S_k^{\varepsilon,0}$ be the price of the bond, and $S_k^{\varepsilon,\imath}$, $\imath = 1, \ldots, d$, be the prices of the stocks at time k, respectively. Suppose that $r(\cdot,\cdot)$, $b^{\imath}(\cdot,\cdot)$, $\sigma^{\imath\jmath}(\cdot,\cdot) : \mathbb{R} \times \mathcal{M} \mapsto \mathbb{R}$, for $\imath, \jmath = 1, \ldots, d$, are some appropriate functions such that corresponding to a market mode $\alpha \in \mathcal{M}$, $r(\cdot, \alpha)$ represents the interest rate of the bond, and $b^{\imath}(\cdot, \alpha)$ and $\sigma^{\imath\jmath}(\cdot, \alpha)$ are the appreciation and volatility rates of the stocks, respectively. For $(t, \alpha) \in \mathbb{R} \times \mathcal{M}$, define

$$c^{\imath}(t,\alpha) = b^{\imath}(t,\alpha) - \frac{1}{2}\sum_{\jmath=1}^{d}(\sigma^{\imath\jmath}(t,\alpha))^2, \; \imath = 1,\ldots,d. \qquad (11.1)$$

Then $S_k^{\varepsilon,\imath}$ satisfies the following system of equations:

$$\begin{cases} S_{k+1}^{\varepsilon,0} = S_k^{\varepsilon,0} + \varepsilon r(\varepsilon k, \alpha_k^{\varepsilon})S_k^{\varepsilon,0} \\ S_0^{\varepsilon,0} = S_0^0 > 0, \\ S_{k+1}^{\varepsilon,\imath} = S_k^{\varepsilon,\imath}\exp\left(\varepsilon c^{\imath}(\varepsilon k, \alpha_k^{\varepsilon}) + \sqrt{\varepsilon}\sum_{\jmath=1}^{d}\sigma^{\imath\jmath}(\varepsilon k, \alpha_k^{\varepsilon})\xi_k^{\jmath}\right), \; \imath = 1,\ldots,d, \\ S_0^{\varepsilon,\imath} = S_0^{\imath} > 0, \end{cases}$$

$$(11.2)$$

where $\{\xi_k^{\imath}\}$, $\imath = 1, \ldots, d$, are sequences of independent and identically distributed random variables. We use the multiplicative model, which is an analogous to the geometric Brownian motion model in continuous time to ensure the nonnegativity of the stock prices.

Suppose that at time k, an investor with an initial endowment $x_0^{\varepsilon} = x_0$ holds $N^{\imath}(\varepsilon k)$ shares of the $\imath$th asset for $\imath = 0, \ldots, d$, during the time interval k to $k+1$. Then his or her wealth at time k is x_k^{ε} given by the equation

$$x_{k+1}^{\varepsilon} - x_k^{\varepsilon} = \sum_{\imath=0}^{d}N^{\imath}(\varepsilon k)(S_{k+1}^{\varepsilon,\imath} - S_k^{\varepsilon,\imath}). \qquad (11.3)$$

Equation (11.3) is based on the premise of self-financing. For a self-financed portfolio (no infusion or withdrawal of funds during the indicated time interval), the difference in total wealth between two consecutive times is purely due to the change in the prices of the stocks (see, e.g., Karatzas and Shreve [77, equation (2.2) in p. 6]). Therefore, we have

$$x_k^{\varepsilon} = \sum_{\imath=0}^{d}u_k^{\varepsilon,\imath} \text{ where } u_k^{\varepsilon,\imath} = N^{\imath}(\varepsilon k)S_k^{\varepsilon,\imath} \text{ for } \imath = 0,\ldots,d. \qquad (11.4)$$

Note that the amount allocated to the bond is completely determined by the selection strategies for the stocks

$$u_k^{\varepsilon,0} = x_k^{\varepsilon} - \sum_{\imath=1}^{d}u_k^{\varepsilon,\imath}.$$

Thus, the control variables are those associated with the stocks, i.e., $u_k^{\varepsilon,\imath}$, for $\imath = 1, \ldots, d$. Given a mean terminal wealth $Ex_{T/\varepsilon}^\varepsilon$, our objective is to minimize the variance of the terminal wealth,

$$E(x_{T/\varepsilon}^\varepsilon - Ex_{T/\varepsilon}^\varepsilon)^2 = E(x_{T/\varepsilon}^\varepsilon)^2 - (Ex_{T/\varepsilon}^\varepsilon)^2.$$

Denote by $\mathcal{F}_k^\varepsilon$, the σ-algebra generated by $\{\alpha_{k_1}^\varepsilon, \xi_{k_1} : 0 \leq k_1 < k\}$, where $\xi_k = (\xi_k^1, \ldots, \xi_k^d)'$. A portfolio

$$u_\cdot^\varepsilon = \{u_k^\varepsilon = (u_k^{\varepsilon,1}, u_k^{\varepsilon,2}, \ldots, u_k^{\varepsilon,d}) : k = 0, 1, \ldots, T/\varepsilon\}$$

is admissible if u_k^ε is $\mathcal{F}_k^\varepsilon$-measurable for each $0 \leq k \leq T/\varepsilon$ and (11.4) has a unique solution

$$x_\cdot^\varepsilon = \{x_k^\varepsilon : k = 0, 1, \ldots, T/\varepsilon\}$$

corresponding to $u_\cdot^\varepsilon$ We also call $(x_\cdot^\varepsilon, u_\cdot^\varepsilon)$ an admissible wealth-portfolio pair. Denote the class of admissible wealth-portfolio pairs by $\mathcal{A}^\varepsilon$.

The objective of the discrete-time mean-variance portfolio selection problem is to find an admissible portfolio $(x_\cdot^\varepsilon, u_\cdot^\varepsilon) \in \mathcal{A}^\varepsilon$, for an initial wealth $x_0^\varepsilon = x_0$ and an initial market mode $\alpha_0^\varepsilon = \ell_0$, such that the terminal wealth is $Ex_{T/\varepsilon}^\varepsilon = z$ for a given $z \in \mathbb{R}$, and the risk in terms of the variance of the terminal wealth, $E[x_{T/\varepsilon}^\varepsilon - z]^2$, is minimized. The problem to be solved is

$$\begin{cases} \text{Minimize } J^\varepsilon(x_0, \ell_0, u_\cdot^\varepsilon) = E[x_{T/\varepsilon}^\varepsilon - z]^2 \\ \text{subject to: } x_0^\varepsilon = x_0, \ \alpha_0^\varepsilon = \ell_0, \ Ex_{T/\varepsilon}^\varepsilon = z \text{ and} \\ (x_\cdot^\varepsilon, u_\cdot^\varepsilon) \in \mathcal{A}^\varepsilon. \end{cases} \quad (11.5)$$

We assume the following conditions.

(A11.1) Condition (HR) of Chapter 6 holds.

(A11.2) For each $\alpha \in \mathcal{M}$, $\imath, \jmath = 1, \ldots, d$, $r(\cdot, \alpha)$, $b^\imath(\cdot, \alpha)$, $\sigma^{\imath\jmath}(\cdot, \alpha)$ are real-valued continuous functions defined on $[0, T]$.

(A11.3) For each $\imath = 1, \ldots, d$, $\{\xi_k^\imath\}$ is a sequence of independent and identically distributed (i.i.d.) random variables that are independent of α_k^ε and that have mean 0 and variance 1. Moreover, for $\imath \neq \jmath$, $\{\xi_k^\imath\}$ and $\{\xi_k^\jmath\}$ are independent.

Remark 11.1. The sequences $\{\xi_k^\imath\}$ are commonly referred to as white noise. Under (A11.3), the Donsker's invariance principle implies that

$$\sqrt{\varepsilon} \sum_{k=0}^{t/\varepsilon-1} \xi_k^\imath \text{ converges weakly to a standard Brownian motion as } \varepsilon \to 0.$$

In fact, correlated noise may be dealt with. What is essential is that a functional central limit theorem holds for the scaled sequence. Sufficient conditions guaranteeing the convergence include, for example, ϕ-mixing processes with appropriate mixing rates. However, as in the hybrid filtering problem, assumption (A11.3) allows us to simplify much of the subsequent discussion.

When the state space $\mathcal{M}$ is large, the cardinality of $\mathcal{M}$ is a large number. To reduce complexity in solving the portfolio selection problem, we aggregate the states in each recurrent class $\mathcal{M}_i$ into one state, resulting in an aggregated process. Then the aggregated process has only l_0 states instead of m_0 states. If $l_0 \ll m_0$, the possible number of regimes is substantially smaller for the aggregated process. The intuitive notion of aggregation, in fact, is used in practice, where we typically merge all the states with similar properties (such as transition rates) into one "super" state; see the schematic illustration given in Figure 1.1. For example, we often put all the states together where the market is up (respectively, down) to form a "bullish" (respectively, "bearish") mode of the market.

Our effort is to show that suitably interpolated processes of the prices converge to limit processes that are switching diffusions. The limit problem is a continuous-time mean-variance portfolio selection problem with regime switching, whose optimal controls have been found in Zhou and Yin [186]. This will, in turn, enable us to construct an asymptotically optimal strategy for our original problem (11.5) based on the optimal strategy of the limit problem. In the remainder of this section, we will establish the connection of the discrete-time and continuous-time portfolio selection problems.

11.2.1   Limit Results

Using aggregation techniques, define an aggregated discrete-time process $\overline{\alpha}_k^\varepsilon$ by

$$\overline{\alpha}_k^\varepsilon = i \ \text{ if } \ \alpha_k^\varepsilon \in \mathcal{M}_i, \ i = 1, 2, \ldots, l_0.$$

Define the interpolated processes

$$S^{\varepsilon,\imath}(t) = S_k^{\varepsilon,\imath}, \ \imath = 0, \ldots, d, \ \ \overline{\alpha}^\varepsilon(t) = \overline{\alpha}_k^\varepsilon \ \text{ for } \ t \in [\varepsilon k, \varepsilon k + \varepsilon). \tag{11.6}$$

They are piecewise-constant interpolations on the interval of length ε. Similarly, for any admissible wealth-portfolio pair $(u_\cdot^\varepsilon, x_\cdot^\varepsilon)$, we define the corresponding interpolated processes

$$u^\varepsilon(t) = u_k^\varepsilon, \ x^\varepsilon(t) = x_k^\varepsilon, \ \text{ for } \ t \in [\varepsilon k, \varepsilon k + \varepsilon). \tag{11.7}$$

Due to its multiplicative nature, the $S^{\varepsilon,\imath}(\cdot)$ is not easy to work with in deriving the weak convergence result for $(S^{\varepsilon,\imath}(\cdot), \overline{\alpha}^\varepsilon(\cdot))$. To overcome these difficulties, we define auxiliary processes $y_k^{\varepsilon,\imath}$ for $\imath = 1, \ldots, d$ as

$$y_{k+1}^{\varepsilon,\imath} = y_k^{\varepsilon,\imath} + \varepsilon c^\imath(\varepsilon k, \alpha_k^\varepsilon) + \sqrt{\varepsilon} \sum_{\jmath=1}^{d} \sigma^{\imath\jmath}(\varepsilon k, \alpha_k^\varepsilon)\xi_k^\jmath, \ \ y_0^{\varepsilon,\imath} = 0, \tag{11.8}$$

where $c^i(\cdot, \cdot)$ is given by (11.1).

Using $\{y_k^{\varepsilon,i}\}$, the price for the ith stock can be written as

$$S_k^{\varepsilon,i} = S_0^i \exp(y_k^{\varepsilon,i}), \quad S_0^{\varepsilon,i} = S_0^i > 0. \tag{11.9}$$

As with the interpolation $S^{\varepsilon,i}(\cdot)$, define the interpolated processes

$$y^{\varepsilon,i}(t) = y_k^{\varepsilon,i} \text{ for } t \in [\varepsilon k, \varepsilon k + \varepsilon), \ i = 1, \ldots, d. \tag{11.10}$$

It follows that

$$S^{\varepsilon,i}(t) = S_0^i \exp(y^{\varepsilon,i}(t)).$$

That is, $S^{\varepsilon,i}(\cdot)$ is related to the interpolation of $y_k^{\varepsilon,i}$ through the exponential mapping.

To proceed, by utilizing the auxiliary process $y^{\varepsilon,i}(\cdot)$, we first show that $(y^{\varepsilon,i}(\cdot), \overline{\alpha}^{\varepsilon}(\cdot))$ converges weakly to certain limit processes, in which the appreciation and volatility rates are averaged out with respect to the stationary measures of the Markov chain. Then by using the continuous mapping theorem (see Theorem 14.21 in the appendix), we obtain the desired result.

11.2.2   Weak Convergence

We shall derive the weak convergence of interpolated processes (11.6) by first verifying the tightness of the underlying sequences and then characterizing the limits as solutions of certain martingale problems having appropriate generators. For each $i = 0, 1, \ldots, d$, we work with a pair of processes $(y^{\varepsilon,i}(\cdot), \overline{\alpha}^{\varepsilon}(\cdot))$. Let $D([0,T]; \mathbb{R} \times \overline{\mathcal{M}})$ denote the space of functions defined on $[0,T]$, taking values in $\mathbb{R} \times \overline{\mathcal{M}}$ that are right continuous, have left-hand limits, endowed with the Skorohod topology (see Section 14.2). The first result is concerned with tightness.

Theorem 11.2 *Under* (A11.1)–(A11.3), *the pair* $\{y^{\varepsilon,i}(\cdot), \overline{\alpha}^{\varepsilon}(\cdot)\}$ *is tight on* $D([0,T]; \mathbb{R} \times \overline{\mathcal{M}})$ *for each* $i = 1, \ldots, d$, *so is* $\{S^{\varepsilon,0}(\cdot), \overline{\alpha}^{\varepsilon}(\cdot)\}$.

Since for each $i = 0, 1, \ldots, d$, $\{S^{\varepsilon,0}(\cdot), \overline{\alpha}^{\varepsilon}(\cdot)\}$ and $\{y^{\varepsilon,i}(\cdot), \overline{\alpha}^{\varepsilon}(\cdot)\}$ are tight by Theorem 11.2, Prohorov's theorem (see Theorem 14.4) enables us to extract weakly convergent subsequences as $\varepsilon \to 0$. Select such convergent subsequences with limits $(S^0(\cdot), \overline{\alpha}(\cdot))$ and $(y^i(\cdot), \overline{\alpha}(\cdot))$, respectively. For notational simplicity, still use ε as the index of the subsequence. We proceed to characterize the limit processes. By virtue of the Skorohod representation (with a slight abuse of notation), we may assume without loss of generality that $(S^{\varepsilon,0}(\cdot), \overline{\alpha}^{\varepsilon}(\cdot)) \to (S^0(\cdot), \overline{\alpha}(\cdot))$ and $y^{\varepsilon,i}(\cdot) \to y^i(\cdot)$ with probability one (w.p.1), and the convergence is uniform on any compact time interval.

Denote the $d \times d$ matrix $(\sigma^{ij}(t, \alpha))$ by $\Sigma(t, \alpha)$. Define

$$
\left.
\begin{aligned}
\bar{r}(t, i) &= \sum_{j=1}^{m_i} \nu^{ij} r(t, s_{ij}), \\
\bar{b}^i(t, i) &= \sum_{j=1}^{m_i} \nu^{ij} b^i(t, s_{ij}), \\
\bar{c}^i(t, i) &= \sum_{j=1}^{m_i} \nu^{ij} c^i(t, s_{ij}),
\end{aligned}
\right\} \quad \text{for } i \in \overline{\mathcal{M}} = \{1, \ldots, l_0\}, \quad (11.11)
$$

where ν^{ij} denotes the jth component of the stationary distribution ν^i corresponding to P^i. Let $\overline{\Sigma}(t, \alpha) = (\bar{\sigma}^{ij}(t, \alpha))$ be such that

$$
\overline{\Sigma}(t, i) \overline{\Sigma}'(t, i) = \sum_{j=1}^{m_i} \nu^{ij} \Sigma(t, s_{ij}) \Sigma'(t, s_{ij}). \quad (11.12)
$$

We aim to show that the limit processes are solutions of

$$
\begin{aligned}
\frac{dS^0(t)}{dt} &= \bar{r}(t, \bar{\alpha}(t)) S^0(t), \\
S^0(0) &= S_0^0, \\
dy^i(t) &= \bar{c}^i(t, \bar{\alpha}(t)) dt + \sum_{j=1}^{d} \bar{\sigma}^{ij}(t, \bar{\alpha}(t)) dw^j(t), \\
y^i(0) &= 0, \ i = 1, \ldots, d,
\end{aligned}
\quad (11.13)
$$

respectively, where $w^i(\cdot)$ for $i = 1, \ldots, d$ are independent, scalar, standard Brownian motions. Equivalently, it suffices to show that for each $i = 0, \ldots, d$, $(y^i(\cdot), \bar{\alpha}(\cdot))$ is a solution of the martingale problem with operator $(\partial/\partial t) + \mathcal{L}^i$, where

$$
\begin{aligned}
\mathcal{L}^0 f(t, y, i) &= f_y(t, y, i) \bar{r}(t, i) y, \\
\mathcal{L}^i f(t, y, i) &= f_y(t, y, i) \bar{c}^i(t, i) + \frac{1}{2} f_{yy}(t, y, i) [\overline{\Sigma}(t, i) \overline{\Sigma}'(t, i)]^{ii} \quad (11.14) \\
&\quad + \overline{Q} f(t, y, \cdot)(i), \ 1 \le i \le d,
\end{aligned}
$$

where $f(\cdot, \cdot, i)$ is a suitable function defined on $\mathbb{R} \times \mathbb{R}$, f_y and f_{yy} denote the first and the second derivatives of f,

$$
\overline{Q} = (\bar{q}_{ij}) = \text{diag}(\nu^1, \ldots, \nu^{l_0}) Q \widetilde{\mathbb{1}},
$$

$$
[\overline{\Sigma}(t, i) \overline{\Sigma}'(t, i)]^{ii} = \sum_{j=1}^{m_i} \nu^{ij} \sum_{j=1}^{d} (\sigma^{ij}(t, s_{ij}))^2,
$$

and

$$\overline{Q}f(t,y,\cdot)(i) = \sum_{j=1}^{l_0} \overline{q}^{ij} f(t,y,j) = \sum_{j\neq i}^{l_0} \overline{q}^{ij}(f(t,y,j) - f(t,y,i)). \quad (11.15)$$

Theorem 11.3 *Assume* (A11.1)–(A11.3). *Then for each $\imath = 1,\ldots,d$, the weak limits of $(y^{\varepsilon,\imath}(\cdot), \overline{\alpha}^{\varepsilon}(\cdot))$ and $(S^{\varepsilon,0}(\cdot), \overline{\alpha}^{\varepsilon}(\cdot))$ are the unique solutions of the martingale problems with the operators $(\partial/\partial t) + \mathcal{L}^{\imath}$ for $\imath = 1,\ldots,d$ and $\imath = 0$, respectively.*

Since $S^{\varepsilon,\imath}(t) = S_0^{\imath} \exp(y^{\varepsilon,\imath}(t))$, and $\exp(\cdot)$ is a continuous function, by the continuous mapping theorem (see Theorem 14.21), $S^{\varepsilon,\imath}(\cdot)$ converges to $S^{\imath}(\cdot)$ such that $S^{\imath}(t) = S_0^{\imath} \exp(y^{\imath}(t))$, where $y^{\imath}(\cdot)$ is the limit of $y^{\varepsilon,\imath}(\cdot)$. Applying Itô's formula and noticing the relations (11.1) and (11.11), we obtain the desired result.

Corollary 11.4. *Under the conditions of Theorem 11.3, $(S^{\varepsilon,\imath}(\cdot), \overline{\alpha}^{\varepsilon}(\cdot))$ converges weakly to $(S^{\imath}(\cdot), \overline{\alpha}(\cdot))$ such that they are solutions of*

$$\frac{dS^0(t)}{dt} = \overline{r}(t,\overline{\alpha}(t))S^0(t),$$
$$S^0(0) = S_0^0,$$
$$dS^{\imath}(t) = \overline{b}^{\imath}(t,\overline{\alpha}(t))S^{\imath}(t)dt + \sum_{j=1}^{d} \overline{\sigma}^{\imath j}(t,\overline{\alpha}(t))S^{\imath}(t)dw^j(t),$$
$$S^{\imath}(0) = S_0^{\imath},\ \imath = 1,\ldots,d,$$

(11.16)

respectively, where $w^{\imath}(\cdot)$ for $\imath = 1,\ldots,d$ are independent, real-valued, standard Brownian motions.

Denote by $\mathcal{F}_t$ the σ-algebra generated by $\{\overline{\alpha}(s), w(s) : 0 \leq s \leq t\}$, where $w(s) = (w^1(s),\ldots,w^d(s))'$. A control $u(\cdot) = (u^1(\cdot),\ldots,u^d(\cdot))$ is admissible for the limit problem if $u(\cdot)$ is $\mathcal{F}_t$-adapted and

$$dx(t) = \left[\overline{r}(t,\overline{\alpha}(t))x(t) + \sum_{\imath=1}^{d}[\overline{b}^{\imath}(t,\overline{\alpha}(t)) - \overline{r}(t,\overline{\alpha}(t))u^{\imath}(t)]\right]dt$$
$$+ \sum_{j=1}^{d}\sum_{\imath=1}^{d} \overline{\sigma}^{\imath j}(t,\overline{\alpha}(t))u^{\imath}(t)dw^j(t)$$

(11.17)

has a unique solution $x(\cdot)$ corresponding to $u(\cdot)$. The $(x(\cdot), u(\cdot))$ is termed an admissible wealth-portfolio pair (for the limit problem). Denote the class of admissible wealth-portfolio pairs by $\overline{\mathcal{A}}$. The limit mean-variance portfolio selection problem is

$$\begin{cases} \text{Minimize } J(x_0, i_0, u(\cdot)) = E[x(T) - z]^2, \\ \text{subject to: } x(0) = x_0, \quad \bar{\alpha}(0) = i_0, \\ Ex(T) = z \text{ and } (x(\cdot), u(\cdot)) \in \overline{\mathcal{A}}, \end{cases} \qquad (11.18)$$

where i_0 is such that $\ell_0 \in \mathcal{M}_{i_0}$ (recall that ℓ_0 is the initial condition of the Markov chain α_0^ε in the original discrete-time problem (11.5)).

For any $(x^\varepsilon, u^\varepsilon) \in \mathcal{A}^\varepsilon$, there are the corresponding interpolated processes $(x^\varepsilon(\cdot), u^\varepsilon(\cdot))$ determined by (11.7). With a slight abuse of notation, we do not distinguish between $(x^\varepsilon, u^\varepsilon)$ and $(x^\varepsilon(\cdot), u^\varepsilon(\cdot))$, but write $(x^\varepsilon(\cdot), u^\varepsilon(\cdot)) \in \mathcal{A}^\varepsilon$ in what follows.

Corollary 11.5. *Under the conditions of Theorem* 11.3, *any sequence of admissible wealth-portfolio pairs* $(x^\varepsilon(\cdot), u^\varepsilon(\cdot)) \in \mathcal{A}^\varepsilon$ *converges weakly to* $(x(\cdot), u(\cdot))$ *that belongs to* $\overline{\mathcal{A}}$. *Moreover,*

$$Ex^\varepsilon(T) \to Ex(T) \quad and \quad E[x^\varepsilon(T) - z]^2 \to E[x(T) - z]^2 \quad as \ \varepsilon \to 0. \ (11.19)$$

11.3   Near-Efficient Portfolio

Having established that associated with the original mean-variance control problem, there is a limit problem, we can construct near-efficient portfolios for (11.5) based on the optimal portfolio for (11.18).

Recall that (11.18) is feasible (for fixed x_0 and i_0) if there is at least one portfolio satisfying all the constraints. It is finite if it is feasible and the infimum of the cost is finite. An optimal portfolio for a given z, if it exists, is called an efficient portfolio, and the corresponding (variance, expected return) pair (Var $x(T), z$) is called an efficient point. The set of all efficient points as z varies is termed efficient frontier. As can be seen, in terms of the standard control terminology, an efficient portfolio is an optimal control policy (corresponding to a particular z).

For a continuous-time mean-variance portfolio selection problem (11.18), a necessary and sufficient condition for the feasibility of the limit problem (11.18) was derived. In addition, it was proved that if it is feasible, then indeed the efficient portfolio corresponding to z exists, which can be expressed in a feedback form, i.e., a function of the time t, the wealth level x, and the market mode i. Using an efficient portfolio of the limit problem, we construct a near-efficient portfolio (i.e., a near-optimal control strategy) for the original problem. In what follows, we simply refer to the optimal portfolio selection strategy of the limit as $u^o(\cdot)$, without specifying its explicit form. For further details and explicit representation of the optimal solution, we refer the reader to Zhou and Yin [186].

Using the optimal control of the limit problem $u^\circ(t, x, i)$, we construct portfolios for the original problem as

$$\widetilde{u}^\varepsilon(t, x, \alpha) = \sum_{i=1}^{l_0} u^\circ(t, x, i) I_{\{\alpha \in \mathcal{M}_i\}}. \tag{11.20}$$

That is, $\widetilde{u}^\varepsilon(t, x, \alpha) = u^\circ(t, x, i)$ if $\alpha \in \mathcal{M}_i$, for $i = 1, 2, \ldots, l_0$. Let x_k^ε be the wealth trajectory of the original problem (11.5) under the feedback control $\widetilde{u}(\varepsilon k, x, \alpha)$. Recall that $x^\varepsilon(t)$ is the continuous-time interpolation of x_k^ε and denote the continuous-time interpolation of $\widetilde{u}(\varepsilon k, x_k^\varepsilon, \alpha_k^\varepsilon)$ by $\widetilde{u}^\varepsilon(t)$. Using such a control leads to near-optimality of (11.5). Write

$$v^\varepsilon(x_0, \ell_0) = \inf_{(x^\varepsilon, u^\varepsilon) \in \mathcal{A}} J^\varepsilon(x_0, \ell_0, u_\cdot^\varepsilon),$$

$$v(x_0, i_0) = \inf_{(x(\cdot), u(\cdot)) \in \overline{\mathcal{A}}} J(x_0, i_0, u(\cdot)).$$

Theorem 11.6 *Suppose that the conditions of Theorem 11.3 are satisfied. Then*

$$\lim_{\varepsilon \to 0} |J^\varepsilon(x_0, \ell_0, \widetilde{u}^\varepsilon(\cdot)) - v^\varepsilon(x_0, i_0)| = 0. \tag{11.21}$$

Thus, the constructed controls based on the optimal solution of the limit problem is asymptotically optimal. That is, based on the optimal portfolio selection rules of the limit problem, we devise a portfolio selection strategy for the original problem and such strategies are asymptotically optimal in the sense of (11.21).

11.4   Inclusion of Transient States

As for the hybrid filtering problem, we can also consider mean-variance control problems modulated by a Markov chain, in which the Markov chain includes transient states in addition to the l_0 ergodic classes. In this case we need to replace (A11.1) by conditions (HT) in Chapter 6.

Theorem 11.7. *Under the conditions of Theorem 11.3 with (A11.1) replaced by (HT) of Chapter 6, the conclusions of Theorem 11.3 and Corollary 11.5 continue to hold with $\overline{Q}$ replaced by $\overline{Q}_*$ defined in (6.16).*

11.5   Continuous-Time Problems

Assuming that the trading of shares takes place continuously, we can then formulate a continuous-time version of the mean-variance control problem under Markovian switching regime. Again, suppose that there are $d + 1$

assets in the underlying market. One of which is the bond and the rest of them are the stock holdings. Let $S^{\varepsilon,0}(\cdot)$ be the price of the bond, and $S^{\varepsilon,\imath}(\cdot)$, $\imath = 1, \ldots, d$, be the prices of the stocks at time t, respectively. Replace the discrete-time price model (11.2) by

$$
\begin{cases}
dS^{\varepsilon,0}(t) = r(t, \alpha^{\varepsilon}(t))S^{\varepsilon,0}(t)dt, \\
S^{\varepsilon,0}(0) = S^0 > 0, \\
dS^{\varepsilon,\imath}(t) = b^{\imath}(t, \alpha^{\varepsilon}(t))S^{\varepsilon,\imath}(t)dt + \sum_{\jmath=1}^{d} \sigma^{\imath\jmath}(t, \alpha^{\varepsilon}(t))dw^{\jmath}(t), \ \imath = 1, \ldots, d, \\
S^{\varepsilon,\imath}(0) = S^{\imath} > 0,
\end{cases}
$$
(11.22)

where $w^{\imath}(\cdot)$ are independent standard Brownian motions, $r(t, \alpha)$ is the interest rate of the bond, and $b^{\imath}(t, \alpha)$ and $\sigma^{\imath\jmath}(t, \alpha)$ represent the rates of stock appreciation and volatility of the $\imath$th stock, corresponding to a market mode $\alpha^{\varepsilon}(t) = \alpha$.

Our main concern is still the reduction of complexity. The techniques to be used are the averaging methods, e.g., in Yin, Zhang, and Badowski [162, 163]. Similar to the continuous-time hybrid filtering problem, we can deal with nonsmooth generators and carry out the averaging procedure analogues to the discrete-time case. Suppose the generator is given by

$$
Q^{\varepsilon}(t) = \frac{\widetilde{Q}(t)}{\varepsilon} + \widehat{Q}(t),
$$
(11.23)

with $\widetilde{Q}(t)$ given by

$$
\widetilde{Q}(t) = \mathrm{diag}(\widetilde{Q}^1(t), \ldots, \widetilde{Q}^{l_0}(t)).
$$
(11.24)

Define the wealth process

$$
x^{\varepsilon}(t) = \sum_{\imath=0}^{d} N^{\imath}(t)S^{\varepsilon,\imath}(t).
$$

We seek to solve the mean-variance problem:

$$
\begin{cases}
\text{Minimize } J^{\varepsilon}(x_0, l_0, u_{\cdot}^{\varepsilon}) = E[x^{\varepsilon}(T) - z]^2, \\
\text{subject to: } x^{\varepsilon}(0) = x_0, \ \alpha^{\varepsilon}(0) = l_0, \ Ex^{\varepsilon}(T) = z, \\
(x^{\varepsilon}(\cdot), u^{\varepsilon}(\cdot)) \text{ is admissible.}
\end{cases}
$$
(11.25)

Suppose that $\widetilde{Q}(t)$ and $\widehat{Q}(t)$ are bounded Borel measurable, that $\widetilde{Q}^i(t)$ for $i = 1, \ldots, l_0$ are weakly irreducible, that (A11.2) holds, and that $w^{\jmath}(\cdot)$ are standard Brownian motions independent of $\alpha^{\varepsilon}(\cdot)$. Then the conclusions of Theorem 11.3 and Corollary 11.5 continue to hold with ν^i replaced by the time-varying quasi-stationary distribution $\nu^i(t)$ for $i = 1 \ldots, l_0$ and $\overline{Q}$ replaced by $\overline{Q}(t)$. Moreover, this result can also be extended to treat the continuous-time problem when in addition to the l_0 weakly irreducible classes, transient states are included. In this case, we obtain the continuous-time analogue of Theorem 11.7 with ν, $\overline{Q}$, and $\overline{Q}_*$ replaced by $\nu(t)$, $\overline{Q}(t)$, and $\overline{Q}_*(t)$ respectively. We omit the details.

11.6   Proofs of Results

Proof of Theorem 11.2. Proposition 6.2-(c) implies that $\{\overline{\alpha}^\varepsilon(\cdot)\}$ is tight. To prove the tightness of $\{y^{\varepsilon,\imath}(\cdot), \overline{\alpha}^\varepsilon(\cdot)\}$, it suffices to show that $\{y^{\varepsilon,\imath}(\cdot)\}$ is tight.

Denote by E_t^ε the conditional expectation with respect to $\mathcal{F}_t^\varepsilon$, the σ-algebra generated by $\{\alpha_{k_1}^\varepsilon, \xi_{k_1}^\imath : k_1 < t/\varepsilon, \imath = 0, \ldots, d\}$. Since $\{\xi_k\}$ and $\{\alpha_k^\varepsilon\}$ are independent, and since $E_t^\varepsilon[\xi_k^\imath \xi_{k_1}^\imath] = 0$ for $k_1 \neq k$ and $t/\varepsilon \leq k_1, k \leq (t+s)/\varepsilon$, owing to the boundedness of $c^\imath(\cdot)$,

$$\varepsilon^2 E_t^\varepsilon \left| \sum_{k=t/\varepsilon}^{(t+s)/\varepsilon-1} \sum_{k_1=t/\varepsilon}^{(t+s)/\varepsilon-1} c^\imath(\varepsilon k, \alpha_k^\varepsilon) c^\imath(\varepsilon k_1, \alpha_{k_1}^\varepsilon) \right|$$
$$\leq K\varepsilon^2 \left(\frac{t+s}{\varepsilon} - \frac{t}{\varepsilon} \right)^2 \leq Ks^2.$$

Similarly, using the boundedness of $\sigma^{\imath j}(\cdot)$,

$$\varepsilon E_t^\varepsilon \sum_{k=t/\varepsilon}^{(t+s)/\varepsilon-1} \left(\sum_{j=1}^d \sigma^{\imath j}(\varepsilon k, \alpha_k^\varepsilon) \xi_k^j \right)^2 \leq Ks.$$

Consequently, for any $\eta > 0$, $t \geq 0$ and $0 \leq s \leq \eta$, and for each $\imath = 1, \ldots, d$,

$$E_t^\varepsilon[y^{\varepsilon,\imath}(t+s) - y^{\varepsilon,\imath}(t)]^2$$
$$\leq KE_t^\varepsilon \left| \varepsilon \sum_{k=t/\varepsilon}^{(t+s)/\varepsilon-1} c^\imath(\varepsilon k, \alpha_k^\varepsilon) \right|^2 + KE_t^\varepsilon \left| \sqrt{\varepsilon} \sum_{k=t/\varepsilon}^{(t+s)/\varepsilon-1} \sum_{j=1}^d \sigma^{\imath j}(\varepsilon k \alpha_k^\varepsilon) \xi_k^j \right|^2$$
$$\leq Ks \leq K\eta.$$

(11.26)

Recall that we use K to represent a generic positive real number; its values may be different for different uses. Note that t/ε and $(t+s)/\varepsilon$ in the summation limits of (11.26) are understood to be the integer parts. Consequently,

$$\lim_{\eta \to 0} \limsup_{\varepsilon \to 0} E \left(E_t^\varepsilon[y^{\varepsilon,\imath}(t+s) - y^{\varepsilon,\imath}(t)]^2 \right) = 0.$$

The desired result follows from the tightness criterion (Lemma 14.12). Likewise,

$$\lim_{\eta \to 0} \limsup_{\varepsilon \to 0} E \left(E_t^\varepsilon [S^{\varepsilon,0}(t+s) - S^{\varepsilon,0}(t)]^2 \right) = 0.$$

Thus $\{S^{\varepsilon,0}(\cdot)\}$ is also tight.   □

Proof of Theorem 11.3. We will mainly work with the stock price processes $(y^{\varepsilon,\imath}(\cdot), \overline{\alpha}^\varepsilon(\cdot))$ for $\imath = 1, \ldots, d$. To proceed, fix $\imath \in \{1, \ldots, d\}$. The uniqueness of the solution of the martingale problem can be verified; see Lemma 14.20.

To characterize the limit, it suffices that for each $i \in \overline{\mathcal{M}}$ and $f(\cdot, \cdot, i) \in C_0^{1,2}$ (the collection of functions that have compact support and that are continuously differentiable with respect to the first variable, and twice continuously differentiable with respect to the second variable),

$$f(t, y^\imath(t), \overline{\alpha}(t)) - f(0, y^\imath(0), \overline{\alpha}(0)) - \int_0^t \left(\frac{\partial}{\partial \tau} + \mathcal{L}^\imath \right) f(\tau, y^\imath(\tau), \overline{\alpha}(\tau)) d\tau \tag{11.27}$$

is a martingale. To this end, it suffices that for any positive integer κ, bounded and continuous functions $h_{j_1}(\cdot)$ with $j_1 \le \kappa$, and any $t, s, t_{j_1} \ge 0$ satisfying $t_{j_1} \le t < t + s \le T$,

$$E \prod_{j_1=1}^{\kappa} h_{j_1}(y^\imath(t_{j_1}), \overline{\alpha}(t_{j_1})) \left[f(t+s, y^\imath(t+s), \overline{\alpha}(t+s)) - f(t, y^\imath(t), \overline{\alpha}(t)) \right.$$
$$\left. - \int_t^{t+s} \left(\frac{\partial}{\partial \tau} + \mathcal{L}^\imath \right) f(\tau, y^\imath(\tau), \overline{\alpha}(\tau)) d\tau \right] = 0. \tag{11.28}$$

To verify (11.28), we begin with the pre-limit processes indexed by ε. The weak convergence of $(y^{\varepsilon,\imath}(\cdot), \overline{\alpha}^\varepsilon(\cdot))$ to $(y^\imath(\cdot), \overline{\alpha}(\cdot))$, the Skorohod representation (without changing notation), the continuity of $f(\cdot, \cdot, \alpha)$, and the dominated convergence theorem lead to

$$E \prod_{j_1=1}^{\kappa} h_{j_1}(y^{\varepsilon,\imath}(t_{j_1}), \overline{\alpha}^\varepsilon(t_{j_1}))$$
$$\times [f(t+s, y^{\varepsilon,\imath}(t+s), \overline{\alpha}^\varepsilon(t+s)) - f(t, y^{\varepsilon,\imath}(t), \overline{\alpha}^\varepsilon(t))]$$
$$\to E \prod_{j_1=1}^{\kappa} h_{j_1}(y^\imath(t_{j_1}), \overline{\alpha}(t_{j_1}))[f(t+s, y^\imath(t+s), \overline{\alpha}(t+s)) - f(t, y^\imath(t), \overline{\alpha}(t))], \tag{11.29}$$

as $\varepsilon \to 0$.

Define

$$\overline{f}(t, y, \alpha) = \sum_{i=1}^{l} f(t, y, i) I_{\{\alpha \in \mathcal{M}_i\}} \quad \text{for each } \alpha \in \mathcal{M}. \tag{11.30}$$

Then $\overline{f}(\cdot,\cdot,\alpha) \in C_0^{1,2}$. Pick $\{n_\varepsilon\}$, a sequence of positive integers satisfying $n_\varepsilon \to \infty$ as $\varepsilon \to 0$ and $\varepsilon n_\varepsilon = \delta_\varepsilon \to 0$ as $\varepsilon \to 0$. Noting that

$$\overline{f}(t, y_k^{\varepsilon,\imath}, \alpha_k^\varepsilon) = f(t, y_k^{\varepsilon,\imath}, \overline{\alpha}_k^\varepsilon),$$

we have

$$
\begin{aligned}
f(t+s, y^{\varepsilon,\imath}(t+s), &\overline{\alpha}^\varepsilon(t+s)) - f(t, y^{\varepsilon,\imath}(t), \overline{\alpha}^\varepsilon(t)) \\
&= \int_t^{t+s} \frac{\partial}{\partial\tau} f(\tau, y^{\varepsilon,\imath}(\tau), \overline{\alpha}^\varepsilon(\tau)) d\tau \\
&+ \sum_{l\delta_\varepsilon=t}^{t+s} [\overline{f}(l\delta_\varepsilon, y_{ln_\varepsilon}^{\varepsilon,\imath}, \alpha_{ln_\varepsilon+n_\varepsilon}^\varepsilon) - \overline{f}(l\delta_\varepsilon, y_{ln_\varepsilon}^{\varepsilon,\imath}, \alpha_{ln_\varepsilon}^\varepsilon)] \\
&+ \sum_{l\delta_\varepsilon=t}^{t+s} [\overline{f}(l\delta_\varepsilon, y_{ln_\varepsilon+n_\varepsilon}^{\varepsilon,\imath}, \alpha_{ln_\varepsilon+n_\varepsilon}^\varepsilon) - \overline{f}(l\delta_\varepsilon, y_{ln_\varepsilon}^{\varepsilon,\imath}, \alpha_{ln_\varepsilon+n_\varepsilon}^\varepsilon)] + o(1),
\end{aligned}
$$

$$(11.31)$$

where $o(1) \to 0$ in probability as $\varepsilon \to 0$ uniformly in $t \in [0, T]$.

By virtue of the weak convergence, the Skorohod representation (without changing notation), and the continuity and hence the boundedness of $(\partial/\partial\tau)f(\cdot, \cdot, \alpha)$, as $\varepsilon \to 0$,

$$
\begin{aligned}
E \prod_{j_1=1}^\kappa h_{j_1}(y^{\varepsilon,\imath}(t_{j_1}), \overline{\alpha}^\varepsilon(t_{j_1})) &\left(\int_t^{t+s} \frac{\partial}{\partial\tau} f(\tau, y^{\varepsilon,\imath}(\tau), \overline{\alpha}^\varepsilon(\tau)) d\tau \right) \\
\to E \prod_{j_1=1}^\kappa h_{j_1}(y^\imath(t_{j_1}), \overline{\alpha}(t_{j_1})) &\left(\int_t^{t+s} \frac{\partial}{\partial\tau} f(\tau, y^\imath(\tau), \overline{\alpha}(\tau)) d\tau \right).
\end{aligned}
$$

$$(11.32)$$

Define

$$\widehat{I}(\alpha) = (I_{\{\alpha=\varsigma_{ij}\}}, 1 \le i \le l_0, 1 \le j \le m_i),$$

and

$$\overline{F}(t,y) = \begin{pmatrix} f(t,y,1)\mathbb{1}_{m_1} \\ \vdots \\ f(t,y,l)\mathbb{1}_{m_l} \end{pmatrix}.$$

For the next to the last line of (11.31), $y_{ln_\varepsilon}^{\varepsilon,\imath}$ is $\mathcal{F}_{ln_\varepsilon}^\varepsilon$-measurable. Assumption (A11.1) and the orthogonality $(P-I)\overline{F}(t,x) = 0$, yield

$$
\begin{aligned}
E_t^\varepsilon \sum_{l\delta_\varepsilon=t}^{t+s} &[\overline{f}(l\delta_\varepsilon, y_{ln_\varepsilon}^{\varepsilon,\imath}, \alpha_{ln_\varepsilon+n_\varepsilon}^\varepsilon) - \overline{f}(l\delta_\varepsilon, y_{ln_\varepsilon}^{\varepsilon,\imath}, \alpha_{ln_\varepsilon}^\varepsilon)] \\
&= \varepsilon E_t^\varepsilon \sum_{l\delta_\varepsilon=t}^{t+s} \sum_{k=ln_\varepsilon}^{ln_\varepsilon+n_\varepsilon-1} \widehat{I}(\alpha_k^\varepsilon) Q \overline{F}(l\delta_\varepsilon, y_{ln_\varepsilon}^{\varepsilon,\imath}) \\
&= \varepsilon E_t^\varepsilon \sum_{l\delta_\varepsilon=t}^{t+s} \sum_{k=ln_\varepsilon}^{ln_\varepsilon+n_\varepsilon-1} Q\overline{F}(l\delta_\varepsilon, y_{ln_\varepsilon}^{\varepsilon,\imath}, \cdot)(\alpha_k^\varepsilon)
\end{aligned}
$$

$$= \varepsilon E_t^{\varepsilon} \sum_{l\delta_\varepsilon=t}^{t+s} \sum_{i=1}^{l_0} \sum_{j=1}^{m_i} Q\overline{f}(l\delta_\varepsilon, y_{ln_\varepsilon}^{\varepsilon,\imath}, \cdot)(\zeta_{ij}) \delta_\varepsilon \frac{1}{n_\varepsilon} \sum_{k=ln_\varepsilon}^{ln_\varepsilon+n_\varepsilon-1} \nu^{ij} I_{\{\alpha_k^\varepsilon \in \mathcal{M}_i\}}$$

$$+ \varepsilon E_t^{\varepsilon} \sum_{l\delta_\varepsilon=t}^{t+s} \sum_{i=1}^{l_0} \sum_{j=1}^{m_i} Q\overline{f}(l\delta_\varepsilon, y_{ln_\varepsilon}^{\varepsilon,\imath}, \cdot)(\zeta_{ij})$$

$$\times \frac{\delta_\varepsilon}{n_\varepsilon} \sum_{k=ln_\varepsilon}^{ln_\varepsilon+n_\varepsilon-1} [I_{\{\alpha_k^\varepsilon=\zeta_{ij}\}} - \nu^{ij} I_{\{\overline{\alpha}_k=i\}}].$$

Lemma 6.2 implies that as $\varepsilon \to 0$,

$$\varepsilon E_t^{\varepsilon} \sum_{l\delta_\varepsilon=t}^{t+s} \sum_{i=1}^{l_0} \sum_{j=1}^{m_i} Q\overline{f}(l\delta_\varepsilon, y_{ln_\varepsilon}^{\varepsilon,\imath}, \cdot)(\zeta_{ij}) \frac{\delta_\varepsilon}{n_\varepsilon} \sum_{k=ln_\varepsilon}^{ln_\varepsilon+n_\varepsilon-1} [I_{\{\alpha_k^\varepsilon=\zeta_{ij}\}} - \nu^{ij} I_{\{\overline{\alpha}_k=i\}}]$$

$\to 0$ in probability uniformly in $t \in [0, T]$.

In view of (6.6),

$$E \left| \sum_{l\delta_\varepsilon=t}^{t+s} \delta_\varepsilon \frac{1}{n_\varepsilon} \sum_{k=ln_\varepsilon}^{ln_\varepsilon+n_\varepsilon-1} [Q\overline{f}(l\delta_\varepsilon, y_{ln_\varepsilon}^{\varepsilon,\imath}, \cdot)(\alpha_k^\varepsilon) - \overline{Q}f(l\delta_\varepsilon, y_{ln_\varepsilon}^{\varepsilon,\imath}, \cdot)(\overline{\alpha}_k^\varepsilon)] \right|$$

$\to 0$ as $\varepsilon \to 0$ uniformly in t.

Putting the above estimates together, as $\varepsilon \to 0$,

$$E \prod_{j_1=1}^{\kappa} h_{j_1}(y^{\varepsilon,\imath}(t_{j_1}), \overline{\alpha}^\varepsilon(t_{j_1})) \left[\sum_{l\delta_\varepsilon=t}^{t+s} \delta_\varepsilon \frac{1}{n_\varepsilon} \sum_{k=ln_\varepsilon}^{ln_\varepsilon+n_\varepsilon-1} Q\overline{f}(l\delta_\varepsilon, y_{ln_\varepsilon}^{\varepsilon,\imath}, \cdot)(\alpha_k^\varepsilon) \right]$$

$$\to E \prod_{j_1=1}^{\kappa} h_{j_1}(y^{\imath}(t_{j_1}), \overline{\alpha}(t_{j_1})) \left[\int_t^{t+s} \overline{Q}f(\tau, y^{\imath}(\tau), \cdot)(\overline{\alpha}(\tau)) d\tau \right].$$

(11.33)

Using a truncated Taylor expansion and denoting

$$\overline{f}_y = (\partial/\partial y)\overline{f} \quad \text{and} \quad \overline{f}_{yy} = (\partial^2/\partial y^2)\overline{f},$$

we have

$$\sum_{l\delta_\varepsilon=t}^{t+s} [\overline{f}(l\delta_\varepsilon, y_{ln_\varepsilon+n_\varepsilon}^{\varepsilon,\imath}, \alpha_{ln_\varepsilon+n_\varepsilon}^{\varepsilon}) - \overline{f}(l\delta_\varepsilon, y_{ln_\varepsilon}^{\varepsilon,\imath}, \alpha_{ln_\varepsilon+n_\varepsilon}^{\varepsilon})]$$

$$= \sum_{l\delta_\varepsilon=t}^{t+s} \overline{f}_y(l\delta_\varepsilon, y_{ln_\varepsilon}^{\varepsilon,\imath}, \alpha_{ln_\varepsilon+n_\varepsilon}^{\varepsilon})[y_{ln_\varepsilon+n_\varepsilon}^{\varepsilon,\imath} - y_{ln_\varepsilon}^{\varepsilon,\imath}]$$

$$+ \frac{1}{2} \sum_{l\delta_\varepsilon=t}^{t+s} \overline{f}_{yy}(l\delta_\varepsilon, y_{ln_\varepsilon}^{\varepsilon,\imath}, \alpha_{ln_\varepsilon+n_\varepsilon}^{\varepsilon})[y_{ln_\varepsilon+n_\varepsilon}^{\varepsilon,\imath} - y_{ln_\varepsilon}^{\varepsilon,\imath}]^2$$

(11.34)

$$+ \frac{1}{2} \sum_{l\delta_\varepsilon=t}^{t+s} [\overline{f}_{yy}(l\delta_\varepsilon, y_{ln_\varepsilon}^{\varepsilon,\imath,+}, \alpha_{ln_\varepsilon+n_\varepsilon}^{\varepsilon})$$

$$- \overline{f}_{yy}(l\delta_\varepsilon, y_{ln_\varepsilon}^{\varepsilon,\imath}, \alpha_{ln_\varepsilon+n_\varepsilon}^{\varepsilon})][y_{ln_\varepsilon+n_\varepsilon}^{\varepsilon,\imath} - y_{ln_\varepsilon}^{\varepsilon,\imath}]^2,$$

where $y_{ln_\varepsilon}^{\varepsilon,\imath,+}$ is on the line segment joining $y_{ln_\varepsilon}^{\varepsilon,\imath}$ and $y_{ln_\varepsilon+n_\varepsilon}^{\varepsilon,\imath}$. Substituting (11.8) in (11.34), we can write

$$
\sum_{l\delta_\varepsilon=t}^{t+s} \overline{f}_y(l\delta_\varepsilon, y_{ln_\varepsilon}^{\varepsilon,\imath}, \alpha_{ln_\varepsilon+n_\varepsilon}^\varepsilon)[y_{ln_\varepsilon+n_\varepsilon}^{\varepsilon,\imath} - y_{ln_\varepsilon}^{\varepsilon,\imath}]
$$
$$
= \sum_{l\delta_\varepsilon=t}^{t+s} \overline{f}_y(l\delta_\varepsilon, y_{ln_\varepsilon}^{\varepsilon,\imath}, \alpha_{ln_\varepsilon+n_\varepsilon}^\varepsilon) \sum_{k=ln_\varepsilon}^{ln_\varepsilon+n_\varepsilon-1} \varepsilon c^\imath(\varepsilon k, \alpha_k^\varepsilon) \tag{11.35}
$$
$$
+ \sum_{l\delta_\varepsilon=t}^{t+s} \overline{f}_y(l\delta_\varepsilon, y_{ln_\varepsilon}^{\varepsilon,\imath}, \alpha_{ln_\varepsilon+n_\varepsilon}^\varepsilon) \sum_{k=ln_\varepsilon}^{ln_\varepsilon+n_\varepsilon-1} \sqrt{\varepsilon} \sum_{j=1}^d \sigma^{\imath j}(\varepsilon k, \alpha_k^\varepsilon)\xi_k^\jmath.
$$

To proceed, we first replace $\alpha_{ln_\varepsilon+n_\varepsilon}^\varepsilon$ in the argument of $\overline{f}_y(\cdot)$ above by $\alpha_{ln_\varepsilon}^\varepsilon$, which adds a term with 0 limit. Letting $l\delta_\varepsilon \to \tau$ as $\varepsilon \to 0$, and using Lemma 6.2-(c), for all $ln_\varepsilon \le k \le ln_\varepsilon + n_\varepsilon - 1$, $\varepsilon k \to \tau$,

$$
E \prod_{j_1=1}^\kappa h_{j_1}(y^{\varepsilon,\imath}(t_{j_1}), \overline{\alpha}^\varepsilon(t_{j_1})) \sum_{l\delta_\varepsilon=t}^{t+s} \overline{f}_y(l\delta_\varepsilon, y_{ln_\varepsilon}^{\varepsilon,\imath}, \alpha_{ln_\varepsilon}^\varepsilon)\frac{\delta_\varepsilon}{n_\varepsilon} \sum_{k=ln_\varepsilon}^{ln_\varepsilon+n_\varepsilon-1} c^\imath(\varepsilon k, \alpha_k^\varepsilon)
$$
$$
= E \prod_{j_1=1}^\kappa h_{j_1}(y^{\varepsilon,\imath}(t_{j_1}), \overline{\alpha}^\varepsilon(t_{j_1})) \sum_{l\delta_\varepsilon=t}^{t+s} \overline{f}_y(l\delta_\varepsilon, y_{ln_\varepsilon}^{\varepsilon,\imath}, \alpha_{ln_\varepsilon}^\varepsilon)
$$
$$
\times \frac{\delta_\varepsilon}{n_\varepsilon} \sum_{k=ln_\varepsilon}^{ln_\varepsilon+n_\varepsilon-1} \sum_{i=1}^{l_0} \sum_{j=1}^{m_i} c^\imath(\varepsilon k, s_{ij})\nu^{ij} I_{\{\alpha_k^\varepsilon \in \mathcal{M}_i\}}
$$
$$
+ E \prod_{j_1=1}^\kappa h_{j_1}(y^{\varepsilon,\imath}(t_{j_1}), \overline{\alpha}^\varepsilon(t_{j_1})) \sum_{l\delta_\varepsilon=t}^{t+s} \overline{f}_y(l\delta_\varepsilon, y_{ln_\varepsilon}^{\varepsilon,\imath}, \alpha_{ln_\varepsilon}^\varepsilon)
$$
$$
\times \delta_\varepsilon \frac{1}{n_\varepsilon} \sum_{k=ln_\varepsilon}^{ln_\varepsilon+n_\varepsilon-1} \sum_{i=1}^{l_0} \sum_{j=1}^{m_i} c^\imath(\varepsilon k, s_{ij})[I_{\{\alpha_k^\varepsilon=s_{ij}\}} - \nu^{ij} I_{\{\alpha_k^\varepsilon \in \mathcal{M}_i\}}]
$$
$$
\to E \prod_{j_1=1}^\kappa h_{j_1}(y^\imath(t_{j_1}), \overline{\alpha}(t_{j_1})) \int_t^{t+s} f_y(\tau, y^\imath(\tau), \overline{\alpha}(\tau))\overline{c}^\imath(\tau, \overline{\alpha}(\tau))d\tau.
$$

By virtue of the independence of $\{\xi_k^\imath\}$, inserting E_t^ε and then E_k, and using $E_k \xi_k^\jmath = 0$, we have

$$
E \prod_{j_1=1}^\kappa h_{j_1}(y^{\varepsilon,\imath}(t_{j_1}), \overline{\alpha}^\varepsilon(t_{j_1})) \left(\sum_{l\delta_\varepsilon=t}^{t+s} \overline{f}_y(l\delta_\varepsilon, y_{ln_\varepsilon}^{\varepsilon,\imath}, \alpha_{ln_\varepsilon}^\varepsilon) \sum_{k=ln_\varepsilon}^{ln_\varepsilon+n_\varepsilon-1} \sqrt{\varepsilon}\sigma^{\imath j}(\varepsilon k, \alpha_k^\varepsilon)\xi_k^\jmath \right)
$$
$$
= 0.
$$

Likewise, detailed estimates reveal that as $\varepsilon \to 0$,

$$
E \prod_{j_1=1}^{\kappa} h_{j_1}(y^{\varepsilon,\imath}(t_{j_1}), \overline{\alpha}^{\varepsilon}(t_{j_1})) \left(\sum_{l\delta_\varepsilon=t}^{t+s} \overline{f}_{yy}(l\delta_\varepsilon, y_{ln_\varepsilon}^{\varepsilon,\imath}, \alpha_{ln_\varepsilon}^{\varepsilon})(y_{ln_\varepsilon+n_\varepsilon}^{\varepsilon,\imath} - y_{ln_\varepsilon}^{\varepsilon,\imath})^2 \right)
$$

$$
= E \prod_{j_1=1}^{\kappa} h_{j_1}(y^{\varepsilon,\imath}(t_{j_1}), \overline{\alpha}^{\varepsilon}(t_{j_1})) \left(\sum_{l\delta_\varepsilon=t}^{t+s} \delta_\varepsilon \overline{f}_{yy}(l\delta_\varepsilon, y_{ln_\varepsilon}^{\varepsilon,\imath}, \alpha_{ln_\varepsilon}^{\varepsilon}) \right.
$$

$$
\left. \times \frac{1}{n_\varepsilon} \sum_{k=ln_\varepsilon}^{ln_\varepsilon+n_\varepsilon-1} \sum_{i=1}^{l_0} \sum_{j=1}^{m_i} E_k \left[\sum_{j=1}^{d} \sigma^{\imath j}(\varepsilon k, s_{ij}) \xi_k^j \right]^2 \nu^{ij} I_{\{\alpha_k^\varepsilon \in \mathcal{M}_i\}} \right) + o(1)
$$

$$
\to E \prod_{j_1=1}^{\kappa} h_{j_1}(y^{\imath}(t_{j_1}), \overline{\alpha}(t_{j_1})) \left(\int_t^{t+s} f_{yy}(\tau, y^{\imath}(\tau), \overline{\alpha}(\tau)) \right.
$$

$$
\left. \times [\overline{\Sigma}(\tau, \overline{\alpha}(\tau)) \overline{\Sigma}'(\tau, \overline{\alpha}(\tau))]^{\imath\imath} d\tau \right),
$$

$$(11.36)$$

where $\overline{\Sigma}(t, \alpha)$ and $[\overline{\Sigma}(t, \alpha) \overline{\Sigma}'(t, \alpha)]^{\imath\imath}$ are given by (11.12) and (11.15), respectively. In addition, similar to the estimates for the term on the third line of (11.31),

$$
E \prod_{j_1=1}^{\kappa} h_{j_1}(y^{\varepsilon,\imath}(t_{j_1}), \overline{\alpha}^{\varepsilon}(t_{j_1})) \sum_{l\delta_\varepsilon=t}^{t+s} [\overline{f}_y(l\delta_\varepsilon, y_{ln_\varepsilon}^{\varepsilon,\imath}, \alpha_{ln_\varepsilon+n_\varepsilon}^{\varepsilon})
$$

$$
- \overline{f}_y(l\delta_\varepsilon, y_{ln_\varepsilon}^{\varepsilon,\imath}, \alpha_{ln_\varepsilon}^{\varepsilon})] \sum_{k=ln_\varepsilon}^{ln_\varepsilon+n_\varepsilon-1} \varepsilon c^{\imath}(\varepsilon k, \alpha_k^\varepsilon) \to 0,
$$

$$(11.37)$$

as $\varepsilon \to 0$ and

$$
E \prod_{j_1=1}^{\kappa} h_{j_1}(y^{\varepsilon,\imath}(t_{j_1}), \overline{\alpha}^{\varepsilon}(t_{j_1})) \sum_{l\delta_\varepsilon=t}^{t+s} [\overline{f}_{yy}(l\delta_\varepsilon, y_{ln_\varepsilon}^{\varepsilon,\imath,+}, \alpha_{ln_\varepsilon+n_\varepsilon}^{\varepsilon})
$$

$$
- \overline{f}_{yy}(l\delta_\varepsilon, y_{ln_\varepsilon}^{\varepsilon,\imath}, \alpha_{ln_\varepsilon}^{\varepsilon})][y_{ln_\varepsilon+n_\varepsilon}^{\varepsilon,\imath} - y_{ln_\varepsilon}^{\varepsilon,\imath}]^2 \to 0.
$$

$$(11.38)$$

Similar to (11.37) and (11.38), as $\varepsilon \to 0$,

$$
E \prod_{j_1=1}^{\kappa} h_{j_1}(y^{\varepsilon,\imath}(t_{j_1}), \overline{\alpha}^{\varepsilon}(t_{j_1})) \left[\sum_{l\delta_\varepsilon=t}^{t+s} [\overline{f}_y(l\delta_\varepsilon, y_{ln_\varepsilon}^{\varepsilon,\imath}, \alpha_{ln_\varepsilon+n_\varepsilon}^{\varepsilon}) \right.
$$

$$
\left. - \overline{f}_y(l\delta_\varepsilon, y_{ln_\varepsilon}^{\varepsilon,\imath}, \alpha_{ln_\varepsilon}^{\varepsilon})] \sum_{k=ln_\varepsilon}^{ln_\varepsilon+n_\varepsilon-1} \sqrt{\varepsilon} \sigma^{\imath j}(\varepsilon k, \alpha_k^\varepsilon) \xi_k^j \right] \to 0.
$$

$$(11.39)$$

Combining the estimates obtained thus far and using (11.34) in conjunction with (11.31), we arrive at

$$E \prod_{j_1=1}^{\kappa} h_{j_1}(y^{\varepsilon,\imath}(t_{j_1}), \overline{\alpha}^{\varepsilon}(t_{j_1}))\overline{\alpha}^{\varepsilon}(t+s)) - f(t, y^{\varepsilon,\imath}(t), \overline{\alpha}^{\varepsilon}(t))]$$

$$\to E \prod_{j_1=1}^{\kappa} h_{j_1}(y^{\imath}(t_{j_1}), \overline{\alpha}(t_{j_1}))$$

$$\times \int_t^{t+s} \left(\left(\frac{\partial}{\partial\tau} + \mathcal{L}^{\imath}\right) f(\tau, y^{\imath}(\tau), \overline{\alpha}(\tau)) + \overline{Q}f(\tau, y^{\imath}(\tau), \cdot)(\overline{\alpha}(\tau)) \right) d\tau.$$

(11.40)

Equation (11.40) together with (11.29) then yields the desired assertion.

The same method works for the proof of $(S^{\varepsilon,0}(\cdot), \overline{\alpha}^{\varepsilon}(\cdot))$ to $(S^0(\cdot), \overline{\alpha}(\cdot))$. The argument is even simpler since no diffusion is involved.   □

Proof of Corollary 11.5. Set

$$\widetilde{S}^{\varepsilon}(\cdot) = (S^{\varepsilon,0}(\cdot), S^{\varepsilon,1}(\cdot), \dots, S^{\varepsilon,d}(\cdot))'.$$

Theorems 11.2 and 11.3 imply that

$(\widetilde{S}^{\varepsilon}(\cdot), \overline{\alpha}^{\varepsilon}(\cdot))$ converges weakly to $(\widetilde{S}(\cdot), \overline{\alpha}(\cdot)) = (S^0(t), \dots, S^d(t), \overline{\alpha}(\cdot))$

that satisfies

$$d\widetilde{S}(t) = \begin{pmatrix} \overline{r}(t, \overline{\alpha}(t)) & 0 \\ 0 & \mathrm{diag}(\overline{b}^1(t, \overline{\alpha}(t)), \dots, \overline{b}^d(t, \overline{\alpha}(t))) \end{pmatrix} \widetilde{S}(t)dt$$

$$+ \begin{pmatrix} 0 & 0 \\ 0 & \overline{\Sigma}(t, \overline{\alpha}(t))\mathrm{diag}(dw^1(t), \dots, dw^d(t)) \end{pmatrix} \widetilde{S}(t).$$

(11.41)

Recall that $u^{\varepsilon}(t) = u_k^{\varepsilon} = (N^1(\varepsilon k)S^{\varepsilon,1}(t), \dots, N^d(\varepsilon k)S^{\varepsilon,d}(t))$ for $t \in [\varepsilon k, \varepsilon k + \varepsilon)$. The weak convergence of $(S^{\varepsilon,1}(\cdot), \dots, S^{\varepsilon,d}(\cdot))$ yields that $u^{\varepsilon}(\cdot)$ converges weakly to $u(\cdot)$ with $u(t) = (N^1(t)S^1(t), \dots, N^d(t)S^d(t))$. As a consequence, $x^{\varepsilon}(\cdot)$ converges weakly to $x(\cdot)$, which satisfies (11.17). This also implies $(x(\cdot), u(\cdot)) \in \bar{\mathcal{A}}$. Finally, the interpolation of x_k^{ε}, the weak convergence of $x^{\varepsilon}(\cdot)$ to $x(\cdot)$, the Skorohod representation (without changing notation), and the dominated convergence theorem lead to (11.19). Therefore, associated with a discrete-time problem (11.5) there is a continuous-time counterpart (11.18) that serves as a limit problem in the sense of Corollary 11.5.   □

Proof of Theorem 11.6. In view of the construction of $\widetilde{u}^{\varepsilon}(\cdot)$ in (11.20), the weak convergence argument as in the proof of Theorem 11.3 yields that $\widetilde{u}^{\varepsilon}(\cdot)$ converges weakly to $u^o(\cdot)$, and $(x^{\varepsilon}(\cdot), \widetilde{u}^{\varepsilon}(\cdot), \overline{\alpha}^{\varepsilon}(\cdot))$ converges weakly to $(x(\cdot), u^o(\cdot), \overline{\alpha}(\cdot))$. Then $J^{\varepsilon}(x_0, \ell_0, \widetilde{u}^{\varepsilon}(\cdot)) \to J(x_0, i_0, u^o(\cdot)) = v(x_0, i_0)$ as $\varepsilon \to 0$. Therefore, $J^{\varepsilon}(x_0, \ell_0, \widetilde{u}^{\varepsilon}(\cdot)) = v(x_0, i_0) + \Delta_1(\varepsilon)$, where $\Delta_1(\varepsilon) \to 0$ as $\varepsilon \to 0$. Select an admissible control $\widehat{u}^{\varepsilon}(\cdot) \in \mathcal{A}^{\varepsilon}$ such that

$$J^{\varepsilon}(x_0, \ell_0, \widehat{u}^{\varepsilon}(\cdot)) \le v^{\varepsilon}(x_0, \ell_0) + \varepsilon.$$

Define

$$\overline{u}(t, x, \alpha) = \sum_{i=1}^{l_0} \widehat{u}^{\varepsilon}(t, x, i) I_{\{\alpha \in \mathcal{M}_i\}},$$

set $\overline{x}_k^{\varepsilon}$ as in (11.4) but with α_k^{ε} replaced by $\overline{\alpha}_k^{\varepsilon}$, and let $\overline{x}^{\varepsilon}(\cdot)$ and $\overline{u}^{\varepsilon}(\cdot)$ be the piecewise constant interpolations of $\overline{x}_k^{\varepsilon}$ and $\overline{u}(\varepsilon k, \overline{x}_k^{\varepsilon}, \overline{\alpha}_k^{\varepsilon})$, respectively. Then $J^{\varepsilon}(x_0, \ell_0, \overline{u}^{\varepsilon}(\cdot)) = E[\overline{x}^{\varepsilon}(T) - z]^2$. Similar to the argument used in Chapter 10, using the mean squares estimate for the occupation measure (see Chapter 6), the wealth equation (11.4), the definition of $\overline{x}_k^{\varepsilon}$, and Gronwall's inequality, we can show that $E|x^{\varepsilon}(t) - \overline{x}^{\varepsilon}(t)|^2 \to 0$ as $\varepsilon \to 0$ for $t \in [0, T]$. This implies that

$$\begin{aligned} J^{\varepsilon}(x_0, \ell_0, \overline{u}^{\varepsilon}(\cdot)) &\leq J^{\varepsilon}(x_0, \ell_0, \widehat{u}^{\varepsilon}(\cdot)) + \Delta_2(\varepsilon) \\ &\leq v^{\varepsilon}(x_0, \ell_0) + \varepsilon + \Delta_2(\varepsilon), \end{aligned} \tag{11.42}$$

where $\Delta_2(\varepsilon) \to 0$ as $\varepsilon \to 0$. The tightness of $(x^{\varepsilon}(\cdot), \overline{u}^{\varepsilon}(\cdot), \overline{\alpha}^{\varepsilon}(\cdot))$ implies that we can extract a convergent subsequence, still denoted by $(x^{\varepsilon}(\cdot), \overline{u}^{\varepsilon}(\cdot), \overline{\alpha}^{\varepsilon}(\cdot))$ for simplicity, such that $J^{\varepsilon}(x_0, \ell_0, \overline{u}^{\varepsilon}(\cdot)) \to J(x_0, i_0, \overline{u}(\cdot))$. It follows that

$$v(x_0, i_0) \leq J(x_0, i_0, \overline{u}(\cdot)) = J^{\varepsilon}(x_0, \ell_0, \overline{u}^{\varepsilon}(\cdot)) + \Delta_3(\varepsilon), \tag{11.43}$$

where $\Delta_3(\varepsilon) \to 0$ as $\varepsilon \to 0$. Combining (11.42) and (11.43),

$$\begin{aligned} v^{\varepsilon}(x_0, \ell_0) &\leq J^{\varepsilon}(x_0, \ell_0, \widehat{u}^{\varepsilon}(\cdot)) \\ &= v(x_0, i_0) + \Delta_1(\varepsilon) \\ &\leq J(x_0, i_0, \overline{u}(\cdot)) + \Delta_1(\varepsilon) \\ &= J^{\varepsilon}(x_0, \ell_0, \overline{u}^{\varepsilon}(\cdot)) + \Delta_1(\varepsilon) + \Delta_3(\varepsilon) \\ &\leq v^{\varepsilon}(x_0, \ell_0) + \varepsilon + \Delta_1(\varepsilon) + \Delta_2(\varepsilon) + \Delta_3(\varepsilon). \end{aligned} \tag{11.44}$$

Subtracting $v^{\varepsilon}(x_0, \ell_0)$ from (11.44) and taking the limit as $\varepsilon \to 0$, we arrive at (11.21). $\square$

11.7   Notes

Concerning the hybrid mean-variance control problem, there have been continuing efforts in extending portfolio selection from the single-period model to multi-period or continuous-time models. Much of the research work on dynamic portfolio selections focused on maximizing expected utility functions of the terminal wealth, which is in spirit different from the original Markowitz's model. For the continuous-time cases, the mean-variance hedging problem was studied by Duffie and Richardson [51] and Schweizer [133], where optimal dynamic strategies were sought to hedge contingent claims in an imperfect market based on the so-called projection theorem. More recently, using the stochastic linear-quadratic (LQ) theory developed in Chen, Li, and Zhou [35] and Yong and Zhou [170], Zhou and Li [185] introduced a stochastic LQ control framework to study the continuous-time version of Markowitz's problem. Within this framework, they derived closed-form efficient policies (in the Markowitz sense) along with an explicit expression of the efficient frontier. For some of the recent development on multi-period, discrete-time Markowitz's portfolio selection problems, see Li and Ng [102], in which efficient strategies were derived together with the efficient frontier.

The switching mean-variance models have been mainly used in the literature in dealing with options; see Barone-Adesi and Whaley [10], and Di Masi, Kabanov, and Runggaldier [47]. An investment-consumption model with regime switching was studied in Zariphopoulou [171]. More recently, in Zhou and Yin [186], the continuous-time version of Markowitz's mean-variance portfolio selection with regime switching was treated where efficient portfolios and the efficient frontier were derived explicitly. The results presented here are based on Yin and Zhou [169]. The main techniques used are the weak convergence methods and martingale averaging; see Ethier and Kurtz [55] and Kushner [96]. There has been a growing interest in solving problems in financial engineering using Markov-modulated models. The recent work of Zhang [176] treated an optimal stock selling rule for a Markov-modulated Black-Scholes model; it showed that an optimal liquidation rule can be obtained via optimal threshold levels by solving a set of two-point boundary value problems. The subsequent work of Yin, Liu, and Zhang [157] provided an enticing alternative using stochastic approximation methods.

12
Production Planning

12.1 Introduction

This chapter is concerned with near-optimal production planning for a class of discrete-time manufacturing problems. Specifically, we consider a manufacturing system consisting of a number of machines that produce a number of parts. Assuming that the machines are subject to breakdown and repair and that the capacity of the machines is a finite-state Markov chain, our objective is to choose the production rates over time to minimize a cost function. A commonly used machinery in solving control problems is the dynamic programming (DP) approach, which leads to a system of DP equations corresponding to the Markovian states. Optimal control can be obtained by solving the systems of equations. Since the state spaces of the Markov chains in these manufacturing systems are often very large, a large number of DP equations have to be solved, which is often computationally infeasible. To resolve this problem, using time-scale separation, we introduce a small parameter in the underlying Markov chain to reflect the different rates of changes of different states, resulting in a two-time-scale formulation.

In this chapter, we first decompose the state space of the underlying Markov chain into a number of recurrent classes and a group of transient states according to their different jump rates. We then aggregate the states and replace the original system with its "average." Under suitable scaling, we obtain a limit control system that has fewer DP equations to be solved. As in Chapter 10, although the original problem is in discrete time, its limit problem is a continuous time one. By using optimal controls of the limit system, we construct controls and show their near optimality.

The rest of the chapter is organized as follows. In the next section, we set up the problem, derive basic properties of the value functions, and present the corresponding limit problem. Using these results, we demonstrate that the value function of the original problem converges to that of the limit problem, construct a near-optimal control policy using the solution of the limit problem, and verify its near optimality. A simple example is provided in Section 12.3 for illustration. Proofs of results are given in Section 12.4. The chapter concludes with some discussions in Section 12.5.

12.2   Main Results

Consider a manufacturing system consisting of a number of machines and producing several types of products. The machines are failure-prone. That is, they are subject to breakdown and repair. For simplicity, we impose no conditions on internal buffers.

Let $x_n \in \mathbb{R}^{n_1}$ be the surplus (inventory/shortage) and $u_n \in \Gamma \subset \mathbb{R}^{n_2}$ the rate of production. Let $\{\alpha_n^\varepsilon : n \geq 0\}$ be a finite-state Markov chain with state space $\mathcal{M}$ and transition matrix $P^\varepsilon = (p^{\varepsilon,ij}) \in \mathbb{R}^{m_0 \times m_0}$, where $P^\varepsilon = P + \varepsilon Q$ with P given in (6.3) and the corresponding $\mathcal{M} = \mathcal{M}_1 \cup \mathcal{M}_2 \cup \cdots \cup \mathcal{M}_{l_0} \cup \mathcal{M}_*$.

The discrete-time control system is governed by

$$x_{n+1} = x_n + \varepsilon(A(\alpha_n^\varepsilon)u_n + B(\alpha_n^\varepsilon)), \ x_0 = x, \ n = 0, 1, \ldots, \qquad (12.1)$$

where $A(\alpha) \in \mathbb{R}^{n_1 \times n_2}$ and $B(\alpha) \in \mathbb{R}^{n_1 \times 1}$, for each $\alpha \in \mathcal{M}$.

Let $u. = \{u_0, u_1, \ldots\}$ be the control sequence, and

$$J^\varepsilon(x, \alpha, u.) = E\left[\sum_{n=0}^{\infty}(1 - \rho\varepsilon)^n \varepsilon G(x_n, \alpha_n^\varepsilon, u_n)\Big| x_0 = x, \alpha_0^\varepsilon = \alpha\right] \qquad (12.2)$$

be the cost function, where $\rho > 0$ is a constant and $G(x, \alpha, u)$ is the running cost function.

Our objective is to choose $u.$ to minimize J^ε. We use the dynamic programming approach to resolve the problem. Let $v^\varepsilon(x, \alpha)$ be the value function, i.e., $v(x, \alpha) = \inf_{u.} J^\varepsilon(x, \alpha, u.)$. The associated system of DP equations is given by

$$v^\varepsilon(x, \alpha) = \min_{u \in \Gamma}\Big\{\varepsilon G(x, \alpha, u) + (1 - \rho\varepsilon)$$
$$\times \sum_{\beta \in \mathcal{M}} p^{\varepsilon, \alpha\beta} v^\varepsilon(x + \varepsilon(A(\alpha)u + B(\alpha)), \beta)\Big\}. \qquad (12.3)$$

For each x, let $u^*(x, \alpha)$ denote the minimizer of the right-hand side of (12.3). Then $u^*(x, \alpha)$ is optimal; see Bertsekas [13].

For typical manufacturing systems, the number of elements in $\mathcal{M}$ is large, so is the number of equations in (12.3). It is therefore computationally intensive to solve these equations. In the rest of the chapter, we study an approximate optimal scheme that requires solving simpler problems and yields near optimal controls. Let us give two examples as special cases to the general model.

Example 12.1 (Single-machine System). Consider a production system with one machine producing one part type. Let $c_n^\varepsilon \in \{0, 1\}$ denote the machine state where 1 means the machine is up with maximum capacity 1 and 0 means that the machine is down. Use $z_n^\varepsilon \in \{z_1, z_2\}$ to denote the part demand rate. The system is given by

$$x_{n+1} = x_n + \varepsilon(c_n^\varepsilon u_n - z_n), \quad x_0 = x. \tag{12.4}$$

In this case, $n_1 = n_2 = 1$, $A(c, z) = c$, $B(c, z) = -z$, $\Gamma = [0, 1]$, and

$$\mathcal{M} = \{(1, z_1), (1, z_2), (0, z_1), (0, z_2)\}.$$

Example 12.2 (Two-machine Flowshop). Let $x_n = (x_n^1, x_n^2)'$ be the surplus of the first machine and second machine and let $c^{\varepsilon,1} \in \{0, 1\}$ and $c^{\varepsilon,2} \in \{0, 1\}$ be the capacity processes. Then $\alpha_n^\varepsilon = (c_n^{\varepsilon,1}, c_n^{\varepsilon,2}) \in \mathcal{M}$ with

$$\mathcal{M} = \{(0, 0), (0, 1), (1, 0), (1, 1)\}.$$

The system is given by

$$x_{n+1} = x_n + \varepsilon(A(c_n^{\varepsilon,1}, c_n^{\varepsilon,2})u_n + B(c_n^{\varepsilon,1}, c_n^{\varepsilon,2})),$$

with $u_n = (u_n^1, u_n^2) \in \Gamma = [0, 1] \times [0, 1]$ and

$$A(c^1, c^2) = \begin{pmatrix} c^1 & c^2 \\ c^2 & 0 \end{pmatrix} \text{ and } B(c^1, c^2) = \begin{pmatrix} 0 \\ -z \end{pmatrix}, \quad \text{for all } (c^1, c^2) \in \mathcal{M}.$$

Next, let us make the following assumptions, which indicates $G(x, \alpha, u)$ verifies a Lipschitz-like condition in the x variable, and $G(x, \alpha, u)$ has at most polynomial growth rate in x. This condition will be used throughout the chapter.

(A12.1) For each $\alpha \in \mathcal{M}$, $G(x, \alpha, u)$ is jointly convex in (x, u) and in addition,

$$|G(x, \alpha, u) - G(y, \alpha, u)| \leq K(1 + |x|^\kappa + |y|^\kappa)|x - y|,$$

for some positive constants K and κ. Moreover,

$$0 \leq G(x, \alpha, u) \leq K(1 + |x|^\kappa).$$

Lemma 12.3. *There exists a constant K, independent of ε and α such that for all x and y,*

$$|v^\varepsilon(x,\alpha) - v^\varepsilon(y,\alpha)| \le K(1 + |x|^\kappa + |y|^\kappa)|x - y|.$$

Properties of Value Functions. Next, we derive basic properties of the value functions. These results are needed for convergence of value functions.

Lemma 12.4. *If there exists a subsequence of $\varepsilon \to 0$ (still denoted by ε for simplicity) such that $v^\varepsilon(x,\alpha) \to v^0(x,\alpha)$ for $\alpha \in \mathcal{M}$, the following assertions hold:*

(a) *For $\alpha \in \mathcal{M}_k$, the limit function $v^0(x,\alpha)$ depends only on k, i.e., $v^0(x,\alpha) = v(x,k)$ for some function $v(x,k)$.*

(b) *For $j = 1,\ldots,m_*$, denote the limit of $v^\varepsilon(x,s_{*j})$ (i.e., $v^0(x,s_{*j})$) by $v(x,*j)$ and write $v(x,*) = (v(x,*1),\ldots,v(x,*m_*))'$. Then*

$$v(x,*) = a^{l_1}v(x,1) + \cdots + a^{l_0}v(x,l_0), \tag{12.5}$$

where $v(x,i)$, for $i = 1,\ldots,l_0$, is given in Part (a).

Limit Problem. Now we show that there exists a limit problem as $\varepsilon \to 0$, which is a continuous-time control problem. The limit problem is simple to solve. From the optimal controls of the limit problem, we can construct controls for the original problem. We will show that such constructed controls are nearly optimal.

Let Γ_0 denote the control points consisting of $U = (U^1,\ldots,U^{l_0})$ with $U^k = (u^{k1},\ldots,u^{km_k})$ and $u^{kj} \in \Gamma$. Define the running cost function for the limit problem as

$$\overline{G}(x,k,U^k) = \sum_{j=1}^{m_k} \nu^{kj} G(x,s_{kj},u^{kj}).$$

Moreover, define

$$F(x,k,U) = \sum_{j=1}^{m_k} \nu^{kj}(A(s_{kj})u^{kj} + B(s_{kj})).$$

Note that $F(x,k,U)$ depends only on U^k, i.e., $F(x,k,U) = F(x,k,U^k)$.

The DP equation for the limit problem has the following form:

$$\rho v(x,k) = \min_{U^k}\left\{\overline{G}(x,k,U^k) + \left(\frac{\partial v(x,k)}{\partial x}\right) \cdot F(x,k,U^k) + \overline{Q}_* v(x,\cdot)(k)\right\}. \tag{12.6}$$

Let $\overline{\alpha}(t) \in \overline{\mathcal{M}} = \{1, 2, \ldots, l_0\}$ be the Markov chain generated by $\overline{Q}_*$ given in (6.16). Consider a class of controls $\mathcal{A}^0$,

$$\mathcal{A}^0 := \left\{ U(t) = (U^1(t), \cdots, U^{l_0}(t)) \in \Gamma_0 : \right.$$
$$\left. U(t) \text{ is progressively measurable w.r.t. } \sigma\{\overline{\alpha}(s) : s \leq t\} \right\}.$$

The corresponding limit control problem is

$$\mathcal{P}^0 : \begin{cases} \text{minimize: } J(x, k, U(\cdot)) = E \int_0^\infty e^{-\rho t} \overline{G}(\overline{x}(t), \overline{\alpha}(t), U(t)) dt, \\ \text{subject to: } \dfrac{d\overline{x}(t)}{dt} = F(\overline{x}(t), \overline{\alpha}(t), U(t)), \ t \geq 0, \\ \qquad\qquad\qquad \overline{x}_0 = i, \ \overline{\alpha}(0) = k, U(\cdot) \in \mathcal{A}^0, \\ \text{value function: } v(x, k) = \inf_{U(\cdot) \in \mathcal{A}^0} J(x, k, U(\cdot)). \end{cases}$$

Theorem 12.5. *For each* $\alpha \in \mathcal{M}_k$ *and* $k = 1, \ldots, l_0$,

$$\lim_{\varepsilon \to 0} v^\varepsilon(x, \alpha) = v(x, k), \tag{12.7}$$

and for $\alpha = s_{*j}$,

$$\lim_{\varepsilon \to 0} v^\varepsilon(x, *) = a^1 v(x, 1) + \cdots + a^{l_0} v(x, l_0),$$

where $v^\varepsilon(x, *) = (v^\varepsilon(x, s_{1*}), \ldots, v^\varepsilon(x, s_{m_**}))$.

Asymptotic Optimality. We consider the asymptotic optimality of our approximation scheme. Let

$$U^o(x) = (U^{o,1}(x), \ldots, U^{o,l_0}(x)) \text{ with} \tag{12.8}$$
$$U^{o,k}(x) = (u^{o,k1}(x), \ldots, u^{o,km_k}(x))$$

be an optimal control for the limit problem $\mathcal{P}^0$.

(A12.2) The cost function $G(x, \alpha, u)$ is twice differentiable with respect to u such that
$$\frac{\partial^2 G(x, \alpha, u)}{\partial u^2} \geq c_0 I > 0,$$
for some constant c_0, where $(\partial^2/\partial u^2)G(x, \alpha, u)$ denotes the Hessian matrix of $G(\cdot)$ with respect to u. There exists a constant K such that
$$\left| G(x + y, \alpha, u) - G(x, \alpha, u) - \left\langle \frac{\partial}{\partial x} G(x, \alpha, u), y \right\rangle \right|$$
$$\leq K(1 + |x|^\kappa)|y|^2.$$

Under (A12.1) and (A12.2), as in Yin and Zhang [158, Lemma 9.10], we can show that

(a) $v(x, k)$ is convex and continuously differentiable.

(b) $U^\circ(x)$ satisfies

$$|U^\circ(x) - U^\circ(y)| \leq K(1 + |x|^\kappa + |y|^\kappa)|x - y|, \qquad (12.9)$$

for some $K > 0$.

(c) $U^\circ(x)$ is an optimal feedback control, i.e., $U(t) = U^\circ(\bar{x}(t)) \in \mathcal{A}^0$ and $J(x, k, U(\cdot)) = v(x, k)$.

Pick $u^* \in \Gamma$. Construct a control for the original discrete-time control problem

$$u_n^\varepsilon = \sum_{i=1}^{l_0} \sum_{j=1}^{m_k} I_{\{\alpha_n^\varepsilon = s_{ij}\}} u^{\circ,ij}(x_n) + \sum_{j=1}^{m_*} I_{\{\alpha_n^\varepsilon = s_{*j}\}} u^*, \qquad (12.10)$$

where $u^{\circ,ij}(x_n)$ is the optimal control of the limit problem defined in (12.8).

Let $J^\varepsilon(x, \alpha)$ be the cost under this control, i.e.,

$$J^\varepsilon(x, \alpha) = J^\varepsilon(x, \alpha, u^\varepsilon).$$

Let $S(r)$ be a ball with radius r and centered at the origin, i.e., $S(r) = \{x \in \mathbb{R}^{n_1} : |x|^2 = \sum_{j=1}^{n_1} (x^j)^2 \leq r^2\}$.

Lemma 12.6. *Assume (A12.2). Then $J^\varepsilon(x, \alpha)$ is locally uniformly continuous, i.e., for each $r > 0$, given $\eta > 0$, there exists $\delta > 0$ such that for $x, y \in S(r)$ and $|x - y| < \delta$, we have*

$$|J^\varepsilon(x, \alpha) - J^\varepsilon(y, \alpha)| < \eta,$$

for all $\varepsilon > 0$.

Theorem 12.7. *The control u^ε is asymptotically optimal for the original control system (12.1) and (12.2) in the sense that*

$$\lim_{\varepsilon \to 0} |J^\varepsilon(x, \alpha, u^\varepsilon) - v^\varepsilon(x, \alpha)| = 0, \quad \text{for all } \alpha \in \mathcal{M}. \qquad (12.11)$$

12.3   Examples

We continue our study of Example 12.1. Our objective is to choose a control u to minimize the surplus costs

$$J^\varepsilon(x, \alpha, u) = E \sum_{n=0}^{\infty} (1 - \rho\varepsilon)^n \varepsilon \left(c^+ x_n^+ + c^- x_n^-\right),$$

where c^+ and c^- are positive constants, $x^+ = \max\{0, x\}$, and $x^- = \max\{0, -x\}$.

Consider the case in which the demand fluctuates more rapidly than the capacity process. In this case, z_n^ε is the fast-changing process, and $c_n^\varepsilon = c_n$ is the slowly varying capacity process being independent of ε. The idea is to derive a limit problem in which the fast-fluctuating demand is replaced by its average. Thus one may ignore the detailed changes in the demand when making an average production planning decision.

Let

$$\mathcal{M} = \{s_{11}, s_{12}, s_{21}, s_{22}\} = \{(1, z_1), (1, z_2), (0, z_1), (0, z_2)\}.$$

Consider the transition matrix P^ε given by

$$P^\varepsilon = \begin{pmatrix} 1 - \lambda_z & \lambda_z & 0 & 0 \\ \mu_z & 1 - \mu_z & 0 & 0 \\ 0 & 0 & 1 - \lambda_z & \lambda_z \\ 0 & 0 & \mu_z & 1 - \mu_z \end{pmatrix} + \varepsilon \begin{pmatrix} -\lambda_c & 0 & \lambda_c & 0 \\ 0 & -\lambda_c & 0 & \lambda_c \\ \mu_c & 0 & -\mu_c & 0 \\ 0 & \mu_c & 0 & -\mu_c \end{pmatrix},$$

where $0 < \lambda_z < 1$ is the jump rate of the demand from z_1 to z_2 and $0 < \mu_z < 1$ is the rate from z_2 to z_1; λ_c and μ_c are the breakdown and repair rates, respectively.

In this example,

$$\overline{Q}_* = \begin{pmatrix} -\lambda_c & \lambda_c \\ \mu_c & -\mu_c \end{pmatrix}.$$

Moreover, the control set for the limit problem

$$\{(u^{11}, u^{12}, 0, 0) : 0 \le u^{11}, u^{12} \le 1\},$$

since when $c_n = 0$ the system is independent of the values of u^{21} and u^{22}. Furthermore, since G is independent of u, we have $\overline{G} = G$. Therefore, the system of equations in the limit problem $\mathcal{P}^0$ is given by

$$\frac{dx(t)}{dt} = c(t)u(t) - \overline{z}, \quad x(0) = x,$$

where $\overline{a}(t) = c(t)$ is a Markov chain generated by $\overline{Q}_*$, and $\overline{z} = \nu^{11}z_1 + \nu^{12}z_2$ with

$$(\nu^{11}, \nu^{12}) = \left(\frac{\mu_z}{\lambda_z + \mu_z}, \frac{\lambda_z}{\lambda_z + \mu_z} \right).$$

Theorem 12.5 implies that $v^\varepsilon(x, \alpha) \to v^0(x, k)$, for $\alpha \in \mathcal{M}_k$, $k = 1, 2$.

Let

$$A_1 = \begin{pmatrix} -\dfrac{\gamma + \mu_c}{\overline{z}} & \dfrac{\mu_c}{\overline{z}} \\ -\dfrac{\lambda_c}{1 - \overline{z}} & \dfrac{\gamma + \lambda_c}{1 - \overline{z}} \end{pmatrix}.$$

It is easy to see that A_1 has two real eigenvalues, one greater than 0 and the other less than 0. Let $a_- < 0$ denote the negative eigenvalue of the matrix A_1 and define

$$\tilde{x} = \max\left(0, \frac{1}{a_-} \log\left[\frac{c^+}{c^+ + c^-}\left(1 + \frac{\gamma\bar{z}}{\lambda_c\bar{z} - (\gamma + \mu_c + \bar{z}a_-)(1 - \bar{z})}\right)\right]\right).$$

The optimal control for $\mathcal{P}^0$ is given by

$$\text{If } c(t) = 0, \ u^o(x) = 0, \text{ and}$$

$$\text{if } c(t) = 1, \ u^o(x) = \begin{cases} 0, & \text{if } x > \tilde{x}, \\ \bar{z}, & \text{if } x = \tilde{x}, \\ 1, & \text{if } x < \tilde{x}. \end{cases}$$

Let

$$U^o(x) = (u^{o,11}(x), u^{o,12}(x), u^{o,21}(x), u^{o,22}(x))$$

be the optimal control for $\mathcal{P}^0$. Note that $(u^{o,11}(x), u^{o,12}(x))$ corresponds to $c(t) = 1$ and $(u^{o,21}(x), u^{o,22}(x))$ corresponds to $c(t) = 0$. Naturally, $(u^{o,21}(x), u^{o,22}(x)) = 0$, since, when $c(t) = 0$, there should be no production. When $c(t) = 1$, let $\nu^{11}u^{o,11}(x) + \nu^{12}u^{o,12}(x) = u^o(x)$. It should be pointed out that in this case the solution $(u^{o,11}(x), u^{o,12}(x))$ is not unique.

Using $u^{o,11}(x)$ and $u^{o,12}(x)$, we construct a control for $\mathcal{P}^0$ as

$$u_n^\varepsilon = u^\varepsilon(x_n, \alpha_n^\varepsilon) = u^\varepsilon(x_n, c_n^\varepsilon, z_n^\varepsilon),$$

where

$$u^\varepsilon(x, c, z) = I_{\{c=1\}}\left(I_{\{z=z_1\}}u^{o,11}(x) + I_{\{z=z_2\}}u^{o,12}(x)\right)$$

$$+ I_{\{c=0\}}\left(I_{\{z=z_1\}}u^{o,21}(x) + I_{\{z=z_2\}}u^{o,22}(x)\right)$$

$$= I_{\{c=1\}}\left(I_{\{z=z_1\}}u^{o,11}(x) + I_{\{z=z_2\}}u^{o,12}(x)\right).$$

Note that in this example, the optimal control $U^o(x)$ is not Lipschitz. Therefore the conditions in Theorem 12.7 are not satisfied. However, noting that

$$|x_n - y_n| = O(\sqrt{n}\varepsilon + |x - y|),$$

where x_n and y_n satisfy (12.4). The foregoing implies

$$|J^\varepsilon(x, \alpha) - J^\varepsilon(y, \alpha)| = O(\varepsilon^{\frac{1}{4}} + |x - y|).$$

As in Theorem 12.7, using Lemma 14.33, we can still show that the constructed control u_n^ε in (12.10) is asymptotically optimal.

One may also consider the case in which the capacity process changes rapidly, whereas the random demand is relatively slowly varying. As in the previous case, assume c_n^ε is the capacity process and $z_n^\varepsilon = z_n$ is the demand. Using exactly the same approach, we may resolve this problem. The discussion is analogous to the previous case; the details are omitted.

12.4   Proofs of Results

Proof of Lemma 12.3. It suffices to show that

$$|v^\varepsilon(x, \alpha) - v^\varepsilon(y, \alpha)| \le K(1 + |x|^\kappa + |y|^\kappa)|x - y|, \text{ for all } x, y \in \mathbb{R}^{n_1}.$$

Given u_n, let x_n and y_n be the corresponding states with $x_0 = x$ and $y_0 = y$, respectively, i.e.,

$$x_{n+1} = x_n + \varepsilon(A(\alpha_n)u_n + B(\alpha_n)), \ x_0 = x,$$
$$y_{n+1} = y_n + \varepsilon(A(\alpha_n)u_n + B(\alpha_n)), \ y_0 = y.$$

Then, we have

$$x_{n+1} - y_{n+1} = x_n - y_n, \ n = 0, 1, \ldots$$

Hence, $x_n - y_n = x - y$, for all n. Moreover, recall that u_n is bounded, which implies there exists some K such that

$$|x_n| \le |x| + Kn\varepsilon, \ |y_n| \le |y| + Kn\varepsilon,$$

for all $\varepsilon > 0$ and $n = 0, 1, \ldots$ It reveals that

$$|J^\varepsilon(x, \alpha, u.) - J^\varepsilon(y, \alpha, u.)|$$
$$\le E \sum_{n=0}^\infty (1 - \rho\varepsilon)^n \varepsilon |G(x_n, \alpha_n, u_n) - G(y_n, \alpha_n, u_n)|$$
$$\le \sum_{n=0}^\infty (1 - \rho\varepsilon)^n \varepsilon K |x - y|(1 + |x_n|^\kappa + |y_n|^\kappa)$$
$$= |x - y| E \sum_{n=0}^\infty (1 - \rho\varepsilon)^n \varepsilon K (1 + |x_n|^\kappa + |y_n|^\kappa).$$

To complete the proof, it suffices to show

$$E \sum_{n=0}^\infty (1 - \rho\varepsilon)^n \varepsilon K (1 + |x_n|^\kappa + |y_n|^\kappa)$$

is bounded. Note that

$$\frac{(|x| + |y|)^\kappa}{1 + |x|^\kappa + |y|^\kappa}$$

is bounded. Therefore,

$$1 + |x_n|^\kappa + |y_n|^\kappa \le 1 + (|x| + Kn\varepsilon)^\kappa + (|y| + Kn\varepsilon)^\kappa$$
$$\le K_0(1 + |x|^\kappa + |y|^\kappa + K(n\varepsilon)^\kappa),$$

for some K_0. It remains to show that

$$\sum_{n=0}^{\infty}(1 - \rho\varepsilon)^n \varepsilon(n\varepsilon)^\kappa < \infty.$$

In fact, note that $(1 - \rho\varepsilon) \leq e^{-\rho\varepsilon}$. Let N be a number large enough such that $x^\kappa e^{-\rho x} \leq e^{-\rho x/2}$, for $x \geq N$. Then we have

$$\sum_{n=0}^{\infty}(1 - \rho\varepsilon)^n \varepsilon(n\varepsilon)^\kappa \leq \sum_{n=0}^{\lfloor N/\varepsilon\rfloor - 1} \varepsilon K + \sum_{n=\lfloor N/\varepsilon\rfloor}^{\infty} \varepsilon^{-\rho n\varepsilon/2} \leq K. \quad \square$$

Proof of Lemma 12.4. Given $u \in \Gamma$, we have, for $\alpha \in \mathcal{M}$,

$$\rho\varepsilon v^\varepsilon(x, \alpha) \leq \varepsilon G(x, \alpha, u)$$
$$+ (1 - \rho\varepsilon)\big(v^\varepsilon(x + \varepsilon(A(\alpha)u + B(\alpha)), \alpha) - v^\varepsilon(x, \alpha)\big)$$
$$+ (1 - \rho\varepsilon)\sum_{\beta \in \mathcal{M}}(p^{\varepsilon,\alpha\beta} - \delta^{\alpha\beta})v^\varepsilon(x + \varepsilon(A(\alpha)u + B(\alpha)), \beta),$$

$$(12.12)$$

where $\delta^{\alpha\beta} = 1$ if $\alpha = \beta$ and 0 otherwise. Using the hypothesis $v^\varepsilon(x, \alpha) \to v^0(x, \alpha)$ and sending $\varepsilon \to 0$ in

$$\sum_{\beta \in \mathcal{M}}(p^{\varepsilon,\alpha\beta} - \delta^{\alpha\beta})v^0(x, \beta) \geq 0,$$

we obtain

$$P^k v^k(x) \geq v^k(x), \text{ for } k = 1, \ldots, l_0,$$

where P^k is the kth block transition matrix defined in (6.3), and $v^k(x) = (v^0(x, s_{k1}), \ldots, v^0(x, s_{km_k}))'$. Now, the irreducibility of P^k and Lemma 14.37 (in Appendix) imply that

$$v(x, k) := v^0(x, s_{k1}) = v^0(x, s_{k2}) = \cdots = v^0(x, s_{km_k}).$$

This proves Part (a).

Next we establish Part (b). Let $u^\varepsilon \in \Gamma$ be an optimal control. Then the equality in (12.12) holds under u^ε. Sending $\varepsilon \to 0$ in the last m_* equations leads to

$$P^{*,1}v^{0,1} + \cdots + P^{*,l_0}v^{0,l_0} + (P^* - I)v(x, *) = 0,$$

where $v^{0,k}(x) = \mathbb{1}_{m_k}v(x, k)$. This yields

$$v(x, *) = -(P^* - I)^{-1}(P^{*,1}v^{0,1}(x) + \cdots + P^{*,l_0}v^{0,l_0}(x)).$$

In view of the definition of a^i, we obtain

$$v(x, *) = \sum_{i=1}^{l_0} -(P^* - I)^{-1}P^{*,i}\mathbb{1}_{m_i}v(x, i)$$
$$= \sum_{i=1}^{l_0} a^i v(x, i),$$

which proves Part (b).   □

Proof of Theorem 12.5. It suffices to show (12.7). By Lemma 12.3, for each sequence of $\{\varepsilon \to 0\}$, there exists a further subsequence (still indexed by ε for notational simplicity) such that $v^\varepsilon(x, \alpha)$ converges. Denote the limit by $v^0(x, \alpha)$. Then by Lemma 12.4, $v^0(x, \alpha) = v(x, k)$. That is, the exact value of α is unimportant and only which subspace $\mathcal{M}_k$ it belongs to matters.

Fix $k = 1, \dots, l_0$. For any $\alpha = s_{kj} \in \mathcal{M}_k$, let $v(x, k)$ be a limit of $v^\varepsilon(x, s_{kj})$ for some subsequence of ε. Given x_0, let a function $\phi(x, k) \in C^1(\mathbb{R}^n)$ such that $v(x, k) - \phi(x, k)$ has a strictly local maximum at x_0 in a neighborhood $N(x_0)$. Choose $x^{\varepsilon, kj} \in N(x_0)$ such that for each $\alpha = s_{kj} \in \mathcal{M}_k$,

$$v^\varepsilon(x^{\varepsilon, kj}, s_{kj}) - \phi(x^{\varepsilon, kj}, k) = \max_{x \in N(x_0)} \{v^\varepsilon(x, s_{kj}) - \phi(x, k)\}.$$

Then it follows that $x^{\varepsilon, kj} \to x_0$ as $\varepsilon \to 0$. Given $u^{kj} \in \Gamma$, let

$$X^{\varepsilon, kj} = x^{\varepsilon, kj} + \varepsilon(A(s_{kj})u^{kj} + B(s_{kj})).$$

Then $X^{\varepsilon, kj} \in N(x_0)$ for ε small enough. We have

$$\rho v^\varepsilon(x^{\varepsilon, kj}, s_{kj}) \leq G(x^{\varepsilon, kj}, s_{kj}, u) + \left(\frac{1 - \rho \varepsilon}{\varepsilon}\right)(v(X^{\varepsilon, kj}, s_{kj}) - v^\varepsilon(x_{kj}^\varepsilon, s_{kj}))$$
$$\times \left(\frac{1 - \rho \varepsilon}{\varepsilon}\right) \sum_{\beta \in \mathcal{M}} (p^{\varepsilon, s_{kj}\beta} - \delta^{s_{kj}\beta})v^\varepsilon(X^{\varepsilon, kj}, \beta).$$

$$(12.13)$$

Noting the definition of $x^{\varepsilon, kj}$, we have

$$v^\varepsilon(X^{\varepsilon, kj}, s_{kj}) - \phi(X^{\varepsilon, kj}, k) \leq v^\varepsilon(x^{\varepsilon, kj}, s_{kj}) - \phi(x^{\varepsilon, kj}, k).$$

Recall, in addition, that $v^\varepsilon \to v$ and $x^{\varepsilon, kj} \to x_0$. It follows that as $\varepsilon \to 0$,

$$\left(\frac{1 - \rho \varepsilon}{\varepsilon}\right) \sum_{j=1}^{m_k} \nu^{kj} \left(v^\varepsilon(X^{\varepsilon, kj}, s_{kj}) - v^\varepsilon(x^{\varepsilon, kj}, s_{kj})\right)$$
$$\leq \left(\frac{1 - \rho \varepsilon}{\varepsilon}\right) \sum_{j=1}^{m_k} \nu^{kj} \left(\phi(X^{\varepsilon, kj}, k) - \phi(x^{\varepsilon, kj}, k)\right)$$
$$\to \frac{\partial \phi(x_0, k)}{\partial x} \cdot \sum_{j=1}^{m_k} \nu^{kj}(A(s_{kj})u^{kj} + B(s_{kj})).$$

Note also that

$$v^\varepsilon(X^{\varepsilon, kj}, s_{kj_1}) - \phi(X^{\varepsilon, kj}, k) \leq v^\varepsilon(x^{\varepsilon, kj_1}, s_{kj_1}) - \phi(x^{\varepsilon, kj_1}, k).$$

We have

$$\sum_{j=1}^{m_k} \nu^{kj} \sum_{j_1=1}^{m_k} (p^{k, s_{kj} s_{kj_1}} - \delta^{s_{kj} s_{kj_1}}) v^\varepsilon (X^{\varepsilon, kj}, s_{kj_1})$$

$$= \sum_{j=1}^{m_k} \nu^{kj} \left(\sum_{j_1=1}^{m_k} p^{k, s_{kj} s_{kj_1}} v^\varepsilon (X^{\varepsilon, kj}, s_{kj_1}) - v^\varepsilon (X^{\varepsilon, kj}, s_{kj}) \right)$$

$$\leq \sum_{j=1}^{m_k} \nu^{kj} \sum_{j_1=1}^{m_k} p^{k, s_{kj} s_{kj_1}} \Big((v^\varepsilon (x^{\varepsilon, kj_1}, s_{kj_1}) - \phi(x^{\varepsilon, kj_1}, k))$$

$$- (v^\varepsilon (X^{\varepsilon, kj}, s_{kj}) - \phi(X^{\varepsilon, kj}, k)) \Big) = 0.$$

We have used the notation $P^k = (p^{k, s_{kj}, s_{kj_1}})$ to denote the kth transition matrix in (6.3). Write (12.13) in vector form, multiply both sides by

$$\nu = \begin{pmatrix} \nu^1 & & & \\ & \ddots & & \\ & & \nu^{l_0} & \\ 0_{m_* \times m_1} & \cdots & 0_{m_* \times m_l} & 0_{m_* \times m_*} \end{pmatrix},$$

and use Lemma 12.4 to obtain that $v(x, k)$ is a viscosity subsolution to (12.6).

Similarly, v is also a viscosity supersolution to (12.6). Moreover, the uniqueness of solution of (12.6) (see Theorem 14.27) implies that $v(x, k)$ is the value function for $\mathcal{P}^0$. Thus, for any subsequence of ε (indexed also by ε), $v^\varepsilon(x, \alpha) \to v^0(x, k)$. The desired result thus follows. $\square$

Proof of Lemma 12.6. Given x and y in $\mathbb{R}^{n_1}$, let x_n and y_n be the corresponding states

$$x_{n+1} = x_n + \varepsilon \bigg\{ A(\alpha_n) \bigg(\sum_{i=1}^{l_0} \sum_{j=1}^{m_k} I_{\{\alpha_n^\varepsilon = s_{ij}\}} u^{o, ij}(x_n) + \sum_{j=1}^{m_*} I_{\{\alpha_n^\varepsilon = s_{*j}\}} u^* \bigg) + B(\alpha_n) \bigg\},$$

and

$$y_{n+1} = y_n + \varepsilon \bigg\{ A(\alpha_n) \bigg(\sum_{i=1}^{l_0} \sum_{j=1}^{m_k} I_{\{\alpha_n^\varepsilon = s_{ij}\}} u^{o, ij}(y_n) + \sum_{j=1}^{m_*} I_{\{\alpha_n^\varepsilon = s_{*j}\}} u^* \bigg) + B(\alpha_n) \bigg\},$$

with $x_0 = x$ and $y_0 = y$. It follows that

$$x_{n+1} - y_{n+1} = x_n - y_n$$
$$+ \varepsilon A(\alpha_n) \left(\sum_{i=1}^{l_0} \sum_{j=1}^{m_k} I_{\{\alpha_n^\varepsilon = s_{ij}\}} (u^{o,ij}(x_n) - u^{o,ij}(y_n)) \right).$$

Let $\phi_n(x) = |x_n - y_n|$. Then, for given $T > 0$ and $n \leq T/\varepsilon$, we have, for $(x, y) \in S(r)$,

$$\phi_{n+1} \leq \phi_n + K\varepsilon(1 + |x|^\kappa + |y|^\kappa + T^\kappa)\phi_n$$
$$\leq \phi_n + K\varepsilon(1 + 2r^\kappa + T^\kappa)\phi_n,$$

with $\phi_0 = |x - y|$. This implies that

$$\phi_n \leq (1 + K\varepsilon)^n |x - y|,$$

where T may depend on r and T. Moreover, recall that

$$G(x_n, \alpha_n, u_n) \leq K(1 + |x_n|^\kappa + (n\varepsilon)^\kappa),$$
$$G(y_n, \alpha_n, u_n) \leq K(1 + |y_n|^\kappa + (n\varepsilon)^\kappa).$$

We have

$$|J^\varepsilon(x, \alpha) - J^\varepsilon(y, \alpha)|$$
$$\leq E \sum_{n=0}^{\lfloor T/\varepsilon \rfloor - 1} (1 - \rho\varepsilon)^n \varepsilon (1 + K\varepsilon)^n |x - y|$$
$$+ \sum_{n=\lfloor T/\varepsilon \rfloor}^{\infty} (1 - \rho\varepsilon)^n \varepsilon K(1 + |x|^\kappa + |y|^\kappa + (n\varepsilon)^\kappa)$$
$$= |x - y| O(e^{(K-\rho)T}) + O(e^{-\rho T/2}).$$

For any $\eta > 0$, choose T large enough such that $O(e^{-\rho T/2}) < \eta/2$ and $\delta = \eta O(e^{(K-\rho)T})/2$. Then whenever $|x - y| < \delta$, we have

$$|J^\varepsilon(x, \alpha) - J^\varepsilon(y, \alpha)| < \eta. \quad \square$$

Proof of Theorem 12.7. As in (12.3), we can show (see Bertsekas [13]) that $J^\varepsilon(x, \alpha)$ satisfies

$$J^\varepsilon(x, s_{ij}) = \varepsilon G(x, s_{ij}, u^{o,ij}(x)) + (1 - \rho\varepsilon)$$
$$\times \sum_{\beta \in \mathcal{M}} p^{\varepsilon, s_{ij}\beta} J^\varepsilon(x + \varepsilon(A(s_{ij})u^{o,ij}(x) + B(s_{ij})), \beta).$$

$$J^\varepsilon(x, s_{*j}) = \varepsilon G(x, s_{*j}, u^*) + (1 - \rho\varepsilon)$$
$$\times \sum_{\beta \in \mathcal{M}} p^{\varepsilon, s_{*j}\beta} J^\varepsilon(x + \varepsilon(A(s_{*j})u^* + B(s_{*j})), \beta).$$

As in Lemma 12.4, we can show that if $v^\varepsilon(x, \alpha) \to v^0(x, \alpha)$ for some sequence of ε, then the limit depends only on k for $\alpha \in \mathcal{M}_k$. It can be shown similarly as in the proof of Theorem 12.5, together with Lemma 12.6, that $J^\varepsilon(x, \alpha)$ convergence to $v^0(x, k)$, for $\alpha \in \mathcal{M}_k$. Therefore,

$$|J^\varepsilon(x, \alpha) - v^\varepsilon(x, \alpha)|$$
$$\leq |J^\varepsilon(x, \alpha) - v^0(x, k)| + |v^0(x, k) - v^\varepsilon(x, \alpha)| \to 0,$$

and the theorem is proved. $\square$

12.5   Notes

This chapter focuses on approximation schemes for a class of discrete-time production planning systems. It provides a systematic approach in reducing complexity of the underlying systems. The computation load is reduced considerably than finding the optimal controls of the original problem directly. This is the most attractive feature of our approach. Furthermore, the asymptotic optimality ensures that such an approximation is almost as good as the optimal one for sufficiently small ε.

This chapter is based on the work of Zhang and Yin [184]. General references on manufacturing systems can be found in the books by Gershwin [61], Buzacott and Shanthikumar [28], Sethi and Zhang [136], and Sethi, Zhang, and Zhang [135]. For structural properties of production policies, see Akella and Kumar [3], Bielecki and Kumar [16], and Zhang and Yin [178], among others. For manufacturing systems involving preventive maintenance, see Boukas [25] and Boukas and Haurie [26].

13
Stochastic Approximation

13.1 Introduction

This chapter is concerned with a class of stochastic approximation (SA) problems with regime switching. Originating from discrete stochastic optimization and time-varying parameter tracking, our study focuses on analyzing the performance of the algorithm in which the time-varying parameter process is a Markov chain.

One of the motivations of our study stems from the following discrete stochastic optimization problem. Let $\mathcal{S}$ be a finite set. For convenience, let $\mathcal{S} = \{e_1, e_2, \ldots, e_S\}$, where e_i, $i = 1, 2, \ldots, S$, are the standard unit vectors in $\mathbb{R}^S$. Consider

$$\min_{y \in \mathcal{S}} EJ(y), \qquad (13.1)$$

where the objective function $J(\cdot)$ is defined on $\mathcal{S}$. One of the difficulties in solving the minimization problem is that the notion of gradient of J is not applicable since $\mathcal{S}$ consists of isolated points only. As a result, we cannot use the gradient information to identify a descent search direction. Let $\mathcal{K} \subset \mathcal{S}$ denote the set of global minimizers for (13.1). If the expected value in (13.1) could be calculated in closed form, one would use stochastic integer programming methods to solve the problem. Nevertheless, $EJ(y)$ cannot be evaluated analytically and only $\{J_n(y)\}$, a sequence of independent and identically distributed (i.i.d.) random variables (with finite second moments), can be measured or simulated. To solve this discrete stochastic optimization problem, Pflug [124, Chapter 5.3] designed a procedure based on an exhaustive enumeration. It proceeds as follows: For each possible candidate $y \in \mathcal{S}$, compute the sample average

$$\widehat{J}_n(y) = \frac{1}{n} \sum_{l=1}^{n} J_l(y),$$

via simulation for large n, and pick $y^* = \arg\min_{y \in \mathcal{S}} \widehat{J}_n(y)$. By virtue of the well-known strong law of large numbers, $\widehat{J}_n(y) \to EJ_1(y) = EJ(y)$ w.p.1, as $n \to \infty$. This and the finiteness of $\mathcal{S}$ imply that as $n \to \infty$,

$$\arg\min \widehat{J}_n(y) \to \arg\min EJ(y) \quad \text{w.p.1}. \tag{13.2}$$

However, this procedure is inefficient since many unnecessary calculations are wasted at those irrelevant (non-optimal) points. Recently, Andradottir considered a combination of discrete stochastic optimization and random search procedure in [5]. It begins with an assumption of stochastic ordering: For each $x, y \in \mathcal{S}$, there exists some random variable $Y^{x,y}$ that is used as a measure to compare two points x and y (If $Y^{x,y} > 0$, it is deemed that y is better than x). Using $Y^{x,y}$ to decide the point to choose in the next move generates a sequence of estimates recursively, which is essentially a variation of the construction of the empirical measures. Effectively, the algorithm generates a homogeneous Markov chain taking values in $\mathcal{S}$ that spends more time at the global optimum than at any other points of $\mathcal{S}$.

The essence is to design an algorithm that is both consistent (i.e., the algorithm converges to the true parameter) and attractive to the minimum (i.e., the iterates move toward the minimizer in each step). The one to be examined in this chapter is a regime-switching stochastic approximation algorithm. We analyze the performance of the algorithm, which is an optimization problem with the underlying parameter (minimum) being time varying. Such algorithms are also useful in wireless communications and blind multiuser detection. The algorithm uses a constant-step size, and updates a sequence of occupation measures; we focus on tracking invariant measures of Markovian parameters. Due to the limitation of tracking capability, in the traditional setup of tracking analysis of time-varying parameters, it is often assumed that the magnitude of the parameter variations is small (so the parameter is regarded as slowly varying). In the Markovian setup, this small variation assumption may be violated. Here, we allow the parameter to have discontinuity with large and infrequent jumps. Consider a class of adaptive algorithms for tracking the invariant distribution of a conditional Markov chain (conditioned on another Markov chain whose transition probability matrix is "near" identity). We evaluate the tracking capability of the stochastic approximation algorithm in terms of mean squares tracking error, mean system of switching ordinary differential equations (ODEs), and limit switching diffusions of associated scaled tracking errors.

Contemporary theory of stochastic approximation relies heavily on the so-called limit mean ODE, while the analysis of rates of convergence depends on a limit diffusion process. Such results are not applicable to the tracking problems or discrete stochastic optimization. Due to the time-varying nature and random jumps, one cannot use the classical SA techniques to analyze the algorithms. Nevertheless, by a combined use of SA methods and two-time-scale Markov chains, asymptotic properties of the algorithm are obtainable.

First, using the perturbed Liapunov function methods of Kushner and Yin [100], we derive mean squares type error bounds for the tracking error based on stability analysis. With the mean squares estimates available, a natural question would be whether an associated limit ODE can be derived via the ODE methods as in the traditional analysis of stochastic approximation and stochastic optimization type algorithms. It turns out that standard ODE method cannot be carried over due to the jump and the time-varying nature of the current system. Nevertheless, a limit system can still be obtained. Distinct from the traditional stochastic approximation results, the limit system is no longer a single ODE but a system of ODEs modulated by a continuous-time Markov chain. Such systems are referred to as ODEs with regime switching. Based on the switching ODEs obtained, we further examine a sequence of suitably normalized errors. Again, in contrast to the classical stochastic approximation method, the limit is not a diffusion but rather a system of diffusions with regime switching. In the system, the diffusion coefficient depends on the modulating Markov chain, which reveals the distinctive time-varying nature of the underlying systems and provides insight into the Markov-modulated stochastic approximation problems.

The rest of the chapter is arranged as follows. Section 13.2 gives the formulation of the problem. Section 13.3 presents the algorithm. Section 13.4 is concerned with a number of asymptotic properties of the algorithm. We first obtain mean squares tracking error bounds, then proceed with a weak convergence analysis for an interpolated sequence of the iterates; further we examine a suitably scaled tracking error sequence of the iterates and derive a switching diffusion limit. The detailed proofs and technical development are in Section 13.5. Additional notes and remarks are included in Section 13.6.

13.2 Problem Formulation

Let $\{\alpha_n^\varepsilon\}$ be a discrete-time Markov chain with finite state space

$$\mathcal{M} = \{z^1, \ldots, z^{m_0}\}. \tag{13.3}$$

We use the following conditions throughout the chapter. Condition (A13.1) characterizes the time-varying parameter as a Markov chain with infrequent transitions, while condition (A13.2) describes the observed signal.

(A13.1) The transition probability matrix of α_n^ε is given by

$$P^\varepsilon = I + \varepsilon Q, \tag{13.4}$$

where $\varepsilon > 0$ is a small parameter, I is an $\mathbb{R}^{m_0 \times m_0}$ identity matrix, and $Q = (q^{ij}) \in \mathbb{R}^{m_0 \times m_0}$ is an irreducible generator of a continuous-time Markov chain. For simplicity, the initial distribution $P(\alpha_0^\varepsilon = z^i) = p_{0,i}$ is independent of ε for each $i = 1, \ldots, m_0$, where $p_{0,i} \geq 0$ and $\sum_{i=1}^{m_0} p_{0,i} = 1$.

(A13.2) Let $\{X_n\}$ be an S-state conditional Markov chain (conditioned on the parameter process). The state space of $\{X_n\}$ is

$$S = \{e_1, \ldots, e_S\}.$$

For each $\alpha \in \mathcal{M}$, $A(\alpha) = (a^{ij}(\alpha)) \in \mathbb{R}^{S \times S}$, the transition probability matrix of X_n, is defined by

$$a^{ij}(\alpha) = P(X_{n+1} = e_j | X_n = e_i, \alpha_n^\varepsilon = \alpha)$$
$$= P(X_1 = e_j | X_0 = e_i, \alpha_0^\varepsilon = \alpha),$$

where $i, j \in \{1, \ldots, S\}$. For each $\alpha \in \mathcal{M}$, the matrix $A(\alpha)$ is irreducible and aperiodic.

Remark 13.1. The parameter ε in (13.4) is sufficiently small so that the entries of the transition probability matrix are nonnegative. The main idea is that although the true parameter is time varying, it is piecewise constant. Moreover, due to the dominating identity matrix in (13.4), $\{\alpha_n^\varepsilon\}$ varies infrequently in time. The time-varying parameter takes a constant value z^i for a random duration and jumps to another state z^j with $j \neq i$ at random time. Note that $\mu'(\alpha)$ (the transpose of the vector $\mu(\alpha)$) may be written as $\nu(\alpha)$ as elsewhere in this book. However, in the stochastic approximation literature, one usually works with column vectors. Thus we use the notation $\mu(\alpha)$ throughout this chapter and also call it a stationary distribution with a slight abuse of notation.

The assumptions on irreducibility and aperiodicity of $A(\alpha)$ imply that for each $\alpha \in \mathcal{M}$, there exists a unique stationary distribution $\mu'(\alpha) \in \mathbb{R}^{1 \times S}$ satisfying

$$\mu'(\alpha) = \mu'(\alpha)A(\alpha), \quad \text{and} \quad \mu'(\alpha)\mathbb{1}_S = 1,$$

where $\mathbb{1}_\ell \in \mathbb{R}^{\ell \times 1}$, with all entries being equal to 1. Our aim is to use a stochastic approximation algorithm to track the time-varying distribution $\mu'(\alpha_n^\varepsilon)$ that depends on the underlying Markov chain α_n^ε.

13.3   Algorithm

The adaptive algorithm is of LMS (least mean squares) type with constant step size. We construct a sequence of estimates $\{\widehat{\mu}_n\}$ for tracking the time-varying distribution $\mu(\alpha_n^\varepsilon)$

$$\widehat{\mu}_{n+1} = \widehat{\mu}_n + \Delta(X_{n+1} - \widehat{\mu}_n), \qquad (13.5)$$

where Δ denotes the step size. The dynamics of the underlying parameter α_n^ε, termed a hypermodel by Benveniste, Metivier, and Priouret [14], are used in the analysis, but they do not explicitly enter the implementation of algorithm (13.5).

In what follows, we first derive a mean squares error bound. Then we proceed with the examination of an interpolated sequence of the iterates. Next, we derive a limit result for a scaled sequence. These three steps are realized in the following section.

13.4   Asymptotic Properties

This section presents several results regarding asymptotic properties of the SA algorithm under consideration. It is divided into three parts. The first part establishes a mean squares estimate for $E|\widehat{\mu}_n - \mu(\alpha_n^\varepsilon)|^2$. The second part derives a limit system of the corresponding switching ODEs. The third part obtains a system of switching diffusions for a suitably scaled sequence of tracking errors. Main results are presented here, whereas the proofs are given in the next section. Throughout the rest of the chapter, we often need to use the notion of fixed-α processes. Given T, by a fixed-α process $X_j(\alpha)$ for $n \leq j \leq T/\varepsilon$, we mean a process in which $\alpha_j^\varepsilon = \alpha$ is fixed for all j with $n \leq j \leq T/\varepsilon$.

13.4.1   Mean Squares Error

Analyzing SA algorithms often requires using Liapunov type functions to prove stability; see Chen [34], and Kushner and Yin [100]. In what follows, we obtain the desired estimate via a stability argument using the perturbed Liapunov function method [100]. Use E_n to denote the conditional expectation with respect to

$$\mathcal{H}_n, \text{ the } \sigma\text{-algebra generated by } \{X_k, \alpha_k^\varepsilon : k \leq n\}. \qquad (13.6)$$

Theorem 13.2. *Assume (A13.1) and (A13.2). In addition, suppose that* $\varepsilon^2 \ll \Delta$ *(i.e.,* $\varepsilon^2/\Delta \to 0$ *as* $\Delta \to 0$*). Then for sufficiently large n,*

$$E|\widehat{\mu}_n - E\mu(\alpha_n^\varepsilon)|^2 = O(\Delta + \varepsilon + \varepsilon^2/\Delta). \qquad (13.7)$$

Remark 13.3 By "for sufficiently large n, (13.7) holds" we mean that there is an n_0 such that for all $n \geq n_0$, the bound (13.7) holds. Note that $\widehat{\mu}_n$ are column vectors taking values in $\mathbb{R}^{S \times 1}$.

In view of Theorem 13.2, to enable the adaptive algorithm to track the time-varying parameter, the ratio ε/Δ cannot be too large. Given the order of magnitude estimate $O(\Delta + \varepsilon + \varepsilon^2/\Delta)$, to balance the two terms Δ and ε^2/Δ, we need to choose $\varepsilon = O(\Delta)$. Therefore, we obtain the following result.

Corollary 13.4. *Under the conditions of Theorem* 13.2, *if* $\varepsilon = O(\Delta)$, *then for sufficiently large* n, $E|\widehat{\mu}_n - E\mu(\alpha_n^\varepsilon)|^2 = O(\Delta)$.

13.4.2   Limit Switching ODEs

Our objective in this section is to derive a limit system for an interpolated sequence of the iterates. In the literature, a usual assumption used for the step size of a tracking algorithm is that $\varepsilon = o(\Delta)$, which means that the true optimum (time-varying parameter) evolves at a much slower rate than the adaptation speed of the stochastic recursive algorithm. However, applications such as those arising from CDMA (code division multiple access) systems require taking into consideration of algorithms with step size $\varepsilon = O(\Delta)$. As can be expected, the analysis for the case of $\varepsilon = O(\Delta)$ is much more difficult. In the rest of this chapter, we consider the case $\varepsilon = O(\Delta)$. For ease of presentation, we choose $\varepsilon = \Delta$ for simplicity henceforth. To proceed, we first examine the asymptotics of the Markov chain, which is essentially an application of the results in Chapter 6. Then we study interpolated sequence of the iterates. The proofs of the results are postponed until Section 13.4.

Limit of the Modulating Markov Chain

Consider the Markov chain α_n^ε. Regarding the probability vector and the n-step transition probability matrix, we have the following approximation results.

Proposition 13.5. *Assume* (A13.1). *Choose* $\varepsilon = \Delta$ *and consider the Markov chain* α_n^Δ. *Then the following assertions hold:*

(a) *Denote* $p_n^\Delta = (P(\alpha_n^\Delta = z^1), \ldots, P(\alpha_n^\Delta = z^{m_0}))$. *Then*

$$
\begin{aligned}
p_n^\Delta &= \widetilde{z}(t) + O(\Delta + e^{-k_0 t/\Delta}), \quad \widetilde{z}(t) \in \mathbb{R}^{1 \times m_0}, \\
\frac{d\widetilde{z}(t)}{dt} &= \widetilde{z}(t)Q, \widetilde{z}(0) = p_0, \\
(P^\Delta)^n &= \widetilde{Z}(t) + O(\Delta + e^{-k_0 t/\Delta}), \\
\frac{d\widetilde{Z}(t)}{dt} &= \widetilde{Z}(t)Q, \ \widetilde{Z}(0) = I.
\end{aligned}
\tag{13.8}
$$

(b) *Define the continuous-time interpolation of α_n^Δ by $\alpha^\Delta(t) = \alpha_n^\Delta$ if $t \in [n\Delta, n\Delta + \Delta)$. Then $\alpha^\Delta(\cdot)$ converges weakly to $\alpha(\cdot)$, which is a continuous-time Markov chain generated by Q.*

Tightness of Interpolated Iterates

For $0 < T < \infty$, we construct a sequence of piecewise constant interpolation of the iterates $\widehat{\mu}_n$ as

$$x^\Delta(t) = \widehat{\mu}_n, \ t \in [\Delta n, \Delta n + \Delta). \tag{13.9}$$

The process $x^\Delta(\cdot)$ so defined is in $D([0, T]; \mathbb{R}^S)$, which is the space of functions defined on $[0, T]$ taking values in $\mathbb{R}^S$ that are right continuous, have left limits, and are endowed with the Skorohod topology. We use weak convergence methods to carry out the analysis. First a tightness result is given in the following lemma.

Lemma 13.6 *Under conditions (A13.1) and (A13.2), $\{x^\Delta(\cdot)\}$ is tight in $D([0, T]; \mathbb{R}^S)$.*

Characterization of the Limit

Consider the pair of processes $(x^\Delta(\cdot), \alpha^\Delta(\cdot))$. This sequence is tight in $D([0, T]; \mathbb{R}^S \times \mathcal{M})$ for $T > 0$, by virtue of Lemma 13.6 and Proposition 13.5. It follows from Prohorov's theorem that we can extract a convergent subsequence, and still index the subsequence by Δ for notational simplicity. Denote the limit of the subsequence by $x(\cdot)$. By virtue of the Skorohod representation (with a slight abuse of notation), we may assume that $x^\Delta(\cdot)$ converges to $x(\cdot)$ w.p.1 and the convergence is uniform on any compact set. We proceed to characterize the limit $x(\cdot)$. The result is stated in the following theorem. Unlike the usual SA approach, the limit is not a deterministic ODE but rather a system of ODEs modulated by a continuous-time Markov chain.

Theorem 13.7. *Under conditions (A13.1) and (A13.2), $(x^\Delta(\cdot), \alpha^\Delta(\cdot))$ converges weakly to $(x(\cdot), \alpha(\cdot))$, which is a solution of the following system of switching ODEs*

$$\frac{d}{dt}x(t) = \mu(\alpha(t)) - x(t), \ x(0) = \widehat{\mu}_0. \tag{13.10}$$

Remark 13.8. The system (13.10) is different from the existing literature on stochastic approximation methods. For SA algorithms, the ODE methods (see Ljung [106] and Kushner and Clark [98]) are now standard and widely used in various applications. The rationale is that the discrete iterations are compared with the continuous dynamics given by a limit ODE. The ODE is then used to analyze the asymptotic properties of the recursive algorithms. Dealing with tracking algorithms for time-varying systems, one may sometimes obtain a non-autonomous differential equation, but the systems are still purely deterministic. Unlike those mentioned above, the limit dynamic system in Theorem 13.7 is only piecewise deterministic due to the underlying Markov chain. In lieu of one ODE, we have a number of ODEs modulated by a continuous-time Markov chain. In any given instance, the Markov chain dictates which regime the system belongs to, and the corresponding system then follows one of the ODEs until the modulating Markov chain jumps into a new location, which explains the time-varying nature of the systems under consideration.

13.4.3   Switching Diffusion Limit

Define $\{v_n\}$, a suitably scaled sequence of tracking errors, and its continuous-time interpolation $v^\Delta(\cdot)$ by

$$v_n = \frac{\widehat{\mu}_n - E\mu(\alpha_n^\Delta)}{\sqrt{\Delta}}, \quad \text{for } n \ge n_0,$$
$$v^\Delta(t) = v_n \text{ for } t \in [n\Delta, n\Delta + \Delta), \tag{13.11}$$

respectively, where n_0 is given in Theorem 13.2 (see Remark 13.3). It follows immediately from Theorem 13.2, $\{v_n\}$ is tight.

By virtue of Proposition 13.5,

$$E\mu(\alpha_n^\Delta) = \overline{\mu}(\Delta n) + O(\Delta + e^{-k_0 n}), \tag{13.12}$$

where

$$\overline{\mu}(\Delta n) \overset{\text{def}}{=} \sum_{i=1}^{m_0} \widetilde{z}^i(\Delta n)\mu(z^i),$$

and $\widetilde{z}^i(t)$ is the ith component of $\widetilde{z}(t)$ given in Proposition 13.5. By (A13.1), $\{\alpha_n^\Delta\}$ is a Markov chain with stationary (time-invariant) transition probabilities. In view of (13.5),

$$v_{n+1} = v_n - \Delta v_n + \sqrt{\Delta}(X_{n+1} - E\mu(\alpha_n^\Delta)) + \frac{E[\mu(\alpha_n^\Delta) - \mu(\alpha_{n+1}^\Delta)]}{\sqrt{\Delta}}. \tag{13.13}$$

Our task in what follows is to figure out the asymptotic properties of $v^\Delta(\cdot)$. We aim to show that it leads to a switching diffusion limit via martingale problem formulation.

Truncation and Tightness

According to the definition (13.11), $\{v_n\}$ is not *a priori* bounded. A convenient way to circumvent this difficulty is to use a truncation device. Let $N > 0$ be a fixed but otherwise arbitrary real number, $B_N(z) = \{z \in \mathbb{R}^S :$

$|z| \leq N\}$ be the spheres with radius N, and $\tau^N(z)$ be a smooth function satisfying

$$\tau^N(z) = \begin{cases} 1, & \text{if } |z| \leq N, \\ 0, & \text{if } |z| \geq N+1. \end{cases}$$

Note that $\tau^N(z)$ is "smoothly" connected between the sphere B_N and B_{N+1}. Now define

$$v_{n+1}^N = v_n^N - \Delta v_n^N \tau^N(v_n^N) + \sqrt{\Delta}(X_{n+1} - E\mu(\alpha_n^\Delta)) \\ + \frac{E[\mu(\alpha_n^\Delta) - \mu(\alpha_{n+1}^\Delta)]}{\sqrt{\Delta}}, \tag{13.14}$$

and define $v^{\Delta,N}(\cdot)$ to be the continuous-time interpolation of v_n^N. It then follows that

$$\lim_{k_0 \to \infty} \limsup_{\Delta \to 0} P(\sup_{0 \leq t \leq T} |v^{\Delta,N}(t)| \geq k_0) = 0 \quad \text{for each } T < \infty$$

and that $v^{\Delta,N}(\cdot)$ is a process that is equal to $v^\Delta(\cdot)$ up until the first exit from B_N, and hence an N-truncation process of $v^\Delta(\cdot)$ (see Definition 14.15). To proceed, we work with $\{v^{\Delta,N}(\cdot)\}$ and derive its tightness and weak convergence first. Then, we let $N \to \infty$ to conclude the proof.

Lemma 13.9. *Under Conditions* (A13.1) *and* (A13.2), $\{v^{\Delta,N}(\cdot)\}$ *is tight in* $D([0,T]; \mathbb{R}^S)$ *and the pair* $\{v^{\Delta,N}(\cdot), \alpha^\Delta(\cdot)\}$ *is tight in* $D([0,T]; \mathbb{R}^S \times \mathcal{M})$.

Representation of Covariance

The results to follow, Lemma 13.10 and Corollary 13.12 for the switching diffusion limit, require representation of the covariance of the conditional Markov chain $\{X_k\}$. This is worked out via the use of a fixed-α process $X_k(\alpha)$ (see also Section 13.4, in particular (13.59)). That is, for any integer $m \geq 0$ satisfying $m \leq k \leq O(1/\Delta)$, with α_k^ε fixed at α, $X_{k+1}(\alpha)$ is a "fixed-α" process (a finite-state Markov chain with 1-step irreducible transition matrix $A(\alpha)$ and stationary distribution $\mu'(\alpha)$). Thus Example 14.11 implies that $\{X_{k+1}(\alpha) - EX_{k+1}(\alpha)\}$ is a ϕ-mixing sequence with zero mean and exponential mixing rate, and hence it is strongly ergodic. As will be seen in Section 13.4, specifically, (13.59), $X_{k+1} - EX_{k+1}$ can be approximated by a fixed-α process $X_{k+1}(\alpha) - EX_{k+1}(\alpha)$. Taking $n = n_\Delta \leq O(1/\Delta)$, as $\Delta \to 0$, $n \to \infty$, and for $m = 1, 2, \ldots$, the strong ergodicity implies

$$\frac{1}{n} \sum_{k_1=m}^{n+m-1} \sum_{k=m}^{n+m-1} (X_{k+1}(\alpha) - EX_{k+1}(\alpha))(X_{k_1+1}(\alpha) - EX_{k_1+1}(\alpha))'$$

$$\to \Sigma(\alpha) \quad \text{w.p.1,}$$

$$\tag{13.15}$$

where $\Sigma(\alpha)$ is an $S \times S$ deterministic matrix and the corresponding expected value

$$E\left(\frac{1}{n}\sum_{k_1=m}^{n+m-1}\sum_{k=m}^{n+m-1}(X_{k+1}(\alpha) - EX_{k+1}(\alpha))(X_{k_1+1}(\alpha) - EX_{k_1+1}(\alpha))'\right)$$
$$\to \Sigma(\alpha).$$

(13.16)

Note that (13.15) is a consequence of ϕ-mixing and strong ergodicity, and (13.16) follows from (13.15) by means of dominated convergence theorem. Clearly, $\Sigma(\alpha)$ is symmetric and nonnegative definite.

Weak Limit via Martingale Problem Solution

To obtain the desired weak convergence result, we first work with the truncated processes $(v^{\Delta,N}(\cdot), \alpha^{\Delta}(\cdot))$. By virtue of the tightness and Prohorov's theorem, we can extract a weakly convergent subsequence (still denoted by $(v^{\Delta,N}(\cdot), \alpha^{\Delta}(\cdot))$ for simplicity) with limit $(v^N(\cdot), \alpha(\cdot))$. We will show that the limit is a switching diffusion.

To proceed with the diffusion approximation, similar to the proof of Theorem 13.7, we will use the martingale problem formulation to derive the desired result. For $v \in \mathbb{R}^S$, $\alpha \in \mathcal{M}$, and any twice continuously differentiable function $f(\cdot, \alpha)$ with compact support, consider the operator $\mathcal{L}$ defined by

$$\mathcal{L}f(v,\alpha) = -f_v'(v,\alpha)v + \frac{1}{2}\text{tr}[f_{vv}(v,\alpha)\Sigma(\alpha)] + Qf(v,\cdot)(\alpha), \qquad (13.17)$$

where $\Sigma(\alpha)$ is given by (13.16), and $f_{vv}(v,\alpha)$ denotes $(\partial^2/\partial v_i \partial v_j)f(v,\alpha)$, the mixed second-order partial derivatives.

Lemma 13.10. *Assume that the conditions of Lemma 13.9 are satisfied. In addition, assume that $(v^{\Delta,N}(0), \alpha^{\Delta}(0))$ converges to $(v^N(0), \alpha(0))$. Then $(v^{\Delta,N}(\cdot), \alpha^{\Delta}(\cdot))$ converges weakly to $(v^N(\cdot), \alpha(\cdot))$, which is a solution of the martingale problem with operator $\mathcal{L}^N$ given by*

$$\mathcal{L}^N f(v,\alpha) = -f_v'(v^N,\alpha)v^N \tau^N(v^N) + \frac{1}{2}\text{tr}[f_{vv}(v^N,\alpha)\Sigma(\alpha)] + Qf(v^N,\cdot)(\alpha),$$

(13.18)

or equivalently $v^N(\cdot)$ satisfies

$$dv^N(t) = -v^N(t)\tau^N(v^N(t))dt + \Sigma^{1/2}(\alpha(t))dw, \qquad (13.19)$$

where $w(\cdot)$ is a standard S-dimensional Brownian motion, and $\Sigma(\alpha)$ is given by (13.16).

Remark 13.11. Note that $(v^{\Delta,N}(0), \alpha^{\Delta}) = (v_{n_0}, \alpha_{n_0}^{\Delta})$. Theorem 13.2 implies that v_{n_0} is tight. In addition, $\alpha^{\Delta}(\cdot)$ converges weakly to $\alpha(\cdot)$. In the above, we simply assume the weak limit is $(v^N(0), \alpha(0))$.

Corollary 13.12. *Under the conditions of Lemma* 13.10, *the untruncated process* $(v^\Delta(\cdot), \alpha^\Delta(\cdot))$ *converges weakly to* $(v(\cdot), \alpha(\cdot))$, *satisfying the switching diffusion equation*

$$dv(t) = -v(t)dt + \Sigma^{1/2}(\alpha(t))dw. \tag{13.20}$$

Combining Lemmas 13.9 and 13.10 and Corollary 13.12, we obtain the following result.

Theorem 13.13. *Assume that conditions* (A13.1) *and* (A13.2) *are satisfied. Then* $(v^\Delta(\cdot), \alpha^\Delta(\cdot))$ *converges weakly to* $(v(\cdot), \alpha(\cdot))$, *which is the solution of the martingale problem with operator defined by* (13.17) *or equivalently, it is the solution of the system of diffusions with regime switching* (13.20).

Occupation Measures for Hidden Markov Model

So far, we have focused on recursive estimations of the occupation measure $\mu(\alpha_n^\Delta)$ given the conditional Markov sequence $\{X_n\}$. The results obtained can be extended to the hidden Markov models (HMMs), where the process $\{X_n\}$ cannot be observed and only noise-corrupted observation $\{Y_n\}$ is available with

$$Y_n = X_n + \zeta_n. \tag{13.21}$$

Assume that $\{\zeta_n\}$ satisfies the standard noise assumptions of a hidden Markov model (see for example, Krishnamurthy and Yin [90]), i.e., it is an i.i.d. noise process independent of X_n and α_n^Δ. Then given $\{Y_n\}$, to estimate $\mu(\alpha_n^\Delta)$ recursively, a modified version of the LMS algorithm (13.5) can be used, which requires the replacement of X_{n+1} in algorithm (13.5) by Y_{n+1}. The mean squares error analysis and switching ODE and switching diffusion results of the previous sections carry over. As a result, the following theorem holds.

Theorem 13.14. *Consider algorithm* (13.5), *where* X_{n+1} *is replaced by the HMM observation* Y_{n+1} *defined in* (13.21). *Assume that* (A13.1) *and* (A13.2) *hold, that* $\{\zeta_n\}$ *is a sequence of i.i.d. random variables with zero mean and* $E|\zeta_1|^2 < \infty$, *and that* $\{\zeta_n\}$ *is independent of* $\{X_n\}$ *and* $\{\alpha_n^\Delta\}$. *Then the conclusions of Theorems* 13.2, 13.7, *and* 13.13 *continue to hold.*

13.4.4   Probability Error Bounds

To estimate probabilities of tracking errors, we may use the asymptotic results to obtain bounds on

$$P(|\mu(\alpha_n^\Delta) - \widehat{\mu}_n| \ge K_0), \tag{13.22}$$

for sufficiently large n and some $K_0 > 0$. In view of (13.20), the leading negative term $-v(t)$ will ensure the stability of the system. Since for large

n, the process v_n is a Gaussian mixture and the limiting process $v(t)$ is a switching diffusion, it is difficult to compute the desired probabilities. Nevertheless, certain approximation can be carried out. We exploit this idea below.

By virtue of an argument from Kushner and Yin [100, p. 323], the covariance of the switching diffusion is given by

$$Ev(t)v'(0) = E\left(\int_{-\infty}^{t} e^{-(t-s)}\Sigma(\alpha(s))dw(s) \int_{-\infty}^{0} e^{-s}\Sigma(\alpha(s))dw(s) \right)'.$$

However, since it is a Gaussian mixture, the covariance is not easy to compute, although some Monte Carlo techniques can be used to assist the computation due to the Markov chain. Thanks to the asymptotic distribution, for n large enough, v_n can be approximated by a normal distribution with mean 0 and appropriate stationary covariance.

To exploit this further, we proceed to obtain the bounds by use of ordering of the covariance matrices. Without loss of generality, we may order the states $z^i \in \mathcal{M}$ so that the covariances $\Sigma(\alpha)$ for $\alpha \in \mathcal{M}$ are in ascending order

$$\Sigma(z^1) \le \Sigma(z^2) \le \cdots \le \Sigma(z^{m_0}), \tag{13.23}$$

where $\Sigma(z^i) \le \Sigma(z^j)$ is in the sense of the order of symmetric definite matrices. That is, $\Sigma(z^j) - \Sigma(z^i)$ is nonnegative definite. To approximate the switching diffusion or the Gaussian mixture, we consider the Gaussian diffusion processes $\underline{v}(\cdot)$ and $\overline{v}(\cdot)$ given by

$$d\underline{v} = -\underline{v}dt + \Sigma^{1/2}(z^1)dw(t), \quad \underline{v}(0) = v_0,$$
$$d\overline{v} = -\overline{v}dt + \Sigma^{1/2}(z^{m_0})dw(t), \quad \overline{v}(0) = v_0.$$

To proceed, first consider a diffusion process

$$d\xi(t) = -\xi(t)dt + \widetilde{\Sigma}^{1/2}dw(t), \ \widetilde{\Sigma} > 0.$$

Owing to its stability, the stationary covariance is given by

$$\int_0^{\infty} e^{-2t}\widetilde{\Sigma}dt = \frac{\widetilde{\Sigma}}{2}.$$

The foregoing implies that the stationary covariance of $\underline{v}(\cdot)$ and $\overline{v}(\cdot)$ are given by $\Sigma(z^1)/2$ and $\Sigma(z^{m_0})/2$, respectively. Note that for any $t > 0$,

$$\Sigma(z^1) \le \Sigma(\alpha(t)) \le \Sigma(z^{m_0}). \tag{13.24}$$

The above lower and upper bounds of the covariance matrix can then be used to find the desired probability bounds.

13.5   Proofs of Results

Proof of Theorem 13.2. The proof is divided into two steps. The first step is a preparation and the second step uses perturbed Liapunov function method.

Step 1: This is a preparation step. Recall the definitions of σ-algebra $\mathcal{H}_n$ in (13.6) and E_n the conditional expectation with respect to $\mathcal{H}_n$. In what follows, we often need to estimate certain quantities such as $E_n[\mu(\alpha_n^\varepsilon) - \mu(\alpha_{n+1}^\varepsilon)]$ etc. We show how such an estimate can be obtained. In view of the Markovian assumption and the structure of the transition probability matrix given by (13.4), we have

$$
\begin{aligned}
&E_n[\mu(\alpha_n^\varepsilon) - \mu(\alpha_{n+1}^\varepsilon)] \\
&= E(\mu(\alpha_n^\varepsilon) - \mu(\alpha_{n+1}^\varepsilon)|\alpha_n^\varepsilon) \\
&= \sum_{i=1}^{m_0} E[\mu(z^i) - \mu(\alpha_{n+1}^\varepsilon)|\alpha_n^\varepsilon = z^i] I_{\{\alpha_n^\varepsilon = z^i\}} \\
&= \sum_{i=1}^{m_0} [\mu(z^i) - \sum_{j=1}^{m_0} \mu(z^j) p^{\varepsilon,ij}] I_{\{\alpha_n^\varepsilon = z^i\}} \\
&= -\varepsilon \sum_{i=1}^{m_0} \sum_{j=1}^{m_0} \mu(z^j) q^{ij} I_{\{\alpha_n^\varepsilon = z^i\}} \\
&= O(\varepsilon).
\end{aligned}
\tag{13.25}
$$

Likewise similar estimates yield

$$
E_n|\mu(\alpha_n^\varepsilon) - \mu(\alpha_{n+1}^\varepsilon)| = \sum_{i=1}^{m_0} \sum_{j=1}^{m_0} |\mu(z^i) - \mu(z^j)| p^{\varepsilon,ij} I_{\{\alpha_n^\varepsilon = z^i\}} = O(\varepsilon), \text{ and}
$$

$$
E_n|\mu(\alpha_n^\varepsilon) - \mu(\alpha_{n+1}^\varepsilon)|^2 = O(\varepsilon).
\tag{13.26}
$$

Such estimates and the calculation will be used in what follows.

Step 2: Define

$$
\tilde{\mu}_n = \hat{\mu}_n - E\mu(\alpha_n^\varepsilon).
$$

Then (13.5) can be rewritten as

$$
\tilde{\mu}_{n+1} = \tilde{\mu}_n - \Delta\tilde{\mu}_n + \Delta(X_{n+1} - E\mu(\alpha_n^\varepsilon)) + E(\mu(\alpha_n^\varepsilon) - \mu(\alpha_{n+1}^\varepsilon)). \tag{13.27}
$$

Define $V(x) = (x'x)/2$. Direct calculations lead to

$$
\begin{aligned}
&E_n V(\tilde{\mu}_{n+1}) - V(\tilde{\mu}_n) \\
&= E_n \tilde{\mu}_n'[-\Delta\tilde{\mu}_n + \Delta(X_{n+1} - E\mu(\alpha_n^\varepsilon)) + E(\mu(\alpha_n^\varepsilon) - \mu(\alpha_{n+1}^\varepsilon))] \\
&\quad + E_n| - \Delta\tilde{\mu}_n + \Delta(X_{n+1} - E\mu(\alpha_n^\varepsilon)) + E(\mu(\alpha_n^\varepsilon) - \mu(\alpha_{n+1}^\varepsilon))|^2.
\end{aligned}
\tag{13.28}
$$

Recall that $\tilde{\mu}'_n$ denotes the transpose of $\tilde{\mu}_n$.

Because of (13.4), the transition probability is independent of time n. Thus, the k-step transition probability depends only on the time lags and can be denoted by $(P^\varepsilon)^k$. By an elementary inequality, we have

$$|\tilde{\mu}_n| = |\tilde{\mu}_n| \cdot 1 \le (|\tilde{\mu}_n|^2 + 1)/2.$$

Therefore,

$$O(\varepsilon)|\tilde{\mu}_n| \le O(\varepsilon)(V(\tilde{\mu}_n) + 1).$$

Since the sequence of signals $\{X_n\}$ is bounded, the boundedness of $\{\hat{\mu}_n\}$ and (13.25) yield

$$E_n| - \Delta\tilde{\mu}_n + \Delta(X_{n+1} - E\mu(\alpha_n^\varepsilon)) + E(\mu(\alpha_n^\varepsilon) - \mu(\alpha_{n+1}^\varepsilon))|^2$$
$$\le KE_n\Big[\Delta^2|\tilde{\mu}_n|^2 + \Delta^2|X_{n+1} - E\mu(\alpha_n^\varepsilon)|^2 + |E(\mu(\alpha_n^\varepsilon) - \mu(\alpha_{n+1}^\varepsilon)|^2\Big]$$
$$= O(\Delta^2 + \varepsilon^2)(V(\tilde{\mu}_n) + 1),$$

$$\text{(13.29)}$$

and

$$E_n\tilde{\mu}'_n[-\Delta\tilde{\mu}_n] = -2\Delta V(\tilde{\mu}_n). \tag{13.30}$$

Using (13.29) and (13.30) in (13.28), we obtain

$$E_nV(\tilde{\mu}_{n+1}) - V(\tilde{\mu}_n)$$
$$= -2\Delta V(\tilde{\mu}_n) + \Delta E_n\tilde{\mu}'_n(X_{n+1} - E\mu(\alpha_n^\varepsilon))$$
$$+ \tilde{\mu}_n E[\mu(\alpha_n^\varepsilon) - \mu(\alpha_{n+1}^\varepsilon)]$$
$$+ O(\Delta^2 + \varepsilon^2)(V(\tilde{\mu}_n) + 1). \tag{13.31}$$

To obtain the desired estimate, we need to "average out" the second and the third terms on the right side of the equality sign of (13.31). To do so, we define the following perturbation

$$V_1^\varepsilon(\tilde{\mu}, n) = \Delta\sum_{j=n}^\infty \tilde{\mu}'E_n(X_{j+1} - E\mu(\alpha_j^\varepsilon)). \tag{13.32}$$

For $V_1^\varepsilon(\tilde{\mu}, n)$ defined in (13.32),

$$\Big|\sum_{j=n}^\infty E_n(X_{j+1} - E\mu(\alpha_j^\varepsilon))\Big| \le \Big|\sum_{j=n}^\infty E_n[X_{j+1} - EX_{j+1}]\Big|$$
$$+ \Big|\sum_{j=n}^\infty [EX_{j+1} - E\mu(\alpha_j^\varepsilon)]\Big|. \tag{13.33}$$

By using the well-known result of mixing processes (see Example 14.11), the Markov property of $\{X_k\}$ implies that it is ϕ-mixing with an exponential mixing rate. Thus, the familiar mixing inequality leads to

$$\left| \sum_{j=n}^{\infty} E_n[X_{j+1} - EX_{j+1}] \right| < \infty.$$

By virtue of Condition (A13.2), for each α, $A(\alpha)$ is irreducible and aperiodic. As a result, we have

$$EX_{j+1} = \sum_{i=1}^{m_0} E\big[E[X_{j+1}|X_n, \alpha_n^{\varepsilon} = z^i] I_{\{\alpha_n^{\varepsilon} = z^i\}}\big]$$

$$= \sum_{i=1}^{m_0} \sum_{i_1=1}^{S} e_{i_1}[A(z^i)]^{j+1-n} P(\alpha_n^{\varepsilon} = z^i).$$

For each z^i, the ergodicity implies that $[A(z^i)]^{j+1-n} \to \mu(z^i)\mathbb{1}_S$ as $(j-n) \to \infty$, and the convergence takes place exponentially. Moreover,

$$E\mu(\alpha_n^{\varepsilon}) = \sum_{i=1}^{m_0} \mu(z^i) P(\alpha_n^{\varepsilon} = z^i).$$

Thus, the foregoing yields

$$\left| \sum_{j=n}^{\infty} [EX_{j+1} - E\mu(\alpha_j^{\varepsilon})] \right| < \infty.$$

Therefore, for each $\widetilde{\mu}$,

$$|V_1^{\varepsilon}(\widetilde{\mu}, n)| \leq O(\Delta)(V(\widetilde{\mu}) + 1). \tag{13.34}$$

Next, define

$$V_2^{\varepsilon}(\widetilde{\mu}, n) = \sum_{j=n}^{\infty} \widetilde{\mu}' E(\mu(\alpha_j^{\varepsilon}) - \mu(\alpha_{j+1}^{\varepsilon})). \tag{13.35}$$

Recall the irreducibility of $(I + \varepsilon Q)$. There is an N_{ε} such that for all $n \geq N_{\varepsilon}$, $|(I + \varepsilon Q)^n - \mathbb{1}_{m_0}\nu_{\varepsilon}| \leq K\varepsilon$, where ν_{ε} denotes the stationary distribution associated with the transition matrix $I + \varepsilon Q$. By telescoping and using the above estimates, we have that for all $\widehat{N} \geq n \geq N_{\varepsilon}$,

$$\left| \sum_{j=n}^{\widehat{N}} \widetilde{\mu}' E[\mu(\alpha_j^{\varepsilon}) - \mu(\alpha_{j+1}^{\varepsilon})] \right| = \left| \widetilde{\mu}' E[\mu(\alpha_n^{\varepsilon}) - \mu(\alpha_{\widehat{N}}^{\varepsilon})] \right|$$

$$\leq |\widetilde{\mu}| O(\varepsilon).$$

Thus,

$$|V_2^{\varepsilon}(\widetilde{\mu}, n)| \leq O(\varepsilon)(V(\widetilde{\mu}) + 1). \tag{13.36}$$

We next show that these perturbations defined in (13.32) result in the desired cancellations in the error estimates. Note that

$$E_n V_1^\varepsilon(\tilde{\mu}_{n+1}, n+1) - V_1^\varepsilon(\tilde{\mu}_n, n)$$
$$= E_n[V_1^\varepsilon(\tilde{\mu}_{n+1}, n+1) - V_1^\varepsilon(\tilde{\mu}_n, n+1)] \qquad (13.37)$$
$$+ E_n V_1^\varepsilon(\tilde{\mu}_n, n+1) - V_1^\varepsilon(\tilde{\mu}_n, n),$$

and that

$$E_n V_1^\varepsilon(\tilde{\mu}_n, n+1) - V_1^\varepsilon(\tilde{\mu}_n, n) = -\Delta E_n \tilde{\mu}_n'(X_{n+1} - E\mu(\alpha_n^\varepsilon)). \qquad (13.38)$$

In addition,

$$E_n V_1^\varepsilon(\tilde{\mu}_{n+1}, n+1) - E_n V_1^\varepsilon(\tilde{\mu}_n, n+1)$$
$$= \Delta \sum_{j=n+1}^{\infty} E_n(\tilde{\mu}_{n+1} - \tilde{\mu}_n)' E_{n+1}(X_{j+1} - E\mu(\alpha_j^\varepsilon))$$
$$= \Delta \sum_{j=n+1}^{\infty} E_n[-\Delta\tilde{\mu}_n + \Delta(X_{n+1} - E\mu(\alpha_n^\varepsilon)) + E(\mu(\alpha_n^\varepsilon) - \mu(\alpha_{n+1}^\varepsilon))]'$$
$$\times E_{n+1}[X_{j+1} - E\mu(\alpha_j^\varepsilon)]$$
$$= O(\Delta^2 + \varepsilon^2)(V(\tilde{\mu}_n) + 1).$$

$$(13.39)$$

To arrive at (13.39), we have used (13.27) and (13.28) to obtain

$$|E_n[\tilde{\mu}_{n+1} - \tilde{\mu}_n]| \le \Delta|\tilde{\mu}_n| + \Delta E_n|X_{n+1} - E\mu(\alpha_n^\varepsilon)|$$
$$+ |E[\mu(\alpha_n^\varepsilon) - \mu(\alpha_{n+1}^\varepsilon)]| \qquad (13.40)$$
$$= O(\Delta + \varepsilon)(V(\tilde{\mu}_n) + 1).$$

We also used $O(\varepsilon\Delta) = O(\varepsilon^2 + \Delta^2)$ via the elementary inequality $2ab \le (a^2 + b^2)$ for $a, b > 0$.

In view of the above estimates,

$$E_n V_1^\varepsilon(\tilde{\mu}_{n+1}, n+1) - V_1^\varepsilon(\tilde{\mu}_n, n)$$
$$= -\Delta E_n \tilde{\mu}_n'(X_{n+1} - E\mu(\alpha_n^\varepsilon)) + O(\Delta^2 + \varepsilon^2)(V(\tilde{\mu}_n) + 1). \qquad (13.41)$$

Analogously, we have

$$E_n V_2^\varepsilon(\tilde{\mu}_{n+1}, n+1) - E_n V_2^\varepsilon(\tilde{\mu}_n, n+1)$$
$$= \sum_{j=n+1}^{\infty} E_n(\tilde{\mu}_{n+1} - \tilde{\mu}_n)' E(\mu(\alpha_j^\varepsilon) - \mu(\alpha_{j+1}^\varepsilon)) \qquad (13.42)$$
$$= O(\varepsilon^2 + \Delta^2)(V(\tilde{\mu}_n) + 1),$$

and

$$E_n V_2^\varepsilon(\widetilde{\mu}_n, n+1) - V_2^\varepsilon(\widetilde{\mu}_n, n) = -\widetilde{\mu}_n' E(\mu(\alpha_n^\varepsilon) - \mu(\alpha_{n+1}^\varepsilon)). \qquad (13.43)$$

Define the perturbed Liapunov function $W(\cdot)$ by

$$W(\widetilde{\mu}, n) = V(\widetilde{\mu}) + V_1^\varepsilon(\widetilde{\mu}, n) + V_2^\varepsilon(\widetilde{\mu}, n).$$

Using the estimates obtained thus far, we deduce

$$\begin{aligned}
E_n W&(\widetilde{\mu}_{n+1}, n+1) - W(\widetilde{\mu}_n, n) \\
&= E_n V(\widetilde{\mu}_{n+1}) - V(\widetilde{\mu}_n) + E_n[V_1^\varepsilon(\widetilde{\mu}_{n+1}, n+1) - V_1^\varepsilon(\widetilde{\mu}_n, n)] \\
&\quad + E_n[V_2^\varepsilon(\widetilde{\mu}_{n+1}, n+1) - V_2^\varepsilon(\widetilde{\mu}_n, n)] \\
&= -2\Delta V(\widetilde{\mu}_n) + O(\Delta^2 + \varepsilon^2)(V(\widetilde{\mu}_n) + 1).
\end{aligned} \qquad (13.44)$$

The above inequality together with (13.34) and (13.36) yields

$$\begin{aligned}
E_n W&(\widetilde{\mu}_{n+1}, n+1) - W(\widetilde{\mu}_n, n) \\
&\leq -2\Delta W(\widetilde{\mu}_n, n) + O(\Delta^2 + \varepsilon^2)(W(\widetilde{\mu}_n, n) + 1).
\end{aligned} \qquad (13.45)$$

Choose Δ and ε small enough so that there is a $\lambda_0 > 0$ satisfying

$$-2\Delta + O(\varepsilon^2) + O(\Delta^2) \leq -\lambda_0 \Delta.$$

Thus, we obtain

$$E_n W(\widetilde{\mu}_{n+1}, n+1) \leq (1 - \lambda_0 \Delta) W(\widetilde{\mu}_n, n) + O(\Delta^2 + \varepsilon^2). \qquad (13.46)$$

By taking the expectation and iterating on the resulting inequality, we have

$$\begin{aligned}
E W(\widetilde{\mu}_{n+1}, n+1) \ &\leq (1 - \lambda_0 \Delta)^{n - N_\varepsilon} E W(\widetilde{\mu}_{N_\varepsilon}, N_\varepsilon) \\
&\quad + \sum_{j=N_\varepsilon}^{n} (1 - \lambda_0 \Delta)^{j - N_\varepsilon} O(\Delta^2 + \varepsilon^2).
\end{aligned} \qquad (13.47)$$

For n large enough, we can make $(1 - \lambda_0 \Delta)^{n - N_\varepsilon} = O(\Delta)$. Then

$$E W(\widetilde{\mu}_{n+1}, n+1) \leq O(\Delta + \varepsilon^2 / \Delta). \qquad (13.48)$$

By applying (13.34) and (13.36) again, replacing $W(\widetilde{\mu}, n)$ by $V(\widetilde{\mu})$ adds another $O(\varepsilon)$ term. Thus we obtain

$$E V(\widetilde{\mu}_{n+1}) \leq O(\Delta + \varepsilon + \varepsilon^2 / \Delta). \qquad (13.49)$$

This concludes the proof. $\square$

Proof of Proposition 13.5. Note that the identity matrix in (13.4) can be written as $I = \text{diag}(1, \ldots, 1) \in \mathbb{R}^{m_0 \times m_0}$. Each of the 1's can be thought of as a 1×1 "transition matrix." Observe that under the conditions for the Markov chain α_n^ε, the $\text{diag}(\nu^1, \ldots, \nu^l)$ defined in (6.6) becomes $I \in \mathbb{R}^{m_0 \times m_0}$, and $\text{diag}(\mathbb{1}_{m_1}, \ldots, \mathbb{1}_{m_l})$ in (6.6) is also I. Moreover, the $\overline{Q}$ defined in (6.6) is now simply Q. Straightforward applications of Proposition 6.2 then yield the desired results. □

Proof of Lemma 13.6. By using the tightness criteria (Lemma 14.12), it suffices to verify that for any $\delta > 0$, $t \in [0, T]$, $0 < s \le \delta$, and $t + \delta \le T$,

$$\lim_{\delta \to 0} \limsup_{\Delta \to 0} E|x^\Delta(t + s) - x^\Delta(t)|^2 = 0. \tag{13.50}$$

To begin, note that

$$\begin{aligned}
x^\Delta(t + s) - x^\Delta(t) &= \widehat{\mu}_{(t+s)/\Delta} - \widehat{\mu}_{t/\Delta} \\
&= \Delta \sum_{k=t/\Delta}^{(t+s)/\Delta - 1} (X_{k+1} - \widehat{\mu}_k).
\end{aligned} \tag{13.51}$$

Note also that both the iterates and the observations are bounded uniformly. Then the boundedness of $\{X_k\}$ and $\{\widehat{\mu}_k\}$ implies that

$$\begin{aligned}
E&|x^\Delta(t + s) - x^\Delta(t)|^2 \\
&= E\left[\Delta \sum_{k=t/\Delta}^{(t+s)/\Delta - 1} (X_{k+1} - \widehat{\mu}_k)'\right]\left[\Delta \sum_{k=t/\Delta}^{(t+s)/\Delta - 1} (X_{k+1} - \widehat{\mu}_k)\right] \\
&= \Delta^2 \sum_{k=t/\Delta}^{(t+s)/\Delta - 1} \sum_{j=t/\Delta}^{(t+s)/\Delta - 1} E(X_{k+1} - \widehat{\mu}_k)'(X_{j+1} - \widehat{\mu}_j) \tag{13.52} \\
&\le K\Delta^2 \left(\frac{t+s}{\Delta} - \frac{t}{\Delta}\right)^2 \\
&= K((t+s) - t)^2 = O(s^2).
\end{aligned}$$

First taking $\limsup_{\Delta \to 0}$ and then $\lim_{\delta \to 0}$ in (13.52), (13.50) is obtained, so the desired tightness follows. □

Proof of Theorem 13.7. To obtain the desired result, we show that the limit $(x(\cdot), \alpha(\cdot))$ is the solution of the martingale problem with operator L_1 given by

$$L_1 f(x, z^i) = f_x'(x, z^i)(\mu(z^i) - x) + Q f(x, \cdot)(z^i), \quad \text{for each } z^i \in \mathcal{M}, \tag{13.53}$$

where for each $z^i \in \mathcal{M}$,

$$\begin{aligned}
Q f(x, \cdot)(z^i) &= \sum_{j \in \mathcal{M}} q^{ij} f(x, z^j) \\
&= \sum_{j \ne i} q^{ij}[f(x, z^j) - f(x, z^i)],
\end{aligned}$$

and $f(\cdot, z^i)$ is twice continuously differentiable with compact support. In the above, $f_x(x, z^i)$ denotes the gradient of $f(x, z^i)$ with respect to x. Using Lemma 14.20, it can be shown that the martingale problem associated with the operator L_1 has a unique solution. Thus, it remains to show that the limit $(x(\cdot), \alpha(\cdot))$ is the solution of the martingale problem. To this end, we need only show that for any positive integer ℓ_0, any $t > 0$, $s > 0$, and $0 < t_j \le t$, and any bounded and continuous function $h_j(\cdot, z^i)$ for each $z^i \in \mathcal{M}$ with $j \le \ell_0$,

$$E \prod_{j=1}^{\ell_0} h_j(x(t_j), \alpha(t_j))$$

$$\times \left[f(x(t+s), \alpha(t+s)) - f(x(t), \alpha(t)) - \int_t^{t+s} L_1 f(x(u), \alpha(u)) du \right]$$

$$= 0.$$

(13.54)

To verify (13.54), we work with the processes indexed by Δ and show that the above equation holds as $\Delta \to 0$.

Since

$$(x^\Delta(\cdot), \alpha^\Delta(\cdot)) \quad \text{converges to} \quad (x(\cdot), \alpha(\cdot)) \quad \text{weakly},$$

using the Skorohod representation (without changing notation) leads to

$$\lim_{\Delta \to 0} E \prod_{j=1}^{\ell_0} h_j(x^\Delta(t_j), \alpha^\Delta(t_j))$$

$$\times \left[f(x^\Delta(t+s), \alpha^\Delta(t+s)) - f(x^\Delta(t), \alpha^\Delta(t)) \right]$$

$$= E \prod_{j=1}^{\ell_0} h_j(x(t_j), \alpha(t_j)) \left[f(x(t+s), \alpha(t+s)) - f(x(t), \alpha(t)) \right].$$

(13.55)

On the other hand, choose a sequence n_Δ such that $n_\Delta \to \infty$, as $\Delta \to 0$, but $\Delta n_\Delta \to 0$. Divide $[t, t+s]$ into intervals of width $\delta_\Delta = \Delta n_\Delta$. Direct calculation shows that

$$\lim_{\Delta \to 0} E \prod_{j=1}^{\ell_0} h_j(x^\Delta(t_j), \alpha^\Delta(t_j))$$

$$\times \left[f(x^\Delta(t+s), \alpha^\Delta(t+s)) - f(x^\Delta(t), \alpha^\Delta(t)) \right]$$

$$= \lim_{\Delta \to 0} E \prod_{j=1}^{\ell_0} h_j(x^\Delta(t_j), \alpha^\Delta(t_j))$$

$$\times \left[\sum_{l n_\Delta = t/\Delta}^{(t+s)/\Delta - 1} [f(\widehat{\mu}_{l n_\Delta + n_\Delta}, \alpha_{l n_\Delta + n_\Delta}^\Delta) - f(\widehat{\mu}_{l n_\Delta + n_\Delta}, \alpha_{l n_\Delta}^\Delta)] \right]$$

$$+ \sum_{l n_\Delta = t/\Delta}^{(t+s)/\Delta - 1} [f(\widehat{\mu}_{l n_\Delta + n_\Delta}, \alpha_{l n_\Delta}^\Delta) - f(\widehat{\mu}_{l n_\Delta}, \alpha_{l n_\Delta}^\Delta)]\Bigg].$$

By virtue of the smoothness and boundedness of $f(\cdot, \alpha)$ for each α, it can be seen that

$$\lim_{\Delta \to 0} E \prod_{j=1}^{\ell_0} h_j(x^\Delta(t_j), \alpha^\Delta(t_j))$$
$$\times \left[\sum_{l n_\Delta = t/\Delta}^{(t+s)/\Delta - 1} [f(\widehat{\mu}_{l n_\Delta + n_\Delta}, \alpha_{l n_\Delta + n_\Delta}^\Delta) - f(\widehat{\mu}_{l n_\Delta + n_\Delta}, \alpha_{l n_\Delta}^\Delta)] \right] \tag{13.56}$$
$$= \lim_{\Delta \to 0} E \prod_{j=1}^{\ell_0} h_j(x^\Delta(t_j), \alpha^\Delta(t_j))$$
$$\times \left[\sum_{l n_\Delta = t/\Delta}^{(t+s)/\Delta - 1} [f(\widehat{\mu}_{l n_\Delta}, \alpha_{l n_\Delta + n_\Delta}^\Delta) - f(\widehat{\mu}_{l n_\Delta}, \alpha_{l n_\Delta}^\Delta)] \right].$$

Thus we need only work with the latter term. Letting $\Delta \to 0$ and $\Delta l n_\Delta \to u$, we denote

$$T_1 = \lim_{\Delta \to 0} E \prod_{j=1}^{\ell_0} h_j(x^\Delta(t_j), \alpha^\Delta(t_j))$$
$$\times \left[\sum_{l n_\Delta = t/\Delta}^{(t+s)/\Delta - 1} [f(\widehat{\mu}_{l n_\Delta}, \alpha_{l n_\Delta + n_\Delta}^\Delta) - f(\widehat{\mu}_{l n_\Delta}, \alpha_{l n_\Delta}^\Delta)] \right]. \tag{13.57}$$

We proceed to figure out the value of T_1. Using the Markov chain α_n^ε and its transition probabilities, we obtain

$$T_1 = \lim_{\Delta \to 0} E \prod_{j=1}^{\ell_0} h_j(x^\Delta(t_j), \alpha^\Delta(t_j))$$
$$\times \left[\sum_{l n_\Delta = t/\Delta}^{(t+s)/\Delta - 1} \sum_{k = l n_\Delta}^{l n_\Delta + n_\Delta - 1} E_k[f(\widehat{\mu}_{l n_\Delta}, \alpha_{k+1}^\Delta) - f(\widehat{\mu}_{l n_\Delta}, \alpha_k^\Delta)] \right]$$
$$= \lim_{\Delta \to 0} E \prod_{j=1}^{\ell_0} h_j(x^\Delta(t_j), \alpha^\Delta(t_j))$$
$$\times \left[\sum_{l n_\Delta = t/\Delta}^{(t+s)/\Delta - 1} \sum_{j_1 = 1}^{m_0} \sum_{i=1}^{m_0} \sum_{k = l n_\Delta}^{l n_\Delta + n_\Delta - 1} [f(\widehat{\mu}_{l n_\Delta}, z^i) P(\alpha_{k+1}^\Delta = z^i | \alpha_k^\Delta = z^{j_1}) \right.$$
$$\left. - f(\widehat{\mu}_{l n_\Delta}, z^{j_1})] I_{\{\alpha_k^\Delta = z^{j_1}\}} \right]$$

$$= E \prod_{j=1}^{\ell_0} h_j(x(t_j), \alpha(t_j)) \left[\int_t^{t+s} Qf(x(u), \alpha(u)) du \right].$$

Since $\widehat{\mu}_{ln_\Delta}^\Delta$ and $\alpha_{ln_\Delta}^\Delta$ are $\mathcal{H}_{ln_\Delta}$-measurable, by virtue of the continuity and boundedness of $f_x(\cdot, \alpha)$,

$$E \prod_{j=1}^{\ell_0} h_j(x^\Delta(t_j), \alpha^\Delta(t_j)) \sum_{ln_\Delta=t/\Delta}^{(t+s)/\Delta-1} [f(\widehat{\mu}_{ln_\Delta+n_\Delta}, \alpha_{ln_\Delta}^\Delta) - f(\widehat{\mu}_{ln_\Delta}, \alpha_{ln_\Delta}^\Delta)]$$

$$= E \prod_{j=1}^{\ell_0} h_j(x^\Delta(t_j), \alpha^\Delta(t_j))$$

$$\times \sum_{ln_\Delta=t/\Delta}^{(t+s)/\Delta-1} \left[\Delta f_x'(\widehat{\mu}_{ln_\Delta}, \alpha_{ln_\Delta}^\Delta) \sum_{k=ln_\Delta}^{ln_\Delta+n_\Delta-1} E_{ln_\Delta}(X_{k+1} - \widehat{\mu}_k) \right] + o(1),$$

where $o(1) \to 0$ as $\Delta \to 0$. Consider the term

$$\lim_{\Delta \to 0} E \prod_{j=1}^{\ell_0} h_j(x^\Delta(t_j), \alpha^\Delta(t_j))$$

$$\times \left[\sum_{ln_\Delta=t/\Delta}^{(t+s)/\Delta-1} \delta_\Delta f_x'(\widehat{\mu}_{ln_\Delta}, \alpha_{ln_\Delta}^\Delta) \left[\frac{1}{n_\Delta} \sum_{k=ln_\Delta}^{ln_\Delta+n_\Delta-1} E_{ln_\Delta} X_{k+1} \right] \right], \tag{13.58}$$

where $\delta_\Delta = \Delta n_\Delta \to 0$. Again, let us use a fixed-α process $X_k(\alpha)$, which is a process with α_k^Δ fixed at $\alpha_{ln_\Delta}^\Delta = \alpha$ for $k \leq O(1/\Delta)$. A close scrutiny of the inner summation in (13.58) shows that

$$\frac{1}{n_\Delta} \sum_{k=ln_\Delta}^{ln_\Delta+n_\Delta-1} E_{ln_\Delta} X_{k+1} \tag{13.59}$$

can be approximated by

$$\frac{1}{n_\Delta} \sum_{k=ln_\Delta}^{ln_\Delta+n_\Delta-1} E_{ln_\Delta} X_{k+1}(\alpha)$$

such that the approximation error goes to 0 in probability, since

$$E_{ln_\Delta}[X_{k+1} - X_{k+1}(\alpha)] = O(\varepsilon) = O(\Delta)$$

by use of the transition matrix (13.4). Thus we have

$$\frac{1}{n_\Delta} \sum_{k=ln_\Delta}^{ln_\Delta+n_\Delta-1} E_{ln_\Delta} X_{k+1}$$

$$= \sum_{j_1=1}^{m_0} \frac{1}{n_\Delta} \sum_{k=ln_\Delta}^{ln_\Delta+n_\Delta-1} E(X_{k+1}(z^{j_1}) I_{\{\alpha_{ln_\Delta}^\Delta=z^{j_1}\}} | \alpha_{ln_\Delta}^\Delta = z^{j_1}) + o(1)$$

$$= \sum_{j_1=1}^{m_0} \frac{1}{n_\Delta} \sum_{k=ln_\Delta}^{ln_\Delta+n_\Delta-1} \sum_{j_2=1}^{S} (e_{j_2}'[A(z^{j_1})]^{k+1-ln_\Delta})' I_{\{\alpha_{ln_\Delta}^\Delta=z^{j_1}\}} + o(1),$$

where e_{j_2} is the standard unit vector and $o(1) \to 0$ in probability as $\Delta \to 0$. Note that for each $j_1 = 1, \ldots, m_0$, as $n_\Delta \to \infty$,

$$\frac{1}{n_\Delta} \sum_{k=ln_\Delta}^{ln_\Delta+n_\Delta-1} [A(z^{j_1})]^{k+1-ln_\Delta} \to \mathbb{1}_S \mu'(z^{j_1}).$$

Note also that $I_{\{\alpha_{ln_\Delta}^\Delta=z^{j_1}\}}$ can be written as $I_{\{\alpha^\Delta(l\delta_\Delta)=z^{j_1}\}}$. As $\Delta \to 0$, and $l\delta_\Delta \to u$, by the weak convergence of $\alpha^\Delta(\cdot)$ to $\alpha(\cdot)$, $I_{\{\alpha^\Delta(\Delta ln_\Delta)=z^{j_1}\}}$ converges in distribution to $I_{\{\alpha(u)=z^{j_1}\}}$. Using the Skorohod representation (with a slight abuse of notation), we may assume that the convergence is w.p.1. Since $\mathbb{1}_S \mu'(z^{j_1})$ has identical rows,

$$\frac{1}{n_\Delta} \sum_{k=ln_\Delta}^{ln_\Delta+n_\Delta-1} E_{ln_\Delta} X_{k+1} \to \sum_{j=1}^{m_0} \mu(z^{j_1}) I_{\{\alpha(u)=z^{j_1}\}} \tag{13.60}$$

$$= \mu(\alpha(u)).$$

This reveals that the limit does not depend on the value of the initial state, a salient feature of Markov chains. As a result,

$$\lim_{\Delta\to 0} E \prod_{j=1}^{\ell_0} h_j(x^\Delta(t_j), \alpha^\Delta(t_j))$$

$$\times \left[\sum_{ln_\Delta=t/\Delta}^{(t+s)/\Delta-1} \delta_\Delta f_x'(\widehat{\mu}_{ln_\Delta}, \alpha_{ln_\Delta}^\Delta) \frac{1}{n_\Delta} \sum_{k=ln_\Delta}^{ln_\Delta+n_\Delta-1} E_{ln_\Delta} X_{k+1} \right]$$

$$= E \prod_{j=1}^{\ell_0} h_j(x(t_j), \alpha(t_j)) \left[\sum_{j_1=1}^{m_0} \int_t^{t+s} f_x'(x(u), z^{j_1}) \mu(z^{j_1}) I_{\{\alpha(u)=z^{j_1}\}} \right]$$

$$= E \prod_{j=1}^{\ell_0} h_j(x(t_j), \alpha(t_j)) \left[\int_t^{t+s} f_x'(x(u), \alpha(u)) \mu(\alpha(u)) du \right].$$

$$\tag{13.61}$$

Likewise, it can be shown that as $\Delta \to 0$,

$$\lim_{\Delta \to 0} E \prod_{j=1}^{\ell_0} h_j(x^\Delta(t_j), \alpha^\Delta(t_j)) \left[\sum_{l_{n_\Delta}=t/\Delta}^{(t+s)/\Delta-1} f'_x(\widehat{\mu}_{l_{n_\Delta}}, \alpha^\Delta_{l_{n_\Delta}}) \frac{\delta_\Delta}{n_\Delta} \sum_{k=l_{n_\Delta}}^{l_{n_\Delta}+n_\Delta-1} \widehat{\mu}_k \right]$$

$$= E \prod_{j=1}^{\ell_0} h_j(x(t_j), \alpha(t_j)) \left[\int_t^{t+s} f'_x(x(u), \alpha(u)) x(u) \, du \right].$$

$$(13.62)$$

Combining (13.55), (13.57), (13.61), and (13.62), the desired result follows. The proof is thus completed. □

Proof of Lemma 13.9. We only verify the first assertion. In view of (13.14), for any $\delta > 0$, and $t, s \geq 0$ with $s \leq \delta$ and $t + \delta \leq T$,

$$v^{\Delta,N}(t+s) - v^{\Delta,N}(t) = -\Delta \sum_{k=t/\Delta}^{(t+s)/\Delta-1} v_k^N \tau^N(v_k^N)$$

$$+ \sqrt{\Delta} \sum_{k=t/\Delta}^{(t+s)/\Delta-1} (X_{k+1} - E\mu(\alpha_k^\Delta)) \qquad (13.63)$$

$$+ \frac{1}{\sqrt{\Delta}} \sum_{k=t/\Delta}^{(t+s)/\Delta-1} E(\mu(\alpha_k^\Delta) - \mu(\alpha_{k+1}^\Delta)).$$

Owing to the N-truncation,

$$\left| \Delta \sum_{k=t/\Delta}^{(t+s)/\Delta-1} v_k^N \tau^N(v_k^N) \right| \leq Ks,$$

so

$$\lim_{\delta \to 0} \limsup_{\Delta \to 0} E \left| \Delta \sum_{k=t/\Delta}^{(t+s)/\Delta-1} v_k^N \tau^N(v_k^N) \right|^2 = 0. \qquad (13.64)$$

Next, by virtue of (A13.1), the irreducibility of the conditional Markov chain $\{X_n\}$ implies that it is ϕ-mixing with an exponential mixing rate; see Example 14.11. Moreover, $E\mu(\alpha_k^\Delta) - EX_{k+1} \to 0$ exponentially fast. Consequently,

$$E \left| \Delta \sum_{k=t/\Delta}^{(t+s)/\Delta-1} (X_{k+1} - E\mu(\alpha_k^\Delta)) \right|^2$$

$$= E \left| \Delta \sum_{k=t/\Delta}^{(t+s)/\Delta-1} [(X_{k+1} - EX_{k+1}) - (E\mu(\alpha_k^\Delta) - EX_{k+1})] \right|^2 = O(s).$$

Thus,

$$\lim_{\delta \to 0} \limsup_{\Delta \to 0} E \left| \Delta \sum_{k=t/\Delta}^{(t+s)/\Delta-1} (X_{k+1} - E\mu(\alpha_k^\Delta)) \right|^2 = 0. \qquad (13.65)$$

In addition, similar to (13.25),

$$\frac{1}{\sqrt{\Delta}} E \sum_{k=t/\Delta}^{(t+s)/\Delta-1} [\mu(\alpha_k^\Delta) - \mu(\alpha_{k+1}^\Delta)]$$
$$= \frac{1}{\sqrt{\Delta}} [E\mu(\alpha_{t/\Delta}^\Delta) - E\mu(\alpha_{(t+s)/\Delta}^\Delta)] \quad\quad (13.66)$$
$$= O(\sqrt{\Delta}).$$

Combining (13.64)–(13.66), we have

$$\lim_{\delta\to0} \limsup_{\Delta\to0} E|v^{\Delta,N}(t+s) - v^{\Delta,N}(t)|^2 = 0,$$

and hence the tightness criterion (see Lemma 14.12), implies that $\{v^{\Delta,N}(\cdot)\}$ is tight.   □

Proof of Lemma 13.10. Again, we use martingale problem formulation. For any positive integer ℓ_0, any $t > 0$, $s > 0$, any $0 < t_j \le t$ with $j \le \ell_0$, and any bounded and continuous function $h_j(\cdot, \alpha)$ for each $\alpha \in \mathcal{M}$, we aim to derive an equation like (13.54) with the operator L_1 replaced by $\mathcal{L}$. As in the proof of Theorem 13.7, we work with the sequence indexed by Δ. Choose n_Δ such that $n_\Delta \to \infty$ but $\delta_\Delta = \Delta n_\Delta \to 0$. Using the tightness of $\{v^{\Delta,N}(\cdot), \alpha^\Delta(\cdot)\}$ and the Skorohod representation (without changing notation), we have that (13.55)–(13.56) hold with $\hat{\mu}^\Delta(\cdot)$ and $\hat{\mu}(\cdot)$ replaced by $v^{\Delta,N}(\cdot)$ and $v^N(\cdot)$, respectively.

In view of (13.66), the term

$$\frac{1}{\sqrt{\Delta}} \sum_{k=t/\Delta}^{(t+s)/\Delta-1} [E\mu(\alpha_k^\Delta) - E\mu(\alpha_{k+1}^\Delta)]$$

contributes nothing to the limit process. Moreover,

$$\sqrt{\Delta} \sum_{k=t/\Delta}^{(t+s)/\Delta-1} [X_{k+1} - E\mu(\alpha_k^\Delta)]$$
$$= \sqrt{\Delta} \sum_{k=t/\Delta}^{(t+s)/\Delta-1} (X_{k+1} - EX_{k+1})$$
$$+ \sqrt{\Delta} \sum_{k=t/\Delta}^{(t+s)/\Delta-1} (EX_{k+1} - E\mu(\alpha_k^\Delta)).$$

Since

$$EX_{k+1} - E\mu(\alpha_k^\Delta) \to 0 \quad \text{exponentially fast},$$

owing to the elementary properties of Markov chain,

$$\sqrt{\Delta} \sum_{k=t/\Delta}^{(t+s)/\Delta-1} (EX_{k+1} - E\mu(\alpha_k^\Delta))$$

only produces an additional term of order $O(\sqrt{\Delta})$. Thus,

$$v^{\Delta,N}(t+s) - v^{\Delta,N}(t)$$

$$= -\Delta \sum_{k=t/\Delta}^{(t+s)/\Delta-1} v_k^N \tau^N(v_k^N)$$

$$+\sqrt{\Delta} \sum_{k=t/\Delta}^{(t+s)/\Delta-1} (X_{k+1} - EX_{k+1}) + O(\sqrt{\Delta}).$$

Like the argument in the proof of Theorem 13.7,

$$\lim_{\Delta \to 0} E \prod_{j=1}^{\ell_0} h_j(v^{\Delta,N}(t_j), \alpha^{\Delta}(t_j))$$

$$\times \left[\sum_{l n_\Delta = t/\Delta}^{(t+s)/\Delta-1} [f(v_{ln_\Delta}^N, \alpha_{ln_\Delta+n_\Delta}^{\Delta}) - f(v_{ln_\Delta}^N, \alpha_{ln_\Delta}^{\Delta})] \right] \qquad (13.67)$$

$$= E \prod_{j=1}^{\ell_0} h_j(v^N(t_j), \alpha(t_j)) \left[\int_t^{t+s} Qf(v^N(u), \alpha(u)) du \right]$$

In addition, with a quantity $o(1) \to 0$ uniform in t as $\Delta \to 0$,

$$E \prod_{j=1}^{\ell_0} h_j(v^{\Delta,N}(t_j), \alpha^{\Delta}(t_j))$$

$$\times \left[-\sum_{ln_\Delta=t/\Delta}^{(t+s)/\Delta-1} \delta_\Delta \frac{1}{n_\Delta} \sum_{k=ln_\Delta}^{ln_\Delta+n_\Delta-1} f_v'(v_{ln_\Delta}^N, \alpha_{ln_\Delta}^{\Delta}) v_k^N \tau^N(v_k^N) \right]$$

$$= E \prod_{j=1}^{\ell_0} h_j(v^{\Delta,N}(t_j), \alpha^{\Delta}(t_j))$$

$$\qquad (13.68)$$

$$\times \left[-\sum_{ln_\Delta=t/\Delta}^{(t+s)/\Delta-1} \delta_\Delta f_v'(v_{ln_\Delta}^N, \alpha_{ln_\Delta}^{\Delta}) v_{ln_\Delta}^N \tau^N(v_{ln_\Delta}^N) \right] + o(1)$$

$$\to E \prod_{j=1}^{\ell_0} h_j(v^N(t_j), \alpha(t_j))$$

$$\times \left[-\int_t^{t+s} f_v'(v^N(u), \alpha(u)) v^N(u) \tau^N(v^N(u)) du \right].$$

Denote

$$\hat{\rho}^{\Delta} = \left| E \prod_{j=1}^{\ell_0} h_j(v^{\Delta,N}(t_j), \alpha^{\Delta}(t_j)) \right.$$

$$\left. \times \left[\sqrt{\Delta} \sum_{ln_\Delta=t/\Delta}^{(t+s)/\Delta-1} f_v'(v_{ln_\Delta}^N, \alpha_{ln_\Delta}^{\Delta}) \sum_{k=ln_\Delta}^{ln_\Delta+n_\Delta-1} [X_{k+1} - EX_{k+1}] \right] \right|.$$

Then, owing to the boundedness of $h(\cdot)$, $v_{ln_\Delta}^N$, and $f_v(\cdot)$,

$$
\widehat{\rho}^\Delta = \left| E \prod_{j=1}^{\ell_0} h_j(v^{\Delta,N}(t_j), \alpha^\Delta(t_j)) \right.
$$

$$
\left. \times \left[\sqrt{\Delta} \sum_{ln_\Delta=t/\Delta}^{(t+s)/\Delta-1} f_v'(v_{ln_\Delta}^N, \alpha_{ln_\Delta}^\Delta) \sum_{k=ln_\Delta}^{ln_\Delta+n_\Delta-1} E_{t/\Delta}[X_{k+1} - EX_{k+1}] \right] \right|
$$

$$
\leq K\sqrt{\Delta} \sum_{k=t/\Delta}^{(t+s)/\Delta-1} E|E_{t/\Delta} X_{k+1} - EX_{k+1}|
$$

$$
\leq K\sqrt{\Delta} \to 0 \quad \text{as} \quad \Delta \to 0.
$$

To arrive at the last line above, we have used the mixing inequality given in Lemma 14.10.

Finally, define

$$
g_{ln_\Delta} g_{ln_\Delta}'
$$
$$
= \frac{1}{n_\Delta} \sum_{k=ln_\Delta}^{ln_\Delta+n_\Delta-1} \sum_{k_1=ln_\Delta}^{ln_\Delta+n_\Delta-1} E_{ln_\Delta}[X_{k+1} - EX_{k+1}][X_{k_1+1} - EX_{k_1+1}]'.
$$

It follows that

$$
E \prod_{j=1}^{\ell_0} h_j(v^{\Delta,N}(t_j), \alpha^\Delta(t_j))
$$

$$
\times \left[\sum_{ln_\Delta=t/\Delta}^{(t+s)/\Delta-1} \text{tr}\left[f_{vv}(v_{ln_\Delta}^N, \alpha_{ln_\Delta}^\Delta)(v_{ln_\Delta+n_\Delta}^N - v_{ln_\Delta}^N)(v_{ln_\Delta+n_\Delta}^N - v_{ln_\Delta}^N)' \right] \right]
$$

$$
= E \prod_{j=1}^{\ell_0} h_j(v^{\Delta,N}(t_j), \alpha^\Delta(t_j)) \left[\sum_{j_1=1}^{m_0} \sum_{ln_\Delta=t/\Delta}^{(t+s)/\Delta-1} I_{\{\alpha_{ln_\Delta}^\Delta = z^{j_1}\}} \right.
$$

$$
\left. \times \text{tr}\left[f_{vv}(v_{ln_\Delta}^N, \alpha_{ln_\Delta}^\Delta)(v_{ln_\Delta+n_\Delta}^N - v_{ln_\Delta}^N)(v_{ln_\Delta+n_\Delta}^N - v_{ln_\Delta}^N)' \right] \right]
$$

$$
= E \prod_{j=1}^{\ell_0} h_j(v^{\Delta,N}(t_j), \alpha^\Delta(t_j))
$$

$$
\times \left[\sum_{j_1=1}^{m_0} \sum_{ln_\Delta=t/\Delta}^{(t+s)/\Delta-1} \delta_\Delta \text{tr}\left[f_{vv}(v_{ln_\Delta}^N, \alpha_{ln_\Delta}^\Delta) E_{ln_\Delta} g_{ln_\Delta} g_{ln_\Delta}' \right] I_{\{\alpha_{ln_\Delta}^\Delta = z^{j_1}\}} \right].
$$

Since conditioned on $\alpha_{ln_\Delta}^\Delta = z^j$, $X_{k+1} - EX_{k+1}$ can be approximated by the fixed-z^{j_1} process

$$
X_{k+1}(z^{j_1}) - EX_{k+1}(z^{j_1}),
$$

and since $X_{k+1}(z^{j_1})$ is a finite-state Markov chain with irreducible transition matrix $A(z^{j_1})$,

$$X_{k+1}(z^{j_1}) - EX_{k+1}(z^{j_1})$$

is ϕ-mixing. Since $\alpha^\Delta(\cdot)$ converges weakly to $\alpha(\cdot)$, sending $\Delta \to 0$ and $l\delta_\Delta = l\Delta n_\Delta \to u$ leads to $\alpha^\Delta(\Delta l n_\Delta)$ converges in distribution to $\alpha(u)$. Moreover, by Skorohod representation (without changing notation), we may assume the above convergence and the convergence of $I_{\{\alpha^\Delta(l\delta_\Delta)=z^{j_1}\}}$ to $I_{\{\alpha(u)=z^{j_1}\}}$ are w.p.1. Therefore, the argument in (13.16) implies that for each $z^{j_1} \in \mathcal{M}$ with $j_1 = 1, \ldots, m_0$,

$$E_{l n_\Delta}[g_{l n_\Delta} g'_{l n_\Delta}] \to \Sigma(z^j) I_{\{\alpha(u)=z^{j_1}\}} \quad \text{as } \Delta \to 0, \tag{13.69}$$

where $\Sigma(z^{j_1})$ is defined in (13.16). It follows that

$$
E \prod_{j=1}^{\ell_0} h_j(v^{\Delta,N}(t_j), \alpha^\Delta(t_j))
$$
$$
\times \left[\sum_{l n_\Delta = t/\Delta}^{(t+s)/\Delta - 1} \operatorname{tr}\left[f_{vv}(v^N_{l n_\Delta}, \alpha^\Delta_{l n_\Delta})(v^N_{l n_\Delta + n_\Delta} - v^N_{l n_\Delta})(v^N_{l n_\Delta + n_\Delta} - v^N_{l n_\Delta})' \right] \right]
$$
$$
\to E \prod_{j=1}^{\ell_0} h_j(v^N(t_j), \alpha(t_j))
$$
$$
\times \left[\int_t^{t+s} \sum_{j_1=1}^{m_0} \operatorname{tr}\left[f_{vv}(v^N(u), z^{j_1}) \Sigma(z^{j_1}) \right] I_{\{\alpha(u)=z^{j_1}\}} du \right]
$$
$$
= E \prod_{j=1}^{\ell_0} h_j(v^N(t_j), \alpha(t_j)) \left[\int_t^{t+s} \operatorname{tr}\left[f_{vv}(v^N(u), \alpha(u)) \Sigma(\alpha(u)) \right] du \right]. \tag{13.70}
$$

In view of (13.67)–(13.70), the proof is completed. □

Proof of Corollary 13.12. The uniqueness of the solution of the associated martingale problem can be proved in a manner similar to that of Lemma 4.9 (see also Lemma 14.20, and Yin and Zhang [158, Lemma 7.18]). The rest of the proof follows from an argument like that in Kushner [96, p. 46] (see also Kushner and Yin [100, Step 4, p. 285]). Let $P^{v(0)}(\cdot)$ and $P^N(\cdot)$ be the measures induced by $v(\cdot)$ and $v^N(\cdot)$ on the Borel subsets of $D([0, T] : \mathbb{R}^S)$. The measure $P^{v(0)}(\cdot)$ is unique by the uniqueness of the martingale problem to $(v(\cdot), \alpha(\cdot))$, which indicates that $P^{v(0)}(\cdot)$ coincides with $P^N(\cdot)$ on all Borel subsets of the set of paths in $D([0, T] : \mathbb{R}^S)$ with values in B_N. As $N \to \infty$, $P^{v(0)}(\sup_{v(t) \leq T} |v(t)| \leq N) \to 1$. The foregoing together with the weak convergence of $(v^{\Delta,N}(\cdot), \alpha^\Delta(\cdot))$ to $(v^N(\cdot), \alpha(\cdot))$ implies that $(v^\Delta(\cdot), \alpha^\Delta(\cdot))$ converges weakly to $(v(\cdot), \alpha(\cdot))$. Some details are omitted. □

13.6   Notes

Stochastic approximation methods were initiated in the work of Robbins and Monro [129] in the early 1950s. The original motivation stemmed from searching for roots of a continuous function $F(\cdot)$, where either the precise form of the function is unknown, or it is too complicated to compute; only "noisy" measurements at desired design points are available. A classical example is to find an appropriate dosage level of a drug, provided that only $\{F(x) + \text{noise}\}$ is available, where x is the level of dosage and $F(x)$ is the probability of success (leading to the recovery of the patient) at the dosage level x. The classical Kiefer–Wolfowitz (KW) algorithm introduced by Kiefer and Wolfowitz [87] concerns the minimization of a real-valued function using only noisy functional measurements. The main issues in the analysis of iteratively defined stochastic processes and applications focus on the basic paradigm of stochastic difference equations. Much of the development has been accompanied by a wide range of applications in optimization, control theory, economic systems, signal processing, communication theory, learning, pattern classification, neural network, and other related fields. In addition, emerging applications have also been found in, for example, wireless communication and financial engineering. The original stochastic approximation algorithms can be considered to be static. That is, they mainly deal with root finding or optimization of a function with a fixed parameter. This has been substantially generalized to treat dynamic systems with varying parameters. For an up-to-date account of stochastic approximation, the reader is referred to Kushner and Yin [100].

The algorithms studied in this chapter are also useful in performance analysis of adaptive discrete stochastic optimization problems, for example, adaptive coding in wireless CDMA communication systems. Recently, in [89], Krishnamurthy, Wang, and Yin treated spreading code optimization of the CDMA system at the transmitter, formulated it as a discrete stochastic optimization problem (since the spreading codes are finite-length and finite-state sequences), and used random search based discrete stochastic optimization algorithm to compute the optimal spreading code. In addition to the random-search-type algorithms, they also designed adaptive step-size stochastic approximation algorithms with both fixed and adaptive step sizes to track slowly time-varying optimal spreading codes originating from fading characteristics of the wireless channel. The numerical results reported in the aforementioned paper have shown remarkable improvement as compared with several heuristic algorithms.

While there are several papers that analyze tracking properties of stochastic approximation algorithms when the underlying parameter varies according to a slow random walk (see Benveniste, Metivier, and Priouret [14], and Solo and Kong [141]), there are few papers examining cases where parameter variations follow a Markov chain with infrequent jumps. Such Markovian models do arise from real applications, however.

Discrete stochastic optimization problems were formulated in Yan and Mukai [151], and subsequently considered in Andradottir [5], and Gong, Ho, and Zhai [63], among others. These discrete stochastic optimization algorithms include selection and multiple comparison methods, multi-armed bandits, a stochastic ruler in [151], nested partition methods and discrete stochastic optimization algorithms based on simulated annealing in Andradottir [5], and Gelfand and Mitter [60].

This chapter presents the results based on the work of Yin, Krishnamurthy, and Ion [156], where error bounds for the adaptive discrete stochastic optimization algorithm were derived by using asymptotic results. To select the step size in the actual computation is an important issue; one enticing alternative is to design a tracking algorithm with step-size adaptation. That is, replace the fixed step size Δ by a time-varying step-size sequence $\{\Delta_n\}$ and adaptively adjust Δ_n as the dynamics evolve. An interested reader may consult Kushner and Yin [100, Section 3.2].

14
Appendix

14.1 Introduction

To make the book reasonably self-contained and to provide the reader with a quick reference and further reading materials, we collect a number of topics here, which are either not in Chapter 2 or more advanced than those covered in Chapter 2. This chapter includes short reviews on systems with regime switching, weak convergence, optimal control and HJB equations, and other related topics.

14.2 Sample Paths of Markov Chains

A central topic discussed in this book is a dynamic system with regime switching, where the switching is modeled by a Markov chain. Perhaps the simplest of such system is a Markov chain itself. This section illustrates how to construct sample paths of Markov chains in both discrete and continuous times. The material presented here is classical and standard in stochastic processes. It is useful for numerical experiments and simulations of stochastic systems.

Discrete-Time Chains

To construct a Markov chain α_k in discrete time requires first prescribing its transition probability matrix $P = (p^{ij})$, then building its sample path. Suppose that $\alpha_k \in \mathcal{M} = \{1, \ldots, m_0\}$, for $k \geq 0$. The sample paths are constructed via comparison determined by the transition probability matrix P. At any time $k \geq 0$, given $\alpha_k = i$, the chain's next move is specified by

$$
\alpha_{k+1} = \begin{cases} 1, & \text{if } U \leq p^{i1}, \\ 2, & \text{if } p^{i1} < U \leq p^{i1} + p^{i2}, \\ \vdots & \vdots \\ m_0, & \text{if } p^{i1} + \cdots + p^{i,m_0-1} < U \leq 1, \end{cases} \tag{14.1}
$$

where U is a random variable following a uniform distribution in $(0,1)$ (i.e., $U \sim U(0,1)$).

Continuous-Time Chains

Suppose that $\alpha(t)$ is a continuous-time Markov chain with state space $\mathcal{M} = \{1, \ldots, m_0\}$ and generator $Q = (q^{ij})$. To construct the sample paths of $\alpha(t)$ amounts to determining its sojourn time at each state and its subsequent moves. The chain sojourns in any given state i for a random length of time, S_i, which has an exponential distribution with parameter $(-q^{ii})$. Subsequently, the process will enter another state. Each state j $(j = 1, \ldots, m_0, \ j \neq i)$ has a probability $q^{ij}/(-q^{ii})$ of being the chain's next residence. The post-jump location is determined by a discrete random variable X^i taking values in $\{1, 2, \ldots, i-1, i+1, \ldots, m_0\}$. Its value is specified by

$$
X^i = \begin{cases} 1, & \text{if } U \leq q^{i1}/(-q^{ii}), \\ 2, & \text{if } q^{i1}/(-q^{ii}) < U \leq (q^{i1} + q^{i2})/(-q^{ii}), \\ \vdots & \vdots \\ m_0, & \text{if } \sum_{j \neq i, j < m_0} q^{ij}/(-q^{ii}) \leq U. \end{cases} \tag{14.2}
$$

where U is a random variable uniformly distributed in $(0,1)$. Thus, the sample path of $\alpha(t)$ is constructed by sampling from exponential and $U(0,1)$ random variables alternately. This section only concerns with stationary Markov chains. Construction for general Markov chain with time-dependent generator $Q = Q(t)$ can be proceeded as in Section 2.4.

14.3   Weak Convergence

The notion of weak convergence is a generalization of convergence in distribution from elementary probability theory. In what follows, we present definitions and results, including tightness, tightness criteria, the martingale problem, Skorohod representation, and Prohorov's theorem, etc.

Definition 14.1 (Weak Convergence). Let P and P_k, $k = 1, 2, \ldots$, be probability measures defined on a metric space $\mathbb{S}$. The sequence $\{P_k\}$ converges weakly to P if

$$\int f dP_k \to \int f dP$$

for every bounded and continuous function $f(\cdot)$ on $\mathbb{S}$. Suppose that $\{x_k\}$ and x are random variables associated with P_k and P, respectively. The sequence x_k converges to x weakly if for any bounded and continuous function $f(\cdot)$ on $\mathbb{S}$, $Ef(x_k) \to Ef(x)$ as $k \to \infty$.

Let $D([0,\infty); \mathbb{R}^r)$ be the space of $\mathbb{R}^r$-valued functions defined on $[0,\infty)$ that are right-continuous and have left-hand limits; let $\mathbb{L}$ be a set of strictly increasing Lipschitz continuous functions $\zeta(\cdot) : [0,\infty) \mapsto [0,\infty)$ such that the mapping is surjective with $\zeta(0) = 0$, $\lim_{t\to\infty} \zeta(t) = \infty$, and

$$\gamma(\zeta) := \sup_{0 \le t < s} \left| \log\left(\frac{\zeta(s) - \zeta(t)}{s - t} \right) \right| < \infty.$$

Similar to $D([0,\infty); \mathbb{R}^r)$, we also use the notation $D([0,T]; \mathbb{F})$ to denote the D-space of functions that take values in $\mathbb{F}$.

Definition 14.2 (Skorohod Topology). For $\xi, \eta \in D([0,\infty); \mathbb{R}^r)$, the *Skorohod topology* $d(\cdot, \cdot)$ on $D([0,\infty); \mathbb{R}^r)$ is defined as

$$d(\xi, \eta) = \inf_{\zeta \in \mathbb{L}} \left\{ \gamma(\zeta) \vee \int_0^\infty e^{-s} \sup_{t \ge 0} \left(1 \wedge |\xi(t \wedge s) - \eta(\zeta(t) \wedge s)|\right) ds \right\}.$$

Analogous definitions and results are available for $D([0,T]; \mathbb{F})$; see Ethier and Kurtz [55] and Billingsley [18] for related references. Although we often work with $D([0,T]; \mathbb{R}^r)$ in this book, the results to follow are often stated with respect to the space $D([0,\infty); \mathbb{R}^r)$. This enables us to apply them to $t \in [0,T]$ for any $T > 0$.

Definition 14.3 (Tightness). A family of probability measures $\mathcal{P}$ defined on a metric space $\mathbb{S}$ is *tight* if for each $\delta > 0$, there exists a compact set $K_\delta \subset \mathbb{S}$ such that

$$\inf_{P \in \mathcal{P}} P(K_\delta) \ge 1 - \delta.$$

The notion of tightness is closely related to compactness. The following theorem, known as Prohorov's theorem, gives such an implication. A complete proof can be found in Ethier and Kurtz [55].

Theorem 14.4 (Prohorov's Theorem). *If $\mathcal{P}$ is tight, then $\mathcal{P}$ is relatively compact. That is, every sequence of elements in $\mathcal{P}$ contains a weakly convergent subsequence. If the underlying metric space is complete and separable, the tightness is equivalent to relative compactness.*

Although weak convergence techniques usually allow one to use weaker conditions and lead to a more general setup, it is often more convenient to work with probability one convergence for purely analytic reasons, however. The Skorohod representation, provides us with such opportunities.

Theorem 14.5 (The Skorohod Representation (Ethier and Kurtz [55])).
*Let x_k and x be random elements belonging to $D([0, \infty); \mathbb{R}^r)$ such that x_k
converges weakly to x. Then there exists a probability space $(\widetilde{\Omega}, \widetilde{\mathcal{F}}, \widetilde{P})$ on
which are defined random elements $\widetilde{x}_k$, $k = 1, 2, \ldots$, and $\widetilde{x}$ in $D([0, \infty); \mathbb{R}^r)$
such that for any Borel set B and all $k < \infty$,*

$$\widetilde{P}(\widetilde{x}_k \in B) = P(x_k \in B), \quad \text{and} \quad \widetilde{P}(\widetilde{x} \in B) = P(x \in B)$$

satisfying

$$\lim_{k \to \infty} \widetilde{x}_k = \widetilde{x} \quad w.p.1.$$

Elsewhere in the book, when we use the Skorohod representation, with a
slight abuse of notation, we often omit the tilde notation for convenience
and notational simplicity.

Let $C([0, \infty); \mathbb{R}^r)$ be the space of $\mathbb{R}^r$-valued continuous functions equipped
with the sup-norm topology, and C_0 be the set of real-valued continuous
functions on $\mathbb{R}^r$ with compact support. Let C_0^l be the subset of C_0 functions
that have continuous partial derivatives up to the order l.

Definition 14.6. Let $\mathbb{S}$ be a metric space and A be a linear operator on
$B(\mathbb{S})$ (the set of all Borel measurable functions defined on $\mathbb{S}$). Let $x(\cdot) =
\{x(t) : t \geq 0\}$ be a right-continuous process with values in $\mathbb{S}$ such that for
each $f(\cdot)$ in the domain of A,

$$f(x(t)) - \int_0^t Af(x(s))ds$$

is a martingale with respect to the filtration $\sigma\{x(s) : s \leq t\}$. Then $x(\cdot)$ is
called a *solution of the martingale problem* with operator A.

Theorem 14.7 (Ethier and Kurtz [55, p. 174]). *A right-continuous process
$x(t)$, $t \geq 0$, is a solution of the martingale problem for the operator A if
and only if*

$$E\left(\prod_{j=1}^i h_j(x(t_j)) \left(f(x(t_{i+1})) - f(x(t_i)) - \int_{t_i}^{t_{i+1}} Af(x(s))ds \right) \right) = 0$$

*whenever $0 \leq t_1 < t_2 < \cdots < t_{i+1}$, $f(\cdot)$ in the domain of A, and $h_1, \ldots, h_i \in
\mathcal{B}(\mathbb{S})$, the Borel field of $\mathbb{S}$.*

Theorem 14.8 (Uniqueness of Martingale Problems, Ethier and Kurtz
[55, p. 184]). *Let $x(\cdot)$ and $y(\cdot)$ be two stochastic processes whose paths are
in $D([0, T]; \mathbb{R}^r)$. Denote an infinitesimal generator by A. If for any function
$f \in A$ (the domain of A),*

$$f(x(t)) - f(x(0)) - \int_0^t Af(x(s))ds, \ t \geq 0,$$

and

$$f(y(t)) - f(y(0)) - \int_0^t Af(y(s))ds, \ t \geq 0$$

are martingales and $x(t)$ and $y(t)$ have the same distribution for each $t \geq 0$, $x(\cdot)$ and $y(\cdot)$ have the same distribution on $D([0,\infty);\mathbb{R}^r)$.

Theorem 14.9. *Let $x^\varepsilon(\cdot)$ be a solution of the differential equation*

$$\frac{dx^\varepsilon(t)}{dt} = F^\varepsilon(t),$$

and for each $T < \infty$, $\{F^\varepsilon(t) : 0 \leq t \leq T\}$ be uniformly integrable. If the set of initial values $\{x^\varepsilon(0)\}$ is tight, then $\{x^\varepsilon(\cdot)\}$ is tight in $C([0,\infty);\mathbb{R}^r)$.

Proof: The proof is essentially in Billingsley [18, Theorem 8.2] (see also Kushner [96, p. 51, Lemma 7]). $\square$

Define the notion of "p-lim" and an operator A^ε as in Ethier and Kurtz [55]. Suppose that $z^\varepsilon(\cdot)$ are defined on the same probability space. Let $\mathcal{F}_t^\varepsilon$ be the minimal σ-algebra over which $\{z^\varepsilon(s), \xi^\varepsilon(s) : s \leq t\}$ is measurable and let E_t^ε denote the conditional expectation given $\mathcal{F}_t^\varepsilon$. Denote

$$\overline{M}^\varepsilon = \{f : f \text{ is real valued with bounded support and is}$$

$$\text{progressively measurable w.r.t. } \{\mathcal{F}_t^\varepsilon\}, \ \sup_t E|f(t)| < \infty\}.$$

Let $g(\cdot), f(\cdot), f^\delta(\cdot) \in \overline{M}^\varepsilon$. For each $\delta > 0$ and $t \leq T < \infty$, $f = p - \lim_\delta f^\delta$ if

$$\sup_{t,\delta} E|f^\delta(t)| < \infty,$$

then

$$\lim_{\delta \to 0} E|f(t) - f^\delta(t)| = 0 \text{ for each } t.$$

The function $f(\cdot)$ is said to be in the domain of A^ε, that is, $f(\cdot) \in \mathcal{D}(A^\varepsilon)$, and $A^\varepsilon f = g$, if

$$p - \lim_{\delta \to 0} \left(\frac{E_t^\varepsilon f(t+\delta) - f(t)}{\delta} - g(t) \right) = 0.$$

If $f(\cdot) \in \mathcal{D}(A^\varepsilon)$, then Ethier and Kurtz [55] or Kushner [96, p. 39] implies that

$$f(t) - \int_0^t A^\varepsilon f(u)du \text{ is a martingale,}$$

and

$$E_t^\varepsilon f(t+s) - f(t) = \int_t^{t+s} E_t^\varepsilon A^\varepsilon f(u)du \quad \text{w.p.1.}$$

In applications, ϕ-mixing processes frequently arise; see [55] and [96]. The assertion below presents a couple of inequalities for uniform mixing processes. Further results on various mixing processes are in [55].

Lemma 14.10 (Kushner [96, Lemma 4.4]). *Let $\xi(\cdot)$ be a ϕ-mixing process with mixing rate $\phi(\cdot)$ and let $h(\cdot)$ be $\mathcal{F}_t^\infty$-measurable and $|h| \leq 1$. Then*

$$\left| E(h(\xi(t+s))|\mathcal{F}_0^t) - Eh(\xi(t+s)) \right| \leq 2\phi(s).$$

If $t < u < v$, and $Eh(\xi(s)) = 0$ for all s, then

$$\left| E(h(\xi(u))h(\xi(v))|\mathcal{F}_0^t) - Eh(\xi(u))h(\xi(v)) \right| \leq 4\left(\phi(v-u)\phi(u-t) \right)^{\frac{1}{2}},$$

where $\mathcal{F}_\tau^t = \sigma\{\xi(s) : \tau \leq s \leq t\}$.

Example 14.11. A useful example of mixing process is a function of a stationary Markov chain with finite state space. Let α_k be such a Markov chain with state space $\mathcal{M} = \{1, \ldots, m_0\}$. Let $\xi_k = g(\alpha_k)$, where $g(\cdot)$ is a real-valued function defined on $\mathcal{M}$. Suppose the Markov chain or equivalently, its transition probability matrix is irreducible and aperiodic. Then as proved in Billingsley [18, pp. 167–169], ξ_k is a mixing process with the mixing measure decaying to 0 exponentially fast.

A crucial step in obtaining many limit problems depends on the verification of tightness of the sequences of interest. A sufficient condition known as Kurtz's criterion appears to be rather handy to use.

Lemma 14.12 (Kushner [96, Theorem 3, p. 47]). *Suppose that $\{y^\varepsilon(\cdot)\}$ is a process with paths in $D([0,\infty); \mathbb{R}^r)$, and suppose that*

$$\lim_{K_1 \to \infty} \left\{ \limsup_{\varepsilon \to 0} P\left(\sup_{0 \leq t \leq T} |y^\varepsilon(t)| \geq K_1 \right) \right\} = 0 \text{ for each } T < \infty, \quad (14.3)$$

and for all $0 \leq s \leq \delta, t \leq T$,

$$E_t^\varepsilon \min\left(1, |y^\varepsilon(t+s) - y^\varepsilon(t)|^2\right) \leq E_t^\varepsilon \gamma_\varepsilon(\delta),$$

$$\lim_{\delta \to 0} \limsup_{\varepsilon \to 0} E\gamma_\varepsilon(\delta) = 0. \quad (14.4)$$

Then $\{y^\varepsilon(\cdot)\}$ is tight in $D([0,\infty); \mathbb{R}^r)$.

Remark 14.13. In lieu of (14.3), one may verify the following condition (see Kurtz [93, Theorem 2.7, p. 10]): Suppose that for each $\eta > 0$ and rational $t \geq 0$ there is a compact set $\Gamma_{t,\eta} \subset \mathbb{R}^r$ such that

$$\inf_\varepsilon P\left(y^\varepsilon(t) \in \Gamma_{t,\eta}\right) > 1 - \eta. \quad (14.5)$$

To deal with singularly perturbed stochastic systems, the perturbed test function method is useful. The next lemma, due to Kushner, gives a criterion for tightness of singularly perturbed systems via perturbed test function methods. Note that in the lemma, the perturbed test functions $f^\varepsilon(\cdot)$ are so constructed that they are close to $f(z^\varepsilon(\cdot))$, and that they result in desired cancellation in the averaging.

Lemma 14.14 (Kushner [96, Theorem 3.4]). *Let $z^\varepsilon(\cdot) \in D([0,\infty); \mathbb{R}^r)$ for an appropriate r, $z^\varepsilon(0) = z_0$, and*

$$\lim_{\kappa_1 \to \infty} \left\{ \limsup_{\varepsilon \to 0} P\left(\sup_{t \le T_1} |z^\varepsilon(t)| \ge \kappa_1 \right) \right\} = 0 \qquad (14.6)$$

for each $T_1 < \infty$. For each $f(\cdot) \in C_0^2$ and $T_1 < \infty$, let there be a sequence $\{f^\varepsilon(\cdot)\}$ such that $f^\varepsilon(\cdot) \in \mathcal{D}(A^\varepsilon)$ and that $\{A^\varepsilon f^\varepsilon(t) : \varepsilon > 0, t < T_1\}$ is uniformly integrable and

$$\lim_{\varepsilon \to 0} P\left(\sup_{t \le T_1} |f^\varepsilon(t) - f(z^\varepsilon(t))| \ge \kappa_2 \right) = 0$$

holds for each $T_1 < \infty$ and each $\kappa_2 > 0$. Then $\{z^\varepsilon(\cdot)\}$ is tight in the space $D([0,\infty); \mathbb{R}^r)$.

To apply Lemma 14.14 for proving tightness, one needs to verify (14.6). Such verifications are usually nontrivial and involve complicated calculations. To overcome the difficulty, one uses the so-called N-truncation device, which is defined as follows.

Definition 14.15 (N-truncation, see Kushner and Yin [100, p. 284]). For each $N > 0$, let $B_N = \{z : |z| \le N\}$ be the sphere with radius N, let $z^{\varepsilon,N}(0) = z^\varepsilon(0)$, $z^{\varepsilon,N}(t) = z^\varepsilon(t)$ up until the first exit from B_N, and

$$\lim_{\kappa_1 \to \infty} \limsup_{\varepsilon \to 0} P\left(\sup_{t \le T_1} |z^{\varepsilon,N}(t)| \ge \kappa_1 \right) = 0 \qquad (14.7)$$

for each $T_1 < \infty$. Then $z^{\varepsilon,N}(t)$ is said to be the N-truncation of $z^\varepsilon(\cdot)$.

Using the perturbed test function techniques, the lemma to follow provides sufficient conditions for weak convergence. Its proof is in Kushner [96].

Lemma 14.16. *Suppose that $\{z^\varepsilon(\cdot)\}$ is defined on $[0,\infty)$. Let $\{z^\varepsilon(\cdot)\}$ be tight on $D([0,\infty); \mathbb{R}^r)$. Suppose that for each $f(\cdot) \in C_0^2$, and each $T_1 < \infty$, there exist $f^\varepsilon(\cdot) \in \mathcal{D}(A^\varepsilon)$ such that*

$$p - \lim_{\varepsilon \to 0} (f^\varepsilon(\cdot) - f(z^\varepsilon(\cdot))) = 0 \qquad (14.8)$$

and

$$p - \lim_{\varepsilon \to 0} (A^\varepsilon f^\varepsilon(\cdot) - Af(z^\varepsilon(\cdot))) = 0. \qquad (14.9)$$

Then $z^\varepsilon(\cdot) \Rightarrow z(\cdot)$.

The theorem to follow is useful in characterizing certain limit processes (in weak convergence analysis). Its proof is in Kushner and Yin [100, Theorem 4.1.1].

Theorem 14.17. *Let $m(t)$ be a continuous-time martingale whose paths are Lipschitz continuous with probability one on each bounded time interval. Then $m(t)$ is a constant with probability one.*

The well-known result due to Cramér and Wold states that problems involving finite-dimensional random vectors can be reduced to problems involving scalar random variables (see Billingsley [18, p. 48]). We present the result in the following lemma.

Lemma 14.18. *Let $x_k = (x_k^1, \ldots, x_k^r)' \in \mathbb{R}^r$ and $x = (x^1, \ldots, x^r)' \in \mathbb{R}^r$. Suppose*

$$\sum_{j=1}^r t^j x_k^j \quad \text{converges in distribution to} \quad \sum_{j=1}^r t^j x^j \quad \text{as} \ \ k \to \infty,$$

for each $t = (t^1, \ldots, t^r)' \in \mathbb{R}^r$. Then x_k converges in distribution to x.

Next, we state a weak convergence result due to Kushner and Huang. For a proof, see Kushner and Huang [99, Theorem 1] and Kushner [96, Theorem 3.2, p. 44].

Theorem 14.19. *Let x_k^ε be a stochastic process in discrete time and $x^\varepsilon(\cdot)$ be its piecewise constant interpolation on the interval $[\varepsilon k, \varepsilon k + \varepsilon)$. Suppose that $\{x^\varepsilon(\cdot)\}$ is tight in $D([0, T]; \mathbb{R}^r)$ and $x^\varepsilon(0) \Rightarrow x_0$, that the martingale problem with operator $\mathcal{L}$ has a unique solution $x(\cdot)$ in $D([0, T]; \mathbb{R}^r)$ for each initial condition, and that for each $g(\cdot) \in C_0^2$, there exists a sequence $\{g^\varepsilon(\cdot)\}$ such that $g^\varepsilon(\cdot)$ is a constant on each interval $[\varepsilon k, \varepsilon k + \varepsilon)$, which is measurable (at $k\varepsilon$) with respect to the σ-algebra induced by $\{x_j^\varepsilon : j \le k\}$. Moreover, suppose that*

$$\sup_{0 \le k \le T/\varepsilon, \varepsilon} E|g^\varepsilon(k\varepsilon)| + \sup_{0 \le k \le T/\varepsilon, \varepsilon} \frac{1}{\varepsilon} |E_k g^\varepsilon(k\varepsilon + \varepsilon) - g^\varepsilon(k\varepsilon)| < \infty, \quad (14.10)$$

and as $\varepsilon \to 0$ with $k\varepsilon \to t$,

$$E|g^\varepsilon(k\varepsilon) - g(x^\varepsilon(k\varepsilon))| \to 0, \quad (14.11)$$

and

$$E \left| \frac{E_k g^\varepsilon(k\varepsilon + \varepsilon) - g^\varepsilon(k\varepsilon)}{\varepsilon} - \mathcal{L}g(x^\varepsilon(k\varepsilon)) \right| \to 0. \quad (14.12)$$

Then $x^\varepsilon(\cdot)$ converges weakly to $x(\cdot)$, the unique solution to the martingale problem with initial condition x_0.

In this book, we focus on hybrid systems or systems with Markov regime switching. Let $\alpha(t)$ be a continuous-time Markov chain with state space $\mathcal{M} = \{1, \ldots, m_0\}$ and let $w(\cdot)$ be an $\mathbb{R}^r$-valued standard Brownian motion that is independent of $\alpha(\cdot)$. Often, we need to examine the following (or its discrete-time counterpart) system of stochastic differential equations modulated by a Markov chain,

$$dx(t) = f(t, \alpha(t), x(t))dt + g(t, \alpha(t), x(t))d\widehat{w}(t), \tag{14.13}$$

where $t \in [0, T]$ for some $T > 0$, $t \in [0, T]$, $f(\cdot, \cdot, \cdot) : [0, T] \times \mathcal{M} \times \mathbb{R}^r \mapsto \mathbb{R}^r$, and $g(\cdot, \cdot, \cdot) : [0, T] \times \mathcal{M} \times \mathbb{R}^r \mapsto \mathbb{R}^r$. It is readily seen that the pair of processes $y(\cdot) = (x(\cdot), \alpha(\cdot))$ is a Markov process. Corresponding to this process $y(\cdot)$, we define an operator $\mathcal{L}$ as follows. For each $i \in \mathcal{M} = \{1, \ldots, m_0\}$ and each $\psi(\cdot, i, \cdot) \in C_0^{1,2}$ ($C_0^{1,2}$ represents the class of functions that have compact support and that are continuously differentiable with respect to t and twice continuously differentiable with respect to x), let

$$\mathcal{L}\psi(t, i, x) = \psi_x'(t, i, x)f(t, i, x) + \frac{1}{2}\text{tr}\left(\psi_{xx}(t, i, x)\Xi(t, i, x)\right) \\ + Q\psi(t, \cdot, x)(i), \tag{14.14}$$

where $\Xi(t, i, x) = g(t, i, x)g'(t, i, x)$. Associated with (14.13), there is a martingale problem with operator $(\partial/\partial t) + \mathcal{L}$. The next lemma establishes the uniqueness of such martingale problem.

Lemma 14.20. *Assume that the following conditions hold: The initial condition x_0 satisfies $E|x_0|^2 < \infty$. The functions $f(\cdot)$ and $g(\cdot)$ satisfy the conditions that for each $\alpha \in \mathcal{M}$, $f(\cdot, \alpha, \cdot)$ and $g(\cdot, \alpha, \cdot)$ are defined and Borel measurable on $[0, T] \times \mathbb{R}^r$; in addition, for each $(t, \alpha, x) \in [0, T] \times \mathcal{M} \times \mathbb{R}^r$,*

$$|f(t, \alpha, x)| \le K(1 + |x|) \quad and \quad |g(t, \alpha, x)| \le K(1 + |x|);$$

for each $x_1, x_2 \in \mathbb{R}^r$,

$$|f(t, \alpha, x_1) - f(t, \alpha, x_2)| \le K|x_1 - x_2|,$$

and

$$|g(t, \alpha, x_1) - g(t, \alpha, x_2)| \le K|x_1 - x_2|.$$

In addition, the Markov chain $\alpha(\cdot)$ and the Brownian motion $w(\cdot)$ are independent. Then the solution of the martingale problem with operator $(\partial/\partial t) + \mathcal{L}$ for $\mathcal{L}$ given by (14.14) is unique.

Proof. The proof follows the same line of argument presented in Yin and Zhang [158, Lemma 7.18] via the use of characteristic functions. □

A useful device known as the continuous mapping theorem indicates that the weak convergence is preserved under nonlinear (continuous) transformation. This enables us to obtain the weak convergence of a desired sequence through the techniques of transformations. We state the following result. The proof can be found in Billingsley [18, p. 31].

Theorem 14.21. *Suppose that x_k is a sequence of random variables (real-valued, or vector-valued, or living in a more general function space), and that x_k converges weakly to x an element in the same space as that of x_k. Then*

(a) *for every real and continuous function $h(\cdot)$, $h(x_k)$ converges weakly to $h(x)$;*

(b) *for every real, measurable function $h(\cdot)$ whose set of discontinuous points has probability measure 0, $h(x_k)$ converges weakly to $h(x)$.*

14.4 Optimal Controls and HJB Equations

To solve optimal control problems involving deterministic or stochastic systems, a commonly used technique is the dynamic programming method; see Fleming and Rishel [57]. Such an approach yields to a partial differential equation (known as the Hamilton–Jacobi–Bellman (HJB) equation) satisfied by the value function. When the dynamic system involves a Markovian switching process, the basic technique still carries over, but a coupling term will be added; see Yin and Zhang [158, Appendix]. In what follows, we present some properties of the value functions and optimal feedback controls arising in dynamic systems with regime switching. Simple but representative examples are treated; models having more complex structures can be dealt with similarly.

Let $(\Omega, \mathcal{F}, P)$ be a probability space. Let $\alpha(t) \in \mathcal{M} = \{1, \ldots, m_0\}$, for $t \geq 0$, denote a Markov chain with generator Q. Assume that $b(x, u, \alpha)$ satisfies the following conditions: There exist bounded functions $b_1(x, \alpha) \in \mathbb{R}^n$ and $b_2(x, \alpha) \in \mathbb{R}^{n \times n_1}$ on $\mathbb{R}^n \times \mathcal{M}$ such that

$$b(x, u, \alpha) = b_1(x, \alpha) + b_2(x, \alpha)u,$$

where $u \in \Gamma$, a convex and compact subset of $\mathbb{R}^{n_1}$. Moreover, for $i = 1, 2$, $b_i(x, \alpha)$ are Lipschitz in that

$$|b_i(x, \alpha) - b_i(y, \alpha)| \leq K|x - y|, \text{ for all } \alpha \in \mathcal{M},$$

for a constant K.

The system of states and the control constraints are

$$\frac{dx(t)}{dt} = b(x(t), u(t), \alpha(t)),$$
$$x(0) = x, \ \alpha(0) = \alpha, \text{ and } u(t) \in \Gamma,$$

where Γ is a convex and compact subset of $\mathbb{R}^{n_1}$.

The problem of interest is to choose an admissible control $u(\cdot)$ so as to minimize an expected cost function

$$J(x, u(\cdot), \alpha) = E \int_0^\infty e^{-\rho t} G(x(t), u(t), \alpha(t)) dt,$$

where x and α are the initial values of $x(t)$ and $\alpha(t)$, respectively, and $G(x, u, \alpha)$ is a running cost function.

Definition 14.22. A control $u(\cdot) = \{u(t) \in \mathbb{R}^{n_1} : t \geq 0\}$ is called *admissible* with respect to the initial α if

(a) $u(\cdot)$ is progressively measurable with respect to the filtration $\{\mathcal{F}_t\}$, where $\mathcal{F}_t = \sigma\{\alpha(s) : 0 \leq s \leq t\}$ and

(b) $u(t) \in \Gamma$ for all $t \geq 0$.

Let $\mathcal{A}$ denote the set of all admissible controls.

Definition 14.23. A function $u(x, \alpha)$ is called an *admissible feedback control*, or simply *feedback control*, if

(a) for any given initial data x, the equation

$$\frac{dx(t)}{dt} = b(x(t), u(x(t), \alpha(t)), \alpha(t)), \quad x(0) = x$$

has a unique solution;

(b) $u(\cdot) = \{u(t) = u(x(t), \alpha(t)), \ t \geq 0\} \in \mathcal{A}$.

Let $v(x, \alpha)$ denote the value function of the problem

$$v(x, \alpha) = \inf_{u(\cdot) \in \mathcal{A}} J(x, u(\cdot), \alpha).$$

As elsewhere in the book, K is a generic positive constant, whose values may be different for different uses.

Lemma 14.24. *The following assertions hold:*

(a) *If $G(x, u, \alpha)$ is locally Lipschitz in that*

$$|G(x_1, u, \alpha) - G(x_2, u, \alpha)| \leq K(1 + |x_1|^\kappa + |x_2|^\kappa)|x_1 - x_2|$$

for some constants K and κ, then $v(x, \alpha)$ is also locally Lipschitz in that

$$|v(x_1, \alpha) - v(x_2, \alpha)| \leq K(1 + |x_1|^\kappa + |x_2|^\kappa)|x_1 - x_2|.$$

(b) *If $G(x, u, \alpha)$ is jointly convex and $b(x, u, \alpha)$ is independent of x, then $v(x, \alpha)$ is convex in x for each $\alpha \in \mathcal{M}$.*

The next lemma presents the dynamic programming principle. A proof of the result can be found in Yin and Zhang [158, Appendix].

Lemma 14.25. *Let τ be an $\{\mathcal{F}_t\}$-stopping time. Then*

$$v(x, \alpha) = \inf_{u(\cdot) \in \mathcal{A}} E\left\{ \int_0^\tau e^{-\rho t} G(x(t), u(t), \alpha(t)) dt + e^{-\rho \tau} v(x(\tau), \alpha(\tau)) \right\}.$$

To characterize further the value functions, let us first recall the notion of viscosity solutions. Let $v(\cdot) : \mathbb{R}^n \times \mathcal{M} \to \mathbb{R}^1$ be a given function and $H(\cdot)$ be a real-valued function on

$$\Omega_H := \mathbb{R}^n \times \mathcal{M} \times \mathbb{R}^{m_0} \times \mathbb{R}^n.$$

Consider the following equation

$$v(x, \alpha) - H\left(x, \alpha, v(x, \cdot), \frac{\partial v(x, \alpha)}{\partial x} \right) = 0. \tag{14.15}$$

Definition 14.26 (Viscosity Solution). $v(x, \alpha)$ is a *viscosity solution* of Equation (14.15) if the following hold:

(a) $v(x, \alpha)$ is continuous in x and $|v(x, \alpha)| \leq K(1 + |x|^\kappa)$ for some $\kappa \geq 0$;

(b) for any $\alpha_0 \in \mathcal{M}$,

$$v(x_0, \alpha_0) - H\left(x_0, \alpha_0, v(x_0, \cdot), \frac{\partial \phi(x_0)}{\partial x} \right) \leq 0,$$

whenever $\phi(x) \in C^1$ (i.e., continuously differentiable) and $v(x, \alpha_0) - \phi(x)$ has a local maximum at $x = x_0$; and

(c) for any $\alpha_0 \in \mathcal{M}$,

$$v(x_0, \alpha_0) - H\left(x_0, \alpha_0, v(x_0, \cdot), \frac{\partial \psi(x_0)}{\partial x} \right) \geq 0,$$

whenever $\psi(x) \in C^1$ and $v(x, \alpha_0) - \psi(x)$ has a local minimum at $x = x_0$.

If (a) and (b) (resp. (a) and (c)) hold, we say that v is a *viscosity subsolution* (resp. *viscosity supersolution*).

In this book, when we consider certain optimal control problems, we obtain a system of HJB equations of the form

$$\rho v(x, \alpha) = \min_{u \in \Gamma} \left\{ b(x, u, \alpha) \frac{\partial v(x, \alpha)}{\partial x} + G(x, u, \alpha) + Q v(x, \cdot)(\alpha) \right\}, \tag{14.16}$$

where $b(\partial/\partial x)v$ is understood to be the inner product of b and $(\partial/\partial x)v$. For proofs and references of the next a few results, we refer the reader to Yin and Zhang [158, A.5 in Appendix].

Theorem 14.27 (Uniqueness Theorem). *Assume that for some positive constants K and κ,*

$$|G(x, u, \alpha)| \leq K(1 + |x|^{\kappa}),$$

and

$$|G(x_1, u, \alpha) - G(x_2, u, \alpha)| \leq K(1 + |x_1|^{\kappa} + |x_2|^{\kappa})|x_1 - x_2|,$$

for all $\alpha \in \mathcal{M}$, x, x_1, and $x_2 \in \mathbb{R}^n$, and u in a compact and convex set Γ. Let Q be the generator of a Markov chain with state space $\mathcal{M}$. Then the HJB equation (14.16) has a unique viscosity solution.

Theorem 14.28. *The value function $v(x, \alpha)$ is the unique viscosity solution to the HJB equation*

$$\rho v(x, \alpha) = \min_{u \in \Gamma} \left\{ b(x, u, \alpha) \frac{\partial v(x, \alpha)}{\partial x} + G(x, u, \alpha) \right\} + Qv(x, \cdot)(\alpha). \quad (14.17)$$

Theorem 14.29 (Verification Theorem). *Let $w(x, \alpha) \in C^1$ such that*

$$|w(x, \alpha)| \leq K(1 + |x|^{\kappa})$$

and

$$\rho w(x, \alpha) = \min_{u \in \Gamma} \left\{ b(x, u, \alpha) \frac{\partial w(x, \alpha)}{\partial x} + G(x, u, \alpha) + Qw(x, \cdot)(\alpha) \right\}.$$

Then the following assertions hold:

(a) *$w(x, \alpha) \leq J(x, u(\cdot), \alpha)$ for any $u(t) \in \Gamma$.*

(b) *Suppose that there are $u^\circ(t)$ and $x^\circ(t)$ satisfying*

$$\frac{dx^\circ(t)}{dt} = b(x^\circ(t), u^\circ(t), \alpha(t))$$

with $x^\circ(0) = x$, $r^\circ(t) = (\partial/\partial x)v(x^\circ(t), \alpha(t))$, and

$$\min_{u \in \Gamma} \left\{ b(x, u, \alpha)r^\circ(t) + G(x^\circ(t), u, \alpha) + Qw(x^\circ(t), \cdot)(\alpha(t)) \right\}$$

$$= b(x^\circ(t), u^\circ(t), \alpha(t))r^\circ(t) + G(x^\circ(t), u^\circ(t), \alpha(t))$$

$$+ Qw(x^\circ(t), \cdot)(\alpha(t))$$

almost everywhere in t with probability one. Then

$$w(x, \alpha) = v(x, \alpha) = J(x, u^\circ(\cdot), \alpha).$$

Lemma 14.30. *Let $c(u)$ be twice differentiable such that $(\partial^2/\partial u^2)c(u) > 0$, $V(x)$ be locally Lipschitz, i.e.,*

$$|V(x_1) - V(x_2)| \leq K(1 + |x_1|^\kappa + |x_2|^\kappa)|x_1 - x_2|,$$

and $u^\circ(x)$ be the minimum of $F(x, u) := uV(x) + c(u)$. Then $u^\circ(x)$ is locally Lipschitz in that

$$|u^\circ(x_1) - u^\circ(x_2)| \leq K(1 + |x_1|^\kappa + |x_2|^\kappa)|x_1 - x_2|.$$

14.5   Miscellany

This section consists of a number of miscellaneous results used in this book. They include the notion of convex functions, equiconitinuity in the extended sense and the corresponding Arzelá-Ascoli theorem, and the Fredholm alternative, among others.

Definition 14.31 (Convex sets and convex functions). A set $S \subset \mathbb{R}^r$ is convex if for any x and $y \in S$, $\gamma x + (1 - \gamma)y \in S$ for any $0 \leq \gamma \leq 1$. A real-valued function $f(\cdot)$ on S is convex if for any $x_1, x_2 \in S$ and $\gamma \in [0, 1]$,

$$f(\gamma x_1 + (1 - \gamma)x_2) \leq \gamma f(x_1) + (1 - \gamma)f(x_2).$$

If the above inequality holds in the strict sense whenever $x_1 \neq x_2$ and $0 < \gamma < 1$, then $f(\cdot)$ is strictly convex.

The definition above can be found in, for example, the work of Fleming [56]. The next lemma establishes the connection of convex functions with Lipschitz continuity and differentiability.

Lemma 14.32 (Clarke [39, Theorem 2.5.1]). *Let $f(\cdot)$ be a convex function on $\mathbb{R}^r$. Then*

(a) *$f(\cdot)$ is locally Lipschitz and therefore continuous, and*

(b) *$f(\cdot)$ is differentiable a.e.*

We present a variation of the Arzelà–Ascoli theorem. Its ingredient is that the family of functions is almost Lipschitz with an extra factor tending to 0 as $k \to \infty$. Note that the sequence $1/k$ in (14.18) can be replaced by a sequence of positive real numbers a_k satisfying $a_k \to 0$.

Lemma 14.33. *Let $\{f_k(x)\}$ be a sequence of functions on a compact subset of $\mathbb{R}^{n_1}$. Assume that $\{f_k(x)\}$ is uniformly bounded and that there exists a constant K such that*

$$|f_k(x) - f_k(y)| \leq K\left(\frac{1}{k} + |x - y|\right), \quad \text{for all } x, y. \tag{14.18}$$

Then there exists a uniformly convergent subsequence.

Proof. The proof is a slight variation of that given in Strichartz [142, p. 312] with the equicontinuous condition replaced by the near Lipschitz condition. □

The following definition of equicontinuity in the extended sense is in Kushner and Yin [100, p. 102]. Theorem 14.35 presents another variation of the Arzelà–Ascoli theorem, namely, Arzelà–Ascoli theorem in the extended sense, which is handy to use for treating interpolated processes from discrete iterations.

Definition 14.34. *Let $\{f_k(t)\}$ be a sequence of $\mathbb{R}^r$-valued measurable function defined on $\mathbb{R}^1$, with $\{f_k(0)\}$ bounded. If for each T_0 and $\eta > 0$, there is a $\delta > 0$ such that*

$$\limsup_{n} \sup_{0 \leq t-s \leq \delta, |t| \leq T_0} |f_k(t) - f_k(s)| \leq \eta,$$

then we say that $\{f_k(\cdot)\}$ is equicontinuous in the extended sense.

Theorem 14.35. *Let $\{f_k(\cdot)\}$ be defined on $\mathbb{R}$ and be equicontinuous in the extended sense. Then there exists a subsequence converging to some continuous limit uniformly on each bounded interval.*

The following Fredholm alternative, which provides a powerful method for establishing existence and uniqueness of solutions for various systems of equations, can be found in, for example, Hutson and Pym [70, p. 184].

Lemma 14.36 (Fredholm Alternative). *Let $\mathbb{B}$ be a Banach space and A a linear compact operator defined on it. Let $I : \mathbb{B} \to \mathbb{B}$ be the identity operator. Assume $\gamma \neq 0$. Then one of the two alternatives holds:*

(a) *The homogeneous equation $(\gamma I - A)f = 0$ has only the zero solution, in which case $\gamma \in \rho(A)$-the resolvent set of A, $(\gamma I - A)^{-1}$ is bounded, and the inhomogeneous equation $(\gamma I - A)f = g$ has also one solution $f = (\gamma I - A)^{-1}g$, for each $g \in \mathbb{B}$.*

(b) *The homogeneous equation $(\gamma I - A)f = 0$ has a nonzero solution, in which case the inhomogeneous equation $(\gamma I - A)f = g$ has a solution iff $\langle g, f^* \rangle = 0$ for every solution f^* of the adjoint equation $\gamma f^* = A^* f^*$.*

Note that in (b) above, $\langle g, f^* \rangle$ is a pairing defined on $\mathbb{B} \times \mathbb{B}^*$ (with $\mathbb{B}^*$ denoting the dual of $\mathbb{B}$). This is also known as an "outer product" (see [70, p. 149]), whose purpose is similar to the inner product in a Hilbert space. If we work with a Hilbert space, this "outer product" is identical to the usual inner product. When one considers linear systems of algebraic equations, the lemma above can be rewritten in a simpler form.

Let B denote an $m_0 \times m_0$ matrix. For any $\gamma \neq 0$, define an operator $A : \mathbb{R}^{m_0 \times m_0} \to \mathbb{R}^{m_0 \times m_0}$ as

$$Ay = y(\gamma I - B).$$

Note that in this case, I is just the $m_0 \times m_0$ identity matrix I. Then the adjoint operator $A^* : \mathbb{R}^{m_0 \times m_0} \to \mathbb{R}^{m_0 \times m_0}$ is

$$A^* x = (\gamma I - B)x.$$

Suppose that b and $y \in \mathbb{R}^{1 \times m_0}$. Consider the system $yB = b$. If the adjoint system $Bx = 0$ where $x \in \mathbb{R}^{m_0 \times 1}$ has only the zero solution, then $yB = b$ has a unique solution given by $y = bB^{-1}$. If $Bx = 0$ has a nonzero solution x, then $yB = b$ has a solution iff $\langle b, x \rangle = 0$.

Lemma 14.37. *If a transition probability matrix $P = (p^{ij})_{m_0 \times m_0}$ is irreducible and aperiodic, and $f(i) \le \sum_{j=1}^{m_0} p^{ij} f(j)$ for a function $f(i)$, $i = 1, \ldots, m_0$, then $f(1) = f(2) = \cdots = f(m_0)$.*

Proof. Note that $(P - I)$ is a generator and is irreducible. A direct application of Lemma A.39 in Yin and Zhang [158] yields this result.   □

The following lemma reveals the connection between the weak irreducibility of a generator Q and the rank of the matrix Q. Its proof is in Yin and Zhang [158, Lemma A.5].

Lemma 14.38. *Let Q be an $m_0 \times m_0$ generator. If $\mathrm{rank}(Q) = m_0 - 1$, then Q is weakly irreducible.*

The next lemma is needed in treating filtering problems.

Lemma 14.39. *Let $w(\cdot)$ and $v(\cdot)$ be independent r-dimensional standard Brownian motions and $\alpha(\cdot)$ be a continuous-time Markov chain with a finite state space $\mathcal{M} = \{1, \ldots, l_0\}$, which is independent of $w(\cdot)$ and $v(\cdot)$. For each $i \in \mathcal{M}$, $A(i)$, $C(i)$ $\sigma_w(i)$, and $\sigma_v(i)$ are $r \times r$ matrices such that $\Sigma(i) = \sigma_w(i)\sigma'_w(i)$ and $C(i)C'(i)$ are positive definite. For each $t \in [0, T]$, $x(t)$ and $y(t)$ are $\mathbb{R}^r$-valued processes such that $x(t)$ is the state and $(y(t), \alpha(t))$ are the observations. That is, $(y(s), \alpha(s))$ is observable for $s \le t$. They satisfy the equations*

$$dx = A(\alpha(t))x dt + \sigma_w(\alpha(t))dw, \quad x(0) = x,$$
$$dy = C(\alpha(t))x dt + \delta dv, \quad y(0) = 0,$$

respectively. In the above, $\delta > 0$ is a small parameter representing the magnitude of observation noise. Then the corresponding Kalman filter is given as follows:

$$d\hat{x} = A(\alpha(t))\hat{x} dt + \frac{1}{\delta^2} R(t)C'(\overline{\alpha}(t))(dy - C(\alpha(t))\hat{x} dt),$$
$$\dot{R} = A(\alpha(t))R(t) + R(t)A'(\alpha(t))$$
$$\qquad - \frac{1}{\delta^2} R(t)C(\alpha(t))C(\alpha(t))R(t) + \Sigma_w^{-1}(\alpha(t)),$$
$$\hat{x}(0) = x, \quad \text{and} \quad R(0) = 0,$$

where $\widehat{x}(t) \in \mathbb{R}^r$ and $R(t) \in \mathbb{R}^{r \times r}$. Moreover,

$$E \int_0^T |x(t) - \widehat{x}(t)|^2 dt = O(\delta).$$

Proof. The proof is given in Zhang [174]. □

It is well known that any random variable X on a probability space $(\Omega, \mathcal{F}, P)$ has an associated characteristic function defined by

$$\phi(t) = E \exp(itX).$$

There is a one-to-one correspondence between the random variable and its characteristic function. A result, known as inversion formula (see Chow and Teicher [37, p. 287]), is stated next.

Theorem 14.40. *Suppose that X is a random variable with characteristic function $\phi(t)$. Then for $-\infty < a < b < \infty$,*

$$\lim_{M \to \infty} \frac{1}{2\pi} \int_{-M}^{M} \frac{e^{-ita} - e^{-itb}}{it} \phi(t) dt$$
$$= P(a < X < b) + \frac{P(X = a) + P(X = b)}{2}.$$

In studying differential equations, a device known as Gronwall's inequality is quite handy; see Hale [66, Corollary 6.5, p. 36]. The following lemma is a discrete-time counterpart and is useful for establishing bounds in difference equations.

Lemma 14.41. *Let $\{\phi_k\}$ be a nonnegative sequence satisfying*

$$\phi_{k+1} \leq C_0 + \varepsilon C_1 \sum_{j=0}^{k} \phi_j, \quad k = 0, 1, 2, \dots, T/\varepsilon, \tag{14.19}$$

for some positive constants C_0 and C_1, and a parameter $\varepsilon > 0$. Then, for $k = 0, 1, 2, \dots, T/\varepsilon$,

$$\phi_k \leq C_0 (1 + \varepsilon C_1)^{T/\varepsilon}.$$

Moreover,

$$\phi_k \leq C_0 \exp(C_1 T). \tag{14.20}$$

Proof: Let $\xi_k = C_0 + \varepsilon C_1 \sum_{j=0}^{k-1} \phi_j$. Then $\phi_k \leq \xi_k$, $\xi_0 = C_0$, and

$$\xi_{k+1} = \xi_k + \varepsilon C_1 \phi_k \leq \xi_k + \varepsilon C_1 \xi_k, \quad k \geq 1.$$

Solving the above inequality yields

$$\phi_k \leq \xi_k \leq C_0 (1 + \varepsilon C_1)^k \leq C_0 (1 + \varepsilon C_1)^{T/\varepsilon}.$$

In addition, noting that $(1 + \varepsilon C_1) \leq \exp(\varepsilon C_1)$, (14.20) follows. □

Remark 14.42. As mentioned elsewhere in the book, T/ε above is meant to be the integer part $\lfloor T/\varepsilon \rfloor$. Taking $\varepsilon = 1$, in (14.19), we obtain the conventional Gronwall's inequality. In fact, Lemma 14.41 can be stated as

$$\sup_{0 \le k \le T/\varepsilon} \phi_k \le C_0 \exp(C_1 T).$$

It is interesting to note that the right-hand side above does not depend on ε. Thus, it also holds uniformly in ε.

There is also a version of the inequality corresponding to the generalized Gronwall's inequality in Hale [66, Lemma 6.2, p. 36], which is stated as follows. Suppose that $\{\phi_k\}$ is a nonnegative sequence of real numbers satisfying

$$\phi_{k+1} \le \psi_{k+1} + \varepsilon \sum_{j=0}^{k} C_j \phi_j, \quad k = 0, 1, 2, \dots, T/\varepsilon$$

for some $\psi_k \ge 0$ and $C_k \ge 0$ and a parameter $\varepsilon > 0$. Then

$$\phi_k \le \psi_k + \varepsilon \sum_{j=0}^{k-1} \prod_{i=j+1}^{k-1} (1 + \varepsilon C_i) C_j \psi_j, \quad k = 0, 1, 2, \dots, T/\varepsilon.$$

The proof is a modification of Lemma 14.41.

14.6   Notes

This chapter reviews a number of technical results. Further reading and details in stochastic processes can be found in Billingsley [18], Ethier and Kurtz [55], and Kushner [96]. Related topics in control theory and singularly perturbed Markov chains can be found in Fleming and Rishel [57], Yin and Zhang [158], and the references therein.

References

[1] M. Abbad and J.A. Filar, Perturbation and stability theory for Markov control problems, *IEEE Trans. Automat. Control*, **37**, (1992), 1415–1420.

[2] M. Abbad, J.A. Filar, and T.R. Bielecki, Algorithms for singularly perturbed limiting average Markov control problems, *IEEE Trans. Automat. Control*, **37** (1992), 1421–1425.

[3] R. Akella and P.R. Kumar, Optimal control of production rate in a failure-prone manufacturing system, *IEEE Trans. Automat. Control*, **31** (1986), 116–126.

[4] E. Altman and V. Gaitsgory, Control of a hybrid stochastic system, *Systems Control Lett.*, **20** (1993), 307–314.

[5] S. Andradottir, A global search method for discrete stochastic optimization, *SIAM J. Optim.*, **6** (1996), 513–530.

[6] V.V. Anisimov, Switching processes: Averaging principle, diffusion approximation and applications, *Acta Appl. Math.*, **40** (1995), 95–141.

[7] K.E. Avrachenkov, J.A. Filar, and M. Haviv, Singular perturbations of Markov chains and decision processes, in *Handbook of Markov Decision Processes: Methods and Applications*, E.A. Feinberg and A. Shwartz Eds., 113–153, Kluwer, Boston, 2002.

[8] B. Avramovic, J. Chow, P.V. Kokotovic, G. Peponides, and J.R. Winkelman, *Time-scale Modeling of Dynamic Networks with Applications to Power Systems*, Lecture Notes in Control & Inform. Sci., Vol. 46, Springer-Verlag, New York, NY, 1982.

334   References

[9] G. Badowski and G. Yin, Stability of hybrid dynamic systems containing singularly perturbed random processes, *IEEE Trans. Automat. Control*, **47** (2002), 2021–2032.

[10] G. Barone-Adesi and R. Whaley, Efficient analytic approximation of American option values, *J. Finance*, **42** (1987), 301–320.

[11] Y. Bar-Shalom and X.R. Li, *Estimation and Tracking: Principles, Techniques, and Software*, Artech House Publishers, Norwood, MA, 1996.

[12] C.M. Bender and S.A. Orszag, *Advanced Mathematical Methods for Scientists and Engineers*, McGraw-Hill, New York, NY, 1978.

[13] D. Bertsekas, *Dynamic Programming: Deterministic and Stochastic Models*, Prentice-Hall, Englewood Cliffs, NJ, 1987.

[14] A. Benveniste, M. Metivier, and P. Priouret, *Adaptive Algorithms and Stochastic Approximations*, Springer-Verlag, Berlin, 1990.

[15] T.R. Bielecki and J.A. Filar, Singularly perturbed Markov control problem: Limiting average cost, *Ann. Oper. Res.*, **28**, (1991), 153–168.

[16] T.R. Bielecki and P.R. Kumar, Optimality of zero-inventory policies for unreliable manufacturing systems, *Oper. Res.*, **36**, (1988), 532–541.

[17] T.R. Bielecki and L. Stettner, Ergodic control of a singularly perturbed Markov process in discrete time with general state and compact action spaces, *Appl. Math. Optim.*, **38** (1998), 261–281.

[18] P. Billingsley, *Convergence of Probability Measures*, J. Wiley, New York, NY, 1968.

[19] T. Björk, Finite dimensional optimal filters for a class of Ito processes with jumping parameters, *Stochastics*, **4** (1980), 167–183.

[20] W.P. Blair and D.D. Sworder, Feedback control of a class of linear discrete systems with jump parameters and quadratic cost criteria, *Internat. J. Control*, **21** (1986), 833–841.

[21] G. Blankenship, Singularly perturbed difference equations in optimal control problems, *IEEE Trans. Automat. Control*, **26** (1981), 911–917.

[22] G.B. Blankenship and G. C. Papanicolaou, Stability and control of stochastic systems with wide band noise, *SIAM J. Appl. Math.*, **34** (1978), 437–476.

[23] H.A.P. Blom, Implementable differential equations for nonlinear filtering, National Aerospace Lab., the Netherlands, NLR MP 81037, 1981.

[24] N.N. Bogoliubov and Y.A. Mitropolskii, *Asymptotic Methods in the Theory of Nonlinear Oscillator*, Gordon and Breach, New York, 1961.

[25] E.K. Boukas, Techniques for flow control and preventive maintenance manufacturing systems, *Control Dynamic Syst.*, **48**, (1991), 327-366.

[26] E.K. Boukas and A. Haurie, Manufacturing flow control and preventive maintenance: A stochastic control approach, *IEEE Trans. Automat. Control*, **35**, (1990), 1024–1031.

[27] L. Breiman, *Probability*, SIAM, Philadelphia, PA, 1992.

[28] J.A. Buzacott and J.G. Shanthikumar, *Stochastic Models of Manufacturing Systems*, Prentice-Hall, Englewood Cliffs, NJ, 1993.

[29] P.E. Caines and H.-F. Chen, Optimal adaptive LQG control for systems with finite state process parameters, *IEEE Trans. Automat. Control*, **30** (1985), 185–189.

[30] X.R. Cao, The relations among potentials, perturbation analysis, and Markov decision processes, *Discrete Event Dyn. Syst.*, **8** (1998), 71–87.

[31] X.R. Cao, The Maclaurin series for performance function of Markov chains, *Adv. in Appl. Prob.*, **30** (1998), 676–692.

[32] X.R. Cao, A unified approach to Markov decision problems and performance sensitivity analysis, *Automatica*, **36** (2000), 771–774.

[33] H.S. Chang, P. Fard, S.I. Marcus, and M. Shayman, Multi-time scale Markov decision processes, preprint, Univ. of Maryland, 2002.

[34] H.-F. Chen, *Stochastic Approximation and Its Applications*, Kluwer Academic, Dordrecht, Netherlands, 2002.

[35] S. Chen, X. Li, and X.Y. Zhou, Stochastic linear quadratic regulators with indefinite control weight costs, *SIAM J. Control Optim.*, **36** (1998), 1685–1702.

[36] J. Chow, J.R. Winkelman, M.A. Pai, and P.W. Sauer, Singular perturbation analysis of large scale power systems, *J. Electric Power Energy Syst.*, **12** (1990), 117–126.

[37] Y.S. Chow and H. Teicher, *Probability Theory*, Springer-Verlag, New York, NY, 1978.

[38] K.L. Chung, *Markov Chains with Stationary Transition Probabilities*, 2nd Ed., Springer-Verlag, New York, NY, 1967.

[39] F. Clarke, *Optimization and Non-smooth Analysis*, Wiley Interscience, New York, NY, 1983.

[40] O.L.V. Costa, Linear minimum mean square error estimation for discrete-time Markov jump linear systems, *IEEE Trans. Automat. Control*, **39** (1994), 1685–1689.

[41] P.J. Courtois, *Decomposability: Queueing and Computer System Applications*, Academic Press, New York, NY, 1977.

[42] M.H.A. Davis, *Markov Models and Optimization*, Chapman & Hall, London, UK, 1993.

[43] F. Delebecque, A reduction process for perturbed Markov chains, *SIAM J. Appl. Math.*, **48** (1983), 325–350.

[44] F. Delebecque and J. Quadrat, Optimal control for Markov chains admitting strong and weak interactions, *Automatica*, **17** (1981), 281–296.

[45] C. Derman, *Finite State Markovian Decision Processes,* Academic Press, New York, NY, 1970.

[46] S. Dey, Reduced-complexity filtering for partially observed nearly completely decomposable Markov chains, *IEEE Trans. Signal Process.*, **48**, (2000), 3334–3344.

[47] G.B. Di Masi, Y.M. Kabanov and W.J. Runggaldier, Mean variance hedging of options on stocks with Markov volatility, *Theory Probab. Appl.*, **39** (1994), 173–181.

[48] R.L. Dobrushin, Central limit theorem for nonstationary Markov chains, *Theory Probab. Appl.* **1** (1956), 65–80, 329–383.

[49] J.L. Doob, *Stochastic Processes*, Wiley Classic Library Edition, Wiley, New York, NY, 1990.

[50] A. Doucet, N.J. Gordon and V. Krishnamurthy, Particle filtering for state estimation for jump Markov linear systems, *IEEE Trans. Signal Process.*, **49** (2001), 613–624.

[51] D. Duffie and H. Richardson, Mean-variance hedging in continuous time, *Ann. Appl. Probab.*, **1** (1991), 1–15.

[52] F. Dufour and P. Bertrand, The filtering problem for continuous-time linear systems with Markovian switching coefficients, *Systems Control Lett.*, **23** (1994), 453–461.

[53] F. Dufour and R.J. Elliott, Adaptive control of linear systems with Markov perturbations, *IEEE Trans. Automat. Control*, **43** (1997), 351–372.

[54] R.J. Elliott, *Stochastic Calculus and Applications*, Springer-Verlag, New York, NY, 1982.

[55] S.N. Ethier and T.G. Kurtz, *Markov Processes: Characterization and Convergence*, J. Wiley, New York, NY, 1986.

[56] W.H. Fleming, *Functions of Several Variables*, Addison-Wesley, Reading, MA, 1965.

[57] W.H. Fleming and R.W. Rishel, *Deterministic and Stochastic Optimal Control*, Springer-Verlag, New York, NY, 1975.

[58] W.H. Fleming and Q. Zhang, Nonlinear filtering with small observation noise: Piecewise monotone observations, *Stochastic Analysis: Liber Amicorum for Moshe Zakai*, E. Merzbach, A. Shwartz, and E. Mayer-Wolf Eds., 153–168, Academic Press, Boston, 1991.

[59] M.I. Friedlin and A.D. Wentzel, *Random Perturbations of Dynamical Systems*, Springer-Verlag, New York, NY, 1984.

[60] S.B. Gelfand and S.K. Mitter, Simulated annealing with noisy or imprecise energy measurements, *J. Optim. Theory Appl.*, **62** (1989), 49–62.

[61] S.B. Gershwin, *Manufacturing Systems Engineering*, Prentice Hall, Englewood Cliffs, NJ, 1994.

[62] I.I. Gihman and A.V. Skorohod, *Introduction to the Theory of Random Processes*, W.B. Saunders, Philadelphia, PA, 1969.

[63] W.-B. Gong, Y. Ho, and W. Zhai, Stochastic comparison algorithm for discrete optimization with estimation, *SIAM J. Optim.*, **10** (1999), 384–404.

[64] W.-B. Gong, P. Kelly, and W. Zhai, Comparison schemes for discrete optimization with estimation algorithms, *Proc. IEEE Conf. Decision Control*, 1993, 2211–2216.

[65] A. Graham, *Kronecker Products and Matrix Calculus with Applications*, Ellis Horwood Ltd., Chichester, 1981.

[66] J. Hale, *Ordinary Differential Equations*, 2nd Ed., R.E. Krieger Pub. Co., Malabar, FL, 1980.

[67] P. Hall and C.C. Heyde, *Martingale Limit Theory and Its Application*, Academic Press, New York, NY, 1980.

[68] U.G. Haussmann and Q. Zhang, Stochastic adaptive control with small observation noise, *Stoch. Stoch. Rep.*, **32** (1990), 109–144.

[69] F.C. Hoppensteadt and W.L. Miranker, Multitime methods for systems of difference equations, *Stud. Appl. Math.*, **56** (1977), 273–289.

[70] V. Hutson and J.S. Pym, *Applications of Functional Analysis and Operator Theory*, Academic Press, London, UK, 1980.

[71] A. Il'in, R.Z. Khasminskii, and G. Yin, Singularly perturbed switching diffusions: Rapid switchings and fast diffusions, *J. Optim. Theory Appl.*, **102** (1999), 555–591.

[72] A. Il'in, R.Z. Khasminskii, and G. Yin, Asymptotic expansions of solutions of integro-differential equations for transition densities of singularly perturbed switching diffusions: Rapid switchings, *J. Math. Anal. Appl.*, **238** (1999), 516–539.

[73] M. Iosifescu, *Finite Markov Processes and Their Applications*, Wiley, Chichester, 1980.

[74] Y. Ji and H.J. Chizeck, Controllability, stabilizability, and continuous-time Markovian jump linear quadratic control, *IEEE Trans. Automat. Control*, **35** (1990), 777–788.

[75] Y. Ji and H.J. Chizeck, Jump linear quadratic Gaussian control in continuous-time, *IEEE Trans. Automat. Control*, **37** (1992), 1884–1892.

[76] Yu. Kabanov and S. Pergamenshchikov, *Two-scale Stochastic Systems: Asymptotic Analysis and Control*, Springer, New York, NY, 2003.

[77] I. Karatzas and S.E. Shreve, *Methods of Mathematical Finance*, Springer, New York, NY, 1998.

[78] S. Karlin and H.M. Taylor, *A First Course in Stochastic Processes*, 2nd Edition, Academic Press, New York, NY, 1975.

[79] S. Karlin and H.M. Taylor, *A Second Course in Stochastic Processes*, Academic Press, New York, NY, 1981.

[80] J. Kevorkian and J.D. Cole, *Multiple Scale and Singular Perturbation Methods*, Springer-Verlag, New York, NY, 1996.

[81] R.Z. Khasminskii, On stochastic processes defined by differential equations with a small parameter, *Theory Probab. Appl.*, **11** (1966), 211–228.

[82] R.Z. Khasminskii, On an averaging principle for Ito stochastic differential equations, *Kybernetika*, **4** (1968), 260-279.

[83] R.Z. Khasminskii, *Stochastic Stability of Differential Equations*, Sijthoff & Noordhoff, Alphen aan den Rijn, Netherlands, 1980.

[84] R.Z. Khasminskii and G. Yin, On transition densities of singularly perturbed diffusions with fast and slow components, *SIAM J. Appl. Math.*, **56** (1996), 1794–1819.

[85] R.Z. Khasminskii, G. Yin, and Q. Zhang, Asymptotic expansions of singularly perturbed systems involving rapidly fluctuating Markov chains, *SIAM J. Appl. Math.*, **56** (1996), 277–293.

[86] R.Z. Khasminskii, G. Yin, and Q. Zhang, Constructing asymptotic series for probability distribution of Markov chains with weak and strong interactions, *Quart. Appl. Math.*, **LV** (1997), 177–200.

[87] J. Kiefer and J. Wolfowitz, Stochastic estimation of the maximum of a regression function, *Ann. Math. Statist.* **23** (1952), 462–466.

[88] P.E. Kloeden and E. Platen, *Numerical Solution of Stochastic Differential Equations*, Springer-Verlag, Berlin, 1992.

[89] V. Krishnamurthy, X. Wang, and G. Yin, Spreading code optimization and adaptation in CDMA via discrete stochastic approximation, to appear in *IEEE Trans. Inform. Theory*.

[90] V. Krishnamurthy and G. Yin, Recursive algorithms for estimation of hidden Markov models and autoregressive models with Markov regime, *IEEE Trans. Inform. Theory*, **48** (2002), 458–476.

[91] P.R. Kumar and P. Varaiya, *Stochastic Systems: Estimation, Identification and Adaptive Control*, Prentice-Hall, Englewood Cliffs, NJ, 1984.

[92] H. Kunita and S. Watanabe, *Stochastic Differential Equations and Diffusion Processes*, North-Holland, Amsterdam, 1981.

[93] T.G. Kurtz, *Approximation of Population Processes*, SIAM, Philadelphia, PA, 1981.

[94] H.J. Kushner, On the differential equations satisfied by conditional probability densities of Markov processes with applications, *SIAM J. Control*, **2** (1964), 106–119.

[95] H.J. Kushner, *Stochastic Stability and Control*, Academic Press, New York, NY, 1967.

[96] H.J. Kushner, *Approximation and Weak Convergence Methods for Random Processes, with applications to Stochastic Systems Theory*, MIT Press, Cambridge, MA, 1984.

[97] H.J. Kushner, *Weak Convergence Methods and Singularly Perturbed Stochastic Control and Filtering Problems*, Birkhäuser, Boston, MA, 1990.

[98] H.J. Kushner and D.S. Clark, *Stochastic Approximation Methods for Constrained and Unconstrained Systems*, Springer-Verlag, New York, NY, 1978.

[99] H.J. Kushner and H. Huang, Averaging methods for the asymptotic analysis of learning and adaptive systems with small adjustment rate, *SIAM J. Control Optim.*, **19** (1980), 635–650.

[100] H.J. Kushner and G. Yin, *Stochastic Approximation and Recursive Algorithms and Applications*, 2nd Ed., Springer-Verlag, New York, NY, 2003.

[101] J.P. LaSalle, *The Stability of Dynamical Systems*, SIAM, Philadelphia, PA, 1979.

[102] D. Li and W.L. Ng, Optimal dynamic portfolio selection: Multi-period mean-variance formulation, *Math. Finance*, **10** (2000), 387–406.

[103] R.H. Liu, Q. Zhang, and G. Yin, Nearly optimal control of singularly perturbed Markov decision processes in discrete time, *App. Math. Optim.*, **44** (2001), 105–129.

[104] R.H. Liu, Q. Zhang, and G. Yin, Asymptotically optimal controls of hybrid linear quadratic regulators in discrete time, *Automatica*, **38** (2002), 409–419.

[105] R.S. Liptser and A.N. Shiryayev, *Statistics of Random Processes I & II*, Springer-Verlag, New York, NY, 2001.

[106] L. Ljung, Analysis of recursive stochastic algorithms, *IEEE Trans. Automat. Control*, **22** (1977), 551–575.

[107] W.P. Malcome, R.J. Elliott, J. van der Hoek, On the numerical stability of time-discretized state estimation via Clark transformations, *Proc. 42nd IEEE Conf. Decision Control*, 2003, 1406–1412,

[108] X. Mao, *Exponential Stability of Stochastic Differential Equations*, Marcel Dekker, New York, NY, 1994.

[109] X. Mao, Stability of stochastic differential equations with Markovian switching, *Stochastic Process. Appl.*, **79** (1999), 45–67.

[110] S.I. Marcus and E.K. Westwood, On asymptotic approximations for some nonlinear filtering problems, in *Proc. IFAC Triennial Congress*, Budapest, Hungary, Vol. VII, 36–41.

[111] M. Mariton and P. Bertrand, Robust jump linear quadratic control: A mode stabilizing solution, *IEEE Trans. Automat. Control*, **30** (1985), 1145–1147.

[112] H. Markowitz, Portfolio selection, *J. Finance*, **7** (1952), 77–91.

[113] H. Markowitz, *Portfolio Selection: Efficient Diversification of Investment*, John Wiley & Sons, New York, NY, 1959.

[114] M. Metivier and P. Priouret, Applications of a Kushner and Clark lemma to general classes of stochastic algorithms, *IEEE Trans. Inform. Theory*, **30** (1984), 140–150.

[115] S.P. Meyn and R.L. Tweedie, *Markov Chains and Stochastic Stability*, Springer-Verlag, London, 1993.

[116] B.M. Miller and W.J. Runggaldier, Kalman filtering for linear systems with coefficients driven by a hidden Markov jump process, *Systems Control Lett.*, **31** (1997), 93–102.

[117] D.S. Naidu, *Singular Perturbation Methodology in Control Systems*, Peter Peregrinus Ltd., Stevenage Herts, UK, 1988.

[118] R.E. O'Malley, Jr., *Singular Perturbation Methods for Ordinary Differential Equations*, Springer-Verlag, New York, 1991.

[119] Z.G. Pan and T. Başar, H^∞-control of Markovian jump linear systems and solutions to associated piecewise-deterministic differential games, in *New Trends in Dynamic Games and Applications*, G.J. Olsder Ed., 61–94, Birkhäuser, Boston, MA, 1995.

[120] Z.G. Pan and T. Basar, H^∞ control of large-scale jump linear systems via averaging and aggregation, *Internat. J. Control*, **72** (1999), 866–881.

[121] G.C. Papanicolaou, Some probabilistic problems and methods in singular perturbations, *Rocky Mountain J. Math.*, **6** (1976), 653–674.

[122] G.C. Papanicolaou, D. Stroock, and S.R.S. Varadhan, Martingale approach to some limit theorems, in *Proc. 1976 Duke Univ. Conf. on Turbulence*, Durham, NC, 1976.

[123] A.A. Pervozvanskii and V.G. Gaitsgory, *Theory of Suboptimal Decisions: Decomposition and Aggregation*, Kluwer, Dordrecht, 1988.

[124] G. Pflug, *Optimization of Stochastic Models: The Interface between Simulation and Optimization*, Kluwer, Boston, MA, 1996.

[125] R.G. Phillips and P.V. Kokotovic, A singular perturbation approach to modeling and control of Markov chains, *IEEE Trans. Automat. Control*, **26** (1981), 1087–1094.

[126] J.P. Quadrat, Optimal control of perturbed Markov chains: The multitime scale case. Singular Perturbations in Systems and Control, M.D. Aradema Ed., 216–239, CISM Courses and Lectures, No. 280, Springer, 1983.

[127] D. Revuz, *Markov Chains*, 2nd Ed., North-Holland, Amsterdam, 1984.

[128] R. Rishel, Control of systems with jump Markov disturbances, *IEEE Trans. Automat. Control*, **20** (1975), 241–244.

[129] H. Robbins and S. Monro, A stochastic approximation method, *Ann. Math. Statist.*, **22** (1951), 400–407.

[130] S. Ross, *Stochastic Processes*, J. Wiley, New York, NY, 1983.

[131] S. Ross, *Introduction to Stochastic Dynamic Programming*, Academic Press, New York, NY, 1984.

[132] W.J. Runggaldier and C. Visentin, Combined filtering and parameter estimation: Approximation and robustness, *Automatica*, **26** (1990), 401–404.

[133] M. Schweizer, Approximation pricing and the variance-optimal martingale measure, *Ann. Probab.*, **24** (1996), 206–236.

[134] S.P. Sethi, W. Suo, M.I. Taksar, and Q. Zhang, Optimal production planning in stochastic manufacturing systems with long run average costs, *J. Optim. Theory Appl.*, **92**, (1997), 161–188.

[135] S.P. Sethi, H. Zhang, and Q. Zhang, *Average-Cost Control of Stochastic Manufacturing Systems*, Springer-Verlag, New York, NY, 2004.

[136] S.P. Sethi and Q. Zhang, *Hierarchical Decision Making in Stochastic Manufacturing Systems*, Birkhäuser, Boston, MA, 1994.

[137] O.P. Sharma, *Markovian Queues*, Ellis Horwood, Chichester, UK, 1990.

[138] H.A. Simon and A. Ando, Aggregation of variables in dynamic systems, *Econometrica*, **29** (1961), 111–138.

[139] A.V. Skorohod, *Studies in the Theory of Random Processes*, Dover, New York, NY, 1982.

[140] A.V. Skorohod, *Asymptotic Methods of the Theory of Stochastic Differential Equations*, Trans. Math. Monographs, Vol. 78, Amer. Math. Soc., Providence, 1989.

[141] V. Solo and X. Kong, *Adaptive Signal Processing Algorithms*, Prentice-Hall, Englewood Cliffs, NJ, 1995.

[142] R.S. Strichartz, *The Way of Analysis*, Jones and Bartlett Publishers, Boston, 1995.

[143] D.W. Stroock and S.R.S. Varadhan, *Multidimensional Diffusion Processes*, Springer-Verlag, Berlin, 1979.

[144] D. N. C. Tse, R. G. Gallager, and J. N. Tsitsiklis, Statistical multiplexing of multiple time-scale Markov streams, *IEEE J. Selected Areas Comm.*, **13** (1995), 1028–1038.

[145] A.B. Vasil'eava and V.F. Butuzov, *Asymptotic Expansions of the Solutions of Singularly Perturbed Equations*, Nauka, Moscow, 1973.

[146] H. Wang and P. Chang, On verifying the first-order Markovian assumption for a Rayleigh fading channel model, *IEEE Trans. Vehicle Tech.*, **45** (1996), 353–357.

[147] J.W. Wang, Q. Zhang, and G. Yin, Two-time-scale hybrid filters: Near optimality, preprint.

[148] W. Wasow, *Asymptotic Expansions for Ordinary Differential Equations*, Interscience, New York, 1965.

[149] D.J. White, *Markov Decision Processes*, Wiley, New York, NY, 1992.

[150] W.M. Wonham, Some applications of stochastic differential equations to optimal nonlinear filtering, *SIAM J. Control*, **2** (1965), 347–369.

[151] D. Yan and H. Mukai, Stochastic discrete optimization, *SIAM J. Control Optim.*, **30** (1992), 594–612.

[152] C. Yang, Y. Bar-Shalom, and C.-F. Lin, Discrete-time point process filter for mode estimation, *IEEE Trans. Automat. Control*, **37** (1992), 1812–1816.

[153] H. Yang, G. Yin, K. Yin, and Q. Zhang, Control of singularly perturbed Markov chains: A numerical study, *ANZIAM J.*, **45** (2003), 49–74.

[154] G. Yin, On limit results for a class of singularly perturbed switching diffusions, *J. Theoret. Probab.*, **14** (2001), 673–697.

[155] G. Yin and S. Dey, Weak convergence of hybrid filtering problems involving nearly completely decomposable hidden Markov chains, *SIAM J. Control Optim.*, **41** (2003), 1820–1842.

[156] G. Yin, V. Krishnamurthy, and C. Ion, Regime switching stochastic approximation algorithms with application to adaptive discrete stochastic optimization, to appear in *SIAM J. Optim.*

[157] G. Yin, R. H. Liu, and Q. Zhang, Recursive algorithms for stock liquidation: A stochastic optimization approach, *SIAM J. Optim.*, **13** (2002), 240–263.

[158] G. Yin and Q. Zhang, *Continuous-time Markov Chains and Applications: A Singular Perturbations Approach*, Springer-Verlag, New York, NY, 1998.

[159] G. Yin and Q. Zhang, Singularly perturbed discrete-time Markov chains, *SIAM J. Appl. Math.*, **61** (2000), 834–854.

[160] G. Yin and Q. Zhang, Discrete-time singularly perturbed Markov chains, to appear in *Stochastic Models and Optimization*, D. Yao, H.Q. Zhang, and X.Y. Zhou Eds., 1–42, Springer-Verlag, 2003.

[161] G. Yin and Q. Zhang, Stability of Markov modulated discrete-time dynamic systems, *Automatica*, **39** (2003), 1339–1351.

[162] G. Yin, Q. Zhang, and G. Badowski, Asymptotic properties of a singularly perturbed Markov chain with inclusion of transient states, *Ann. Appl. Probab.*, **10** (2000), 549–572.

[163] G. Yin, Q. Zhang, and G. Badowski, Singularly perturbed Markov chains: Convergence and aggregation, *J. Multivariate Anal.*, **72** (2000), 208–229.

[164] G. Yin, Q. Zhang, and G. Badowski, Occupation measures of singularly perturbed Markov chains with absorbing states, *Acta Math. Sinica*, **16** (2000), 161–180.

[165] G. Yin, Q. Zhang, and G. Badowski, Discrete-time singularly perturbed Markov chains: Aggregation, occupation measures, and switching diffusion limit, *Adv. in Appl. Probab.*, **35** (2003), 449–476.

[166] G. Yin, Q. Zhang, and Q.G. Liu, Error bounds for occupation measure of singularly perturbed Markov chains including transient states, *Probab. Engrg. Inform. Sci.*, **14** (2000), 511–531.

[167] G. Yin, Q. Zhang, and Y.J. Liu, Discrete-time approximation of Wonham filters, to appear in *J. Control Theory Appl.*

[168] G. Yin, Q. Zhang, H. Yang, and K. Yin, Discrete-time dynamic systems arising from singularly perturbed Markov chains, *Nonlinear Anal.*, **47** (2001), 4763–4774.

[169] G. Yin and X.Y. Zhou, Markowitz's mean-variance portfolio selection with regime switching: from discrete-time models to their continuous-time limits, *IEEE Trans. Automat. Control*, **49** (2004), 349–360.

[170] J. Yong and X.Y. Zhou, *Stochastic Controls: Hamiltonian Systems and HJB Equations*, Springer-Verlag, New York, NY, 1999.

[171] T. Zariphopoulou, Investment-consumption models with transactions costs and Markov-chain parameters, *SIAM J. Control Optim.*, **30** (1992), 613–636.

[172] Q. Zhang, Risk sensitive production planning of stochastic manufacturing systems: A singular perturbation approach, *SIAM J. Control Optim.*, **33** (1995), 498–527.

[173] Q. Zhang, Finite state Markovian decision processes with weak and strong interactions, *Stoch. Stoch. Rep.*, **59** (1996), 283–304.

[174] Q. Zhang, Nonlinear filtering and control of a switching diffusion with small observation noise, *SIAM J. Control Optim.*, **36** (1998), 1738–1768.

[175] Q. Zhang, Hybrid filtering for linear systems with non-Gaussian disturbances, *IEEE Trans. Automat. Control*, **45** (2000), 50–61.

[176] Q. Zhang, Stock trading: An optimal selling rule, *SIAM J. Control Optim.*, **40** (2001), 64–87.

[177] Q. Zhang, R.H. Liu, and G. Yin, Nearly optimal controls of Markovian systems, in *Stochastic Models and Optimization*, D. Yao, H.Q. Zhang, and X.Y. Zhou Eds., 43–86, Springer-Verlag, 2003.

[178] Q. Zhang and G. Yin, Turnpike sets in stochastic manufacturing systems with finite time horizon, *Stoch. Stoch. Rep.*, **51**, (1994), 11-40.

[179] Q. Zhang and G. Yin, A central limit theorem for singularly perturbed nonstationary finite state Markov chains, *Ann. Appl. Probab.*, **6** (1996), 650–670.

[180] Q. Zhang and G. Yin, Structural properties of Markov chains with weak and strong interactions, *Stochastic Process. Appl.*, **70** (1997), 181–197.

[181] Q. Zhang and G. Yin, On nearly optimal controls of hybrid LQG problems, *IEEE Trans. Automat. Control*, **44** (1999), 2271–2282.

[182] Q. Zhang and G. Yin, Nearly optimal asset allocation in hybrid stock-investment models, *J. Optim. Theory Appl.*, **121** (2004), 197–222.

[183] Q. Zhang and G. Yin, Exponential bounds for discrete-time singularly perturbed Markov chains, *J. Math. Anal. Appl.*, **293** (2004), 645–662.

[184] Q. Zhang and G. Yin, A two-time-scale approach for production planning in discrete time, in *Analysis, Control, and Optimization of Complex Dynamic Systems*, E.K. Boukas and R. Malhamè Eds., 37–54, Kluwer Academic, Boston, MA, 2004.

[185] X.Y. Zhou and D. Li, Continuous-Time mean-variance portfolio selection: A stochastic LQ framework, *Appl. Math. Optim.*, **42** (2000), 19–33.

[186] X.Y. Zhou and G. Yin, Markowitz's mean-variance portfolio selection with regime switching: A Continuous-time model, *SIAM J. Control Optim.*, **42** (2003), 1466–1482.